Food Science and Technology

Also of Interest

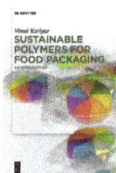

Sustainable Polymers for Food Packaging.
An Introduction
Vimal Katiyar, 2020
ISBN 978-3-11-064453-1, e-ISBN (PDF) 978-3-11-064803-4,
e-ISBN (EPUB) 978-3-11-064463-0

Food Contamination by Packaging.
Migration of Chemicals from Food Contact Materials
Ana Rodríguez Bernaldo de Quirós, Antía Lestido Cardama,
Raquel Sendón, Verónica García Ibarra, 2019
ISBN 978-3-11-064487-6, e-ISBN (PDF) 978-3-11-064806-5,
e-ISBN (EPUB) 978-3-11-064502-6

Recycling of Polyethylene Terephthalate
Martin J. Forrest, 2019
ISBN 978-3-11-064029-8, e-ISBN (PDF) 978-3-11-064030-4,
e-ISBN (EPUB) 978-3-11-064045-8

Industrial Biotechnology
Mark Anthony Benvenuto, 2019
ISBN 978-3-11-053639-3, e-ISBN 978-3-11-053662-1,
e-ISBN (EPUB) 978-3-11-053676-8

Downstream Processing in Biotechnology
Venko N. Beschkov, Dragomir Yankov (Eds.), 2020
ISBN 978-3-11-057395-4, e-ISBN (PDF) 978-3-11-057411-1,
e-ISBN (EPUB) 978-3-11-057400-5

Food Science and Technology

Trends and Future Prospects

Edited by
Oluwatosin Ademola Ijabadeniyi

DE GRUYTER

Editor
Prof. Oluwatosin Ademola Ijabadeniyi
Durban University of Technology
Department of Biotechnology and Food Technology
Faculty of Applied Sciences
PO Box 1334
4000 Durban
South Africa
oluwatosini@dut.ac.za

ISBN 978-3-11-066745-5
e-ISBN (PDF) 978-3-11-066746-2
e-ISBN (EPUB) 978-3-11-066757-8

Library of Congress Control Number: 2020949818

Bibliographic information published by the Deutsche Nationalbibliothek
The Deutsche Nationalbibliothek lists this publication in the Deutsche Nationalbibliografie;
detailed bibliographic data are available on the Internet at http://dnb.dnb.de.

© 2021 Walter de Gruyter GmbH, Berlin/Boston
Cover image: Colin Anderson/Photographer's Choice RF/Getty Images
Typesetting: Integra Software Services Pvt. Ltd.
Printing and binding: CPI books GmbH, Leck

www.degruyter.com

Foreword

At this point of human existence in the twenty-first century, there is an ever-present global food crisis. The Global Report on Food Crises produced by the Food Security Information Network and published by the Food and Agriculture Organization of the United Nations in April 2019 does not make for encouraging reading. According to the report, over 113 million people across 53 countries experienced acute hunger in 2018. The report singles out eight countries, five of them from Africa that accounted for two-thirds of the total number of people facing acute food insecurity, which amounted to about 72 million people.

Clearly, there is a global food crisis that is palpably exhibited in the form of acute food insecurity with Africa (especially sub-Saharan Africa) being the hardest hit. The United Nations has forecast that given the current trajectory, Africa is not on track to realize the Sustainable Development Goal 2 of Zero Hunger by 2030. The statistics for Africa in this respect are dire. About 20% of the African population is undernourished, with sub-Saharan Africa being the worst hit by food insecurity. About one-third of undernourished people in the world today are in Africa, and of these 90% are in sub-Saharan Africa. The main drivers of food insecurity are identified as conflict and insecurity, climate shocks, and economic turbulence.

Food security is not just about access to food. It is also about how nutritious the food is. Micronutrient and macronutrient malnutrition are well-known scourge in the developing world. However, many parts of the developing world such as sub-Saharan Africa are recording significant increases in the occurrence of diet-related noncommunicable diseases. Rapid urbanization, rising incomes, and poor dietary choices due to a demand for convenience have brought on a nutrition transition and change in dietary patterns from nutrient-dense to more energy-dense foods. This has resulted in overnutrition accompanied by susceptibility to conditions such as obesity and noncommunicable diseases.

The gloomy picture presented earlier puts in sharp focus on the pivotal role that food science and technology can play in combating the global food crisis and contributing to attaining food security. It underscores the need to train more food scientists and technologists and, in that regard, the importance of this book *Food Science and Technology: Trends and Future Prospects* cannot be overemphasized.

This book begins with a look at the food industry as a whole and goes on to discuss aspects of postharvest handling and food processing. The processing of the major plant and animal food groups, namely cereals and legumes, meat, poultry, and fish, is discussed. The book then goes on to deal with a broad range of issues within the three major pillars of food science and technology – food microbiology, food chemistry, and product development. To round off, the book includes important chapters that discuss recent innovations in food processing and in the food industry and this is followed by a couple of offerings dealing with food business, entrepreneurship, management, and regulation.

https://doi.org/10.1515/9783110667462-202

This is a much-needed book in light of the prevailing global food crisis and challenges with food insecurity. It will be useful for a wide range of people including academics, researchers, and students in food science and technology, food industry professionals, food entrepreneurs, farmers, policymakers, and the curious consumer. It is envisaged that this book will have a significant impact in the field of food science and technology.

Kwaku G. Duodu
Professor: Food Science
University of Pretoria
Department of Consumer and Food Sciences
Pretoria, South Africa
26 November 2019

Contents

Part I: **Postharvest handling and food processing**

Oluwatosin Ademola Ijabadeniyi, Ibilola Itiolu, Titilayo Adenike Ajayeoba,
Adebola Olubukola Oladunjoye

Oluwafemi James Caleb, Rebogile R.R. Mphahlele, Zinash A. Belay,
Asanda Mditshwa

Betty O. Ajibade, Omotola F. Olagunju, Oluwatosin Ademola Ijabadeniyi

Tremayne S. Naiker, John J. Mellem

Olugbenga Philip Soladoye

AyoJesutomi O. Abiodun-Solanke

Part II: **Food microbiology use of microorganisms**

Titilayo Adenike Ajayeoba, Oluwatosin Mary Kaka, Oluwatosin Akinola
Ajibade

Angela Parry-Hanson Kunadu, Emmanuel Addo-Preko, Nikki Asuming-Bediako

Part III: Food chemistry: analysis and nutrition

Part IV: Product development, sensory evaluation, and packaging

Part V: Food innovations and nonthermal processing

Part VI: Food business: entrepreneurship and regulation

Part VII: Outlook

List of contributors

Abimbola Kemisola
Department of Home Economics and Food
Science
University of Ilorin
Kwara State, Nigeria

Abimbola M. Enitan-Folami
Department of Biotechnology and Food
Technology
Durban University of Technology
Durban, South Africa

Adebola Olubukola Oladunjoye
Department of Food Technology
University of Ibadan
Ibadan, Nigeria

Adeoluwa Iyiade Adetunji
Labworld/Philafrica Foods (Pty) Ltd.
Centurion, South Africa

Ahmad Cheikhyoussef
Science and Technology Division
Multidisciplinary Research Centre (MRC)
University of Namibia
Windhoek, Namibia

Ajibola Bamikole Oyedeji
Department of Biotechnology and Food
Technology
University of Johannesburg,
Gauteng, South Africa

Akintayo Olaide
Department of Home Economics and Food
Science
University of Ilorin
Kwara State, Nigeria

Angela Parry-Hanson Kunadu
University of Ghana
Accra, Ghana

Anna V. Begunova
All-Russian Research Institute of Dairy
Industry
Moscow, Russia

Aribisala Jamiu Olaseni
Department of Biotechnology and Food
Technology
Durban University of Technology
Durban, South Africa
Department of Microbiology
Federal University of Technology
Akure, Nigeria

Aruwa Christiana Eleojo
The School of Sciences
Department of Microbiology
Federal University of Technology
Akure, Nigeria

Asanda Mditshwa
School of Agriculture
Earth and Environmental Sciences
College of Agriculture, Engineering
and Science
University of KwaZulu-Natal (PMB-Campus)
Scottsville, South Africa

Beatrice I.O. Ade-Omowaye
Department of Food Science
Ladoke Akintola University of Technology
Nigeria

Betty O. Ajibade
Durban University of Technology
Durban, South Africa

Brendan A. Niemira
USDA-ARS
Eastern Regional Research Center
Food Safety & Intervention technologies
Research Unit
Wyndmoor, USA

Dele Raheem
Arctic Centre (NIEM)
University of Lapland
Rovaniemi, Finland

Emmanuel Addo-Preko
University of Ghana
Accra, Ghana

https://doi.org/10.1515/9783110667462-204

Faith Ruzengwe
Department of Biotechnology and Food
Technology
Durban University of Technology
Durban, South Africa

Fatima Raji
Food Technology Department
Federal Institute of Industrial Research Oshodi
Lagos, Nigeria

Feroz M. Swalaha
Department of Biotechnology and Food
Technology
Durban University of Technology
Durban, South Africa

Garuba Taofeeq
Plant Biology Department
University of Ilorin
Ilorin, Nigeria

Ibilola Itiolu
Department of Biotechnology and Food
Technology
Durban University of Technology
Durban, South Africa

Irina V. Rozhkova
All-Russian Research Institute of Dairy
Industry
Moscow, Russia

John J. Mellem
Department of Biotechnology and Food
Technology
Durban University of Technology
Durban, South Africa

Konstantin V. Moiseenko
Research Centre of Biotechnology
Moscow, Russia

Madende Moses
The Food Biosciences Department
Teagasc Food Research Centre
Dublin, Ireland

Mlungisi Mtolo
Reckitt Benckiser South Africa (Pty) Ltd.
Elandsfontein, South Africa

Ngcala Mamosa
Department of Molecular and Cell Biology
University of Cape Town
Rondebosch, South Africa

Nikki Asuming-Bediako
University of Ghana
Accra, Ghana

Ocen M. Olanya
USDA-ARS
Eastern Regional Research Center
Food Safety & Intervention technologies
Research Unit
Wyndmoor, PA, USA

Olaide Akande
Food Technology Department
Federal Institute of Industrial Research Oshodi
Lagos, Nigeria

Olayemi Eyituoyo Dudu
Department of Food Science and Engineering
Harbin Institute of Technology
Harbin, China

Olga A. Glazunova
Research Centre of Biotechnology
Moscow, Russia

Olugbenga Philip Soladoye
Food Processing Development Centre
Food and Bio Processing Branch
Alberta Agriculture and Forestry
Leduc, Canada

Oluwafemi Ayodeji Adebo
Department of Biotechnology and Food
Technology
University of Johannesburg,
Gauteng, South Africa

Oluwafemi James Caleb
Agri-Food Systems & Omics
Post-Harvest and Agro-Processing
Technologies (PHATs)
Agricultural Research Council (ARC)
Stellenbosch, South Africa

Oluwatosin Ademola Ijabadeniyi
Department of Biotechnology and Food
Technology
Durban University of Technology
Durban, South Africa

Oluwatosin Mary Kaka
Department of Biotechnology and Food
Technology
Durban University of Technology
Durban, South Africa

Oluwatoyin Oluwole
Food Technology Department
Federal Institute of Industrial Research Oshodi
Lagos, Nigeria

Omotola F. Olagunju
Afe Babalola University
Ekiti State, Nigeria

Oyedeji Ajibola Bamikole
Department of Biotechnology and Food
Technology
University of Johannesburg
Gauteng, South Africa

Oyedeji Amusa Mariam Oyefunke
Department of Botany and Plant
Biotechnology
University of Johannesburg
Auckland Park, South Africa

Pillay Charlene
Department of Biotechnology and Food
Technology
Durban University of Technology
Durban, South Africa

Rebogile R.R. Mphahlele
Agricultural Research Council-Tropical
and Subtropical Crops (ARC-TSC)
Nelspruit, South Africa

Sabiu Saheed
Department of Biotechnology and Food
Technology
Durban University of Technology
Durban, South Africa

Samson A. Oyeyinka
Department of Home Economics and Food
Science
University of Ilorin
Nigeria

Tatyana V. Fedorova
All-Russian Research Institute of Dairy
Industry
Moscow, Russia

Titilayo Adenike Ajayeoba
Department of Microbiology, Faculty of
Science
Adeleke University Ede
Nigeria
Department of Biotechnology and Food
Technology
Durban University of Technology
Durban, South Africa

Tremayne S. Naiker
Department of Biotechnology and Food
Technology
Durban University of Technology
Durban, South Africa

Zinash A. Belay
Agri-Food Systems & Omics
Post-Harvest and Agro-Processing
Technologies (PHATs)
Agricultural Research Council (ARC)
Stellenbosch, South Africa

Introduction

Food is important to humans because it fuels the body and provides energy to carry out daily activities. Hence, a sustainable food supply system is required to ensure that the world's growing population is fed with safe, nutritious, and wholesome foods.

There is a need to focus on sustaining the food supply chain and having a circular food economy underpinned by a cross-disciplinary approach. A cross-disciplinary subject like food science can therefore play an important role in facilitating a healthy lifestyle for consumers while at the same time reducing waste and the environmental impact of food production, processing, distribution, and consumption. This will ensure the availability of food for future generations.

Food Science and Technology: Trends and Future Prospects presents different aspects of food science, that is, food microbiology, food chemistry, nutrition, and process engineering. These can be applied for selection, preservation, processing, packaging, and distribution of quality food. The authors focus on fundamental aspects of food and also highlight emerging technology and innovations that are changing the food industry. Such innovation and smart solutions (blockchain for traceability, big data, sensor technology, superfoods ingredients, personalized nutrition, 3D printing, artificial intelligence, and augmented and visual reality) provide reasons to be optimistic about an increase in food security in the digitalized global food system.

The chapters are written by leading researchers, lecturers, and experts in food chemistry, food microbiology, biotechnology, nutrition, and management. *Food Science and Technology: Trends and Future Prospects* is valuable for researchers and students in food science and technology. It is also useful for food industry professionals, food entrepreneurs, and farmers. The book covers important topics affecting quality and safety of foods from farm to fork.

Oluwatosin Ademola Ijabadeniyi

https://doi.org/10.1515/9783110667462-205

Part I: **Postharvest handling and food processing**

Oluwatosin Ademola Ijabadeniyi, Ibilola Itiolu,
Titilayo Adenike Ajayeoba, Adebola Olubukola Oladunjoye

1 The food industry: yesterday, today, and tomorrow

1.1 Introduction

The term food industry is broad though often interchangeably called food production or manufacturing industry. It is characterized by a complex system of activities concerning supply, consumption, and delivery of food product across the globe (Campbell-Platt, 2011). These activities include agriculture, manufacturing, food processing, marketing, logistics, food services, regulation, research, and development. The Economic Research Service of USDA states that food industries cover a series of industrial activities directed at the processing, conversion, preparation, preservation, and packaging of foodstuffs (Malagie et al. 2011). Food (which water is also a component) is the most basic aspect of survival for any organism though, with an exception of oxygen, it is also an important element of human cultural tradition and social structure as well as well-being (Alifoods, 2016). Food's global diversities are often celebrated as it unites people around the dinner table. This industrial complex that produces, processes, and distributes food is one of the largest industries in the world (Drummond et al. 2014), and its trends and future prospects from farm to fork are vital topics that affect everyone either directly or indirectly. The food industry plays an important role in the economy of any nation in the world, currently food has been a global commodity as they can be transported across nations around the world, and this is because changes in consumer's lifestyle and choices have contributed to the growth and expansion of food industries globally (Campbell-Platt, 2011). In the twentieth century, the industry has become more dynamic compared to decades back due to new trends in research and development, advanced food processing technologies innovations, and obviously consumers' lifestyle modifications and choices.

The food industry is huge and if all its functions are combined it is greater than all than all all other manufacturing sectors, that is, mining, communications, automobile, chemical manufacturing, and so on put together (Potter and Hotchkiss, 1995).

Oluwatosin Ademola Ijabadeniyi, Ibilola Itiolu, Department of Biotechnology and Food
Technology, Durban University of Technology, South Africa
Titilayo Adenike Ajayeoba, Department of Microbiology, Faculty of Science, Adeleke University
Ede, Nigeria
Adebola Olubukola Oladunjoye, Department of Food Technology, University of Ibadan, Nigeria

https://doi.org/10.1515/9783110667462-001

Several millions of personnel are trained to occupy positions within the food production process and thus contributing directly or indirectly to its growth (Harcourt, 2011). In any economy, the quantity of food produced via agriculture has a direct correlation on the price of food and other commodities. Surplus supply of food reflects in lowered prices of items in the market. Food is often consumed in different forms other than the form it was produced, this includes transformation from raw form into finished goods, that is, the potatoes into dehydrated forms in crisps and raw chicken fillets into packaged breaded crumbled chicken bites. In recent times, due to change in consumers lifestyles, religious beliefs, and even dietary modification, the total food consumed outside the home outweighs the ones prepared in the home (Potter and Hotchkiss, 1995). Homemade meals are becoming less popular due to high job demands, unease in preparation, and dynamism in food industry (Campbell-Platt, 2011). The process of meal preparation is often faster and easier for a food enterprise than an individual, which explains fast food joints springing up in every suburb of both developing and developed economy. Most grocery stores have also added a restaurant or food processing facility while food manufacturers are constantly acquiring franchise or restaurant chains and have become the promoter of their products (Drummond et al. 2014). Apart from the fact that more than any other sector of the economy the food industry has gained more international presence with the big players such as Coca-Cola, Nestle, PepsiCo, Kraft, and Friesland have manufacturing facilities in many countries of the world (Drummond et al. 2014). Food has become a global commodity (Potter and Hotchkiss, 1995); hence, it is common to see products made millions of miles away available in a grocery stores near you this is because food products are been merchandised and transported worldwide.

1.2 Historical origin of food industry

The processing of foods dates back to prehistoric ages before the Industrial Revolution which is a major contributing factor to the development of the food industry (Overton, 2010). Prior mid-1700s in Europe and USA, humans use crude and simple methods for the production (preharvest and harvest, animal husbandry) and processing of food but the methods changed to tractors/power tools for production and machines for processing of food after the industrial revolution (in the mid-1700s) (Murano, 2003). Traditional methods such as sun drying, smoking, roasting, and addition of salt were crude methods employed for preserving food until the introduction of canning methods. Canning of foods can be dated back to 1752. Frenchman Nicholas Appert was the first man to "can" foods (that is why canning is also called appertization) (Malagie, 2011). He was able to "can" cooked meat, vegetables, and milk by using bottle then at a later date in 1810 Peter Durand used "tin" instead of bottle for canning and then Louis Pasteur in 1862 introduced pasteurization method, a significant advance in ensuring microbiological safety of food (Rahman, 2007). Canned food has become a staple around the world so more advanced food processing methods were introduced from spray drying to addition of preservative.

Food shortages and starvation became less common as a result of implementation of more efficient mechanized methods of food production, processing, and distribution (Murano, 2003). Prior to the invention of refrigeration technology, houses (in form of icehouse) were specially built for storing and maintaining masses of natural ice collected from cold northern lakes (Murano, 2003). The invention of ice-making technology/machines in the 1800s stopped the storage in icehouses but discovery of refrigeration and canning enhanced the technology of preservation. Advances in food chemistry knowledge and application have also assisted to solve the problems of preservation, especially spoilage, though in recent times consumers began questioning the problem of chemical residues (from fertilizer and preservatives) in food which has given rise to acceptance of organic food. Table 1.1 shows different food processing methodologies and technologies developed from 10,000 BC to 1900s. Recently, technologies such as pulse electric field, cold plasma, high pressure processing, and ultrasound are applied in the food industry. Furthermore, today's food systems have become diverse and complex, involving everything from subsistence farming to multinational food companies (Hueston and McLeod, 2012). Also, because everyone eats; everyone relies on food systems, local and global (Hueston and McLeod, 2012).

Table 1.1: Historical dates of food technology.

Technology developed	Approximate date
Milling of cereal grains into flour	10,000 BC
Baking of unleavened bread	10,000 BC
Meat and fish smoking, salting, drying	4,000 BC
Grape and barley fermentation	3,000 BC
Yogurt fermentation	AD 200
Canning	1800s
Iron roller milling of flour	1800s
Milk pasteurization	1800s
Freeze-drying	1900s
Modified atmosphere packaging	1900s
Food irradiation	1900s

Adapted from Murano (2003).

1.3 The food pipeline

The food pipeline consists of the functional progression of various operations and their complex interaction involved in transforming the food materials from farm during harvesting to the consumption stage. Quality and safety of the final food product can be hampered during any of the stages involved (Figure 1.1).

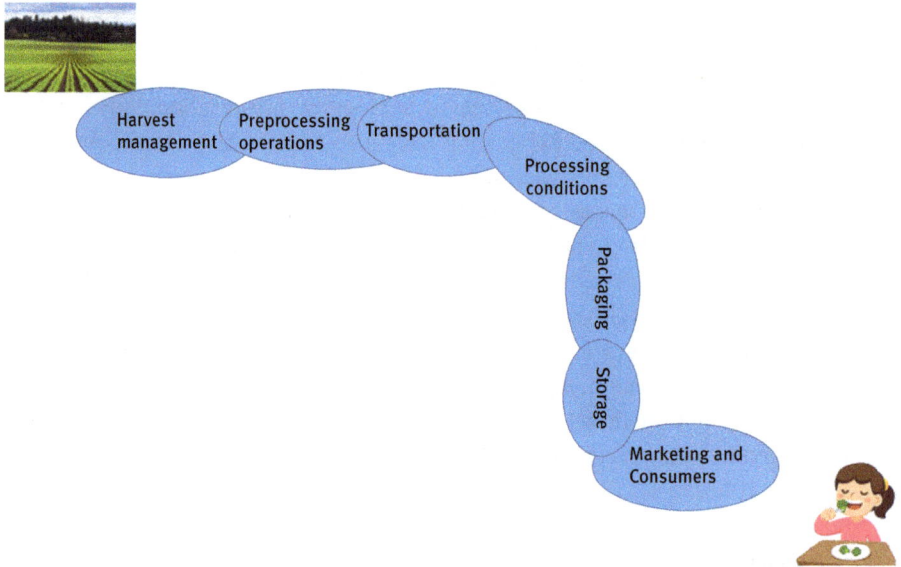

Figure 1.1: The food processing pipeline.

Harvest management: Different harvest management practices are implemented depending on the type of raw material to ensure wholeness and safety to the consumer. The conditions of the raw material during harvesting usually affect the end-product of food. These can be influenced by fertilizer application, pruning, irrigation, cultivar type, and stage of maturity before harvesting (Arah et al. 2015). Although the use of human biosolids and animal waste give a significant economic cost savings, the potential risks and threats to food safety in the manufacture of raw materials must be balanced particularly during irrigation/fertilizing because it can affect the quality and quantity during processing and storage (Wei and Kniel 2010). Therefore, management strategies require multidimensional approaches that mitigate risks from multiple sources such as risk of contamination during crop growth, harvesting, and postharvesting (Falade 2018). The food chain needs to be seen as a whole in order to implement preventive measures in the food chain, from farm to fork. Preharvest has been considered the major contributor to finished product contamination in previous years, which in many countries is a current issue.

Recent methods, however, explain the importance of understanding the entire chain and using scientific evidence to handle and control foodborne pathogens. Food testing should therefore be considered not only for finished products, but also for prefarming and food processing environments (Rivera et al. 2018). A key and crucial aspect of the development of the food value chain also depends on improved postharvest management that, in turn, makes it possible to better and more efficiently meet consumer demand, reduce costs, and increase benefits. The causes of postharvest losses are primarily related to financial administrative and technological shortcomings in harvesting methods, pest control, storage, and refrigeration facilities under challenging climate conditions, infrastructure, packaging, and marketing systems (Falade 2018; Kalaitzis et al. 2019).

Preprocessing operations: Depending on the type of raw materials (perishable, semiperishable, and unperishable), its condition at harvest and the knowledgeable skill of handler, preprocessing operations influence the quality and safety of final product. The mechanical operations employed immediately after harvesting determine the physical structure/integrity of the raw material (Floros et al. 2010). The temperature, water activity, nutrient composition of the raw material (which may initiate microbial contamination), as well as degree of pest infestation are critical factors for the quality and safety of such products (Floros et al. 2010; Kumar and Kalita 2017). Also, the tradition of each society largely depends on the economic, biological, cultural, and political contexts for preprocessing food produce.

Transportation: Moving raw material from one location to another is a critical factor for consideration during the food chain. Depending on the time, the method of transportation, and conditions of raw materials during transit, the quality, quantity, and safety of food can be impaired during this step. Perishable foods are susceptible to natural persistent and permanent biophysicochemical changes (high-water content, protein presence, carbohydrates, and fat causing oxidation and rancidity processes) and the use of properly selected methods of cooling (freezing), storage, and distribution conditions and continuous monitoring and control in compliance with the relevant qualities (Jakubowski 2015). Together with the multitude of container, temperature, and handling criteria for each food product, the sheer quantity and variety of foods transported highlights the food industry's vulnerability to potential contamination during transport (Ackerley, Sertkaya and Lange 2010). Contamination risk factors that may protect raw food products against physical, chemical, and microbial contamination during transportation and holding include abuse of temperature, unsanitary storage fields, unsafe loading or unloading practices, defective packaging, ill-repaired shipping containers, poor employee habits, and road conditions (Ackerley, Sertkaya and Lange 2010; Floros et al. 2010; Arah et al. 2015). The earlier an undetected problem is introduced into the system in the complex food transportation system, the higher the risk (as calculated by the probability and effect of exposure); that is, an issue that is introduced earlier in the supply chain will spread to many manufacturers, retailers, and then customers just because of the system structure. Sources indicate that tampering and sabotage, temperature manipulation, and cross-contamination are the major concerns for food safety during transportation (Ackerley, Sertkaya and Lange 2010).

Processing conditions: Diverse applications have been employed in the processing of food raw materials. The industry location, design, equipment layout, exterior landscaping, drainage, and waste disposal are critical factors essential for food processing. These parameters influence pest control, intercompartment processing, handling patterns, level of contamination on food, storage, and distribution management after processing (Campbell, Arthur and Mullen 2004; Cramer 2013). Raw materials arriving at processing plants may have been exposed to various sources of contaminants, so cleaning, sorting, and grading are the logical order of the preliminary operations. Throughout these operations, emphasis is placed on maintaining sanitary conditions, preventing product losses, ensuring raw material quality, reducing bacterial growth, and scheduling all transfers and deliveries to reduce the holdup time, which can be both expensive and harmful to the quality of the product (Ortega-Rivas 2012). Dehydration is among the most important steps used in food preservation by means of moisture control to protect and add value to food products. Conduction, convection, and radiation are also the essential methods that allow water to vaporize and expel the resulting vapor either naturally or by force that results in dehydration (Eswara and Ramakrishnarao 2013). Successful drying depends heavily on the properties of the food material (Qiu, Boom and Schutyser 2019). Heating conditions, temperature holding patterns, and food additive(s) that may be used enhance organoleptic qualities during processing are also important factors that determine the successful processing operations of food product and type of microbial growth during and after such processing (Tamme et al. 2009; De Silvestri et al. 2018). There may be intrinsic and extrinsic factors which can cause microbial spoilage in foods. The intrinsic properties of foods dictate the expected shelf life or perishability of foods and also influence the form and rate of microbial spoilage. The key intrinsic properties related to food spoilage are endogenous metabolites, substrates, light sensitivity, and oxygen. These properties can be regulated during the formulation of food products in order to monitor food quality and safety. pH, water activity, nutrient content, and oxidation–reduction capacity are intrinsic causes of food spoilage while extrinsic factors include relative humidity, temperature, presence, and activities of other microbes and these parameters determine the type of preservation method employed (Amit et al. 2017). Good manufacturing practices (GMP) are the procedures required to comply with guidelines provided by regulatory agencies that regulate food product production and sale. Those standards include minimum requirements that a manufacturer of food products must meet during manufacture to ensure that the goods are of high quality and present no safety risks (Featherstone 2015).

Jones (1992), in the past, has reported the advantages and disadvantages of food processing:

Advantages
- Food processing has the ability to make food look, taste, and smell as though it had not been preserved but to allow it to be safely eaten at a later time.
- Development of totally new food products, experiences, and levels of convenience
- Food processing frees populations from dependence on geography, climate, and other environmental factors.
- Food processing helps to convert agricultural products into forms more suitable or desirable for human consumption.
- Food processing helps to extract commodities that are not readily available in the food when harvested, for example, hydrogenation of oils into margarines for use as instead of butter.
- Food processing makes food safer by removing or killing pathogenic microorganisms.
- Food processing encourages allow choice. It is possible to choose between full cream, low fat, and no fat milk.
- Food processing and convenience go together, so do fast food and food processing.

Disadvantages
- All forms of food processing cause nutrient loss.
- Some products do not really provide nutritional benefits to the consumer, for example replacing real juice with flavored water.
- Some products are high in constituents such as salt, fat, or cholesterol which can be dangerous to health.
- Processed foods may also be low in certain important constituents of the diet, such as fiber.
- Processed foods also depend greatly on packaging which may create a possible environmental concern.
- Skills of food preparation (cooking) may be lost in a family where there is heavy dependence on processed foods.
- Another demerit of food processing (i.e., fast food) is eating of junks which have led to obesity and overweight.

Packaging: Food packaging is used to safeguard food from contamination of the atmosphere and other effects (such as odors, vibrations, dust, temperature, physical damage, light, microorganisms, and humidity) and is crucial for maintaining the quality and safety of food, while also increasing the shelf life and reducing food loss and waste. Nonetheless, offering environmentally friendly packaging alternatives without compromising the packaging's key features (such as barrier properties, mechanical properties, and extended product shelf life) will require continuous innovation and the development of new sustainable packaging technologies in the decades to come (Han et al. 2018). A variety of packaging material, ranging from plastics (polyvinyl chloride, polyethylene and varieties, and polystyrene), metals (steel, tins, and aluminum), glass, woody materials (cardboard and papers) and other advanced packaging materials such as three-dimensional (3D), nanopackaging, and biomedical packaging have been employed for different food product (Cha and Chinnan 2004; Lu and Wong 2009; Malhotra, Keshwani and Kharkwal 2015; Said and Sarbon 2019).

Storage: Food storage is an important aspect of the food chain, as losses can be attributed mainly to storage methods on-farm and postharvesting. The primary purpose of storing food is to ensure the availability for human consumption of harvested and processed food products (Barba et al. 2019). A wide variety of storage systems are used depending on the type of food to be stored, including wire clothing, polythene containers, metal silos (for imperishable and, in some cases, semiperishable food items), and cold chain for highly perishable food products such as fruit and vegetables, livestock, and fishery products. Nevertheless, by investing in enhanced and user-friendly storage methods for postharvest processing, minimizing food losses increases food safety and reduces economic losses (Owach, Bahiigwa and Elepu 2017). Moisture content and water activity (a_w) are vital parameters that affect most food products' shelf life. Excessive amounts of water will cause spoilage in most cases, while loss of water will make food unacceptable in other cases. Moisture management in foods requires knowledge of their sorption/desorption behavior, essential humidity content, temperature and relative humidity environmental storage conditions, packaging parameters including volume-to-area ratios, and permeability (Mannheim, Liu and Gilbert 1994).

The shelf life of food after storage is a critical factor to the acceptance of such food products to consumers. For most food products, the quality does not improve during storage, hence several parameters must be considered to determine the effect of storage on the flavor, color, texture, nutritional composition, and potential microbial growth of the food material (Tanner 2016). Food storage can change the content of nutrients and bioactive compounds as well as the chemical shape. It also affects their bioavailability (positive or negative) depending on storage conditions (e.g., time and temperature) and the compounds being targeted (Barba et al. 2019).

Marketing and consumers' handling pattern: Efforts are beyond cleaner production to sustainable use in order to achieve sustainable development. It includes consumer preference for greener goods, which during processing leads to less pollution. A coherent marketing strategy is required, based on a deeper and more detailed understanding of food consumers and their purchasing behavior. Consumer behavior is evolving due to quality and environmental awareness of different food products (Basha et al. 2015). Market structure and marketing channels have become more complex and rely on the marketing strategy being followed by the food marketing network actors. The process of exchanging food and agricultural products may occur through institutionalized markets or through a direct relationship between producer and retailer (Meulenberg and Viaene 1998) as these influences the end product of the food material. In addition, marketing is affected by food accessibility, food availability, and food choice, which in turn may be influenced by geography, demography, disposable income, urbanization, globalization, marketing, religion, culture, and consumer attitudes (Kearney 2010). Consumers are expected to have several roles in food safety, because consumers not only purchase and obtain goods but also prepare and provide food for themselves and others. Failure to implement food hygiene practices can increase cases of foodborne diseases and the way customers handle food influences the risk of proliferation of pathogen as well as cross-contamination to other food products (Redmond and Griffith 2003).

Food-borne/related diseases have increased over the years and have had a negative impact on many nations' health and economic well-being, mainly due to contaminated food and drinking water ingestion. Food poisoning is caused by consuming food infected with microorganisms or their metabolites, by contamination caused by insufficient storage procedures, by unhygienic handling practices, by cross-contamination from food contact surfaces, or by persons harboring microorganisms in their nares and eyes (Jay, Comar and Govenlock 1999). Unhygienic practices during the processing, handling, and storage of food create the conditions for the proliferation and transmission of diseases that cause organisms (Akabanda, Hlortsi and Owusu-Kwarteng 2017). Different factors such as the general health requirements, handwashing lavatories, refuse management systems, and dishwashing facilities impact food safety in food establishments. Other important factors affecting food safety are product storage, preparation, and service activities. Cooking utensils conditions, food storage systems (time and temperature), as well as the experience and behaviors of food handlers have a direct or indirect impact on food safety. Food hygiene depends directly on the personal hygiene and practices of the workers at the establishments (Ifeadike et al. 2014).

1.4 Environmental conditions affecting the food industry

A variety of parameters are usually considered as dimensions to food industry within an industrial food web perspective. Multiple environmental variables are involved. The national economies' growth prospects are considered to meet their resilience and competitiveness criteria, including the design and implementation of domestic policies to meet global needs. Globalization, which refers to the way in which changes in one area may easily have a significant impact on the food consumption pattern, health, and well-being of people (Olayiwola, Soyibo and Atinmo 2004) is a complex interrelated phenomenon between the economic, social, cultural, and political environmental pattern and responses (Najam, Runnalls and Halle 2016). The trade and foreign investments of food material/product, which may directly influence the flow of information, people and national policy are critical factors to the

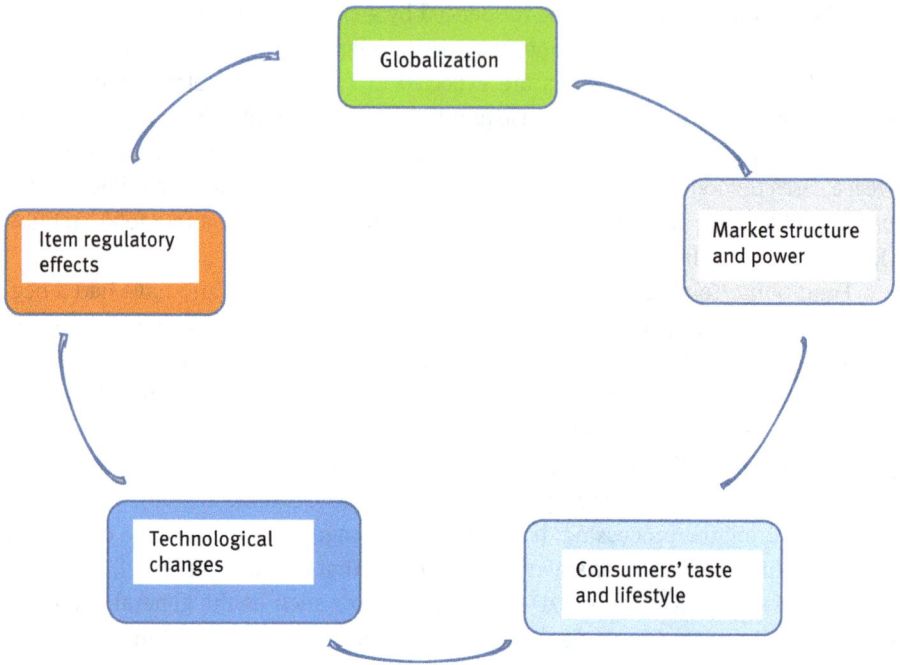

Figure 1.2: Environmental dimensions affecting the food industry.

socioeconomic development of urbanization and food preferences (Cuevas García-Dorado et al. 2019). Establishing a global market for food products has significant effects on food commodities' supply and costs. On the production side, global markets promote specialization in export crops, which tend to create economies of scale in agricultural and food production, leading to higher global performance, but also homogenizing food availability (Khoury et al. 2014; Cuevas García-Dorado et al. 2019). On the demand side, through imports, countries will increase their access to a variety of commodities, including vital food and healthy foods, as well as potentially unhealthy processed and ultraprocessed items. Access to international commodity markets will reduce the volatility of food prices by reducing the impact of local shocks (Baker et al. 2016; Cuevas García-Dorado et al. 2019).

The food choices, taste preferences, personal and cultural ideas, family structures, employment status (wages, length of time devoted to work, work-life stress), acculturation among immigrant populations, and access to personal transportation are outcomes of globalization which will influence the market structure/organization, policies, and legislative power in place (Caswell, Yaktine and Council 2013; Bartkiene et al. 2019). National food control systems such as food legislation and regulations, policy and institutional frameworks, food inspection and monitoring, food laboratory services, involvement of all stakeholders, and dissemination of information are designed to address specific needs and priorities of countries. This involves understanding that food safety is a broadly shared obligation and requires cooperation between all actors in the farm-to-table process, creating a systemic, co-ordinated, and proactive approach to minimizing contamination risks throughout the food chain, the most efficient way of producing safe food, implementing sci-ence-based control strategies and prioritizing activities based on risk investigation and management strategies, creating emergency measures for dealing with specific hazards (Omojokun 2013). Such innovations have enabled rapid advancements in information technology to improve product quality, productivity, and profitability for food service providers with creative technologies such as hardware or software developments. Innovation is occurring at an ever-increasing pace and changing business models in the industry drastically, showing benefits in the areas of qual-ity, cost, speed, reliability, flexibility, and employee training (Pantelidis 2009). Although scientific and technological advances have allowed some nutritional defi-ciencies to be addressed, address food safety and quality, and feed larger popula-tions, more progress is needed to address the challenges of sustainably feeding the increasing future population in developed and developing nations alike. In reality, it is critically important to accelerate and implement scientific and technological developments in both the agricultural and food producing sectors in order to meet the food needs of the future (Floros et al. 2010). To ensure that each food organiza-tion meets the required standard, a regulatory system is generally structured. Different policy and administrative tools governments are used to ensure the safe manufacture, labeling, distribution, and marketing of food products (Buckley and Riviere 2012)

1.5 Design and layout of food processing facilities

Food processing or production is a value-chain system which primarily starts from farming operation to production and eventual consumption. These operations are regarded as pre- and postharvest operations. However, to ensure quality preserva-tion and safety, food processing operations are usually structured or designed into a facility layout to accommodate all that is needed for processing operations, stor-age, and optimal productivity with sanitary and safety principles in mind (Smith 2006; Heinz 2013).

The design and layout of food processing facility is premised to incorporate re-duction of harborages and elimination of food contamination. These facilities in-cludes, among others; processing and/or control equipment, utility, plant building, warehouse, and water and waste treatment sections.

Importantly, safety and hygiene principles represent a significant factor of con-sideration in the design and layout of food processing facility. Hence, critical measures are being put in place to attain safety requirements at maximum production economy (Wijtzes et al. 1998; Holah 2003). These principles include hazard analysis critical control points (HACCP), GMP and good hygiene practices (GHP) (Wanniarachchi et al. 2016). More often, the design food processing layout incorporates various inclusions of materials based on availability, cost, and sanitary qualities in order to meet required outputs (Tak and Yadav 2012). However, other salient parameters which contribute significantly to design and layout of food processing facilities including site of location, materials for internal and external building, and water usage are discussed below.

In the design of food processing facility, the site of location plays a primary role. Facilities should be designed in a location that is quite away from contamination sources such as chemical plants, waste or effluent treatment plants as such could cost the company more capital to maintain process control and hygiene (Saravacos and Kostaropoulos 2002). The orientation of the plant should create ample possibilities of airwaves around the parking lots and administrative building while the loading bay and warehouse located in opposite direction to environmental effect of sunshine, rain, or wind (Moerman and Wouters 2016). Furthermore, access to raw materials, product market, energy and water supply, waste or effluent treatment, labor supply and environmental laws, legislations, and tax systems should also be considered. Plants processing large chunk of raw materials (e.g., cocoa crop) into export products should be located near plantation site, while those using large quantities of imported raw materials (e.g., wheat flour) should be sited near the seaports. Site condition should be thoroughly cleared and cleaned of any potential toxic materials that could impart contamination of food production process, and should be graded for appropriate drainage system which will prevent stagnancy of effluent water. Storm sewers should be designed and located to allow for smooth runoff (Park et al. 2018).

The physical condition of food raw materials and end product largely influences the design and layout of a food processing facility (Wanniarachchi et al. 2016). Hence, qualities such as gravity flow; rheology can be applied in processing liquid and grains. In planning high-moisture foods such as fruit juice beverage and dairy products, high humidity is needed in packaging and storage while low-moisture foods such as wheat flour, low-humidity packaging and storage is needed. Food management principles such as HACCP and GMP have been adopted to enhance effective process control in food processing plants, so that foods produced are safe for consumption while quality assurance measure helps to enhance quality and minimize cost (Van Donk and Gaalman 2004; Fellows 2009).

The food plant is basically designed to accommodate processing and utility equipment with other allied facilities such as storage areas, laboratories, workers' common room, and restaurants. The type and nature of building in a food plant layout is determined by the processing operation. Most building is often one-story building and constructed with steel frame (Sule 2008). The plant foundation should be designed with strong concrete to prevent cracks and harborage site for pests and microbial proliferation. For example, food plant floor should be made of concrete slabs to withstand the weight and vibrations of heavy equipment which can be leveled to accuracy range of ±4 mm in 2 m (normal) to ±1 mm in 2 m (flat). Generally, food plant floors should be durable with high resistance to shock, sound, and vibrations as well changes in temperature, humidity, or chemical spillage. It should also not be too slippery and be easy to clean. Also, floor drains should be designed to prevent movement of pests and rodents using water traps and sewer screens respectively (Moerman and Wouters 2016). Roofs could be an entry point of microbial contamination as a results of bird droppings, air pollutions around processing location, or food waste from air extracts. Hence, this development calls for control during layout design (Holah 2014). Choice of roof for a food plant layout could vary from long aluminum or steel spans of composite fireresistant materials. This roof material could come in different shapes such as saw tooth, monitor, dome, and flat shapes. The roof should be designed to lend itself to regular cleaning. The roofs should be sealed hermetically to prevent leakages from water and pest infestations.

1.6 Food plant water usage

Enormous quantity of water is used in food production and processing. According to IFT (2015), eight hundred gallons of water per day are used to produce enough food for one person. Also, one person eats 2,000 to 5,000 L of virtual water embedded in food per day (IFT, 2015). At the same time, recent factors of climate change, population increase, and economic development has been reported to contribute significantly to water scarcity thereby affecting food value chain both at preharvest (irrigation) and postharvest period (processing) (Meneses et al. 2017). In food processing plant, high-quality water plays a primary role either as part of ingredients or raw food material and sanitation process, hence proper and adequate attention must be paid for its utilization. Table 1.2 shows the water food print/usage of different foods.

Furthermore, because water could serve as vehicle for microbial contamination of foods from the source, plant facilities such as water-holding facilities and distribution channels, and personnel involved in processing strict HACCP safety regulations have been legislated by various public health agencies to maintain a healthy society (Kirby et al. 2003; Casani et al. 2005; EPA 2012). In recent times, most food processing plants have designed various logistic mechanism to reduce water usage and consumption in order to reduce cost of excess effluents and

Table 1.2: Water food print/usage of different foods.

Food	Liters
Chocolate	17,196 L/kg
Beef	15,415 L/kg
Pork	5,988 L/kg
Chicken	4,325 L/kg
Eggs	3,300 L/kg
Cheese	3,178 L/kg
Rice	2,497 L/kg
Wheat bread	1,608 L/kg
Maize	1,222 L/kg
Apple	822 L/kg
Orange	560 L/kg
Potato	287 L/kg
Milk	255 L/glass
Coffee	132 L/cup
Wine	109 L/glass
Beer	74 L/glass

Adapted from IFT (2015).

disposal treatments. These mechanisms include decrease in unregulated use, water reuse, and recycling after treatment and enhanced layout design.

In reducing water usage, food plant managers, being aware of the huge operating cost and financial implication of using water both for production and effluent treatment, continue to adopt veritable total quality management approaches to reduce water usage by relevant public agencies (European Commission 2006). Other approaches in water usage in food processing plants involves water reclamation, reconditioning, recycling, and reuse via application of appropriate water treatment technologies (Table 1.3).

Table 1.3: Some water treatment technologies and their applications (Asano and Levine, 1998).

Process	Description	Application
Solid/liquid separation		
Sedimentation	Gravity sedimentation of particulate matter, chemical floc, and precipitates from suspension by gravity settling	Removal of particles from turbid water that are larger than 30 μm
Filtration	Particle removal by passing water through sand or other porous medium	Removal of particles from water that are larger than about 3 μm. Frequently used after sedimentation or coagulation/flocculation
Biological treatment (wastewater)		
Aerobic biological treatment	Biological metabolism of wastewater by microorganisms in an aeration basin or biofilm process	Removal of dissolved and suspended organic matter from wastewater
Oxidation pond	Ponds up to 1 m in depth for mixing and sunlight penetration	Reduction of suspended solids, BOD, pathogenic bacteria, and ammonia from wastewater
Biological nutrient removal	Combination of aerobic, anoxic, and anaerobic processes to optimize conversion of organic and ammonia nitrogen to molecular nitrogen (N_2) and removal of phosphorus	Reduction of nutrient content of reclaimed water
Waste stabilization ponds	Pond system consisting of anaerobic, facultative, and maturation ponds linked in series to increase retention time	Reduction of suspended solids, BOD, pathogens, and ammonia from wastewater
Disinfection	The inactivation of pathogenic organisms using oxidizing chemicals, ultraviolet light, caustic chemicals, heat, or physical separation processes (e.g., membranes)	Protection of public health by removal of pathogenic organisms
Advanced Treatment		
Activated carbon	Process by which contaminants are physically adsorbed onto the surface of activated carbon	Removal of hydrophobic organic compounds

Table 1.3 (continued)

Process	Description	Application
Ion Exchange	Exchange of ions between an exchange resin and water using a flow through reactor	Effective for removal of cations such as calcium, magnesium, iron, ammonium, and anions such as nitrate
Membrane filtration	Microfiltration, nanofiltration, and ultrafiltration	Removal of particles and microorganisms from water
Reverse Osmosis	Membrane system to separate ions and particles from solution based on reversing osmotic pressure differentials	Removal of dissolved salts and minerals from solution; also effective for pathogen removal

1.7 Food operational management

The operational structure in any food industry varies depending on size, production capacity, equipment, and so on, however, it is important to note that the different departments are working together to achieve a common goal which is to produce a wholesome product that meets the consumers specification. The following units may be in a typical food processing industry:

Engineering department: The engineering department consists of a team of engineering experts and technicians whose roles are daily maintenance and operation of all processing edifice, automation controls, and other technological equipment in the production floor (factory). This unit is usually headed by factory engineer that oversees by ensuring that all equipment and machinery are in a functional state, in addition to this they are in charge of all maintenance plan the entire organization.

Production department: This department deals with all processing activities involving the conversion the raw material to packaging. They create the products of the company with the use of expertise which could be manmade or machine. The unit is overseen by production managers or factory managers who supervise a team of production technicians and shop floor workers to ensure the smooth running of the food production line. Other responsibilities of this unit include ensuring the compliance to GMP and good housekeeping protocols to prevent the production of unwholesome products or those with food safety issue. They liaise with the quality team, warehouse, and distribution units to achieve their objectives.

Technical department: In most world-class food industries, the department houses the quality assurance/control unit, research and development unit, and organoleptic/ sensory evaluation unit. The main facilities that are integral to this department are the microbiology, quality, and sensory laboratories. The research and development unit usually known as R&D are strategically designed for innovations toward promoting a company's brand and their activities including introducing new products, ideas, and processes that supports the continuous improvement of the company visions or goals. They work with all units of the organization, customers, and consumers by collecting useful information and feedbacks to formulate a new idea, product, or process. The quality assurance been part of the technical department is the custodian of the implementation of the quality and food safety management systems within the organization and are often regarded as the police of the system. The major function of the department is to monitor continuously the required parameters from raw material delivery to finish products in order to produce a product that meets consumer's needs. They are accountable for the production of wholesome and safe food products, monitor the entire production process from the receipt of raw materials till product is finally consumed using initiatives like GMP, GHP, and HACCP to prevent any form of contaminants or allergens. Food chemistry, microbiological assessments, and hygiene monitoring applications are employed in carrying out their roles and responsibilities within the organization. A quality assurance manager supervises the entire process and sometimes might be expected to halt production process if the appropriate quality criteria are not maintained; however that can always lead to a bone of contention between the QC manager and production manager due to the fact that their targets are not alignment usually the quality manager is enforcing compliance whereas Production manager is targeting customer orders (i.e., quantity). The validation of processes, construction materials and equipment (e.g., the type of material to be used on food machine must be food grade) used in the production facility are also one of the many responsibilities of this unit. In the food industry, assessment of food contact surfaces is necessary to determine whether equipment is properly cleaned and/or sanitized and whether living pathogenic are present (Fratamico et al., 2009). All these are carried out by quality personnel in accordance to Food safety Laws and regulation. Summarily, quality control or quality assurance section have personnel that

- monitor compositional standards of raw material
- monitor compositional standards of finished products
- ensure that governmental regulations are followed
- enforce compliance with Food safety tools GMP, GHK, and HACCP

The organoleptic/sensory evaluation unit is responsible for assessment of the organoleptic properties of raw materials, intermediate products, and finished goods.

Sales and Marketing: All sales promotion, publicity, and activities are the key responsibilities of this unit of the organization. They monitor how their brands affect the market forces and whether the food industry will make profit or not depends on the sales and marketing department. They liaise often with almost every other unit within the organization for effective achievements of their goals.

Procurement, finance, and human resources department: All these departments are important for proper financing and planning for the overall profitable operation of the company.

Logistics and planning: This unit houses the warehousing, distribution, and fleet management. One important section of the food industry is the warehouse/distribution. The warehouse unit manages the receipt and storage of raw materials, finished goods, packaging material, and auxiliary materials. It is common to have refrigerated storage areas, that is, cold room and transportation vehicles like trucks and tankers, material-handling equipment such as fork-lift trucks, palletizing machinery, and workshop for maintenance of transportation vehicles in this section. Good GHP and GMP are enforced in this section, especially storage condition in the warehouse should be carefully controlled to avoid deterioration of products in quality. Warehouses are separated according to content occupied, that is, some warehouse are designated for the raw materials while others are for the finished goods (food products), auxiliary materials, and packaging materials. Incoming inspection of raw materials, auxiliary materials, and packaging material are done and received by the warehouse by approved suppliers (i.e., suppliers already certified by auditors to provide quality materials) and are used when procuring these materials. This is important because poor-quality raw material will give rise to a poor-quality finished product (remember the saying, garbage in, garbage out).

Environmental sustainability: This could be an adjunct to the quality department in most food industry while in others it is an independent unit which can also be outsourced. They deal with the management of waste such as solid wastes, effluents, and also carry out projects like conservation of water and energy in the food industry. Regulatory bodies responsible for the environmental concerns in most countries have made it mandatory for all wastes from factory facilities into the community environment to control in order to minimize the impact on environment, human health, and ecosystem. A separate unit in manufacturing facility is required to carry out tests of all effluents before they are released into the immediate environment and disposal of solid waste appropriately. There are occasional visits to the factory by regulatory bodies to inspect and check records. If a food facility is found defaulting, it may be shut down. They liaise between the engineering and quality team to achieve their goals. This unit of the food industry ensure that they do not contribute to the pollution of the environment. This is achieved by having an efficient waste treatment facility (effluent point) for the treatment of processing wastes, wash water, discarded whey (if the dairy plant is processing milk to cheese). This is imperative because untreated wastewaters from food industries have high content of biological matter which can serve as nutrient supply for various microorganisms in rivers and other water bodies but also could be detrimental to the environment if not controlled. In recent times, community social responsibility projects like source water protection, recycling, and water conservation are been supported by big food industries.

Health and safety: This is another independent unit within the organization required by regulation. This unit is responsible for the enforcement of occupational health and safety within the food facility. It is spearheaded by a safety manager who supervises safety marshals. The roles and responsibilities expected involves every activity that minimizes risks and accidents. This ranges from planning and execution of fire drills, investigating and recording near misses while considering root causes, and correction action plans to prevent them. This unit works hand in hand with medical personnel in the factory clinic, ambulance, and also admission of staff into hospital in the case of a major accident. The roles of the unit also involve assessment of ergonometric conditions of staff workstations, monitoring the safety condition of all factory machinery, equipment, vehicles, and construction at large, enforcing the use of personal protective equipment. The importance of this unit cannot be overemphasized due to the fact that it is a requirement of the law to monitor fatal accidents and death in factory. Any death occurring must be reported to the appropriate authorities and investigated if the factory found wanted could lead to the closure of the facility.

1.8 The future of food industry

Numerous factors such as technologies, consumer changing food preference and climate change mitigation strategies is changing the way food is produced, processed, distributed, and sold to consumers. GMOs, 3D printing, blockchain, nanotechnology, drone in farming, internet of things, artificial intelligence, big data, and virtual reality are some technologies that will make the food industry more smarter, safer and environmentally friendly. According to Salmon (2017), "food retailing and production are changing around the globe. From how food is designed and where it's grown, to how it's consumed and who is consuming it." Salmon (2017) also reported digital platforms, urban agriculture, food as a service, vertical farming, DNA-based diets, and lab-grown meat to be trends that will change the food industry significantly in the next 10 years than in the last 50. Also because of environmental and health concerns, many consumers are giving up or reducing consumption of meat. There is now huge demand for plant-based diets. Beyond meat and impossible foods are examples of companies in this space. While impossible meat has produced hi-tech vegan burgers from water, soy protein concentrate, coconut oil, sunflower oil, and natural meat flavoring from a molecule called found in animal muscle tissues as well as in small quantities in legumes, Beyond main ingredients are proteins derived from peas, rice, mung beans, canola, and coconut (Goldstuck, 2020). Novameat, a Spanish company, recently announced that it has developed "the most realistic" plant-based steak that has the texture and appearance of the real one. The product, whose ingredients include pea, seaweed, and beetroot juice, is projected to be available in restaurants in Spain and Italy in 2020, and by 2021 it will be introduced into other countries and supermarkets (Mehmet, 2020).

Another food of the future is insect. According to Stull (2019), edible insects such crickets are environmentally friendly protein source. In addition, they require less land, feed, water for their production unlike traditional livestock. In addition, they also emit fewer greenhouse gases (Stull, 2019). Insects and other food products whose production have less impact on the environment will be of high demand in the future. According Tai et al. (2014), "future food production is highly vulnerable to both climate change and air pollution with implications for global food security." Climate change adaptation and ozone regulation will, therefore, be important strategies to protect the global food production and supply (Tai et al., 2014). Example of such strategies or initiatives is the target set by AB InBec, the world's largest brewer, to buy 50% of their electricity from renewable sources by the end of 2020 and all of by 2025 (Furlonger, 2020). It is also important to protect the food supply from existing and emerging food safety risks including toxicological chemical risks, toxicological biological risks (mycotoxins, phytotoxins, phycotoxins), microbiological risks, veterinary drug residue in animals, risky consumer behaviors (e.g., consumption of expired products, eating with unwashed hands), zoonoses, plant diseases, unintended effects of GM, unintended effects of nanotechnology, unintended effects of other new technologies, radioactive contaminations, plant pests, pesticide residues, pollutants unrelated to agricultural production (e.g., contaminants from mining activities), and growth hormones in animal products (European Commission, 2016). For this to take place, food safety must be a global priority and there should be continuous food safety strides through improved research, innovation, leadership, and collaboration.

1.9 Conclusion

The food industry has evolved over the years with industrial revolution being the main contributing factor. This diverse food business from farm to fork will continue to enable advancement through provision of nutritious, safe, and wholesome food. However, problems of food waste, water usage, and other sustainability issues remain a daunting challenge. Also, food production will continue to be vulnerable to food safety risks and climate change. These problems, undoubtedly, can be overcome through adoption of technologies, willingness to change, cooperation among stakeholders and innovation. For example, win–win strategies for climate as well as technologies such as artificial intelligence, blockchain, and internet of things will offer solutions to food insecurity challenges. Furthermore, food industry is becoming more committed to sustainability and also willing to implement circular food economy.

References

Ackerley, N., Sertkaya, A. and Lange, R. (2010). Food transportation safety: characterizing risks and controls by use of expert. *Food Protection Trends*, 30(4), 212–222.

Akabanda, F., Hlortsi, E. H. and Owusu-Kwarteng, J. (2017). Food safety knowledge, attitudes and practices of institutional food-handlers in Ghana. *BMC Public Health*, 17(1), 40.

Alifoods. (2016). *Food and Taste Blog. Historical Origin of Food Preservation*. Global Italian Food Traders. https://www.italian-feelings.com/historical-origins-of-food-preservation-technique/ (Accessed January 13, 2020).

Amit, S. K., Uddin, M. M., Rahman, R., Islam, S. R. and Khan, M. S. (2017). A review on mechanisms and commercial aspects of food preservation and processing. *Agriculture & Food Security*, 6(1), 51.

Arah, I. K., Amaglo, H., Kumah, E. K. and Ofori, H. (2015). Preharvest and postharvest factors affecting the quality and shelf life of harvested tomatoes: a mini review. *International Journal of Agronomy*, 2015, 6.

Asano, T. and Levine, A. D. (1998). Wastewater reclamation, recycling, and reuse: an introduction. In: T. Asano (Ed.), *Wastewater Reclamation and Reuse* (pp. 1–56). Lancaster, PA: Techonomic Publishing Company.

Baker, P., Friel, S., Schram, A. and Labonte, R. (2016). Trade and investment liberalization, food systems change and highly processed food consumption: a natural experiment contrasting the soft-drink markets of Peru and Bolivia. *Globalization and Health*, 12(1), 24.

Banks, G and Overton, J 2010. Old world, new world, third world? Reconceptualising the worlds of wine. *Journal of Wine Research*, 21 (1): 57–75.

Barba, F. J., Munekata, P. E. S., Lorenzo, J. M. and Cilla, A. (2019). *Health Effects of Food Storage*.

Bartkiene, E., Steibliene, V., Adomaitiene, V., Juodeikiene, G., Cernauskas, D., Lele, V., Klupsaite, D., Zadeike, D., Jarutiene, L. and Guin, R. P. (2019). Factors affecting consumer food preferences: food taste and depression-based evoked emotional expressions with the use of face reading technology. *BioMed Research International*, 2019, 10.

Basha, M. B., Mason, C., Shamsudin, M. F., Hussain, H. I., Salem, M. A. and Ali, A. (2015). Consumer acceptance towards organic food. *GJISS*, 4(3), 29–32.

Buckley, G. J. and Riviere, J. E. (2012). *Ensuring Safe Foods and Medical Products Through Stronger Regulatory Systems Abroad*. National Academies Press.

Campbell, J. F., Arthur, F. H. and Mullen, M. A. (2004). Insect management in food processing facilities. *Advances in Food and Nutrition Research*, 48(2), 239–295.

Campbell-Platt, G. (2011). *Food Science and Technology IUFoST*. Wiley-Blackwell, UK.

Casani, S., Rouhany, M. and Knøchel, S. (2005). A discussion paper on challenges and limitations to water reuse and hygiene in the food industry. *Water Research*, 39(6), 1134–1146.

Caswell, J. A., Yaktine, A. L. and Council, N. R. (2013). Individual, household, and environmental factors affecting food choices and access. In: *Supplemental Nutrition Assistance Program: Examining the Evidence to Define Benefit Adequacy*. National Academies Press (US).

Cha, D. S. and Chinnan, M. S. (2004). Biopolymer-based antimicrobial packaging: a review. *Critical Reviews in Food Science and Nutrition*, 44(4), 223–237.

Cramer, M. M. (2013). *Food Plant Sanitation: Design, Maintenance, and Good Manufacturing Practices*. CRC Press.

Cuevas García-Dorado, S., Cornselsen, S. R. and Walls, H. (2019). Economic globalization, nutrition and health: a review of quantitative evidence. *Globalization and Health*, 15(1), 15.

De Silvestri, A., Ferrari, E., Gozzi, S., Marchi, F. and Foschino, R. (2018). Determination of temperature dependent growth parameters in psychotrophic pathogen bacteria and tentative use of mean kinetic temperature for the microbiological control of food. *Frontiers in Microbiology*, 9(3023).

Drummond, E. H. and Goodwin, J. H. (2014). *Agricultural Economics (Third Edition)*. Pearson New International.

EPA. (2012). *Guidelines for Water Reuse*. nepis.epa.gov/Exe/ZyPURL.cgi?Dockey¼P100FS7K.TXT (Accessed January 13, 2020).

Eswara, A. R. and Ramakrishnarao, M. (2013). Solar energy in food processing – a critical appraisal. *Journal of Food Science and Technology*, 50(2), 209–227.

European Commission. (2006). *Integrated Pollution Prevention and Control. Food, Drink and Milk Industries*. http://eippcb.jrc.ec.europa.eu/reference/BREF/fdm_bref_0806.pdf (Accessed January 13, 2020).

European Commission. (2016). *Final Report Summary – COLLAB4SAFETY (Towards Sustainable Global Food Safety Collaboration)*. https://cordis.europa.eu/project/id/311611/reporting (Accessed January 21, 2020).

Falade, T. (2018). Aflatoxin management strategies in sub-saharan Africa. In: *Fungi and Mycotoxins-Their Occurrence, Impact on Health and the Economy as Well as Pre-and Postharvest Management Strategies*. IntechOpen.

Featherstone, S. (2015). 11 – process room operations. In: S. Featherstone (Ed.), *A Complete Course in Canning and Related Processes (Fourteenth Edition)* (pp. 203–238). Woodhead Publishing. http://www.sciencedirect.com/science/article/pii/B9780857096777000116 (Accessed January 14, 2020).

Fellows, P. J. (2009). *Food Processing Technology: Principles and Practice*. Elsevier.

Floros, J. D., Newsome, R., Fisher, W., Barbosa-Cánovas, G. V., Chen, H., Dunne, C. P., German, J. B., Hall, R. L., Heldman, D. R. and Karwe, M. V. (2010). Feeding the world today and tomorrow: the importance of food science and technology: an IFT scientific review. *Comprehensive Reviews in Food Science and Food Safety*, 9(5), 572–599.

Food and Drug Administration. (2010). *Letter to The Honorable Louise M. Slaughter: Sales of Antibacterial Drugs in Kilograms*. Washington, DC. https://www.govinfo.gov/content/pkg/CHRG-111hhrg77921/pdf/CHRG-111hhrg77921.pdf (Accessed January 14, 2020).

Fratamico, P. M., Annous, B. A. and Guenther, N. 2009. *Biofilms in the food and beverage industries*. Elsevier. CRC Press New York, USA.

Furlonger, D. (2020). *Red Tape Hinders SAB Bid for Clean Power*. https://www.businesslive.co.za/bt/business-and-economy/2020-01-19-red-tape-hinders-sab-bid-for-clean-power/ (Accessed January 19, 2020).

Goldstuck, A. (2020). *Price is the Only Beef with Hi-Tech Vegan Burgers*. https://www.timeslive.co.za/sunday-times/business/2020-01-12-price-is-the-only-beef-with-hi-tech-vegan-burgers/ (Accessed January 12, 2020).

Gustafson, R. H. and Bowen, R. E. (1997). Antibiotic use in animal agriculture. *Journal of Applied Microbiology* 83(5), 531–541.

Han, J., Ruiz-Garcia, L., Qian, J. P. and Yang, X. (2018). Food packaging: a comprehensive review and future trends. *Comprehensive Reviews in Food Science and Food Safety*, 17(4), 860–877.

Harcourt. (2011). *The South African Food Processing Industry* (pp. 12, 13, 19). Report by the Agriculture, Nature and Food Quality Department of the Embassy of the Kingdom of the Netherlands, July 2011.

Heinz, H. (2013). *Principles and Practices for the Safe Processing of Foods*. Elsevier.

Holah, J. (2003). *Guidelines for the Hygienic Design, Construction and Layout of Food Processing Factories*. Campden & Chorleywood Food Research Association Group.

Holah, J. (2014). Cleaning and disinfection practices in food processing. *Hygiene in Food Processing (Second Edition)* (pp. 259–304). Woodhead Publishing.

Hueston, W. and McLeod, A. (2012). *Overview of the Global Food System: Changes Over Time/Space and Lessons for Future Food Safety*. https://www.ncbi.nlm.nih.gov/books/NBK114491/ (Accessed January 13, 2020).

Ifeadike, C. O., Ironkwe, O. C., Adogu, P. O. and Nnebue, C. C. (2014). Assessment of the food hygiene practices of food handlers in the Federal Capital Territory of Nigeria. *Tropical Journal of Medical Research*, 17(1), 10.

IFT. (2015). *How Sustainable Water Use can Boost Food Security Worldwide*. https://www.news wise.com/articles/how-sustainable-water-use-can-boost-food-security-worldwide (Accessed January 14, 2020).

Jakubowski, T. (2015). Temperature monitoring in the transportation of meat products. *Journal of Food Processing & Technology*, 6(10), 1.

Jay, L. S., Comar, D. and Govenlock, L. D. (1999). A video study of Australian domestic food-handling practices. *Journal of Food Protection*, 62(11), 1285–1296.

Jones, M. J. (1992). *Effect of Food Processing in Food Safety* (pp. 171–201). St. Paul, MN: Eagan Press.

Kalaitzis, P., Elena, C., Bita, C. and Hilmi, M. (2019). Innovative Postharvest Technologies for Sustainable Value Chain.

Kearney, J. (2010). Food consumption trends and drivers. *Philosophical Transactions of the Royal Society B: Biological Sciences*, 365(1554), 2793–2807.

Khoury, C. K., Bjorkman, A. D., Dempewolf, H., Ramirez-Villegas, J., Guarino, L., Jarvis, A., Rieseberg, L. H. and Struik, P. C. (2014). Increasing homogeneity in global food supplies and the implications for food security. *Proceedings of the National Academy of Sciences*, 111(11), 4001–4006.

Kirby, R. M., Bartram, J. and Carr, R. (2003). Water in food production and processing: quantity and quality concerns. *Food Control*, 14(5), 283–299.

Kumar, D. and Kalita, P. (2017). Reducing postharvest losses during storage of grain crops to strengthen food security in developing countries. *Foods (Basel, Switzerland)*, 6(1), 8.

Lu, D. and Wong, C. (2009). *Materials for Advanced Packaging*. Springer.

Malagié, M., Jensen, G., Graham, J. C. and Donald, L. S. (2011). The food industry processes. In: *Encyclopaedia of Occupational Health & Safety* (chapter 67). https://www.iloencyclopaedia.org/contents/part-x-96841/food-industry (Accessed January 13, 2020).

Malhotra, B., Keshwani, A. and Kharkwal, H. (2015). Antimicrobial food packaging: potential and pitfalls. *Frontiers in Microbiology*, 6(611).

Mannheim, C. H., Liu, J. X. and Gilbert, S. G. (1994). Control of water in foods during storage. *Journal of Food Engineering*, 22(1), 509–532.

Mehmet, S. (2020). *"Most Realistic" Plant-Based Steak Revealed by Spanish Company*. https://www.newfoodmagazine.com/news/102572/most-realistic-plant-based-steak-revealed-by-spanish-company (Accessed January 17, 2020).

Meneses, Y. E., Stratton, J. and Flores, R. A. (2017). Water reconditioning and reuse in the food processing industry: current situation and challenges. *Trends in Food Science & Technology*, 61, 72–79.

Meulenberg, M. and Viaene, J. (1998). Changing food marketing systems in western countries. In: *Innovation of Food Marketing Systems* (pp. 5–36). Wageningen Pers.

Moerman, F. and Wouters, P. C. (2016). Hygiene concepts for food factory design. In: *Innovation and Future Trends in Food Manufacturing and Supply Chain Technologies* (pp. 81–133). Elsevier.

Murano, P. S. (2003). *Understanding Food Science and Technology (1st Edition)*. USA: Thomson Learning.

Najam, A., Runnalls, D. and Halle, M. (2016). Environment and globalization: five propositions (2010). *The Globalization and Environment Reader*, 94.

Olayiwola, K., Soyibo, A. and Atinmo, T. (2004). Impact of globalization on food consumption, health and nutrition in Nigeria. *Globalization of Food Systems in Developing Countries: Impact on Food Security and Nutrition*, 83, 99.

Omojokun, J. (2013). Regulation and enforcement of legislation on food safety in Nigeria. *Mycotoxin and Food Safety in Developing Countries*, 251–268.

Ortega-Rivas, E. (2012). Common preliminary operations: cleaning, sorting, grading. In: *Non-Thermal Food Engineering Operations* (pp. 11–25). Springer.

Overton, M. 1996. *Agricultural Revolution in England. The Transformation of the Agrarian Economy 1500–1850*. UK: Cambridge University Press.

Owach, C., Bahiigwa, G. and Elepu, G. (2017). Factors influencing the use of food storage structures by agrarian communities in Northern Uganda. *Journal of Agriculture, Food Systems, and Community Development*, 7(2), 127–144.

Pantelidis, I. (2009). High tech foodservice; an overview of technological advancements. In: *Proceedings of the Annual Research Conference, Eastbourne, UK.*

Park, J. W., Oh, H. Y., Kim, D. Y. and Cho, Y. J. (2018). Plant location selection for food production by considering the regional and seasonal supply vulnerability of raw materials. *Mathematical Problems in Engineering*, 2018.

Potter, N. N. and Hotchkiss, J. H. (1995). *Food Science (Fifth Edition)*. Springer, UK.

Qiu, J., Boom, R. M. and Schutyser, M. A. (2019). Agitated thin-film drying of foods. *Drying Technology*, 37(6), 735–744.

Rahman, S. M. (2007). *Handbook of Food Preservation*. CRC Press. Taylor and Francis.

Redmond, E. C. and Griffith, C. J. (2003). Consumer food handling in the home: a review of food safety studies. *Journal of food protection*, 66(1), 130–161.

Rivera, D., Toledo, V., Reyes-Jara, A., Navarrete, P., Tamplin, M., Kimura, B., Wiedmann, M., Silva, P. and Moreno Switt, A. I. (2018). Approaches to empower the implementation of new tools to detect and prevent foodborne pathogens in food processing. *Food Microbiology*, 75, 126–132.

Said, N. S. and Sarbon, N. M. (2019). Protein-based active film as antimicrobial food packaging: A. *Active Antimicrobial Food Packaging*, 53.

Salmon, K. (2017). *The Future of Food: New Realities for the Industry*. https://www.accenture.com/us-en/_acnmedia/pdf-70/accenture-future-of-food-new-realities-for-the-industry.pdf (Accessed January 16, 2020).

Saravacos, G. D. and Kostaropoulos, A. E. (2002). *Handbook of Food Processing Equipment*. Springer.

Smith, D. (2006). Design and management concepts for high care food processing. *British Food Journal*, 108(1),54–60.

Stull, V. (2019). *Dissecting the Health Benefits of Edible Insects*. https://www.ift.org/news-and-publications/food-technology-magazine/issues/2019/october/columns/iftnext-dissecting-the-health-benefits-of-edible-insects (Accessed January 16, 2020).

Sule, D. R. (2008). *Manufacturing Facilities: Location, Planning, and Design*. CRC Press.

Tai, A. P. K., Martin, M. V. and Heald. C. L. (2014). Threat to future global food security from climate change and ozone air pollution. *Nature Climate Change*, 4, 817–821.

Tak, C. S. and Yadav, L. (2012). Improvement in layout design using SLP of a small size manufacturing unit: a case study. *IOSR Journal of Engineering*, 2(10),1–7.

Tamme, T., Reinik, M., Roasto, M., Meremäe, K. and Kiis, A. (2009). Impact of food processing and storage conditions on nitrate content in canned vegetable–based infant foods. *Journal of Food Protection*, 72(8), 1764–1768.

Tanner, D. (2016). Impacts of storage on food quality. In: *Reference Module in Food Science.* Elsevier. http://www.sciencedirect.com/science/article/pii/B978008100596503479X.

Van Donk, D. and Gaalman, G. (2004). Food safety and hygiene: systematic layout planning of food processes. *Chemical Engineering Research and Design*, 82(11), 1485–1493.

Wanniarachchi, W., Gopura, R. and Punchihewa, H. (2016). Development of a layout model suitable for the food processing industry. *Journal of Industrial Engineering*, 2016.

Wei, J. and Kniel, K. E. (2010). Pre-harvest viral contamination of crops originating from fecal matter. *Food and Environmental Virology*, 2(4), 195–206.

Wijtzes, T., Van't Riet, K., in't Veld, J. H. and Zwietering, M. (1998). A decision support system for the prediction of microbial food safety and food quality. *International Journal of Food Microbiology*, 42(1–2), 79–90.

Oluwafemi James Caleb, Rebogile R.R. Mphahlele,
Zinash A. Belay, Asanda Mditshwa

2 Postharvest handling of fresh produce

2.1 Introduction

Horticultural commodities are largely susceptible to deterioration as they continue to depend on their internal substrate for maintenance. Postharvest handling practices and conditions during distribution and retail display remain a complex challenge, as this could result in huge losses. Disruptions in optimum cool chain can create accelerated ripening or deterioration of fresh produce. Detrimental impact of suboptimum practices and conditions have been extensively reported in postharvest literature for a whole range of fruit and vegetables. The sensitivity of tropical and subtropical fruit, vegetables, and root and tuber crops to low-storage temperatures is equally a significant issue.

For example, the respiration rate (RR) of selected fruit types may dramatically increase under chilling temperatures or thereafter when the product is transferred to a higher storage temperature. This phenomenon could be due to the cells' response to remove various intermediate metabolites accumulated under chilling conditions, and to repair damaged subcellular and cellular components. Hence, detailed understanding of produce-specific needs is required. There has been considerable research on the use of postharvest physical, chemical, and gaseous treatments. These tools have been applied to maintain freshness and safety with high-nutritional value for fresh produce. These postharvest treatments are often applied in combination with appropriate storage management strategies. This chapter presents the current status of postharvest handling practices that can be used to preserve fresh quality and reduce losses.

Rebogile R.R. Mphahlele, Agricultural Research Council-Tropical and Subtropical Crops (ARC-TSC), Nelspruit, South Africa
Zinash A. Belay, Oluwafemi James Caleb, Agri-Food Systems & Omics, Post-Harvest and Agro-Processing Technologies (PHATs), Agricultural Research Council (ARC) Infruitec-Nietvoorbij, Stellenbosch, South Africa
Asanda Mditshwa, School of Agriculture, Earth and Environmental Sciences, College of Agriculture, Engineering and Science, University of KwaZulu-Natal (PMB-Campus), Scottsville, South Africa

https://doi.org/10.1515/9783110667462-002

2.2 Physiological factors affecting storage and shelf life of fresh produce

Fresh horticultural commodities continue their living processes after harvest and their postharvest life depends on the rate at which they use substrates. An increase in the rate of loss because of normal physiological changes as respiration and transpiration is caused by conditions that increase the rate of deterioration such as exposure to extreme temperatures, relative humidity (RH), atmospheric conditions, and physical injury. Therefore, a thorough understanding of both respiration and transpiration process in fresh produce will facilitate the selection of optimum storage conditions to reduce loss and extend the shelf.

2.2.1 Respiration rate

Fresh produce requires energy to remain alive and the energy is generated by respiration, which is the oxidative catabolism of carbohydrates. Respiratory pathways by which fruit oxidize sugars are glycolysis, oxidative pentose phosphate pathway, and the tricarboxylic acid pathway. During this process, malic acid is used as a respiratory substrate via malic enzyme, which decarboxylates malate to pyruvate. The handling and storing process of fresh produce can increase the rate at which the fresh produce respires, using up sugar and acid substrates and shortening shelf life. Usually, sugars and acids in fresh produce are sequestered in the vacuole but are released periodically or maintained in a separate pool for use in respiration.

Respiration is the main process which leads to deterioration of fresh produce, and it can be considered a metabolic process aiming at the oxidative breakdown of organic substrates into simple molecules (Fagundes et al., 2013). RR of fresh produce can be accelerated or slowed down depending on factors associated with the environment, packaging material, and product characteristics. This includes type of the fresh produce, maturity, storage temperature, humidity, and surrounding atmosphere.

2.2.1.1 Measurement of RR

Respiratory gas exchange is often used as a general measure of the metabolic rate of tissues, since respiration has a central position in the overall metabolism. Reducing the RR slows down the metabolic process and thus reduces quality deterioration. Generally, RR is determined by measuring the O_2 and CO_2 concentrations in the atmosphere surrounding the product. There are various ways of measuring gas exchange, but the three methods used most often are as follows: (1) static method: a specific gas composition is generated around the produce and the gas flow is closed for a specific period of time. Gas composition is measured at the beginning and the end of the period (or regular interval). (2) Flow-through method: a specific gas composition is generated around the produce and the difference in gas composition between the input and effluent gas flow measured provides the RR. Flow-through method provides accurate measurement of gas exchange, but these systems have limitations in tracking rapid changes. (3) Permeable method: using a package with film of known O_2 and CO_2 permeability, the equilibrium concentrations that develop inside the packages are measured (Table 2.1).

Table 2.1: Respiration rate measuring methods, basic equations, and their main characteristics.

Types of system	Basic equations	Characteristics
Static	$RO_2, CO_2 = \dfrac{(YO_{2i}, CO_{2f} - YO_{2f}, CO_{2i}) \times Vf}{M(t_f - t_i)}$	Nondestructive, simple experimental set up, able to test different combination of gases, sensitive to free volume. Concentration cannot be kept approximately constant during the experiment.
Flow through	$RO_2, CO_2 = \dfrac{(YO_{2i}, CO_{2f} - YO_{2f}, CO_{2i}) \times F}{M}$	Nondestructive, complex experimental set up, able to test different combination of gases, sensitive to free flow rate. Concentration is kept approximately constant during the experiment, not suitable for low-respiring products.
Permeable	$RO_2, CO_2 = \dfrac{(YO_2, CO_2 \times A)}{l \times M}(PO_2, CO_{2f} - PO_{2f}, CO_{2i})$	Nondestructive, complex experimental set up, unable to test different combination of gases, suitable for low-respiring products, sensitive to permeability package dimensions and steady-state connections.

Adapted and modified from Belay et al. (2016).

Respiratory quotient (RQ) is the ratio of CO_2 produced per O_2 consumed, and its value is 1 for glucose catabolism, however, this value changes depending on the substrate required during respiration process. The RQ value of less than 1 can be expected when lipids or proteins, molecules often containing less O_2. Therefore, RQ can be used as an indication of which substrates are being used in the respiratory pathway (Fonseca et al., 2002). The fresh product composition determines what type of substrate is available for respiration and consequently the RQ.

2.2.1.2 Factors affecting RR of fresh produce

2.2.1.2.1 Product characteristics

Fresh produce with different metabolic activities, consequently, have different RR. This includes diversity of roots, tubers, seeds, fruit, stems, and leaves that have different metabolic activities. Fresh produce are well equipped to live detached from the plant, especially those designed for vegetative production (carrots, potato, and onion) and generative reproduction (seeds and fruit). These products often contain large amount of carbohydrates that enable the maintenance of respiration and energy production. On the other hand, other fresh produce such as leaves (spinach) or whole plant (lettuce) do not contain much storage material and are susceptible to rapid senescence and wilting. Fresh produce such as citrus, grape, onion, and potato have low-reparation rate compared to cauliflower, avocado, Brussel sprouts, broccoli, mushroom, spinach, and sweet corn, which respires very high during storage. Other fresh produce including cherry, pear, fig, lettuce, and tomato have moderate RR (Caleb et al., 2016a. Apart from different type of fresh produce, even different cultivars of the same fresh produce with different maturity stage can have different RR.

2.2.1.2.2 Temperature

Temperature is the most important external factor influencing RR. The effect of temperature on RR is mostly associated with the variety of enzymatic reactions which are involved in respiration. The rate of all these reactions increases exponentially with increasing temperature within the physiological temperature range. Biological reactions generally increase two- or threefolds for every 10 °C rise in temperature. Extremely high or low temperatures have negative effects on fresh produce, where the first could lead to enzyme denaturation and the later could result in physiological injury, which leads to an increase in RR. The increase in temperature can lead to gradual decrease and increase in O_2 and CO_2 concentrations. Reducing the storage temperature from 30 to 20 °C decreased O_2 consumption by about 13% and increased CO_2 production by about 11.31% over the entire storage period for fresh-cut papaya (Rahman et al., 2013). The RR showed significant differences ($p \leq 0.05$) across the storage temperature of 2, 5, and 7 °C for fresh-cut "Gala" apple (Fagundes et al., 2013). It was observed in this same study that the highest RR was at 7 °C and the lowest at 2 °C. Therefore, the lower the temperature the lower the metabolic process, and the higher the shelf life (Caleb et al., 2013).

2.2.1.2.3 Surrounding atmospheres

The RR of fresh produce depends on the concentration and combination of the atmosphere surrounding the product. The most common atmosphere composition used for fresh produce includes oxygen (O_2), carbon dioxide (CO_2), and nitrogen at different concentration. O_2 availability is a prerequisite for metabolic activity in a commodity or more specifically for RR, and the higher the RR, the lower the shelf life (Belay et al., 2017). Decreasing the O_2 concentration and increasing the CO_2 concentration could significantly decrease the RR (Rahman et al., 2013).

The rapid reduction of RR during postharvest handling of fresh produce could be related to the limited substrate reserves for respiratory metabolism with the storage time. Excessive accumulation of CO_2 can result in cell membrane damage and physiological injuries to the product (Caleb et al., 2012). Under excessively low O_2 concentration (<1%) anaerobic respiration may occur, resulting in the development of off-odors. Furthermore, reduction in RR delays enzymatic degradation of organic substrates, thus, extending the shelf life of the fresh produce. Proper control of the surrounding atmospheres, that is O_2 and CO_2, has been shown to be effective in reducing the RR. In general, storage of fresh produce in an optimal atmosphere delays the RR and extends the shelf life.

2.2.2 Transpiration rate

Transpiration rate (TR) is the physiological process in fresh produce which is related to the mass transfer process whereby water moves from the surface of the plant origin or from the stored commodity into the surrounding air (Caleb, Mahajan, Al-Said et al., 2013). Transpiration is one of the phenomena that affect physiological deterioration of horticultural products. It is considered as the major cause of postharvest losses and poor quality in leaf vegetables, such as lettuce, spinach, and cabbage (Madani et al., 2019). Transpiration induces wilting, shrinkage, and loss of crispness (Ben-Yehoshua and Rodov, 2013). Transpiration involves the interaction between moisture evaporation from the produce surface as a result of water vapor pressure difference and moisture release as a result of metabolic activity (Ngcobo et al., 2012). Transpiration is driven by a concentration difference and can be described in terms of water activity differences across the membrane, moisture concentration, and water vapor pressure differences between a product's surface and its surrounding (Bovi et al., 2016).

This process of moisture loss induces wilting, shrinkage, and loss of firmness and crispness of fresh produce, and thus adversely affects the appearance, texture, flavor, and mass of the product. Conversely, accumulation of water at the product surface favors the growth of spoilage microorganisms (Caleb, Mahajan, Manley et al., 2013). TR during postharvest period can be influenced by fresh product factors such as, morphological characteristics, maturity stage, presence of injuries, surface to volume ratio, as well as environmental factors including, RH, temperature, water vapor pressure difference, and air velocity (Mahajan et al., 2008). Biophysical properties of the skin, air film resistance, respiration heat generation, and vapor as variables also affect the TR (Kang and Lee, 1998). TR increases with increasing storage temperature (Xanthopoulos, Athanasiou, Lentzou, Boudouvis and Lambrino, 2014) and with decreasing RH from 100% to 75% (Tano et al., 2005).

2.2.2.1 TR measurements

TR can be expressed by the difference in weight loss of a produce through time as presented in the following equation.

$$TR = \frac{M_i - M_f}{t - M_i} \qquad (2.1)$$

where, TR, is the transpiration rate (g kg^{-1} h^{-1}), Mi is the initial weight (g), and Mt is the weight of the product (g) at time t (h) (Caleb et al., 2013).

Fresh produce TR has linear relationship with water loss and water vapor pressure deficit (WVPD), where the flow of water vapor through a produce occurs where there is a gradient of water vapor pressure between the produce skin and the surrounding environment (Andongo et al., 2014; Ben-Yehoshua and Rodov, 2002). Thus, the flow of water vapor can be calculated according to Fick's law of diffusion, which is proportional to the difference between humidity of the fresh produce internal atmosphere and the humidity of the surrounding air (Pereira et al., 2018). These relationships can be expressed as a function of water activity by the following equation:

$$TR = k_i \times (a_{wi} - a_w) \qquad (2.2)$$

where k_i is the mass transfer coefficient, a_{wi} and a_w are water activities of the fresh produce and water activity of the container (RH/100), respectively. In order to consider the effect of temperature on the TR of fresh produce the temperature term can be added to eq. (2.2) as follows:

$$TR = k_i \times (a_{wi} - a_w) \times \left(1 - \exp^{-aT}\right) \qquad (2.3)$$

where a is a coefficient constant and T is the temperature (K) (Caleb et al., 2013).

2.2.2.2 Factors affecting TR

2.2.2.2.1 Intrinsic factors

TR of fresh produce during postharvest can be affected by intrinsic factors such as surface to volume ratio, surface injuries, maturity stage, and morphological and anatomical characteristics (Caleb et al., 2013). The rate of water loss is directly proportional to the surface area to volume ratio of the produce. Therefore, the size, shape, and surface of the commodity affects the TR, where large produce have smaller surface to volume ratio, which makes them lose less moisture than small produce. Similarly, produce with thicker skin also lose moisture slowly than thin-skin produce. In addition, the surface structure of the fresh produce including those that contains wax affects the TR by providing extra layer of resistance. In this case, the rate of water loss is lower than those fresh produce without a waxy structure.

Additionally, physiological condition such as the maturity stage of fresh produce after harvest has been shown to significantly influence TR. Furthermore, the end products of fresh produce respiration are water and heat, both of which have influence on the transpiration. The produced water remains in the fresh produce, on the other hand heat can be released through heat transfer process to the environment. This heat transfer raises the temperature of the produce and can create a difference in vapor pressure deficit which may lead to increase in evaporation. Evaporation, which occurs at the produce surface, is an endothermal process which will cool the surface, thus lowering the vapor pressure at the surface and reducing transpiration. On the other hand, respiration within the fresh produce could lead to increase in the produce temperature, thus raising the vapor pressure at the surface and increasing transpiration (Bovi et al., 2016).

2.2.2.2.2 Extrinsic factors

The TR phenomenon is also influenced by factors that affect the mass transfer process including environmental condition during postharvest period such as temperature, RH, vapor pressure, surrounding air, and packaging type and condition. Moisture loss from the fresh produce during postharvest handling and storage is dependent on RH, air velocity, and heat of respiration. The rate of moisture loss can be reduced by increasing the air velocity. Increase in air velocity lowers the produce temperature and the temperature gradient, hence reducing the heat of respiration, eliminating low and saturated RH, and minimizing the rate of moisture loss. Furthermore, the difference in temperature between the product and the surrounding air can affect the equilibrium water vapor pressure, and thus transpiration. Transpiration in fresh produce mainly results from the differences in water vapor pressure between the interior of the produce and the surrounding environment. The product surface may be assumed to be saturated, and thus, the WVP at the fresh produce surface is equal to the water vapor saturation pressure at the produce surface temperature.

In this sense, the vapor pressure within the produce solely depends on the temperature, whereas, the vapor pressure of the surrounding is influenced by the temperature and RH. Practically, WVPD is often expressed in terms of the RH, where saturated RH (100%) is equivalent to equilibrium water vapor pressure, evaporation is then proportional to the difference between RH in the surrounding air at saturation (Tano et al., 2005). In addition, respiratory heat can also significantly influence water loss from fresh produce under water vapor saturated conditions. When the fruit temperature is higher than the surrounding air temperature, there is increase in surface temperature due to heat of respiration (Bovi et al., 2018b). This difference in temperature could lead to an increase in water vapor pressure gradient for the mass transfer between the fruit and the surrounding and then increase mass loss (Bovi et al., 2018b).

The pronounced effect of RH on the TR has been reported for fresh mushrooms (Mahajan et al., 2008), pomegranate (cv. Acco) (Caleb et al., 2013), strawberry (cv. Elsanta) (Sousa-Gallagher et al., 2013; Bovi et al., 2018a), and banana (Murmu and Mishra, 2016). In order to control the significant effect of RH on the mass loss of fresh produce, recent studies introduced humidity regulating packaging system for *Agaricus* mushroom (Singh et al., 2010; Sängerlaub et al., 2011) and button mushroom (Rux et al., 2015). Furthermore, various combinations of packaging films has been used to design a packaging material, which can regulate the RH around the product, thereby reducing the effect of WVPD; such systems were developed for broccoli (Caleb et al., 2016b) and for pomegranate arils (Belay et al., 2018).

Apart from environmental conditions factors, surface injuries, which include cuts, bruises, and scratches on the skin surface of fresh produce, tend to increase the TR as they reduce the tissue resistance to moisture loss due to modification of the skin. In general, optimum storage conditions will extend the shelf life of fresh produce by providing low temperature and high RH environment, which reduces moisture loss and decreases respiratory activity, thereby lowering TR. Furthermore, factors such as skin permeability and airflow can also affect the TR of fresh produce. Therefore, in order to analyze the TR of fresh reduce, the complex heat and mass transfer phenomena should be considered.

2.3 Role of postharvest techniques in prolonging shelf life of fresh produce

Fruit quality is determined by both external and internal attributes. External quality translates to shape, size, skin color, and overall appearance, whereas internal parameters include flesh firmness, soluble solids content, titratable acidity, pH, vitamins, phenolics, carotenoids, and antioxidant activity (Mesa et al., 2016; Navarro et al., 2015). For the quality of the fresh produce to be extended, it is important to reduce the rates at which biochemical, enzymatic, and microbial degradation reactions take place. Proper harvesting, processing, postharvest treatment, temperature, and RH are among the processes that could help preserve the shelf life of fresh produce.

Advanced technological system including heat treatment (Vicente et al., 2006; Yun et al., 2013; Liu et al., 2017), edible coatings (Athmaselv et al., 2013; Fagundes et al., 2014), irradiation (Artés-Hernández et al., 2010; Latorre et al., 2010; Alegria et al., 2012), modified atmosphere packaging (MAP), and controlled atmosphere (CA) techniques have been adopted by fresh produce industry to slow down physiological and biochemical processes and delay senescence of fresh produce (Caleb et al., 2012;). MAP and CA techniques use gas compositions including O_2, CO_2, and N_2 to improve storage quality of fresh produce for as long as possible when combined with proper cold-storage conditions which in turn will minimize waste.

2.3.1 Heat treatments

2.3.1.1 Hot water treatment

The postharvest use of water at temperatures above 35 °C for treating fresh horticultural produce is called hot water treatment (HWT) (Caleb et al, 2016c). Although HWT is not widely used during postharvest handling of fresh produce, it is environmentally friendly and completely safe for humans. HWT has a great potential, particularly for fresh produce destined for organic markets where only residue-free or chemically free commodities are acceptable. Fresh horticultural produce is not homogenous, some fruit and vegetables are soft skinned while some are actually hard. Usall et al. (2016) indicated that growing conditions, harvest maturity, cultivar, as well as the thermal conductivity of fresh produce tissue have an enormous influence on the effect of HWT. If improperly applied, HWT can easily result to irreversible mechanical damage. It is therefore important that each HWT protocol considers the appropriate treatment temperature and duration suitable for each specific produce.

The use of HWT has been demonstrated to be one of the effective postharvest handling techniques for extending the shelf life of fresh produce (Table 2.2). For instance, Lamikanra et al. (2005) investigated the effect of HWT on the quality of fresh-cut cantaloupe melon. The authors found that exposing the whole cantaloupe melon to heat treatment of 50 °C for 60 min before cutting reduced respiration, water loss, and microbial count while it generally maintained or improved sensory attributes of fresh-cut fruit. Similarly, Yun et al. (2013) reported that dipping "Kamei" Satsuma mandarins in warm water at 52 °C for 2 min resulted in lower incidences of pathological and physiological disorders. Using the proteomics and metabolomics analyzes, the authors found that proteins, such as beta-1,3-glucanase (GNS) and class II chitinase, associated with stress resistance were upregulated in the treated pericarp. On the other hand, redox metabolism enzymes such as oxidoreductase, isoflavones reductase, and superoxide dismutase (SOD) were notably downregulated. The efficacy of HWT to extend the shelf life of fresh produce is strongly linked to the increased accumulation of heat shock protein. Moreover, the expression of defense-related genes such as chitinase (CHI), GNS, and phenylalanine ammonia lyase (PAL) has often shown to increase in HWT-treated fresh produce (Klein and Lurie 1991; Liu et al., 2012; Yun et al., 2013). Although HWT is effective in maintaining quality and extending the shelf life of fresh produce, it

Table 2.2: The effect of hot water treatment (HWT) and hot air treatment (HAT) on postharvest quality of fresh produce.

Heat treatment	Commodity	Temperature and duration	Benefit	References
HWT	Cantaloupe melon	50 °C, 60 min	Reduced respiration and water loss, increased sensory attributes	Lamikanra et al. (2005)
	"June Prince" peach	40 °C, 5–10 min	Reduced brown rot	Liu et al. (2012b)
	"Kamei" Satsuma mandarin	52 °C, 2 min	Inhibited disease and decay incidence	Yun et al. (2013)
	Fresh-cut cantaloupe melon	60 °C, 60 min	Increased fruit firmness	Lamikanra and Watson (2007)
	"Nankou'"mume fruit	45 °C, 5 min	Reduced fruit susceptibility to chilling injury	Endo et al. (2019)
HAT	Strawberry	45 °C, 3 h	Lower respiration rate and decay, higher antioxidant capacity	Vicente et al. (2006)
	"Rhapsody" tomato	38 °C, 24 h	Resulted to heat injury, lower lycopene content	Yahia et al. (2007)
	"Baifeng" peaches	38 °C, 12 h	Lower internal browning, higher mealiness	Jin et al. (2009)
	"Cicco" broccoli	48 °C, 3 h	Reduced senescence	Costa et al. (2006)
	"Wumei" bayberries	48 °C, 3 h	Reduced respiration and decay incidence	Wang et al. (2010)
	"Qiandaowuhe" persimmon	52 °C, 3 h	Extended shelf life, reduced weigh loss	Luo (2006)
	"Biqi" Chinese bayberry	50 °C, 3 h	Maintained firmness and reduced disease incidence	Luo et al. (2009)
	"Jiefangzhong" loquat	38 °C, 36 h	Improved chilling resistance	Shao and Tu (2014)
	"Cicco" broccoli	48 °C, 3 h	Delayed senescence	Lemoine et al. (2009)

is critical that product-specific protocols are developed. This is based on the fact that the response of horticultural products to HWT is not homogenous as gas permeability and membrane structure differs. Also, HWT can lead to undesired textural changes especially for fresh produce with soft pericarp.

2.3.1.2 Hot air treatment

Hot air treatment (HAT) is another form of heat treatment in which fresh produce is exposed to air temperature higher than 30 °C for several hours or days. Various studies have shown that HAT is effective in maintaining quality and extending the shelf life of fresh produce (Table 2.2). The influence of HAT on fruit quality, decay incidence, as well as oxidative metabolism was assessed in "Selva" strawberries during cold storage (Vicente et al., 2006). Exposing the fruit to air heat at 45 °C for 3 h did not only reduce RR and decay incidence, but higher activities of ascorbate peroxidase (APX) as well as SOD were noted in treated fruit fundamental changes in oxidative metabolism. Lower incidences of internal browning have been reported in peaches exposed to HAT (38 °C, 12 h) before storage at 0 °C for 3–5 weeks (Jin et al., 2009).

Similar to HWT, HAT has also been associated with negative results including mechanical damage and compromised nutritional quality. For instance, Yahia et al. (2007) reported heat injury in mature green tomatoes exposed to 38 °C and 95% RH (24 h) before being stored at 4 °C for four weeks. The authors also found that nutritional attributes, such as β-carotene and lycopene were slightly reduced compared to fruit exposed to 34 °C. Similarly, Jin et al. (2009) reported a higher incidence of mealiness in HAT-treated "Baifeng" peaches. Reduced weight loss has also been reported in "Qiandaowuhe" persimmon fruits exposed to 52 °C prior to ripening at 20 °C for 14 days (Luo, 2006). Thus, it is critical that HAT should be properly executed and supervised in order to minimize quality loss.

To this effect, combining HAT with other postharvest treatments such as methyl jasmonate (Jin et al., 2009), nanopackaging containing nano-TiO_2 and Ag (Wang et al., 2010), UV-C (Lemoine et al., 2008), and low-concentration acidic electrolyzed water (Liu et al., 2017) has been proposed. In understanding its mechanism of action for extending shelf life, HAT has been demonstrated to increase the accumulation of certain beneficial enzymes such as catalase, APX, and glutathione S-transferase (Yahia et al., 2007). Glutathione S-transferase is well known for protecting the cells from oxidative stress by scavenging reactive O_2 species including superoxide anion, hydroxyl radical, hydrogen peroxide, and singlet O_2. Additionally, HAT has been demonstrated to inhibit pectinmethylesterase (PME) and polygalacturonase (PG) activities (Luo, 2006; Luo et al., 2009). The reduced activities of PME and PG are strongly linked with the delayed depolymerization of pectic substances thereby maintaining pericarp firmness.

2.3.2 Edible coatings

Edible coatings are thin layers of edible material applied on the surface of fresh produce (Dhall, 2013). Various application methods, such as dipping, spraying, and brushing are used for applying the coatings. The use of edible coatings as postharvest treatments has become very popular in the fresh horticultural produce industry. This trend is largely due to the environment and health concerns regarding the application of synthetic and nonedible coatings as well as waxes on food materials (Mditshwa et al., 2017a). Among many of the advantages, edible coatings are known for reducing water and weight loss, maintaining organoleptic attributes and external quality, controlling oxidation processes as well as acting as carriers for antimicrobial agents (Lin and Zhao, 2007; Dhall, 2013; Ncama et al., 2018). Moreover, the potential use of food processing by-products is an innovative strategy for controlling pollution while inexpensively adding value in fresh produce (Otoni et al., 2017).

Table 2.3: The effect of edible coatings on postharvest quality of fresh produce.

Edible coating	Commodity	Benefit	References
Chitosan (1%) + Arabic gum (10%)	"Pisang Berangan" banana	Delayed mass and firmness loss, reduced decay	Maqbool et al. (2010)
Aloe vera	"Ruchi 618"	Inhibited weight loss, delayed ripening and extended shelf life	Athmaselv et al. (2013)
	"Hayward' kiwifruit	Extended shelf life and maintained sensory properties	Benítez et al. (2015)
Carboxymethyl cellulose + chitosan	"Or" and "Mor" mandarins, "Navel" oranges, and "Star Ruby" grapefruit	Increased fruit firmness, no effect on weight loss	Arnon et al. (2014)
Cassava starch (3%)	"Aroma" strawberry	Reduced respiration and retained fruit color	Garcia et al. (2010)
Hydroxypropyl methylcellulose (HPMC), beeswax (BW)	"Josefina" cherry tomato	Inhibited gray mold and extended shelf life	Fagundes et al. (2014)
Stafresh 151™ (Mineral oil)	"Grandela" tomato	Delayed color change, reduced weight loss	De Jesus Dávila-Aviña et al. (2011)
Gelatin; Waxy corn starch	"Red Crimson" grapes	Improved appearance, inhibited weight loss	Fakhouri et al. (2015)
CMC (1%) + moringa leaf extract (2%)	"Hass" and "Gem" avocado	Suppressed disease incidence, lower mass loss, reduced respiration	Tesfay et al. (2017)

Table 2.3 shows the effect of edible coatings on postharvest quality of fresh produce. Studies by Athmaselv et al. (2013) demonstrated that *Aloe vera* based coating prolonged the shelf-life of "Ruchi 618" tomatoes to 39 days compared to the control treatment which had the shelf-life of only 19 days. Similarly, Maqbool et al. (2010) investigated the effect of different edible coatings on postharvest anthracnose of banana fruit stored at 13 °C for 28 days. The authors found that chitosan (CH) at 1% to 1.5% did not only delay ripening but it also inhibited anthracnose. On the other hand, Arabic gum (AG) did not exhibit any antifungal effects. However, incorporating 10% AG with 1% CH did not only reduce weight and firmness loss, but decay incidence was also reduced. In another unrelated study, applying carboxymethyl cellulose and CH bilayer was shown to maintain postharvest quality and sensory attributes in "Navel" oranges and "Star Ruby" grapefruit (Arnon et al., 2014). However, the same coating decreased flavor acceptability and the overall consumer experience in "Or" and "Mor" mandarins after four weeks of cold storage. Edible coatings are very diverse is terms of their properties, they differ significantly in gas permeability and this affects the postharvest quality and nutritional attributes of the coated produce (Mditshwa et al., 2017a). Normally, the coating should not completely seal the fruit surface as this might increase internal CO_2 and ethanol accumulation, thereby, leading to off-flavors. Therefore, the concentration of the edible coatings, and the application methods used, must be product-specific for effectiveness.

2.3.3 Irradiation

Irradiation is one of the novel and innovative techniques used for extending the shelf-life of fresh or minimally processed fresh produce and inactivating postharvest pathogens such as *Salmonella typhimurium* and *Escherichia coli*. It is generally regarded as a safe food sanitization alternative to chemical methods. There are different types of irradiation treatments used in food industry. For this section, the discussion will be limited to only UV-C and gamma irradiation methods.

2.3.3.1 UV-C irradiation

The use of UV-C irradiation as postharvest treatment of fresh produce has gained popularity in recent years (Mditshwa et al., 2017b). There is growing evidence that UV-C irradiation does not only extend shelf life but it also promotes the accumulation of nutritional attributes such as carotenoids and antioxidants (Table 2.4). A study by Alegria et al. (2012) demonstrated that exposing fresh-cut carrots to UV-C

Table 2.4: The effect of UV-C and gamma irradiation on postharvest quality of fresh produce.

Irradiation	Commodity	Dosage	Benefit	References
UV-C	"Nantes" carrot	0.78 ± 0.36 kJ/m^2	Reduced respiration, maintained whiteness index, reduced microbial load	Alegria et al. (2012)
	"Henry's Beauts" tomatoes	969.8 µW/cm^2	Inhibited decay, increased carotenoid content	Khubone and Mditshwa (2018)
	"Zhenfen 202" tomatoes	4 or 8 kJ/ m^2	Increased shelf life, increased the contents of gallic acid, chlorogenic acid	Lui et al. (2012)
	"Silver Whale" spinach	24 kJ/m^2	Reduced microbial counts of Listeria monocytogenes and Salmonella enterica	Escalona et al. (2010)
	"Comte de Paris" pineapple	4.5 kJ/m^2	Reduced firmness loss, maintained titratable acidity	Pan and Zu (2012)
	"Button" mushrooms	0.45–3.15 kJ/m^2	Extended shelf life, reduced microbial count of E. coli O157:	Guan et al. (2012)
Gamma	"Conditiva" red beet	1–2 kGy	No effect on color and firmness	Latorre et al. (2010)
	"Corona" strawberry	1–1.5 kGy	Reduced decay and mass loss, no effect on chemical attributes	Majeed et al. (2014)
	Paprika	4 kGy	Reduced gray mold	Yoon et al. (2014)
	"BC-79" cabbage	2 kGy	Inhibited wound-induced browning	Banerjee et al. (2015)
	"Kurdistan" strawberry	1 kGy	Reduced botrytis fruit rot	Jouki and Khazaei (2014)

maintains the whiteness index values and increases carotenoid content almost threefold compared to the control treatment. Moreover, microbial load was much lower in UV-C-treated produce compared to their untreated counterparts. Escalona et al. (2010) also reported lower microbial counts of *Listeria monocytogenes* and *Salmonella enterica* in "Silver Whale" spinach treated with various doses of UV-C before storage at 5 °C for 13 days. Similarly, Khubone and Mditshwa (2018) found low decay incidence and high antioxidant capacity in "Henry's Beauts" tomatoes exposed to UV-C irradiation before storage. Artés-Hernández et al. (2010) concluded that, regardless of the dose, UV-C increases the antioxidant capacity and maintains the overall quality of fresh-cut watermelon. Clearly, there is enough literature evidence to suggest that UV-C is an effective treatment that can be employed during postharvest handling. Its effectiveness, as a sanitizing treatment, on fresh-cut produce is highly desired and this will greatly benefit the fresh-cut produce industry, particularly for horticultural produce with delicate texture. However, reduced vitamin C content in UV-C-treated fresh produce has been reported (Pan and Zu, 2012). Also, the effect of UV-C irradiation on organoleptic attributes of the treated produce needs further investigation. More research is thus required to establish vitamin C-friendly doses for some types of fresh produce.

2.3.3.2 Gamma irradiation

Gamma irradiation is another nonchemical and nonthermal food preservation technique that has gained popularity in the fresh produce industry. Low doses of gamma irradiation have been shown to influence the postharvest quality and shelf life of various fruits and vegetables (Table 2.4). Loterre et al. (2010) assessed the effect of gamma irradiation on biochemical and physicochemical attributes of fresh-cut red beetroot. Their findings demonstrated that exposing beetroot to 2 kGy reduced the loss of flesh firmness during storage. Irradiation treatments resulted in increased cell-cell adhesion through increasing calcium cross-linking at the middle lamellae region which resulted to high firmness in treated. However, the authors also reported that gamma irradiation (1 and 2 kGy) had no effect on fruit color. A study by Majeed et al. (2014) reported that gamma irradiation doses of 1 and 1.5 kGy reduced mass loss and decay incidence in 'Corona' strawberries stored at room temperature for 9 days. Remarkably, these doses did not cause drastic changes to the chemical profile and organoleptic properties of the treated fruit. However, exposing the fruit to 0.5 kGy irradiation did not offer any benefits during shelf life. It could be argued that, although low doses are highly recommended for safety reasons, in order to ensure superior postharvest quality and longer shelf life, each gamma irradiation protocol for fresh produce should be based on empirical evidence. The combinatory use of low doses of gamma irradiation with other postharvest treatments such as edible coatings (Hussain et al., 2012a), calcium chloride dips (Hussain et al., 2012b) have also been demonstrated to be effective in facilitating the marketing of fresh produce to distant markets. The efficacy of gamma irradiation is closely linked to its ability to influence the regulation and expression of some of the key genes affecting quality and shelf life at postharvest. The downregulation of PAL gene as well as PAL activity has been reported in cabbages (Jouki and Khazaei, 2014). Although its adoption as the food preservation technique remains low, gamma irradiation is globally being recognized as one of the best technologies for maintaining the postharvest quality of fresh produce (Fernandes et al., 2012).

2.3.4 Modified atmosphere packaging

In MAP environment, the gas composition can either be passive or active. In passive atmosphere condition, the CO_2/O_2 gas composition is generated through respiration of the packaged product, its equilibrium in turn depends on the permeability of the plastic film and storage temperature (Lee et al., 1996). On the other hand, active-MAP (A-MAP) involves flushing of desired gases (CO_2/O_2 and N_2) inside the package before sealing and somewhat can be achieved through use of gas absorbers or scavengers to attain desired gas level. The N_2 gas is generally used as a "filler" gas under MAP to balance the volume decrease owing to CO_2 absorption and to inhibit package collapse (Sandhya, 2010). Overall, passive-MAP (P-MAP) is relatively cheaper than the A-MAP.

Schematic presentation of MAP systems is shown in Figure 2.1. For a successful MA, several interrelated factors that should be considered including product type, level of gas composition and the suitability of the film. Moreover, hygiene and temperature at which product are stored play a significant role in making MAP a success. Table 2.5 highlights numerous research studies on the successful use of passive and active MAP on prolonging the shelf life of fresh produce for the past 10 years.

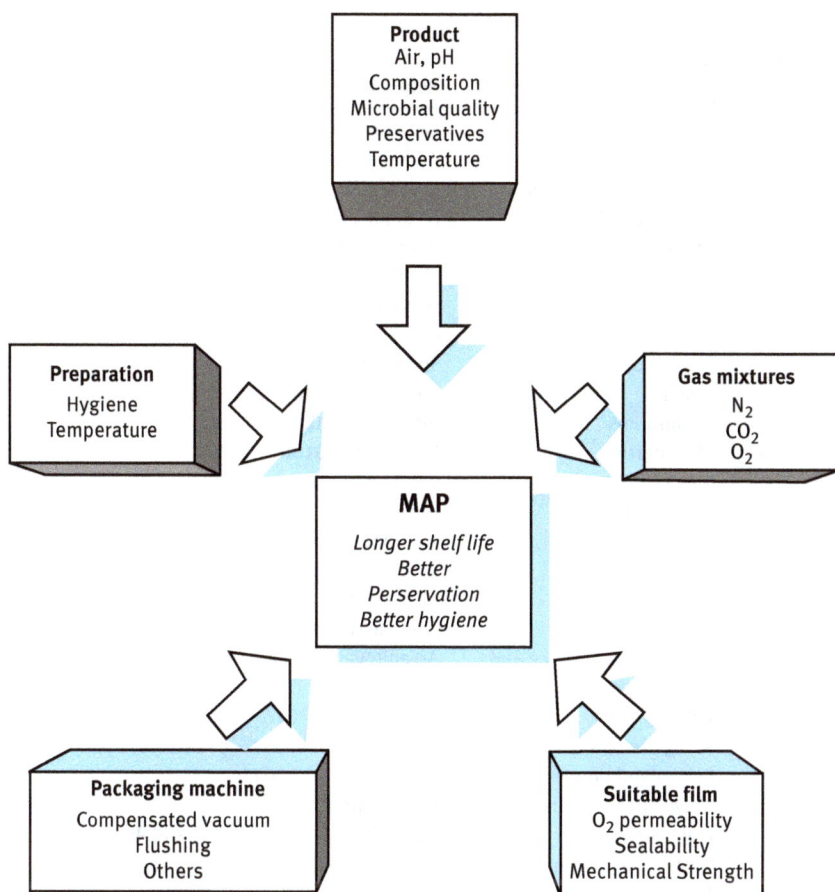

Figure 2.1: Modified atmosphere packaging system (www.Globalspec.com).

2.3.4.1 MAP of fresh produce

Generally, low O_2 (2–6%) and high CO_2 (5–15%) are preferable for extending the shelf life of fresh produce. However, several fresh produce have shown to tolerate high level of O_2 over 7% compared to other fresh fruit. Study conducted by Afifi et al. (2016) showed that the optimum atmosphere composition for strawberry was 7.5% O_2 and 15% CO_2. An internal atmosphere of 12.3% O_2 + 5.6% CO_2 reduced the internal browning and extended the eating quality of "Bartlett" pear for up to 4 months at −1.1 °C (Wang and Sugar, 2013). On the contrary, however, Bhat and Stamminger (2016) reported that MAP (24% CO_2 and 1% O_2) had promising results on the preservation of strawberries. According to Nielsen and Leufvén (2008), the recommended gas composition for strawberry storage varies a great deal between different reports.

Table 2.5: MAP published articles in the past 10 years.

MAP condition	Fruit type	Reference
A-MAP	Cherry tomatoes	Fagundes et al., 2015
P-MAP	Table grapes	Admane et al., 2018
A-MAP	Shiitake mushrooms	Li et al., 2014; Jing-jun et al., 2012
A-MAP	Cherry tomatoes	Fagundes et al., 2015
P-MAP	Blueberry	Rodriguez et al., 2016
P-MAP	Pomegranate	Mphahlele et al., 2016
P-MAP	Rocket leaves	Mastrandrea et al., 2017
P-MAP	Pomegranate	Candir et al., 2018
E-MAP	Cherry tomatoes	Tumwesigye et al., 2017
P-MAP	Guava	Murmu and Mishra, 2017
P-MAP	Mushrooms	Joshi et al., 2018
P-MAP + Humidity	strawberries	Bovi et al., 2018a
P-MAP	Fresh basil	Patiño et al., 2018
P-MAP	Fresh tomatoes	Domínguez et al., 2016
P-MAP	Figs	Villalobos et al., 2017; Villalobos et al., 2018
P-MAP	Cherry tomatoes	D'Aquino et al., 2016
P-MAP	Green chilies	Chitravathi et al., 2015
P-MAP	Bartlett pears	Wang and Sugar, 2013
A-MAP	"Lollo verde" lettuce and rocket	Arvanitoyannis et al., 2011
P-MAP	Baby spinach	Garrido et al., 2016
P-MAP	Red globe grapes	Candir et al., 2012
P-MAP	Yellow and purple plum	Díaz-Mula et al., 2011
P-MAP	"Fuyu" persimmon	Liamnimitr et al., 2018
P-MAP	Fresh strawberry	Matar et al., 2018; Barikloo and Ahmadi, 2018
P-MAP	Oyster mushroom	Jafri et al., 2013
P-MAP + humidity	Tomatoes	Park et al., 2018
P-MAP	"Italia" table grapes	Cefola et al., 2018
P-MAP	Sweet cherry	Wang and Long, 2014; Hayta and Aday, 2015; Colgecen and Aday, 2015; Aglar et al., 2017
P-MAP	Wild rocket	Edelenbos et al., 2017
A-MAP	Guava	Murmu and Mishra, 2018
P-MAP	white mushrooms	Gholami et al., 2017
P-MAP and A-MAP	Prickly pears	Ochoa-Velasco and Guerrero-Beltrán, 2016
E-MAP	Figs	Villalobos et al., 2014
	Strawberry	Afifi et al., 2016; Bhat & Stamminger, 2015
A-MAP	Bell pepper	Singh et al., 2014
A-MAP	Shiitake mushrooms	Jing-jun et al., 2012
P-MAP	Tomatoes	Park et al., 2018

P-MAP: Passive modified atmosphere packaging; A-MAP: Active modified atmosphere packaging; E-MAP: Equilibrium modified atmosphere packaging

This can partly be attributed to different cultivars used in the studies. Importantly, growing conditions, degree of ripeness, storage temperature, and the choice of quality parameter determine the most appropriate storage atmosphere.

Several fruits have shown to tolerate relatively high level of O_2 and low CO_2 in different studies under P-MAP condition. For instance, Gholami et al. (2017) observed that the use of MAP (15% O_2 and 5% CO_2) characterized by O_2 barrier material with good permeability properties against CO_2 and low permeability of H_2O retained the quality of mushrooms in terms of overall appearance and weight loss over the conventional PVC film stored at 4 °C. Similarly, a study conducted by Cefola et al. (2018) reported that low CO_2 (0.03–10%) preserved the quality and sensory parameters, whereas high CO_2 (>20%) caused a fermentative metabolism of Italia grapes stored at 5 °C. A study conducted by Jafri et al. (2013) in oyster mushrooms subjected to chemical treatments (sorbitol (0.05%, w/v), citric acid (3%, w/v), $CaCl_2$ (1%, w/v), and MAP (10% O_2 and 5% CO_2)) retained better/acceptable quality characteristics and had high sensory scores resulting in storage life of 25 days when stored at 4 °C. The study showed that a combinatorial effect of MAP and chemical treats could be beneficial than MAP alone. Fresh chilies packed under antifog (RD45) film (5% O_2 and 3% CO_2 gas composition) maintained pigment stability, retention of phenolics, capsaicin, antioxidants, and ascorbic acid as well as firmness with moderate-weight loss during storage at low temperature (8 ± 1 °C) for 28 days (Chitravathi et al., 2015). D'Aquino et al. (2016) observed that microperforated films which created MAPs with moderate levels of CO_2 (2–4 kPa), O_2 partial pressures of 15–18 kPa O_2, and in-package RH of approximately 100%, contributed to reduced respiration and decrease in the rate of degradation of sugars, organic acids, vitamin C, and lycopene in tomato fruit cultivars ("Trebus" and "Dorotea"). Moreover, in the same study, the MAP conditions conserved the eating quality, freshness, firmness, and with minimal weight loss. On the contrary, Rodriguez and Zoffoli (2016) observed that MAP (two perforations of 3 mm^2) with CO_2 level of 6 KPa was detrimental to blueberry and induced-fruit softening in cvs. "O'Neal," "Duke," and "Legacy." Bovi et al. (2018a) observed no significant changes on the physicochemical quality attributes of strawberries packaged under modified atmosphere and humidity packaging.

MAP has proven to be more beneficial in extending the shelf life of pomegranate fruit but this varied significantly across the cultivars investigated. Mphahlele et al. (2016) reported that pomegranate fruit cv. Wonderful was kept for 3 months at 7 °C under P-MAP alone. Interestingly, Candir et al. (2018) found that P-MAP combined with CH (1%) maintained the quality and extended shelf life of pomegranate fruit cv. "Hicaznar" for 6 months when stored at 6 °C.

2.3.4.2 Active MAP of fresh produce

Active MAP is a system that involves incorporation of several compounds into packaging systems to prolong shelf life and preserve or enhance quality under cold storage (Restuccia et al., 2010). Figure 2.2 shows a schematic presentation of A-MAP systems. Even though it has proven to be beneficial, its use is primarily limited to low adaptability due to high costs. Very few studies have been conducted for the past 10 years mainly for A-MAP for enhancing shelf life of fresh produce. A study conducted by Fagundes et al. (2015) found that atmosphere composition of 5% O_2 + 5% CO_2 extended the shelf life up to 25 days when stored at 5 °C which reveals that gas concentration can influence the postharvest quality of cherry tomatoes. Singh et al. (2014) reported that the shelf life of bell pepper was extended to 49 days in active packages of 4.5% O_2, 7.8% CO_2, and 4.7% O_2, 7.5% CO_2 with or without moisture absorbent. Active atmosphere with gas concentration as high as 100% O_2 and 50% O_2 had a better effect on preserving nutritional quality of shiitake mushrooms stored for 7 days at 10 °C while storage under initial low O_2 (3% O_2 and 5% CO_2) induced fermentative metabolites (Li et al., 2014). However, Jing-jun et al. (2012) was of the observation that the gas components of 2% O_2 + 10–13% CO_2 within the packages is best for preservation of shiitake mushrooms when stored at 4 °C.

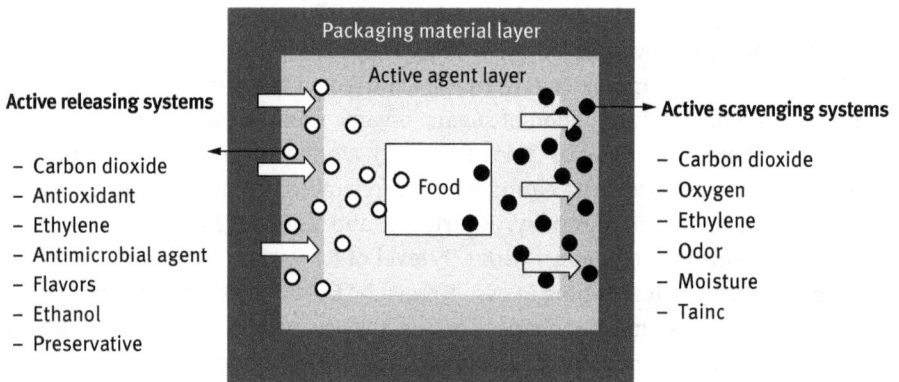

Figure 2.2: Schematic presentation of active modified packaging phenomenon (Lee et al., 2015).

More interest is shifting toward designing package with permeability (O_2 and CO_2) adapted to the product respiration known as equilibrium modified atmosphere packaging (E-MAP). For successful E-MAP, appropriate balance of several parameters influencing the package atmosphere such as gas permeability of the packaging film, RR of the commodity, and commodity weight need to be taken into consideration. Although E-MAP has been studied, very few studies reported on the successful use E-MAP for extending the shelf life of fresh produce. For instance, feijoa fruits packed in perforated polypropylene bags at 6 °C with equilibrium gas levels in the headspace of 8.21 kPa of

O_2 and 5.82 kPa of CO_2 had the lowest decline and change in their quality properties up to 38 days (Castellanos et al., 2016). Tumwesigye et al. (2017) observed that the IBC film attained equilibrium O_2 (2–3%) after 180 h at 10 °C for 75% and 95% RH, with or without perforation prolonged the shelf life of cherry tomatoes.

2.3.5 Overview of CA

Under CA, O_2, CO_2, and N_2 as well as temperature and relatively humidity of the storage are strictly regulated. CA was simply developed for long term storage of various fresh produce without causing significant losses and be able to offer high quality product. Low O_2 concentration of approximately 1 KPa has shown to improve storage term of product and this has led to wide adoption and application of ultra-low O_2 (0.8–1.2 kPa) globally (Dilley, 2006). There are two type of CA storage: static and dynamic controlled atmosphere (DCA). In a static-CA, hypoxic conditions (0.8–1.2 kPa) are strictly maintained for the entire storage duration while dynamic-CA system relies on the O_2 concentration as low as fruit metabolism can withstand. For a successful CA system, the reduction of O_2 to extremely low level should be monitored. Recently, there are three methods that permit the monitoring of low-O_2 level under CA during storage, namely, chlorophyll fluorescence based method (Prange et al., 2007); method based on the anaerobic respiration products (ethanol) (Veltman et al., 2003) and the RQ-based method (Weber et al., 2015a; Brackmann, 2015; Van Schaik et al., 2015). Recent report has shown that chlorophyll fluorescence is widely used compared to other methods around the world (Wright et al., 2010; Thewes et al., 2015; Weber et al., 2017; Mditshwa et al., 2018). Recent published peer-reviewed articles conducted for last 10 years on the successful use of CA on the storage of fresh produce are presented in Table 2.6. This, however, excludes

Table 2.6: Studies conducted for last 10 years on the successful use of CA for storage of fresh fruit.

CA – condition(s)*	Fruit type	Reference
DCA-RQ	Apples	Thewes et al. (2017); Thewes et al. (2018); Stanger et al. (2018)
Static and D-CA	Apples	Brizzoloara et al. (2017); Both et al. (2018)
RQ-DCA	Apples	Bessemans et al. (2016)
Static-CA	Apples	Both et al. 2014; DeEll et al., 2012; Lee et al., 2012; Ho et al., 2013; Lumpkin et al., 2015; DeEll et al., 2016a; DeEll et al., 2016b; Jung et al., 2011; Kweon et al., 2013; Gwanpua et al., 2012; Vermathen et al., 2017; Watkins and Nock, 2012; Lum et al., 2016; Toivonen and Hampson, 2014; East et al., 2013; Nock and Watkins, 2013

Table 2.6 (continued)

CA – condition(s)*	Fruit type	Reference
Static-CA	Hardy-Kiwifruit	Latocha et al., 2014; Li et al., 2017
Static-CA	Guava	Alba-Jiménez et al., 2018; Teixeira et al., 2016
CA-ULO	Grape	Liu et al., 2013
DCA	Apples	Weber et al., 2015b; Mditswa et al., 2017a; Mditshwa et al., 2017b; Weber et al., 2017; dos Santosa et al., 2018
DCA-ULO; CA	Apples	Thewes et al., 2015; Stanger et al., 2018
DCA-CF; DCA-EtOH, CA	Pear	Deuchande et al., 2016; Mattheis et al., 2013
CA	Apples	Bekele et al., 2016; dos Santosa et al., 2018; Thewes et al., 2017
Static-CA	Figs	Bahar and Lichter (2018)
Static-CA	Litchi	Sivakumar and Korsten (2010); Ali et al. (2016)
Static-CA	Sweet cherries	Cozzolino et al. (2019); Serradilla et al. (2013); Yang et al. (2019)
DCA-CF; DCA-RQ	Apples	Both et al. (2017)
Static-CA	Avocados	Alamar et al. (2017); Hernández et al. (2017)
Static-CA	Papaya	Martins and Resende (2015)
Static-CA	Strawberry, Red raspberry, Highbush blueberry	Alamar et al. (2017); Li et al. (2015); Forney et al. (2015); Chiabrando et al. (2011)
DCA-CF	Apples	Tran et al. (2015); Stanger et al. (2018)
CA; CA-CF	Pears	Mattheis et al. (2013)
Static-CA	Plums	Singh and Singh (2012)
Static-CA	Pomegranate	Matityahu et al. (2016)
Static-CA	Pears	Mattheis et al. (2013); Saquet et al. (2017); Vanoli et al. (2016); Yu et al. (2017)

*CA – controlled atmosphere; DCA – dynamic controlled atmosphere; CF – chlorophyll fluorescence; RQ – respiration quotient; EtOH – ethanol; and ULO – ultra low oxygen.

those published in conference proceedings. Figure 2.3 shows the percentage of CA studies for fresh produce. Between the year 2010 and early 2019, the data shows that among the fresh produce subjected to CAs were much higher for apples compared to other crops (Figure 2.3). On the other hand, fruit including pears, litchis, sweet cherries, avocadoes, and sweet cherries are gradually gaining momentum.

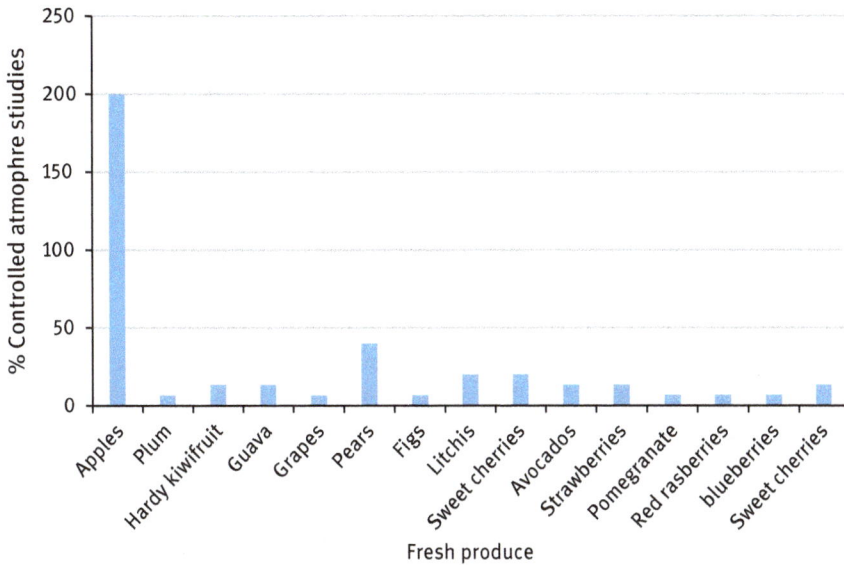

Figure 2.3: The percentage controlled atmosphere studies for fresh produce from 2010 to 2019.

2.3.5.1 Influence of CA on the quality of fruit

Several studies have reported on the effect of CA, dynamic controlled atmosphere (DCA), DCA-RQ, dynamic controlled atmosphere-chlorophyll fluorescence (DCA-CF), DCA-RQ, DCA-chlorophyll fluorescence (DCA-CF) and DAC-ethanol on quality mainte- nance of various fresh produce. Thewes et al. (2015) subjected "Royal Gala" and "Galaxy" apples to both CA (1.2 kPa O_2 + 2.0 kPa CO_2) and 1.2 kPa CO_2 DCA-CF + ultra- low oxygen (ULO) storage with 0.4 kPa O_2 + 1.2 kPa CO_2 storage during 2012 and initial low oxygen stress (0.05 kPa of O_2) + ULO for one day during 2013 seasons. The author found that fruit exposed to CA had higher ethylene followed by fruit under ULO and the lowest for fruit stored under DCA-CF. Moreover, the author showed that fruit stored under ULO and DCA-CF had higher firmness during 2013 seasons. The results from the study show that season has effect on the quality of apple fruit. Weber et al. (2017) found that "Fuji Suprema" apple stored in DCA-RQ1.5 treated with 1-MCP reduced physiological disorders and maintained flesh firmness but induced fermentative me- tabolites which in turn reduced ethylene production and RR. However, the author ob- served that 1-MCP reduced fruit quality due to occurrence of cavities and therefore concluded that it is not recommended for "Fuji Suprema" apple stored in DCA condi- tions. Both et al. (2017) observed that DCA-RQ1.5 has a trend to maintain higher flesh firmness of "Royal Gala" apple compared to DCA-CF when stored for 9 months plus 7 days shelf life due to low-ethylene evolution and RR. More so, the author found that fruit stored under DCA-RQ 2.0 had higher ethanol and ethyl acetate production.

Similarly, Bessemans et al. (2016) reported lower ethylene production rate in "Royal Gala" and "Granny Smith" apples stored in DCA-RQ for 8 to 9 months than fruit stored under CA. Low production of ethylene for fruit stored under DCA-CF and ULO is attributed to lower ACC oxidase enzyme activity.

Recently, Bahara and Lichter (2018) found that CA conditions of $5O_2 + 5CO_2$ kPa retains figs' quality when stored for 30 days at 2 °C. Combined effect of CA (5% O_2 + 10% CO_2 + 85% argon), and pressurized argon (0.5 MPa for 1 h at 0 °C) maintained the better quality sweet cherries during 63 days of storage at 0 °C (Yang et al., 2019). Study conducted by Hernández et al. (2017) reported that heat shock at 38 °C for a 1 h prior to CA storage (4 kPa O_2 and 6 kPa CO_2) for 30 days at 5 °C decreased ripening heterogeneity in early and middle season "Hass" avocado fruit whereas exposure to nitrogen shock + CA did not. Limited O_2 concentration during storage has been alleged to cause synthesis of fermentative metabolites including acetaldehyde and ethanol (Saquet et al., 2000). Papaya fruit stored under CA (1% or 3% $O2$ + 12% CO_2) minimized mass loss and fruit ripening, but affected the sensory attributes of the fruit after the storage (Martins and Resende, 2015). Furthermore, the author indicated that sensorial attributes were best maintained in an atmosphere with 3% O_2 plus 6% CO_2. Litchi fruit stored under CA conditions with gas combination of 1% $O2$ + 5% $CO2$ resulted in reduced weight loss, pericarp browning, membrane leakage and malondialdehyde contents (Ali et al., 2016).

DCA and CA have shown to be beneficial in delaying development of physiological disorders in some cultivars. For instance, pears cv. "Rocha" harvested at optimal maturity stored for 145 days at 0.5 °C and 95% RH showed no signs of internal browning disorders DCA-CF (Deuchande et al., 2016). On the contrary, Mattheis et al. (2013) found higher incidence of IBD in pears "d'Anjou" stored under DCA-CF than under CA. Moreover, the author concluded that IBD development did not trigger any modification in chlorophyll fluorescence signal, suggesting that this sensor cannot detect the stress condition that leads to the development of IBD in pear cultivar investigated. Chilling injury was fully prevented in white mushroom under static CA (80% O_2 + 20% CO_2) at 1 °C (Li et al., 2019). Pomegranate cvs. PG100-1, EVE, PG116-17 stored under CA (CA: 2 kPa O_2 + 5 kPa CO_2) at 7 °C, had lower incidences of husk scald and decay, however; RA prevented off-flavor during 5 months of cold storage (Matityahu et al., 2016).

DCA has shown to reduce superficial scald largely in various apple and pear cultivars at low temperature. Development of superficial scald has been widely associated with the conjugated trienes formation by oxidation of a-farnesene resulting in necrosis of hypodermal cortical tissue (Whitakker et al., 2009). Mditshwa et al. (2017) observed that DCA controlled superficial scald in preoptimally and optimally harvested fruit stored at 0 °C for 5 days to 20 weeks followed by 6 or 10 weeks simulated handling temperature (0.5 °C) for distant market plus 7 days at 20°C. They found that exposing DCA-treated fruit to 10 weeks shipment period increases the possibility of superficial scald development.

2.3.5.2 Influence of CA on the quality of fresh vegetables

From the gathered articles, very few studies report on the exposure of fresh vegetables to CA (Table 2.3). In a study conducted by Simões et al. (2011) who exposed baby carrots to CA of 5 kPa O_2 and 5 kPa CO_2 was of the conclusion that the baby carrots lasted only for 8 days. Naheed et al. (2017) found that storage duration/life of garlic scape cultivars differed even though they were stored under one CA condition ($O_2 = 2–5\%$, $CO_2 = 3–6\%$) at a temperature of 0 ± 0.5 °C. Wang et al. (2019) reported that the suitable gas concentration under CA for storage of broccoli is 50% $O_2 + 50\%$ CO_2, which in turn improved the quality, reduced accumulation of off-flavors. However, the author reported that high CO_2 promoted production of fermentative metabolites. Fernández-León et al. (2013) investigated the effect of 1-MCP and CA on broccoli quality and found that CA storage (10% O_2, 5% CO_2, 1–2 °C and 85–90% RH) preserved the quality and prolonged the shelf life of broccoli head than 1-MCP treatment and when stored in cold storage alone (1–2 °C and 85–90% RH). Conversely, the author observed that loss of chlorophyll was more prevalent at shelf life (2 and 4 days at 20 °C) for both treatments.

Table 2.7: Studies conducted for the last 10 years in the successful use CA for the storage of fresh vegetable.

CA- condition	Fruit type	Reference
Static-CA	Broccoli	Fernández-León et al. (2013); Wang et al. (2019)
Static-CA	Garlic scape	Naheed et al. (2017)
Static-CA	Mushrooms	Li et al. (2019)
Static-CA	Baby carrot	Simões et al. (2011)
Static-CA	Pak-choi	Harbaum-Piayda et al. (2010)
Static-CA	Calçots (anion stems)	Zudaire et al. (2017)
DCA	Spinach	Wright et al. (2012)
Static-CA	Broccoli plant	Fernández-León et al. (2013)

2.3.6 Chemical treatments

Chemical treatments have been at the center of postharvest for many years to curb losses during storage of fresh produce (Table 2.8). Losses may result from fungal and bacterial infections which affect fruit physiology and metabolism (Prusky et al., 2010). Fungi including *Penicillium digitatum*, *P. italicum*, and *Geotrichum citri-aurantii* are the most common postharvest fungi which infect their host through wounds

Table 2.8: Effect of postharvest chemical treatment on the quality of fresh produce.

Fruit/ vegetable	Treatment	Effect/result	Reference
Litchi	6-Benzylaminopurine (BAP)	Inhibited growth of *Peronophythora litchii* and browning incidence	Zhang et al. (2018)
Peach	6-Benzylaminopurine	The treatment directly inhibited *Monilinia fructicola*	Zhang et al. (2015)
Lemons	NaOCl (200 mg L^{-1}), imazalil (50 and 1,000 mg L^{-1}) at 20 or 50 °C, fruit were individually wrapped	Consecutive treatment with NaOCl and at 50 mg L^{-1} at 50°C, was as effective as IMZ at 1,000 mg L^{-1} at 20°C in controlling *Penicillium* decay.	D'Aquino et al. (2017)
Muskmelon	Acetylsalicylic acid (ASA)	ASA concentration of 3.2 mg/mL inhibited *Fusarium* rot development.	Huali et al. (2019)
Carrot	Ammonium sulfate, calcium sulfate, magnesium sulfate, potassium sulfate, sodium metabisulfite, sodium sulfate	Concentration of 50 and 200 mM metabisulfite salts provided 100% inhibition of cavity spot and dry rot, respectively; 50 mM calcium sulfate and sodium sulfate significantly reduced carrot cavity spot lesions; ammonium sulfate, magnesium sulfate, potassium sulfate, and sodium sulfate reduced potato dry rot lesions at 200 mM.	Kolaei et al. (2012)
Papaya	Hydrothermal plus calcium chloride	The treatment reduced anthracnose incidence up to 10 days during storage of papaya at 12°C.	Ayón-Reyna et al. (2017)
Fig	Calcium chloride (CaCl)	Concentration of 4% CaCl was found to be most effective in suppressing growth of mesophilic aerobic bacteria and yeast and molds at 1 ± 0.5°C, 95–98% RH.	Irfana et al. (2013)

Produce	Treatment	Description	Reference
Pears	γ-Aminobutyric acid	γ-Aminobutyric acid (GABA), at 100–1,000 lg/mL, induced strong resistance against blue mold rot caused by *Penicillium expansum*.	Yu et al. (2014)
Peach	Salicylic acid and ultrasound	Combined application of salicylic acid (SA, 0.05 mM) and ultrasound (40 kHz, 10 min) was effective in preventing fungal decay caused by blue mold (*Penicillium expansum*) during storage than the SA treatment alone while ultrasound had no effect.	Yang et al. (2011)
Citrus fruit	Sodium dehydroacetate (SD) and sodium silicate (Na_2SiO_3)	SD and Na_2SiO_3 combination (1:4, w/w) was effective against mycelial growth of *Geotrichum citri-aurantii* compared to single compound, with the minimum inhibitory concentration (MIC) and minimum fungicidal concentration (MFC) being 1.8 and 3.6 g L^{-1}, respectively.	Li et al. (2019)
Plums	Fumaric acid	Fruit treated with 0.5% fumaric acid decreased the population of *L. monocytogenes* by 1.65 log CFU/g.	Kim and Song (2017)
Satsuma mandarin	Ammonium molybdate	Treatment with high doses of ammonium molybdate reduced disease severity caused by *Penicillium digitatum*	Lu et al. (2018)
Jujube, peaches	Trisodium phosphate (TP)	Application of TP reduced disease development of brown rot caused by *Monilinia fructicola* on jujube and peach fruits	Cai et al. (2015)
Citrus	Sodium dehydroacetate	Various concentrations reduced the incidence of green and blue molds 25 ± 2 °C.	Duan et al. (2016)
Plums	Chlorine dioxide and ultrasonic	Combined treatment was more effective in reducing the initial microflora	Chen and Zhu (2011)

(continued)

Table 2.8 (continued)

Fruit/vegetable	Treatment	Effect/result	Reference
Pears	Acibenzolar-S-methyl (ASM)	ASM dipping treatment significantly decreased lesion diameter on the pear fruit inoculated with *Penicillium expansum*.	Ge et al. (2017)
Mango	Bentonite/potassium sorbate	Bentonite or bentonite combined with potassium sorbate reduced decay	Lui et al. (2014)
Loquat	Methyl jasmonite (MeJA)	MeJA treatment significantly reduced decay incidence (anthracnose development) caused by *Colletotrichum acutatum*.	Cao et al. (2014)
Banana	Oxalic acid	Treatment at 8 and 20 mM reduced brown coloring of fruit skin compared to untreated fruit	Huang et al. (2013)
Artichokes	Oxalic acid (OA)	OA significantly reduced mesophilic aerobics and yeast and molds counts stored at 20 °C and 85% RH for 3 days.	Ruíz-Jiménez et al. (2014)
Pears	Pyraclostrobin + boscalid (Pyr + Bosc), cyprodinil + fludioxonil (Cyp + Flu), fludioxonil (Flu), pyrimetanil (PyrM) and myclobutanil (Myc)	Pyraclostrobin + boscalid, cyprodinil + fludioxonil, and fludioxonil + pyrimetanil reduced natural incidence of *Alternaria* and *Cladosporium*.	Lutz et al. (2017)
Apple	Sodium nitroprusside	1.0 mmol L^{-1} SNP significantly inhibited lesion development of apple fruit inoculated with *P. expansum*	Ge et al. (2019)
Plums	Salicylic acid (SA)	Application of SA inhibited disease incidence.	Luo et al. (2011)

Produce	Treatment	Findings	Reference
Strawberry, cherry	Vaporized ethyl pyruvate (EP)	500 µl of EP (192 mL/L in air) gave favorable results for retardation of fungal decay.	Bozkurt et al. (2016)
Litchi	n-Butanol	n-Butanol-treated fruit had a lower browning index and disease index than untreated fruit.	Sun et al. (2011)
Lettuce	Hydrogen peroxide vapor	Vaporized 10% hydrogen peroxide treatment for 10 min was the most effective combination for reducing Salmonella Typhimurium, Escherichia coli O157:H7 and Listeria monocytogenes.	Back et al. (2014)
Apples	Trisodium phosphate (TSP)	TSP at various concentrations (0.25, 0.5, 1.0, 2.0 mg/mL) inhibited mycelia growth of Alternaria alternate.	Ge et al. (2019)
Apples	β-Aminobutyric acid (BABA)	BABA had an effect on controlling blue mold caused Penicillium expansum in apple fruit stored at 25 °C	Zhang et al. (2011)
Pears	Calcium chloride (CaCl₂) dipping and pullulan coating	Both treatments reduced the incidence of brown spots, inhibited the activities of PPO (polyphenol oxidase) and POD (peroxidase), increased the activities of CAT (catalase) and SOD (superoxide dismutase), and delayed the loss of phenolic compounds	Kou et al. (2015)
Blueberries	Aqueous chlorine dioxide (ClO₂) and UV-C	Combined effect of 2 mg L⁻¹ aqueous ClO₂ with 4 kJ m⁻² UV-C irradiation inhibited increase of decay incidence at 4 ± 1 °C	Xu et al. (2016)
Lettuce, spinach, parsley	Chlorine	Chlorine (100 mg L⁻¹) washes resulted in average log reductions of 2.9 ± 0.1 for Escherichia coli in the vegetables tested	Karaca et al. (2014)
Muskmelon	Thiamine	Thiamine (100 mmol L⁻¹) inhibited lesion development of muskmelon fruit inoculated with Trichothecium roseum or Alternaria alternata	Ge et al. (2017)

(continued)

Table 2.8 (continued)

Fruit/vegetable	Treatment	Effect/result	Reference
Apple	Salicylic acid	2.5 mM SA solution controlled blue mold in situ when applied under storage temperatures of 25 and 4 °C.	da Rocha Neto et al. (2016)
Citrus	Electrolyzed sodium bicarbonate (eNaHCO₃)	Application of eNaHCO₃ was able to control *Penicillium digitatum* infections when applied in wounds	Fallanaj et al. (2016)
Oranges	Benzothiadiazole (BTH), β-aminobutyric acid (BABA), 2,6-dichloroisonicotinic acid (INA), sodium silicate (SSi), salicylic acid (SA), acetylsalicylic acid (ASA), and harpin.	Postharvest dip treatments with chemical resistance inducers did not satisfactorily reduce decay caused by *P. digitatum* and *P. italicum* on citrus fruit after harvest.	Moscoso-Ramírez and Palou (2013)
Figs	Sulfur dioxide (SO₂) through fumigation and/or by dual release generating pads	Fumigation of warm fruit at 25 (L/L) h of SO₂ reduced populations of *Alternaria* and *Rhizopus* spp. on fig surface. The treatment was more effective against *Rhizopus* spp. than against *Alternaria* spp., *Botrytis* spp., and *Penicillium* spp. was also reduced treatment with by SO₂.	Cantín et al. (2011)
Citrus	Dehydroacetic acid, dimethyl dicarbonate, ethylene diamine tetraacetic acid, sodium acetate, and sodium benzoate (SB)	Dip treatments for 60 s with 3% (w/v) SB heated above 50 °C resulted in approximately 90% decrease of green and blue mold prevalence on "Valencia" oranges inoculated, treated, and incubated at 20 °C and 90% RH for 7 days. The treatment was also effective on "Lanelate" oranges, "Fino" lemons, and "Ortanique" mandarins, but not on "Clemenules" mandarins.	Montesinos-Herrero et al. (2016)
Figs	Sulfur dioxide (SO₂) through fumigation and/or by dual release generating pads	Fumigation of warm fruit at 25 (L/L) h of SO₂ reduced populations of *Alternaria* and *Rhizopus* spp. on fig surface. The treatment was more effective against *Rhizopus* spp. than against *Alternaria* spp. *Botrytis* spp. and *Penicillium* spp. was also reduced treatment with by SO₂.	Cantín et al. (2011)

Produce	Treatment	Description	Reference
Pear	L-glutamate	The treatment at 1.00 mM induced strong resistance against blue mold rot caused by *P. expansum* under either 25 or 4 °C condition. L-Glutamate reduced spore germination of *P. expansum* both in fruit wounds and *in vitro* after 24 h of treatment.	Jin et al. (2019)
Pear	Indole-3-acetic acid (IAA)	IAA at 100–500 µg mL^{-1} was effective against blue mold rot when the treating interval was between 24 and 48 h. No direct antifungal activities were observed toward *P. expansum in vitro*.	Zhang et al. (2018)
Pear	Sodium bicarbonate (NaHCO$_3$)	Fruits inoculated with NaHCO3-treated *P. expansum* spores showed lower disease incidence and severity.	Lai et al. (2015)
Apple	Methyl thujate	Treatment with 6,000 mg L^{-1} alleviated disease severity in apple fruit.	Ji et al. (2018)
Chinese bayberry	Methyl jasmonate (MeJA)	Treatment with 10 mol L^{-1} of MeJA significantly inhibited green mold rot caused by *P. citrinum*, with the decay incidence being 66.2% lower than that of the control fruit after storage at 1 °C for 8 d.	Wang et al. (2014)
Table grapes	MeJA	MeJA treatment (10 µmol/L) reduced the incidence and growth of gray mold decay in grape caused by *Botrytis cinerea*.	Jiang et al. (2015)
Litchi	Nitric oxide (NO)	Application of 0.5, 1.0, or 2.0 mM NO were effective against pericarp browning and reduced decay incidence up to 8 days at ambient condition (30 ± 2 °C temperature and 85 ± 5% RH).	Barman et al. (2014)
Nectarine	Chlorogenic acid (CHA)	Decay rate could be kept at lower levels both at 25 or 50 mg L^{-1} CHA	Xi et al. (2017)

(continued)

Table 2.8 (continued)

Fruit/ vegetable	Treatment	Effect/result	Reference
Litchi	L-Cysteine	L-Cysteine (0.25%) treated fruit showed reduced disease incidence and browning index	Ali et al. (2016)
Muskmelon	Oxalic acid	50 Mm concentration reduced disease severity on fruit inoculated with *Trichothecium roseum*	Deng et al. (2015)
Citrus	Potassium sorbate (PS)	PS did not provide adequate decay control caused by *Penicillium digitatum* in waxing treatments.	Parra et al. (2014)
Litchi	Kojic acid (KA)	4 mmol L^{-1} KA had reduced fruit decay; while, 6 mmol L^{-1} KA delayed pericarp browning by maintaining higher total anthocyanin when stored at 5 ± 1 °C with 90 ± 5% RH for 20 days.	Shah et al. (2017)
Litchi	Methionine	0.25% methionine inhibited fruit decay and reduced pericarp browning when kept under cold storage at 5 ± 1 °C for 28 days.	Ali et al. (2018)
Strawberry	$CaCl_2$	1% $CaCl_2$ g treated fruit had better decay rate during storage at 4 °C for 15 days	Chen et al. (2011)
Table grapes	Trans-resveratrol and glycine betaine (GB)	Trans-resveratrol (1.6×10^{-5}, 1.6×10^{-4}, and 1.6×10^{-3} M) especially at medium and high rates decreased decay after storage. GB (10, 15, and 20 mM/L) at high rate, decreased decay incidences.	Awad et al. (2015)

Mango	Nitric oxide (NO)	NO (sodium nitroprusside of 0.1 mM) treatment inhibited lesion growth on mango fruit inoculated with *Colletotrichum gloeosporioides* and the lesion diameter at 2AU: Please provide unit for "2" in "*Colletotrichum gloeosporioides* and the lesion diameter at 2 up to". up to 8 days averaged 30% when stored at	Hu et al. (2014)
Satsuma mandarins	Salicylic acid (SA)	The disease incidence of control/SA-treated fruit at 50 and 120 days exposed to 2 mM SA was 23.3%/10% and 67.3%/23.3%	Zhu et al. (2016)
Litchi	*para*-Aminosalicylate (PAS-Na)	Concentration of 0.3 g L^{-1} PAS-Na significantly prevented the development of pericarp browning.	Li et al. (2019)
Grapes	AC concentration of 50 mL L^{-1} was after 4 or 8 weeks; AC concentrations (30, 50 and 75 mL L^{-1}) were repeated 5, 3, and 2 times, respectively	All treatments reduced gray mold incidence after 8 weeks of storage, however, repeated treatments resulted the most effective. Two fumigations at 50 mL L^{-1} or five fumigations at 30 mL L^{-1} reduced gray mold incidence by 63.6% or 57.1% respectively.	Venditti et al. (2017)

inflicted during harvest handling and processing (Wuryatmo et al., 2014; Karim et al., 2016). Generally, control of microbial infection is achieved through application of wide variety of synthetic fungicides. Some fungicides are able to kill pathogens at the fruit/vegetable surface as contact but do not have any protective activity. Moreover, preservation of membrane structure is crucial in maintaining fruit quality and prolonging fruit and vegetable shelf life.

Studies have reported various valuable results on application of fungicides and anti-browning agents on fresh fruit and vegetables. In the recent study, Zhang et al. (2018) found that 6-benylminopurine at 0.1 g L^{-1} did not only act as antifungal agent but also delayed browning incidence in litchi fruit stored at 25 °C for 8 days. Methionine at 0.25% inhibited fruit decay and reduced pericarp browning in litchi when kept under cold storage at 5 ± 1 °C for 28 days (Ali et al., 2018). Application of 6-Benzylaminopurine at concentration ranging from 500 to 1,000 mg L^{-1} of inhibited *Monilinia fructicola*, a common peach fungus (Zhang et al., 2015). Study by Cai et al. (2015) was of the observation that disodium phosphate at 1.5 and 2% (w/v) completely inhibited brown rot caused by *Monilinia fructicola* in jujube and peach fruits, respectively. It could be noted that results from both studies showed antifungal activity is directly correlated to the chemical concentrations used. However, acetylsalicylic acid (ASA) at 3.5 mg L^{-1} was found to be effective against *Fusarium* rot compared to other concentrations (1.6 and 6.4 mg L^{-1}) by inhibiting the growth of *F. sulphureum* on muskmelon after 6 and 8 d of pathogen inoculation (Huali et al., 2019). From these results, it could be suggested that higher concentration of various fungicides may not necessarily be effective across broad spectrum of pathogens. According to Huali et al. (2019), treatment with ASA resulted in the degeneration of fungal hypha and impairment of secondary metabolites of toxin. Treatment with 6,000 mg L^{-1} methyl thujate alleviated disease severity in apple fruit caused by *Botrytis cinerea* than when exposed to lower concentrations (1,000 and 3,000 mg L^{-1}) during storage at room temperature (23 ± 2 °C) (Ji et al., 2018). The author observed that methyl thujate was more active against spore germination than mycelial growth.

Lu et al. (2018) was of the observation that ammonium molybdate at high dose of 5 mmol L^{-1} reduced disease severity caused by *Penicillium digitatum* in satsuma mandarin than when treated with 0.01, 0.1, and 1 mmol L^{-1} doses. Pears treated with *y*-aminobutyric acid at a concentration in a range of 100–1,000 µg/mL, induced strong resistance against blue mold rot caused by *P. expansum* (Yu et al., 2014). In apple fruit, a dose of 1.0 mmol^{-1} sodium nitroprusside inhibited lesion development of the fruit inoculated with *P. expansum* (Ge et al., 2019). On the contrary, Moscoso-Ramírez and Palou (2013) found that orange fruit treated with benzothiadiazole, *β*-aminobutyric acid, 2,6-dichloroisonicotinic acid, sodium silicate, salicylic acid, acetylsalicylic acid, and harpin and stored at 20 °C, 90% RH for 6 days were less effective against *P. digitatum* and *P. italicum* after harvest.

Several combinatorial effects of fungicides against foodborne pathogens have been reported by researchers. For instance, in citrus fruit sodium dehydroacetate combined with sodium silicate (1:4 w/w) inhibited mycelial growth of *Geotrichum cit-riaurantii* with minimum inhibitory concentration and minimum fungicidal concentration being 1.8 and 3.6 g L^{-1}, respectively (Li et al., 2019c). Combination of fungicide pyraclostrobin + boscalid, cyprodinil + fludioxonil and fludioxonil + pyrimethanil reduced natural incidence of *Alternaria–Cladosporium* complex (4 months at – 1/0 °C), between 24% and 35%, whereas on wounds artificially inoculated, 91% and 98% reduction was achieved in pears (Lutz et al., 2017). Lui et al. (2014) reported that bentonite or bentonite combined with potassium sorbate reduced decay incidence in mango fruit stored at room temperature for 16 days. Application of trisodium phosphate at 1.5% and 2% (w/v) reduced disease development of brown rot caused by *Monilinia fructicola* on jujube and peach fruits (Cai et al., 2015).

Microbial loads of various fruit and vegetable during storage can be reduced by postharvest treatment. Irfana et al. (2013) investigated calcium chloride on the quality of fig (*Ficus carica* L.) during storage and found that 4% calcium chloride was effective in suppressing growth of mesophilic aerobic and yeast and molds when stored at 1 ± 0.5 °C, 95–98% RH and during shelf life. In lettuce, spinach, and parsley treated with 100 mg L^{-1} resulted in average log reductions of 2.9 ± 0.1 of *Escherichia coli* (Karaca et al., 2014). Kim and Song (2017) observed that population of *Listeria monocytogenes* was reduced by 1.65 log CFU/g in plums treated with a concentration of 0.5% fumaric acid.

2.4 Conclusion

Maintenance of fresh produce quality is a complex procedure and race against time. Hence, the critical knowledge of the interactions between various factors that impact quality and the role of available tools to maintain desired quality for a longer time is needed in order to design optimal postharvest handling practices. A wide range of physical and chemical treatments have been shown to be applicable to certain types of fresh commodity and the effectiveness of existing treatments is well documented. Postharvest treatments, such as CA and MAP, in combination with other hurdle techniques and under optimal storage conditions have been shown to be effective for maintaining physical, nutritional, and sensory attributes. However, as new cultivars emerge and climate change induce changes on existing fresh produce, research on these technologies remains relevant and important to minimize postharvest losses. In addition, recognizing that the metabolism of the fresh horticultural produce changes in response to the applied postharvest treatments, in depth omics research to understand underlying responses is crucial and this will open new arena within postharvest biology and technology.

Acknowledgements: This work is based on research supported in part by the Agricultural Research Council, South Africa, the National Research Foundation of South Africa, Competitive Support for Unrated Researchers (Grant Number: 116272), and Incentive Funding for Rated Researchers (Grant Number: 119192) awarded to Dr OJ Caleb. The National Research Foundation – Thuthuka Grant awarded to Dr RR Mphahlele is gratefully acknowledged. Sincere gratitude goes to Dr Mduduzi Ngcobo for his guidance and mentorship to Dr RR Mphahlele.

"It does not matter how slowly you go so long as you do not stop." ~ Confucius

References

Admane, N., Genovese, F., Altieri, G., Tauriello, A., Trani, A., Gambacort, G., Verrastroc, V. and Di Renzo, G. C. (2018). Effect of ozone or carbon dioxide pre-treatment during long-term storage of organic table grapes with modified atmosphere packaging. *LWT-Food Science and Technology*, 98, 170–178.

Afifi, E. H., Ragab, M. E., Abd El-Gawad, H. G. and Emam, M. S. (2016). Effect of active and passive modified atmosphere packaging on quality attributes of strawberry fruits during cold storage. Arab Universities Journal of Agricultural Sciences, 24(1), 157–168.

Aglar, E., Ozturk, B., Guler, S. K., Karakay, O., Uzun, S. and Saracoglu, O. (2017). Effect of modified atmosphere packaging and 'Parka' treatments on fruit quality characteristics of sweet cherry fruits (*Prunus avium* L. '0900 Ziraat') during cold storage and shelf life. *Scientia Horticulturae*, 222, 162–168.

Aindongo, W. V., Caleb, O. J., Mahajan, P. V., Manley, M. and Opara, U. L. (2014). Effects of storage conditions on transpiration rate of pomegranate aril-sacs and arils. *South African Journal of Plant and Soil*, 31(1), 7–11.

Alamar, M. C., Collings, E., Cools, K. and Terry, L. A. (2017). Impact of controlled atmosphere scheduling on strawberry and imported avocado fruit. *Postharvest Biology and Technology*, 134, 76–86.

Alba-Jiménez, J. E., Benito-Bautista, P., Nava, G. M., Rivera-Pastrana, D. M., Vázquez-Barrios, M. E. and Mercado-Silva, E. M. 2018. Chilling injury is associated with changes in microsomal membrane lipids in guava fruit (*Psidium guajava* L.) and the use of controlled atmospheres reduce these effects. *Scientia Horticulturae*, 240, 94–101.

Alegria, C., Pinheiro, J., Duthoit, M., Gonçalves, E. M., Moldão-Martins, M. and Abreu, M. (2012). Fresh-cut carrot (cv. Nantes) quality as affected by abiotic stress (heat shock and UV-C irradiation) pre-treatments. *LWT-Food Science and Technology*, 48, 197–203.

Ali, S., Khan, A. S. and Malik, A. U. (2016). Postharvest L-cysteine application delayed pericarp browning, suppressed lipid peroxidation and maintained antioxidative activities of litchi fruit. *Postharvest Biology and Technology*, 121, 135–142.

Ali, S., Khan, A. S., Malik, A. U., Shaheen, T. and Shahid, M. (2018). Pre-storage methionine treatment inhibits postharvest enzymatic browning of cold stored 'Gola' litchi fruit. *Postharvest Biology and Technology*, 140, 100–106.

Arnon, H., Zaitsev, Y., Porat, R. and Poverenov, E. (2014). Effects of carboxymethyl cellulose and chitosan bilayer edible coating on postharvest quality of citrus fruit. *Postharvest Biology and Technology*, 87, 21–26.

Artés-Hernández, F., Robles, P. A., Gómez, P. A., Tomás-Callejas, A. and Artés, F. (2010). Low UV-C illumination for keeping overall quality of fresh-cut watermelon. *Postharvest Biology and Technology*, 55, 114–120.

Arvanitoyannis, I. S., Bouletis, A. D., Papa, E. A., Gkagtzis, D. C., Hadjichristodoulou, C. and Papaloucas, C. (2011). Microbial and sensory quality of "*Lollo verde*" lettuce and rocket salad stored under active atmosphere packaging. *Anaerobe*, 17, 307–309.

Athmaselvi, K. A., Sumitha, P. and Revathy, B. (2013). Development of Aloe vera based edible coating for tomato. *International Agrophysics* 27, 369–375.

Awad, M. A., Al-Qurashi, A. D. and Mohamed, S. A. (2015). Postharvest trans-resveratrol and glycine betaine treatments affect quality, antioxidant capacity, antioxidant compounds and enzymes activities of 'El-Bayadi' table grapes after storage and shelf life. *Scientia Horticulturae*, 197, 350–356.

Ayón-Reyna, L. E., González-Robles, A., Rendón-Maldonado, J. G., Báez-Flores, M. E., López-López, M. E. and Vega-García, M. O. (2017). Application of a hydrothermal-calcium chloride treatment to inhibit postharvest anthracnose development in papaya. *Postharvest Biology and Technology*, 124, 85–90.

Back, K. H., Ha, J. W. and Kang, D. H. (2014). Effect of hydrogen peroxide vapor treatment for inactivating *Salmonella Typhimurium, Escherichia coli* O157:H7 and *Listeria monocytogenes* on organic fresh lettuce. *Food Control*, 44, 78–85.

Bahara, A. and Lichter, A. (2018). Effect of controlled atmosphere on the storage potential of Ottomanit fig fruit. *Scientia Horticulturae*, 227, 196–201.

Banerjee, A., Suprasanna, P., Variyar, P. S. and Sharma, A. (2015). Gamma irradiation inhibits wound induced browning in shredded cabbage. *Food Chemistry*, 173, 38–44.

Barman, K., Siddiqui, M. W., Patel, V. B. and Prasad, M. (2014). Nitric oxide reduces pericarp browning and preserves bioactive antioxidants in litchi. *Scientia Horticulturae* 171, 71–77.

Barikloo, H. and Ahmadi, E. (2018). Shelf life extension of strawberry by temperatures conditioning, chitosan coating, modified atmosphere, and clay and silica nanocomposite packaging. *Scientia Horticulturae*, 240, 496–508.

Belay, Z. A., Caleb, O. J., Mahajan, P. V. and Opara, U. L. (2018). Design of active modified atmosphere and humidity packaging (MAHP) for 'wonderful' pomegranate arils. *Food and Bioprocess Technology*, 11(8), 1478–1494.

Belay, Z. A., Caleb, O. J. and Opara, U. L. (2016). Modelling approaches for designing and evaluating the performance of modified atmosphere packaging (MAP) systems for fresh produce: a review. *Food Packaging and Shelf Life*, 10, 1–15.

Belay, Z. A., Caleb, O. J., Opara, U. L. 2017. Impacts of low and super-atmospheric oxygen concentrations on quality attributes, phytonutrient content and volatile compounds of minimally processed pomegranate arils (cv. Wonderful). *Postharvest Biology Technology*, 124, 119–127.

Bekele, E. A., Ampofo-Asiama, J., Alis, R., Hertog, M. L., Nicolai, B. M. and Geeraerd, A. H. (2016). Dynamics of metabolic adaptation during initiation of controlled atmosphere storage of 'Jonagold' apple: effects of storage gas concentrations and conditioning. *Postharvest Biology Technology*, 117, 9–20.

Ben-Yehoshua, S., Rodov, V. 2002. Transpiration and water stress. In J. Barz, & J. K. Brecht (Eds.), Postharvest physiology and pathology of vegetables (pp. 111e159). NY, USA: Marcel Dekker.

Ben-Yehoshua, S. and Rodov, V. (2013). Transpiration and water stress. In: J. A. Bartz and J. K. Brecht (Eds.), *Postharvest Physiology and Pathology of Vegetables (Second Edition)* (p. 111159). New York, NY: Marcel Dekker.

Benítez, S., Achaerandio, I., Pujolà, M. and Sepulcre, F. (2015). Aloe vera as an alternative to traditional edible coatings used in fresh-cut fruits: a case of study with kiwifruit slices. *LWT-Food Science and Technology*, 61, 184–193.

Bessemans, N., Verboven, P., Verlinden, B. and Nicolaï, B. (2016). A novel type of dynamic controlled atmosphere storage based on the respiratory quotient (RQ-DCA). Postharvest Biology and Technology, 115, 91–102.

Bhat, R. and Stamminger, R. (2015). Impact of combination treatments of modified atmospheric packaging and refrigeration on the status of antioxidants in highly perishable strawberries. *Journal of Food Process Engineering*, 39, 121–131.

Bhat, R. and Stamminger, R. (2015). Impact of combination treatments of modified atmosphere packaging and refrigeration on the status of antioxidants in highly perishable strawberries. *Journal of Food Processing*, 39, 121–131.

Both, V., Brackmann, A., Thewes, F. R., Ferreira, D. and Wagner, F. de, R. (2014). Effect of storage under extremely low oxygen on the volatile composition of Royal Gala apples. *Food Chemistry*, 156, 50–57.

Both, V., Thewes, F. R., Brackmann, A., de Oliveira Anese, R., de Freitas Ferreira, D. and Wagner, R. (2017). Effects of dynamic controlled atmosphere by respiratory quotient on some quality parameters and volatile profile of 'Royal Gala' apple after long-term storage. *Food Chemistry*, 215, 483–492.

Both, V., Brackmann, A., Thewes, F. R., Weber, A., Schultz, E. E. and Ludwig, V. (2018). The influence of temperature and 1-MCP on quality attributes of 'Galaxy' apples stored in controlled atmosphere and dynamic controlled atmosphere. *Postharvest Biology and Technology*, 16, 168–177.

Both, V., Thewes, F. R., Brackmann, A., de Oliveira Anese, R., de Freitas Ferreira, D. and Wagner, R. (2017). Effects of dynamic controlled atmosphere by respiratory quotient on some quality parameters and volatile profile of 'Royal Gala' apple after long-term storage. *Food Chemistry*, 215, 483–492.

Bovi, G. G., Caleb, O. J., Ilte, K., Rauh, C. and Mahajan, P. V. (2018). Impact of modified atmosphere and humidity packaging on the quality, offodour development and volatiles of 'Elsanta' strawberries. *Food Packaging and Shelf Life*, 16, 204–210.

Bovi, G.G., Caleb, O.J., Ilte, K., Rauh, C., Mahajan, P.V. 2018a. Impact of modified atmosphere and humidity packaging on the quality, offodour development and volatiles of 'Elsanta' strawberries. Food Packaging and Shelf Life, 16, 204–210.

Bovi, G. G., Rux, G., Caleb, O. J., Herppich, W. B., Linke, M., Rauh, C., & Mahajan, P. V. 2018b. Measurement and modelling of transpiration losses in packaged and unpackaged strawberries. *Biosystems engineering*, 174, 1–9.

Bovi, G. G., Caleb, O. J., Linke, M., Rauh, C. and Mahajan, P. V. (2016). Transpiration and moisture evolution in packaged fresh horticultural produce and the role of integrated mathematical models: a review. *Biosystems Engineering*, 150, 24–39.

Bovi, G. G., Rux, G., Caleb, O. J., Herppich, W. B., Linke, M., Rauh, C. and Mahajan, P. V. (2018). Measurement and modelling of transpiration losses in packaged and unpackaged strawberries. *Biosystems Engineering*, 174, 1–9.

Bozkurt, F., Tornuk, F., Toker, O. S., Karasu, S., Arici, M. and Durak, M. Z. (2016). Effect of vaporized ethyl pyruvate as a novel preservation agent for control of postharvest quality and fungal damage of strawberry and cherry fruits. *LWT-Food Science and Technology*, 65, 1044–1049.

Brackmann, A. (2015). Control apparatus for controlled atmosphere cells for storing perishable items. U.S. Patent, n.US2015/0257401A1.

Brizzolara, S., Santucci, C., Tenori, L., Hertog, M., Nicolai, B., Stürz, S., Zanella, A. and Tonutti, P. (2017). A metabolomics approach to elucidate apple fruit responses to static and dynamic controlled atmosphere storage. *Postharvest Biology and Technology*, 127, 76–87.

Cai, J., Chen, J., Lu, G., Zhao, Y., Tian, S. and Qin, G. (2015). Control of brown rot on jujube and peach fruits by trisodium phosphate. *Postharvest Biology and Technology*, 99, 93–98.

Caleb, O.J., Herppich, W.B., Mahajan, P.V., 2016a. The basics of respiration for horticultural products. Reference Module in Food Sciences. Elsevier, pp. 1–7.

Caleb, O. J., Ilte, K., Fröhling, A., Geyer, M. and Mahajan, P. V. 2016. Integrated modified atmosphere and humidity package design for minimally processed Broccoli (Brassica oleracea L. var. italica). *Postharvest Biology and Technology*, 121, 87–100.

Caleb, O. J., Mahajan, P. V., Al-Said, F. A. and Opara, U. L. (2013). Transpiration rate and quality of pomegranate arils as affected by storage conditions. *CyTA-Journal of Food*, 11(3), 199–207.

Caleb, O. J., Mahajan, P. V., Manley, M., Opara, U. L. 2013. Evaluation of modified atmosphere packaging engineering design parameters for pomegranate arils. *International Journal of Food Science and Technology*, 48, 2315–2323.

Caleb, O. J., Opara, U. L. and Witthuhn, C. R. Modified atmosphere packaging of pomegranate fruits and arils: a review. *Food Bioprocess Technology*, 5, 15–30.

Caleb, O. J., Opara, U. L., Witthuhn, C. R. 2012. Modified atmosphere packaging of pomegranate fruits and arils: a review. *Food bioprocess Technology*, 5, 15–30.

Caleb, O. J., Wegner, G., Rolleczek, C., Herppich, W. B., Geyer, M. and Mahajan, P. V. (2016). Hot water dipping: impact on postharvest quality, individual sugars, and bioactive compounds during storage of 'Sonata' strawberry. *Scientia Horticulturea*, 210, 150–157.

Candir, E., Ozdemir, A. E. and Aksoy, M. C. (2016). Effects of chitosan coating and modified atmosphere packaging on postharvest quality and bioactive compounds of pomegranate fruit cv. 'Hicaznar'. *Scientia Horticulturae*, 235, 235–243.

Candir, E., Ozdemir, A. E. and Aksoy, M. C. (2018). Effects of chitosan coating and modified atmosphere packaging on postharvest quality and bioactive compounds of pomegranate fruit cv. 'Hicaznar'. *Scientia Horticulturae*, 235, 235–243.

Candir, E., Ozdemir, E. A., Kamiloglu, O., Soylu, E. M., Dilbaz, R., Ustun, D. (2012). Modified atmosphere packaging and ethanol vapor to control decay of 'Red Globe' table grapes during storage. *Postharvest Biology and Technology*, 63, 98–106.

Cantín, C. M., Palou, C. L., Michailides, B. V., Carlos, T. J. and Crisosto, H. (2011). Evaluation of the use of sulfur dioxide to reduce postharvest losses on dark and green figs. *Postharvest Biology and Technology* 59, 150–158.

Cao, S., Cai, Y., Yang, Z., Joyce, D. C., Zheng, Y. (2014). Effect of MeJA treatment on polyamine, energy status and anthracnose rot of loquat fruit. *Food Chemistry*, 145, 86–89.

Castellanos, D. A., Polanía, W. and Herrera, A. O. (2016). Development of an equilibrium modified atmosphere packaging (EMAP) for feijoa fruits and modeling firmness and color evolution. *Postharvest Biology and Technology* 120, 193–203.

Cefola, M., Damascelli, A., Lippolis, V., Cervellieri, S., Linsalata, V., Logrieco, A. F. and Pace, B. (2018). Relationships among volatile metabolites, quality and sensory parameters of 'Italia' table grapes assessed during cold storage in low or high CO_2 modified atmospheres. *Postharvest Biology and Technology*, 142, 124–134.

Chen, Z. and Zhu, C. (2011). Combined effects of aqueous chlorine dioxide and ultrasonic treatments on postharvest storage quality of plum fruit (*Prunus salicina* L.). *Postharvest Biology and Technology*, 61, 117–123.

Chiabrando, V. and Giacalone, G. (2011). Shelf-life extension of highbush blueberry using 1-methylcyclopropene stored under air and controlled atmosphere. *Food Chemistry*, 126, 1812–1816.

Chitravathi, K., Chauhan, O. P. and Raju, P. S. (2015). Influence of modified atmosphere packaging on shelf-life of green chillies (*Capsicum annuum* L.). *Food Packaging and Shelf Life*, 4, 1–9.

Colgecen, I. and Aday, M. S. (2015). The efficacy of the combined use of chlorine dioxide and passive modified atmosphere packaging on sweet cherry quality. *Postharvest Biology and Technology*, 109, 10–19.

Costa, M. L., Civello, P. M., Chaves, A. R. and Martínez, G. A. (2006). Hot air treatment decreases chlorophyll catabolism during postharvest senescence of broccoli (Brassica oleracea L. var. italica) heads. *Journal of the Science of Food and Agriculture*, 86, 1125–1131.

Cozzolino, R., Martignetti, A., Cefola, M., Pace, B., Capotorto, I., De Giulio, B., Montemurro, N. and Pellicano, M. (2019). Volatile metabolites, quality and sensory parameters of "Ferrovia" sweet cherry cold stored in air or packed in high CO_2 modified atmospheres. *Food Chemistry*, 286, 659–668.

da Rocha Neto, A. C., Luiz, C., Maraschin, M., and Di Piero, R. M. (2016). Efficacy of salicylic acid to reduce *Penicillium expansum* inoculum and preserve apple fruits. *International Journal of Food Microbiology*, 221, 54–60.

D'Aquino, S., Dai, S., Deng, Z., Gentile, A., Angioni, A., De Pau, L. and Palma, A. (2017). A sequential treatment with sodium hypochlorite and a reduced dose of imazalil heated at 50 °C effectively control decay of individually film-wrapped lemons stored at 20 °C. *Postharvest Biology and Technology*, 124, 75–84.

D'Aquino, S., Mistriotis, A., Briassoulis, D., Di Lorenzo, M. L., Malinconico, M. and Palma, A. (2016). Influence of modified atmosphere packaging on postharvest quality of cherry tomatoes held at 20 °C. *Postharvest Biology and Technology*, 115, 103–112.

Díaz-Mula, H. M., Zapata, P. J., Guillén, F., Valverde, J. M., Valero, D. and Serrano, M. (2011). Modified atmosphere packaging of yellow and purple plum cultivars. 2. Effect on bioactive compounds and antioxidant activities. *Postharvest Biology and Technology*, 61, 110–116.

de Jesús Dávila-Aviña, J. E., Villa-Rodríguez, J., Cruz-Valenzuela, R., Rodríguez-Armenta, M., Espino-Díaz, M., Ayala-Zavala, J. F. and González-Aguilar, G. (2011). Effect of edible coatings, storage time and maturity stage on overall quality of tomato fruits. *American Journal of Agricultural and Biological Sciences*, 6, 162–171.

DeEll, J. R. and Ehsani-Moghaddam, B. (2012). Delayed controlled atmosphere storage affects storage disorders of 'Empire' apples. *Postharvest Biology and Technology*, 67, 167–171.

DeEll, J. R., Lum, G. B. and Ehsani-Moghaddam, B. (2016a). Effects of multiple 1-methylcyclopropene treatments on apple fruit quality and disorders in controlled atmosphere storage. *Postharvest Biology and Technology*, 111, 93–98.

DeEll, J. R., Lum, G. B. and Ehsani-Moghaddam, B. (2016b). Elevated carbon dioxide in storage rooms prior to establishment of controlled atmosphere affects apple fruit quality. *Postharvest Biology and Technology*, 118, 11–16.

Deng, J., Bi, Y., Zhang, Z., Xie, D., Ge, Y., Li, W., Wang, J. and Wang, Y. (2015). Postharvest oxalic acid treatment induces resistance against pink rot by priming in muskmelon (*Cucumis melo* L.) fruit. *Postharvest Biology and Technology*, 106, 53–61.

Deuchande, T. Carvalho, S. M. P. Guterres, U., Fidalgo, F., Isidoro, N., Larrigaudiere, C. and Vasconcelos, M. W. (2016). Dynamic controlled atmosphere for prevention of internal browning disorders in 'Rocha' pear. *LWT-Food Science and Technology*, 65, 725–730.

Dhall, R. K. (2013). Advances in edible coatings for fresh fruits and vegetables: a review. *Critical Reviews in Food Science and Nutrition*, 53, 435–450.

Dilley, R. D. (2006). Development of controlled atmosphere storage technologies. e storage technologies. *Stewart Postharvest Review*, 2, 1–8. DOI: 10.2212/spr.2006.6.5.

Domínguez, I., Lafuente, M. T., Hernández-Muñoz, P. and Gavara, R. (2016). Influence of modified atmosphere and ethylene levels on quality attributes of fresh tomatoes (*Lycopersicon esculentum* Mill.). *Food Chemistry*, 209, 211–219.

Dos Santos, I. D., Pizzutti, I. R., Dias, J. V., Fontana, M. E., Brackmann, A., Anese, R. O., Thewes, F. R., Marques, L. N. and Cardoso, C. D. (2018). Patulin accumulation in apples under dynamic controlled atmosphere storage. *Food Chemistry*, 255, 275–281.

East, A. R., Smale, N. J. and Trujillo, F.J. (2013). Potential for energy cost savings by utilising alternative temperature control strategies for controlled atmosphere stored apples. *International Journal of Refrigeration*, 36, 1109–1117.

Edelenbos, M., Lokke, M. M. and Seefeldt, H. F. (2017). Seasonal variation in colour and texture of packaged wild rocket (*Diplotaxis tenuifolia* L.). *Food Packaging and Shelf Life*, 14A, 46–51.

Endo, H., Ose, K., Bai, J. and Imahori, Y. (2019). Effect of hot water treatment on chilling injury incidence and antioxidative responses of mature green mume (Prunus mume) fruit during low temperature storage. *Scientia Horticulturae*, 246, 550–556.

Escalona, V. H., Aguayo, E., Martínez-Hernández, G. B. and Artés, F. (2010). UV-C doses to reduce pathogen and spoilage bacterial growth in vitro and in baby spinach. *Postharvest Biology and Technology*, 56, 223–231.

Fagundes, C., Carciofi, B.A.M., Monteiro, A.R. (2013). Estimate of respiration rate and physicochemical changes of fresh-cut apples stored under different temperatures. *Food Science and Technology, Campinas*, 33(1), 60–67.

Fagundes, C., Moraes, K., Pérez-Gago, M. B., Palou, M. L. and Monteiroa, M. A. R. (2015). Effect of active modified atmosphere and cold storage on the postharvest quality of cherry tomatoes. *Postharvest Biology and Technology*, 109, 73–81.

Fagundes, C., Palou, L., Monteiro, A. R. and Pérez-Gago, M. B. (2014). Effect of antifungal hydroxypropyl methylcellulose-beeswax edible coatings on gray mold development and quality attributes of cold-stored cherry tomato fruit. *Postharvest Biology and Technology*, 92, 1–8.

Fakhouri, F. M., Martelli, S. M., Caon, T., Velasco, J. I. and Mei, L. H. I. (2015). Edible films and coatings based on starch/gelatin: film properties and effect of coatings on quality of refrigerated Red Crimson grapes. *Postharvest Biology and Technology*, 109, 57–64.

Fallanaj, F., Ippolito, A., Ligorio, A., Garganese, F., Zavanella, C. and Sanzani, S. M. (2016). Electrolyzed sodium bicarbonate inhibits *Penicillium digitatum* and induces defence responses against green mould in citrus fruit. *Postharvest Biology and Technology*, 115, 18–29.

Fernandes, Â., Antonio, A. L., Oliveira, M. B. P., Martins, A. and Ferreira, I. C. (2012). Effect of gamma and electron beam irradiation on the physico-chemical and nutritional properties of mushrooms: a review. *Food Chemistry*, 135, 641–650.

Fernández-León, M. F., Fernández-León, A. M., Lozano, M., Ayuso, M. C. and González-Gómez, D. (2013). Different postharvest strategies to preserve broccoli quality during storage and shelf life: controlled atmosphere and 1-MCP. *Food Chemistry*, 138, 564–573.

Fagundes, C., Carciofi, B.A.M., Monteiro, A.R. 2013. Estimate of respiration rate and physicochemical changes of fresh-cut apples stored under different temperatures. *Food Science and Technology, Campinas*, 33(1), 60–67.

Fonseca, S. C., Oliveira, F. A. R., & Brecht, J. K. (2002). Modelling respiration rate of fresh fruits and vegetables for modified atmosphere packages: a review. *Journal of Food Engineering*, 52, 99–119.

Forney, C. F., Jamieson, A. R., Pennell, K. D. M., Jordan, M. A., Fillmore. S. A. E. (2015). Relationships between fruit composition and storage life in air or controlled atmosphere of red raspberry. *Postharvest Biology and Technology*, 110, 121–130.

Garcia, L. C., Pereira, L. M., de Luca Sarantópoulos, C. I. and Hubinger, M. D. (2010). Selection of an edible starch coating for minimally processed strawberry. *Food and Bioprocess Technology*, 3, 834–842.

Garrido, Y., Tudela, J. A., Hernández, J. A. and Gil, M. I. (2016). Modified atmosphere generated during storage under light conditions is the main factor responsible for the quality changes of baby spinach. *Postharvest Biology and Technology*, 114, 45–53.

Ge, Y., Li, C., Lü, J., Zhu, D. 2017a. Effects of thiamine on *Trichothecium* and *Alternaria* rots of muskmelon fruit and the possible mechanisms involved. *Journal of Integrative Agriculture*, 16, 2623–2631.

Ge, Y., Chen, Y., Li, C., Wei, M., Li, X., Tanga, Q. and Duan, B. (2019). Effect of trisodium phosphate treatment on black spot of apple fruit and the roles of anti-oxidative enzymes. *Physiological and Molecular Plant Pathology*, 106, 226–231.

Ge, Y., Chen, Y., Li, C., Zhao, J., Wei, M., Lid, X., Yang, S. and Mi, Y. (2019). Effect of sodium nitroprusside treatment on shikimate and phenylpropanoid pathways of apple fruit. *Food Chemistry*, 290, 263–269.

Ge, Y., Chen, Y., Li, C., Wei, M., Li, X., Tanga, Q., Duan, B. 2019a. Effect of trisodium phosphate treatment on black spot of apple fruit and the roles of anti-oxidative enzymes. Physiological and Molecular Plant Pathology, 106, 226–231.

Ge, Y., Chen, Y., Li, C., Zhao, J., Wei, M., Lid, X., Yang, S., Mi, Y. 2019b. Effect of sodium nitroprusside treatment on shikimate and phenylpropanoid pathways of apple fruit. *Food Chemistry*, 290, 263–269.

Ge, Y., Li, C., Lü, J. and Zhu, D. (2017). Effects of thiamine on *Trichothecium* and *Alternaria* rots of muskmelon fruit and the possible mechanisms involved. *Journal of Integrative Agriculture*, 16, 2623–2631.

Ge, Y., Wei, M., Li, C., Chen, Y., Lv, J. and Li, J. (2017). Effect of acibenzolar-S-methyl on energy metabolism and blue mould of Nanguo pear fruit. *Scientia Horticulturae* 225, 221–225.

Gholami, R., Ahmadi, E. and Farris, S. (2017). Shelf life extension of white mushrooms (*Agaricus bisporus*) by low temperatures conditioning, modified atmosphere, and nanocomposite packaging material. *Food Packaging and Shelf Life*, 14, 88–95.

Guan, W., Fan, X. and Yan, R. (2012). Effects of UV-C treatment on inactivation of Escherichia coli O157: H7, microbial loads, and quality of button mushrooms. *Postharvest Biology and Technology*, 64, 119–125.

Gwanpua, S. G., Verlinden, B. E., Hertog, M. L. A. T. M., Bulens, I., Van de Poel, B., Van Impe, J., Nicolaï, B. M. and Geeraerd, A. H. (2012). Kinetic modeling of firmness breakdown in "Braeburn" apples stored under different controlled atmosphere conditions. *Postharvest Biology and Technology*, 67, 68–74.

Hayta, E. and Aday, M. S. (2015). The effect of different electrolyzed water treatments on the quality and sensory attributes of sweet cherry during passive atmosphere packaging storage. *Postharvest Biology and Technology*, 102, 32–41.

Hernández, I., Fuentealba, C., Olaet, J. A., Poblete-Echeverría, C., Defilippic, B. G., González-Agüeroc, M., Campos-Vargas, R., Luriee, S. and Pedreschia, R. (2017). Effects of heat shock and nitrogen shock pre-treatments on ripening heterogeneity of hass avocados stored in controlled atmosphere. *Scientia Horticulturae*, 225, 408–415.

Ho, Q. T., Verboven, P., Verlinden, B. E., Schenk, A. and Nicolaï, B. M. (2013). Controlled atmosphere storage may lead to local ATP deficiency in apple. *Postharvest Biology and Technology*, 78, 103–112.

Huali, X., Yang, B., Yaxuan, S., Raza, H., Hujun, W., Shan, Z., Rui, Z., Haitao, L., Mina, N., Xiaoyan, C. and Calderóon-Urrea, A. (2019). Acetylsalicylic acid treatment reduces *Fusarium* rot development and neosolaniol accumulation in muskmelon fruit. *Food Chemistry*, 289, 278–284.

Huang, H., Zhu, Q., Zhang, Z., Yanga, B., Duan, X. and Jiang, Y. (2013). Effect of oxalic acid on antibrowning of banana (Musa spp., AAA group, cv. 'Brazil') fruit during storage. *Scientia Horticulturae*, 160, 208–212.

Hussain, P. R., Dar, M. A. and Wani, A. M. (2012). Effect of edible coating and gamma irradiation on inhibition of mould growth and quality retention of strawberry during refrigerated storage. *International Journal of Food Science & Technology*, 47, 2318–2324.

Hussain, P. R., Meena, R. S., Dar, M. A. and Wani, A. M. (2012). Effect of post-harvest calcium chloride dip treatment and gamma irradiation on storage quality and shelf-life extension of red delicious apple. *Journal of Food Science and Technology*, 49, 415–426.

Irfana, P. K., Vanjakshi, V., Keshava Prakasha, M. N., Ravi, R. and Kudachikar, V. B. (2013). Calcium chloride extends the keeping quality of fig fruit (*Ficus carica* L.) during storage and shelf-life. *Postharvest Biology and Technology*, 82, 70–75.

Jafri, M., Jha, A., Bunkar, D. S. and Ram, R. C. (2013). Quality retention of oyster mushrooms (*Pleurotus florida*) by a combination of chemical treatments and modified atmosphere packaging. *Postharvest Biology and Technology*, 76, 112–118.

Ji, D., Chen, T., Ma, D., Liu, J., Xu, Y. and Tian, S. (2018). Inhibitory effects of methyl thujate on mycelial growth of *Botrytis cinerea* and possible mechanisms. *Postharvest Biology and Technology*, 142, 46–54.

Jia, L.-E., Liu, S., Duan, X., Zhang, C., Wu, Z., Liu, M., Guo, S., Zuo, J. and Wang, L. (2017). 6-Benzylaminopurine treatment maintains the quality of Chinese chive (*Allium tuberosum* Rottler ex Spreng.) by enhancing antioxidant enzyme activity. *Journal of Integrative Agriculture*, 16, 1968–1977.

Jiang, L., Jin, P., Wang, L., Yu, X., Wang, H. and Zheng, Y. (2015). Methyl jasmonate primes defense responses against *Botrytis cinerea* and reduces disease development in harvested table grapes. *Scientia Horticulturea*, 192, 218–223.

Jing-jun, Y., Jian-rong, L., Xiao-xiang, H., Lei, Z., Tian-jia, J. and Mia, X. (2012). Effects of active modified atmosphere packaging on postharvest quality of shiitake mushrooms (*Lentinula edodes*) stored at cold storage. *Journal of Integrative Agriculture*, 11, 474–482.

Jin, L., Cai, Y., Sun, C., Huang, Y. and Yu, T. (2019). Exogenous L-glutamate treatment could induce resistance against *Penicillium expansum* in pear fruit by activating defense-related proteins and amino acids metabolism. *Postharvest Biology and Technology*, 150, 148–157.

Jin, P., Zheng, Y., Tang, S., Rui, H. and Wang, C. Y. (2009). A combination of hot air and methyl jasmonate vapor treatment alleviates chilling injury of peach fruit. *Postharvest Biology and Technology*, 52, 24–29.

Jung, S. -K. and Watkins, C. B. (2011). Involvement of ethylene in browning development of controlled atmosphere-stored 'Empire' apple fruit. *Postharvest Biology and Technology*, 59, 219–226.

Joshi, K., Warby, J., Valverde, J., Tiwari, B., Cullen, P. J. and Frias, J. M. (2018). Impact of cold chain and product variability on quality attributes of modified atmosphere packed mushrooms (*Agaricus bisporus*) throughout distribution. *Journal of Food Engineering*, 232, 44–55.

Jouki, M. and Khazaei, N. (2014). Effect of low-dose gamma radiation and active equilibrium modified atmosphere packaging on shelf life extension of fresh strawberry fruits. *Food Packaging and Shelf Life*, 1, 49–55.

Kang, J. S., Lee, D. S. 1998. A kinetic model for transpiration of fresh produce in a controlled atmosphere. *Journal of Food Engineering*, 35(1), 65e73.

Karaca, H. and Velioglu, Y. S. (2014). Effects of ozone treatments on microbial quality and some chemical properties of lettuce, spinach, and parsley. *Postharvest Biology and Technology*, 88, 46–53.

Karim, H., Boubaker, H., Askarne, L., Talibi, I., Msanda, F., Boudyach, E. H., Ben, A. and Aoumar, A. (2016). Antifungal properties of organic extracts of eight *Cistus* L. species against postharvest citrus sour rot. *Letter of Applied Microbiology*, 62, 16–22.

Khubone, L. W. and Mditshwa, A. (2018). The effects of UV-C irradiation on postharvest quality of tomatoes (*Solanum lycopersicum*). *Acta Horticulturae*, 1201, 75–82.

Kim, H.-G. and Song, K. B. (2017). Combined treatment with chlorine dioxide gas, fumaric acid, and ultraviolet-C light for inactivating *Escherichia coli* O157:H7 and *Listeria monocytogenes* inoculated on plums. *Food Control*, 71, 371–375.

Klein, J. D. and Lurie, S. (1991). Postharvest heat treatment and fruit quality. *Postharvest News and Information*, 2, 15–19.

Kolaei, E. A., Tweddell, R. J. and Avis, T. J. (2012). Antifungal activity of sulfur-containing salts against the development of carrot cavity spot and potato dry rot. *Postharvest Biology and Technology*, 63, 55–59.

Kou, X., Wu, M., Li, L., Wang, S., Xue, Z., Liu, B. and Fei, Y. (2015). Effects of CaCl₂ dipping and pullulan coating on the development of brown spot on 'Huangguan' pears during cold storage. *Postharvest Biology and Technology*, 99, 63–72.

Kowalczyk, D., Kordowska-Wiater, M., Zieba, E. and Baraniak, B. (2017). Effect of carboxymethylcellulose/candelilla wax coating containing potassium sorbate on microbiological and physicochemical attributes of pears. *Scientia Horticulturae* 218, 326–333.

Kweon, H. J., Kang, I. K., Kim, M. J., Lee, J., Moon, Y. S., Choi, C., Choi, D. G. and Watkins, C. B. (2013). Fruit maturity, controlled atmosphere delays and storage temperature affect fruit quality and incidence of storage disorders of 'Fuji' apples. *Scientia Horticulturae*, 157, 60–64.

Lai, T., Bai, X., Wang, Y., Zhou, J., Shi, N. and Zhou, T. (2015). Inhibitory effect of exogenous sodium bicarbonate on development and pathogenicity of postharvest disease *Penicillium expansum*. *Scientia Horticulturae*, 187, 108–114.

Lamikanra, O., Bett-Garber, K. L., Ingram, D. A. and Watson, M. A. (2005). Use of mild heat pre-treatment for quality retention of fresh-cut cantaloupe melon. *Journal of Food Science*, 70, C53–C57.

Lamikanra, O. and Watson, M. A. (2007). Mild heat and calcium treatment effects on fresh-cut cantaloupe melon during storage. *Food Chemistry*, 102, 1383–1388.

Latocha, P., Krupa, T., Jankowski, P. and Radzanowska, J. (2014). Changes in postharvest physicochemical and sensory characteristics of hardy kiwifruits (*Actinidia arguta* and its hybrid) after cold storage under normal versus controlled atmosphere. *Postharvest Biology Technology*, 88, 21–33.

Latorre, M. E., Narvaiz, P., Rojas, A. M. and Gerschenson, L. N. (2010). Effects of gamma irradiation on bio-chemical and physico-chemical parameters of fresh-cut red beet (*Beta vulgaris* L. var. conditiva) root. *Journal of Food Engineering*, 98, 178–191.

Lee, J., Cheng, L., Rudell, D. R. and Watkins, C. B. (2012). Antioxidant metabolism of 1-methylcyclopropene (1-MCP) treated 'Empire' apples during controlled atmosphere storage. *Postharvest Biology and Technology*, 65, 79–91.

Lee, L., Arul, J., Lencki, R. and Castaigne, F. (1996). A review on modified atmosphere packaging and preservation of fresh fruits and vegetables: physiological basis and practical aspects; part 2. *Packaging Technology and Science*, 9, 1–17.

Lee, S. Y., Lee, S. J., Choi, D. S. and Hur, S. J. (2015). Current topics in active and intelligent food packaging for preservation of fresh foods. *Journal of the Science of Food and Agriculture*, 95, 2799–2810.

Lemoine, M. L., Civello, P. M., Chaves, A. R. and Martínez, G. A. (2008). Effect of combined treatment with hot air and UV-C on senescence and quality parameters of minimally processed broccoli (*Brassica oleracea* L. var. Italica). *Postharvest Biology and Technology*, 48, 15–21.

Lemoine, M. L., Civello, P., Chaves, A., Martínez, G. 2009. Hot air treatment delays senescence and maintains quality of fresh-cut broccoli florets during refrigerated storage. *LWT-Food Science and Technology*, 42, 1076–1081.

Liamnimitr, N., Thammawong, M., Techavuthiporn, C., Fahmy, K., Suzuki, T. and Nakano, K. (2018). Optimization of bulk modified atmosphere packaging for long-term storage of "Fuyu" persimmon fruit. *Postharvest Biology and Technology*, 135, 1–7.

Li, H., Billing, D., Pidakala, P. and Burdon, J. (2017). Textural changes in 'Hayward' kiwifruit during and after storage in controlled atmospheres. *Scientia Horticulturae*, 222, 40–45.

Li, L., Kitazawa, H., Zhang, R., Wang, X., Zhang, L., Yu, S. and Li, Y. (2019). New insights into the chilling injury of postharvest white mushroom (*Agaricus bisporus*) related to mitochondria and electron transport pathway under high O_2/CO_2 controlled atmospheres. *Postharvest Biology and Technology*, 152, 45–53.

Li, L., Luo, Z., Huang, X., Zhang, L., Zhao, P., Ma, H., Li, X., Ban, Z. and Liu, X. (2015). Label-free quantitative proteomics to investigate strawberry fruit proteome changes under controlled atmosphere and low temperature storage. *Journal of Proteomics*, 120, 44–57.

Li, L., Tang, X., Ouyang, Q. and Tao, N. (2019). Combination of sodium dehydroacetate and sodium silicate reduces sour rot of citrus fruit. *Postharvest Biology and Technology*, 151, 19–25.

Li, L., Tang, X., Ouyang, Q., Tao, N. 2019c. Combination of sodium dehydroacetate and sodium silicate reduces sour rot of citrus fruit. *Postharvest Biology and Technology*, 151, 19–25.

Li, L., Kitazawa, H., Zhang, R., Wang, X., Zhang, L., Yu, S., Li, Y. 2019a. New insights into the chilling injury of postharvest white mushroom (*Agaricus bisporus*) related to mitochondria and electron transport pathway under high O_2/CO_2 controlled atmospheres. *Postharvest Biology and Technology*, 152, 45–53.

Li, T., Shi, D., Wu, Q., Zhang, Z., Qu, H., Jiang, Y. 2019b. Sodium para-aminosalicylate delays pericarp browning of litchi fruit by inhibiting ROS-mediated senescence during postharvest storage. *Food Chemistry*, 278, 552–559.

Li, T., Shi, D., Wu, Q., Zhang, Z., Qu, H. and Jiang, Y. (2019). Sodium para-aminosalicylate delays pericarp browning of litchi fruit by inhibiting ROS-mediated senescence during postharvest storage. *Food Chemistry*, 278, 552–559.

Li, L., Tang, X., Ouyang, Q., Tao, N. (2019c). Combination of sodium dehydroacetate and sodium silicate reduces sour rot of citrus fruit. *Postharvest Biology and Technology*, 151, 19–25.

Li, Y., Ishikawa, Y., Satake, T., Kitazawa, H., Qiu X. and Rungchang, S. (2014). Effect of active modified atmosphere packaging with different initial gas compositions on nutritional compounds of shiitake mushrooms (*Lentinus edodes*). *Postharvest Biology and Technology*, 92, 107–113.

Lin, D. and Zhao, Y. (2007). Innovations in the development and application of edible coatings for fresh and minimally processed fruits and vegetables. *Comprehensive Reviews in Food Science and Food Safety*, 6, 60–75.

Liu, C. H., Cai, L. Y., Lu, X. Y., Han, X. X. and Ying, T. J. (2012). Effect of postharvest UV-C irradiation on phenolic compound content and antioxidant activity of tomato fruit during storage. *Journal of Integrative Agriculture*, 11, 159–165.

Liu, C.H., Cai, L.Y., Lu, X.Y., Han, X.X., Ying, T.J., 2012a. Effect of postharvest UV-C irradiation on phenolic compound content and antioxidant activity of tomato fruit during storage. Journal of Integrative Agriculture, 11, 159–165.

Liu, J., Sui, Y., Wisniewski, M., Droby, S., Tian, S., Norelli, J. and Hershkovitz, V. (2012). Effect of heat treatment on inhibition of *Monilinia fructicola* and induction of disease resistance in peach fruit. *Postharvest Biology and Technology*, 65, 61–68.

Liu, J., Sui, Y., Wisniewski, M., Droby, S., Tian, S., Norelli, J., Hershkovitz, V. 2012b. Effect of heat treatment on inhibition of Monilinia fructicola and induction of disease resistance in peach fruit. Postharvest Biology and Technology, 65, 61–68.

Liu, K., Wang, X. and Young, M. (2014). Effect of bentonite/potassium sorbate coatings on the quality of mangos in storage at ambient temperature. *Journal of Food Engineering*, 137, 16–22.

Liu, Q., Tan, C. S. C., Yang, H. and Wang, S. (2017). Treatment with low-concentration acidic electrolysed water combined with mild heat to sanitise fresh organic broccoli (*Brassica oleracea*). *LWT-Food Science and Technology*, 79, 594–600.

Liu, Y. -B. (2013). Controlled atmosphere treatment for control of grape mealybug *Pseudococcus maritimus* (Ehrhorn) (*Hemiptera: Pseudococcidae*), on harvested table grapes. *Postharvest Biology and Technology*, 86, 113–117.

Lu, L., Ji, L., Qiao, L., Zhang, Y., Chen, M., Wang, C., Chen, H. and Zheng, X. (2018). Combined treatment with *Rhodosporidium paludigenum* and ammonium molybdate for the management of green mold in satsuma mandarin (*Citrus unshiu* Marc.). *Postharvest Biology and Technology*, 140, 93–99.

Luo, Z. (2006). Extending shelf-life of persimmon (*Diospyros kaki* L.) fruit by hot air treatment. *European Food Research and Technology*, 222, 149–154.

Luo, Z., Chen, C. and Xie, J. (2011). Effect of salicylic acid treatment on alleviating postharvest chilling injury of 'Qingnai' plum fruit. *Postharvest Biology and Technology*, 62, 115–120.

Luo, Z., Xu, T., Xie, J., Zhang, L. and Xi, Y. (2009). Effect of hot air treatment on quality and ripening of Chinese bayberry fruit. *Journal of the Science of Food and Agriculture*, 89, 443–448.

Lum, G. B., Brikis, C. J., Deyman, K. L., Subedi, S., DeEll, J. R., Shelp, B. J. and Bozzo, G. G. (2016). Pre-storage conditioning ameliorates the negative impact of 1-methylcyclopropene on physiological injury and modifies the response of antioxidants and γ-aminobutyrate in 'Honeycrisp' apples exposed to controlled-atmosphere conditions. *Postharvest Biology and Technology*, 116, 115–128.

Lumpkin, C., Fellman, J. K., Rudell, D. R. and Matheis, J. (2015). 'Fuji' apple (*Malus domestica* Borkh.) volatile production during high pCO2 controlled atmosphere storage. *Postharvest Biology and Technology*, 100, 234–243.

Lutz, M. C., Sosa, M. C. and Colodner, A. D. (2017). Effect of pre and postharvest application of fungicides on postharvest decay of Bosc pear caused by *Alternaria – Cladosporium* complex in North Patagonia, Argentina. *Scientia Horticulturae*, 225, 810–817.

Madani, B., Mirshekari, A., Imahori, Y. 2019. Physiological responses to stress. In: Yahia E. M. (Ed.), Postharvest Physiology and Biochemistry of fruits and vegetables. Elsevier, pp 406–423.

Mahajan, P., Rux, G., Caleb, O., Linke, M., Herppich, W. and Geyer, M. (2015). Mathematical model for transpiration rate at 100% humidity for designing modified humidity packaging. In *III International Conference on Fresh-Cut Produce: Maintaining Quality and Safety 1141* (pp. 269–274).

Mahajan, P. V., Oliveira, F. A. R. and Macedo, I. (2008). Effect of temperature and humidity on the transpiration rate of the whole mushrooms. *Journal of Food Engineering*, 84, 281–288.

Majeed, A., Muhammad, Z., Majid, A., Shah, A. H. and Hussain, M. (2014). Impact of low doses of gamma irradiation on shelf life and chemical quality of strawberry (*Fragaria x ananassa*) cv. 'Corona'. *Journal of Animal and Plant Sciences*, 24, 1531–1536.

Maqbool, M., Ali, A., Ramachandran, S., Smith, D. R. and Alderson, P. G. (2010). Control of postharvest anthracnose of banana using a new edible composite coating. *Crop Protection*, 29, 1136–1141.

Massolo, J. F., Lemoine, M. L., Chaves, A. R., Concellon, A. and Vicente, A. R. (2014). Benzylaminopurine (BAP) treatments delay cell wall degradation and softening, improving quality maintenance of refrigerated summer squash. *Postharvest Biology Technology*, 93, 122–129.

Mastrandrea, L., Amodio, M. L., de Chiara, M. L. V., Pati, S. and Colelli, G. (2017). Effect of temperature abuse and improper atmosphere packaging on volatile profile and quality of rocket leaves. *Food Packaging and Shelf Life*, 14, 59–65.

Matar, C., Gaucel, S., Gontard, N., Guilbert, S. and Guillard, V. (2018). Predicting shelf life gain of fresh strawberries 'Charlotte cv' in modified atmosphere packaging. *Postharvest Biology and Technology*, 142, 28–38.

Martins, D. R. and Resende, E. D. (2015). External quality and sensory attributes of papaya cv: golden stored under different controlled atmospheres. *Postharvest Biology Technology*, 110, 40–42.

Matityahu, I., Marciano, P., Holland, D., Ben-Arie, R. and Amir, R. (2016). Differential effects of regular and controlled atmosphere storage on the quality of three cultivars of pomegranate (*Punica granatum* L.). *Postharvest Biology and Technology*, 115, 132–141.

Mattheis, J., Felicetti, D. and Rudell, D. R. (2013). Pithy brown core in 'd'Anjou' pear (*Pyrus communis* L.) fruit developing during controlled atmosphere storage at pO₂ determined by monitoring chlorophyll fluorescence. *Postharvest Biology and Technology*, 86, 259–264.

Mditshwa, A., Fawole, O.A., Vries, F., van der Merwe, K., Crouch, E., Opara, U.L., 2017a. Minimum exposure period for dynamic controlled atmospheres to control superficial scald in 'Granny Smith' apples for long distance supply chains. *Postharvest Biology and Technology*, 127, 27–34.

Mditshwa, A., Fawole, O. A. and Opara, U. L. (2018). Recent developments on dynamic controlled atmosphere storage of apples – a review. *Food Packaging and Shelf Life*, 16, 59–68.

Mditshwa, A., Fawole, O. A., Vries, F., van der Merwe, K., Crouch, E. and Opara, U. L., Minimum exposure period for dynamic controlled atmospheres to control superficial scald in 'Granny Smith' apples for long distance supply chains. *Postharvest Biology and Technology*, 127, 27–34.

Mditshwa, A., Magwaza, L. S., Tesfay, S. Z. and Mbili, N. C. (2017). Effect of ultraviolet irradiation on postharvest quality and composition of tomatoes: a review. *Journal of Food Science and Technology*, 54, 3025–3035.

Mditshwa, A., Magwaza, L. S., Tesfay, S. Z. and Opara, U. L. (2017). Postharvest factors affecting vitamin C content of citrus fruits: a review. *Scientia Horticulturae*, 218, 95–104.

Mesa, K., Serra, S., Masia, A., Gagliardi, F., Bucci, D. and Musacchi, S. (2016). Seasonal trends of starch and soluble carbohydrates in fruits and leaves of 'Abbé Fétel' pear trees and their relationship to fruit quality parameters. *Scientia Horticulturae*, 211, 60–69.

Montesinos-Herrero, C., Moscoso-Ramíreza, P. A. and Palou, L. (2016). Evaluation of sodium benzoate and other food additives for the control of citrus postharvest green and blue molds. *Postharvest Biology and Technology*, 115, 72–80.

Moscoso-Ramírez, P. A. and Palou, L. (2013). Evaluation of postharvest treatments with chemical resistance inducers to control green and blue molds on orange fruit. *Postharvest Biology and Technology*, 85, 132–135.

Mphahlele, R. R., Fawole, O. A. and Opara, U. L. (2016). Influence of packaging system and long term storage on physiological attributes, biochemical quality, volatile composition and antioxidant properties of pomegranate fruit. *Scientia Horticulturae*, 211, 140–151.

Murmu, S. B. and Mishra, H. N. (2016). Measurement and modelling the effect of temperature, relative humidity and storage duration on the transpiration rate of three banana cultivars. *Scientia Horticulturae*, 209, 124–131.

Murmu, S.B., and Mishra, H.N. (2017). Engineering evaluation of thickness and type of packaging materials based on the modified atmosphere packaging requirements of guava (Cv. Baruipur). *LWT-Food Science and Technology*, 78, 273–280.

Murmu, S. B. and Mishra, H. N. (2018). Selection of the best active modified atmosphere packaging with ethylene and moisture scavengers to maintain quality of guava during low temperature storage. *Food Chemistry*, 253, 55–62.

Naheed, Z., Cheng, Z., Wu, C., Wen, Y. and Ding, H. 2017. Total polyphenols, total flavonoids, allicin and antioxidant capacities in garlic scape cultivars during controlled atmosphere storage. *Postharvest Biology and Technology*, 131, 39–45.

Navarro, J. M., Botía, P. and Pérez-Pérez, J. G. (2015). Influence of deficit irrigation timing on the fruit quality of grapefruit (*Citrus paradisi* Mac.). *Food Chemistry*, 175, 329–336.

Ncama, K., Magwaza, L. S., Mditshwa, A. and Tesfay, S. Z. (2018). Plant-based edible coatings for managing postharvest quality of fresh horticultural produce: a review. *Food Packaging and Shelf life*, 16, 157–167.

Ngcobo, M. E., Delele, M. A., Pathare, P. B., Chen, L., Opara, U. L., & Meyer, C. J. (2012). Moisture loss characteristics of fresh table grapes packed in different film liners during cold storage. *Biosystems Engineering*, 113(4), 363–370.

Nielsen, T., and Leufvén, A. (2008). The effect of modified atmosphere packaging on the quality of Honeoye and Korona strawberries. *Food Chemistry*, 107, 1053–1063.

Ochoa-Velasco, C. E. and Guerrero-Beltrán, J. A. (2016). The effects of modified atmospheres on prickly pear (*Opuntia albicarpa*) stored at different temperatures. *Postharvest Biology Technology*, 111, 314–321.

Otoni, C. G., Avena-Bustillos, R. J., Azeredo, H. M., Lorevice, M. V., Moura, M. R., Mattoso, L. H., and McHugh, T. H. (2017). Recent advances on edible films based on fruits and vegetables – a review. *Comprehensive Reviews in Food Science and Food Safety*, 16, 1151–1169.

Pan, Y. G. and Zu, H. (2012). Effect of UV-C radiation on the quality of fresh-cut pineapples. *Procedia Engineering*, 37, 113–119.

Park, M.-H., Sangwanangkul, P. and Choi, J.-W. (2018). Reduced chilling injury and delayed fruit ripening in tomatoes with modified atmosphere and humidity packaging. *Scientia Horticulturae*, 231, 66–72.

Parra, J., Ripoll, G. and Orihuel-Iranzo, B. (2014). Potassium sorbate effects on citrus weight loss and decay control. *Postharvest Biology and Technology*, 96, 7–13.

Patiño, L. S., Castellanos, D. A. and Herrera, A. O. (2018). Influence of 1-MCP and modified atmosphere packaging in the quality and preservation of fresh basil. *Postharvest Biology and Technology*, 136, 57–65.

Pereira, E., Spagnol, W. A. and Silveira Jr., V. (2018). Water loss in table grapes: model development and validation under dynamic storage conditions. *Food Science and Technology*, 38, 473–479.

Prange, R. K., Delong, J. M., Harrison, P., Mclean, S., Scrutton, J. and Cullen, J. (2007). Method and apparatus for monitoring a condition in chlorophyll containing matter. U.S. Patent, n.WO/ 2002/006795.

Prusky, D., Alkan, N., Miyara, I., Barad, S., Davidzon, M., Kobiler, I., et al. (2010). Mechanisms modulating postharvest pathogen colonization of decaying fruits. In: *Postharvest Pathology, Netherlands* (Chapter 4). Springer.

Rahman, E. A. A., Talib, R. A., Aziz, M. G., Yusof Y. A. 2013. Modelling the effect of temperature on respiration rate of fresh cut papaya (*Carica papaya* L.) fruits. *Food science and biotechnology*, 22(6), 1581–1588.

Restuccia, D., Spizzirri, U. G., Parisi, O. I., Cirillo, G., Curcio, M., Iemma, F., et al., (2010). New EU regulation aspects and global market of active and intelligent packaging for food industry applications. *Food Control*, 21, 1425–1435.

Rodriguez, J. and Zoffoli, J. P. (2016). Effect of sulfur dioxide and modified atmosphere packaging on blueberry postharvest quality. *Postharvest Biology and Technology*, 117, 230–238.

Ruíz-Jiménez, J. M., Zapata, P. J., Serrano, M., Martínez-Romero, D. V. D., Castillo, S. and Guillén, F. (2014). Effect of oxalic acid on quality attributes of artichokes stored at ambient temperature. *Postharvest Biology and Technology*, 95, 60–63.

Rux, G., Mahajan, P. V., Geyer, M., Linke, M., Pant, A., Saengerlaub, S. and Caleb, O. J. (2015). Application of humidity-regulating tray for packaging of mushrooms. *Postharvest Biology and Technology*, 108, 102–110.

Sängerlaub, S., Singh, P., Stramm, C. and Langowski, H. (2011). Shelf-life extension of fresh raw Agaricus mushrooms – a preliminary study with humidity regulating packages. In: *Proceedings of 25th IAPRI Symposium on Packaging, Berlin*.

Sandhya. 2010. Modified atmosphere packaging of fresh produce: current status and future needs. LWT-Food Science and Technology, 43(3), 381–392.

Saquet, A. A., Streif, J., Bangerth, F., 2000. Changes in ATP, ADP and pyridine nucleotide levels related to the incidence of physiological disorders in 'Conference' pears and 'Jonagold' apples during controlled atmosphere storage. *Journal Horticultural Science and Biotechnology*, 75, 243–249.

Saquet, A. A., Streif, J. and Almeida, D. P. F. (2017). Responses of 'Rocha' pear to delayed controlled atmosphere storage depend on oxygen partial pressure. *Scientia horticulturae*, 19, 17–21.

Serradilla, M. J., Villalobos, M. C., Hernández, A., Martín, A., Lozano, M. and Córdoba, M. G. (2013). Study of microbiological quality of controlled atmosphere packaged 'Ambrunés sweet cherries and subsequent shelf-life. *International Journal of Food Microbiology*, 166, 85–92.

Shah, H. M. S., Khan, A. S. and Ali, S. (2017). Pre-storage kojic acid application delays pericarp browning and maintains antioxidant activities of litchi fruit. *Postharvest Biology and Technology*, 132, 154–161.

Shao, X. and Tu, K. (2014). Hot air treatment improved the chilling resistance of loquat fruit under cold storage. *Journal of Food Processing and Preservation*, 38, 694–703.

Simões, A. D. N., Allende, A., Tudela, J. A., Puschmann, R. and Gil. M. I. (2011). Optimum controlled atmospheres minimise respiration rate and quality losses while increase phenolic compounds of baby carrots. *LWT-Food Science and Technology*, 44, 277–283.

Singh, P., Saengerlaub, S., Stramm, C. and Langowski, H. C. (2010). Humidity regulating packages containing sodium chloride as active substance for packing of fresh raw *Agaricus* mushrooms. In: *Procceding of the 4th International Workshop Cold Chain Management, Bonn*.

Singh, R., Giri, S. K. and Kotwaliwale, N. (2014). Shelf-life enhancement of green bell pepper (*Capsicum annuum* L.) under active modified atmosphere storage. *Food Packaging and Shelf Life*, 1, 101–112.

Singh, S. P. and Singh, Z. (2012). Postharvest oxidative behaviour of 1-methylcyclopropene treated Japanese plums (*Prunus salicina* Lindell) during storage under controlled and modified atmospheres. *Postharvest Biology Technology*, 74, 26–35.

Sivakumar, D. and Korsten, L. (2010). Fruit quality and physiological responses of litchi cultivar McLean's Red to 1-methylcyclopropene pre-treatment and controlled atmosphere storage conditions. *LWT-Food Science and Technology*, 43, 942–948.

Sousa-Gallagher, M. J., Mahajan, P. V. and Mezdad, T. (2013). Engineering packaging design accounting for transpiration rate: model development and validation with strawberries. *Journal of Food Engineering*, 119(2), 370–376.

Stanger, M. C., Steffens, C. A., Soethe, C., Moreira, M. A., Amarante, C. V. T., Both, V. and Brackmann, A. (2018). Phenolic compounds content and antioxidant activity of 'Galaxy' apples stored in dynamic controlled atmosphere and ultralow oxygen conditions. *Postharvest Biology and Technology*, 144, 70–76.

Sun, J., You, X., Li, L., Peng, H., Suc, W., Li, C., He, Q. and Liao, F. (2011). Effects of a phospholipase D inhibitor on postharvest enzymatic browning and oxidative stress of litchi fruit. *Postharvest Biology and Technology*, 62, 288–294.

Tano, K., Kamenan, A. and Arul, J. (2005). Respiration and transpiration characteristics of selected fresh fruits and vegetables. *Agronomie Africaine*, 17, 103–115.

Teixeira, G. H. A., Júnior, L. C. C., Ferraudo, A. S. and Durigan, J. F. (2016). Quality of guava (*Psidium guajava* L. cv. Pedro Sato) fruit stored in low-O_2 controlled atmospheres is negatively affected by increasing levels of CO_2. *Postharvest Biology and Technology*, 111, 62–68.

Tesfay, S. Z., Magwaza, L. S., Mbili, N. and Mditshwa, A. (2017). Carboxyl methylcellulose (CMC) containing moringa plant extracts as new postharvest organic edible coating for avocado (*Persea americana* Mill.) fruit. *Scientia Horticulturae*, 226, 201–207.

Thewes, F. R., Both, V., Brackmann, A., Weber, A. and Anese, R. O. (2015). Dynamic controlled atmosphere and ultralow oxygen storage on 'Gala' mutants quality. *Food Chemistry*, 188, 62–70.

Toivonen, P. M. A. and Hampson, C. R. (2014). Relationship of IAD index to internal quality attributes of apples treated with 1-methylcyclopropene and stored in air or controlled atmospheres. *Postharvest Biology and Technology*, 91, 90–95.

Tran, D. T., Verlinden, B. E., Hertog, M. and Nicolaï, B. M. (2015). Monitoring of extremely low oxygen control atmosphere storage of 'Greenstar' apples using chlorophyll fluorescence. *Scientia Horticulturae*, 184, 18–22.

Tumwesigye, K. S., Sousa, A. R., Oliveira, J. C., and Sousa-Gallagher, M. J. (2017). Evaluation of novel bitter cassava film for equilibrium modified atmosphere packaging of cherry tomatoes. *Food Packaging and Shelf Life*, 13, 1–14.

Usall, J., Ippolito, A., Sisquella, M. and Neri, F. (2016). Physical treatments to control postharvest diseases of fresh fruits and vegetables. *Postharvest Biology and Technology*, 122, 30–40.

Ustun, D., Candir, E., Ozdemir, A. E., Kamiloglu, O., Soylu, E. M. and Dilbaz, R. (2012). Effects of modified atmosphere packaging and ethanol vapor treatment on the chemical composition of 'Red Globe' table grapes during storage. *Postharvest biology and Technology*, 68, 8–15.

Vanoli, M., Grassi, M. and Rizzolo, A. (2016). Ripening behavior and physiological disorders of 'Abate Fetel' pears treated at harvest with 1-MCP and stored at different temperatures and atmospheres. *Postharvest Biology and Technology*, 111, 274–285.

Van Schaik, A. C. R., Van de Geijn, F. G., Verschoor, J. A. and Veltman, R. H. (2015). A new interactive storage concept: dynamic control of respiration. *Acta Horticulturea*, 1071, 245–252.

Veltman, R. H., Verschoor, J. A. and Ruijsch van Dugteren, J. H. (2003). Dynamic control system (DCS) for apples (*Malus domestica* Borkh. cv. 'Elstar'): optimal quality through storage based on products response. *Postharvest Biology Technology*, 27, 79–86.

Venditti, T., Ladu, G., Cubaiu, L., Myronycheva, O. and D'hallewin, G. (2017). Repeated treatments with acetic acid vapors during storage preserve table grapes fruit quality. *Postharvest Biology and Technology*, 125, 91–98.

Vermathen, M., Marzorati, M., Diserens, G., Baumgartner, D., Good, C., Gasser, F. and Vermathen, P. (2017). Metabolic profiling of apples from different production systems before and after controlled atmosphere (CA) storage studied by 1H high resolution-magic angle spinning (HR-MAS) NMR. *Food Chemistry*, 233, 391–400.

Vicente, A. R., Martínez, G. A., Chaves, A. R. and Civello, P. M. (2006). Effect of heat treatment on strawberry fruit damage and oxidative metabolism during storage. *Postharvest Biology and Technology*, 40, 116–122.

Villalobos, M. D. C, Serradilla, M. J., Martín, A., Hernandez-Leon, A., Ruiz-Moyano, S. and Cordoba, M. G. (2017). Characterization of microbial population of breba and main crops (*Ficus carica*) during cold storage: influence of passive modified atmospheres (MAP) and antimicrobial extract application. *Food Microbiology*, 63, 35–46.

Villalobos, M. C. Serradilla, M. J., Martín, A., Aranda, E., López-Corrales, M., Córdoba, M. G. (2018). Influence of modified atmosphere packaging (MAP) on aroma quality of figs (*Ficus carica* L.). *Postharvest Biology and Technology*, 136, 145–151.

Villalobos, M.D.C., Serradilla, M.J., Martín, A., Ruíz-Moyano, S., Pereira, C., Cordoba, M.G., 2014. Use of equilibrium modified atmosphere packaging for preservation of 'San Antonio' and 'Banane' breba crops (*Ficus carica* L.). Postharvest Biology Technology, 98, 14–22.

Wang, K., Jin, P., Han, L., Shang, H., Tang, S., Rui, H., Duan, Y., Kong, F., Kai, X. and Zheng, Y. (2014). Methyl jasmonate induces resistance against *Penicillium citrinum* in Chinese bayberry by priming of defense responses. *Postharvest Biology and Technology*, 98, 90–97.

Wang, K., Jin, P., Shang, H., Li, H., Xu, F., Hu, Q. and Zheng, Y. (2010). A combination of hot air treatment and nano-packing reduces fruit decay and maintains quality in postharvest Chinese bayberries. *Journal of the Science of Food and Agriculture*, 90, 2427–2432.

Wang, L., Zhang, Y., Chen, Y., Liu, S., Yun, L., Guo, Y., Zhang, X. and Wang, F. (2019). Investigating the relationship between volatile components and differentially expressed proteins in broccoli heads during storage in high CO_2 atmospheres. *Postharvest Biology and Technology*, 153, 43–51.

Wang, Y. and Sugar, D. (2013). Internal browning disorder and fruit quality in modified atmosphere packaged 'Bartlett' pears during storage and transit. *Postharvest Biology and Technology*, 83, 72–82.

Wang, Y. and Long, L. E. (2014). Respiration and quality responses of sweet cherry to different atmospheres during cold storage and shipping. *Postharvest Biology and Technology*, 92, 62–69.

Watkins, C. B. and Nock, J. F. (2012). Rapid 1-methylcyclopropene (1-MCP) treatment and delayed controlled atmosphere storage of apples. *Postharvest Biology and Technology*, 69, 24–31.

Weber, A., Brackmann, A., Both, V., Pavanello, E. P., Anese, R. O. and Thewes, F. R. (2015). Respiratory quotient: innovative method for monitoring 'Royal Gala' apple storage in a dynamic controlled atmosphere. *Scientia Agricola*, 72, 28–33.

Weber, A., Thewes, F. R., Anese, R. O., Both, V., Pavanello, E. P. and Brackmann, A. (2015). Dynamic controlled atmosphere (DCA): interaction between DCA methods and 1-methylcyclopropene on 'Fuji Suprema' apple quality. *Food Chemistry*, 235, 136–144.

Weber, A., Thewes, F. R., de Oliveira Anese, R., Both, V., Pavanello, E. P. and Brackmann, A. (2017). Dynamic controlled atmosphere (DCA): interaction between DCA methods and 1-methylcyclopropene on 'Fuji Suprema' apple quality. *Food Chemistry*, 235, 136–144.

Whitaker, B. D., Villalobos-Acuna, M., Mitcham, E. J. and Mattheis, J. P. (2009). Superficial scald susceptibility and α-farnesene metabolism in 'Bartlett' pears grown in California and Washington. *Postharvest Biology and Technology*, 53, 43–50.

Wright, H., DeLong, J., Harrison, P. A., Gunawardena, A. H. L. N. and Prange, R. (2010). The effect of temperature and other factors on chlorophyll fluorescence and the lower oxygen limit in apples (*Malus domestica*). *Postharvest Biology Technology*, 55, 21–28.

Wuryatmo, E., Able, A. J., Ford, C. M. and Scott, E. S. (2014). Effect of volatile citral on the development of blue mould, green mould and sour rot on navel orange. *Australasian Plant Pathology*, 43, 403–411.

Xanthopoulos, G., Koronaki, E. D. and Boudouvis, A. G. (2012). Mass transport analysis in perforation-mediated modified atmosphere packaging of strawberries. *Journal of Food Engineering*, 111, 326–335.

Xanthopoulos, G. T., Athanasiou, A. A., Lentzou, D. I., Boudouvis, A. G., Lambrinos, G. P. 2014. Modelling of transpiration rate of grape tomatoes. Semi-empirical and analytical approach. *Biosystems Engineering*, 124, 16e23.

Xi, Y., Fan, X., Zhao, H., Li, X., Cao, J. and Jiang, W. (2017). Postharvest fruit quality and antioxidants of nectarine fruit as influenced by chlorogenic acid. *LWT-Food Science and Technology*, 75, 537–544.

Xu, F., Chen, X. H., Yang, Z. F., Jin, P., Wang, K. T., Shang, H. T., Wang, X. L. and Zheng, Y. H. (2013). Maintaining quality and bioactive compounds of broccoli by combined treatment with 1-methylcyclopropene and 6-benzylaminopurine. *Journal Science Food Agriculture*, 93, 1156–1161.

Yahia, E. M., Soto-Zamora, G., Brecht, J. K. and Gardea, A. (2007). Postharvest hot air treatment effects on the antioxidant system in stored mature-green tomatoes. *Postharvest Biology and Technology*, 44, 107–115.

Yang, Q., Zhang, X., Wang, F. and Zhao, Q. (2019). Effect of pressurized argon combined with controlled atmosphere on the postharvest quality and browning of sweet cherries. *Postharvest Biology and Technology*, 147, 59–67.

Yang, Z., Cao, S., Cai, Y. and Zheng, Y. (2011). Combination of salicylic acid and ultrasound to control postharvest blue mold caused by *Penicillium expansum* in peach fruit. *Innovative Food Science and Emerging Technologies*, 12, 310–314.

Ye, J., Li, J., Han, X., Zhang, L., Jiang, T. and Xia, M. (2012). Effects of active modified atmosphere packaging on postharvest quality of shiitake mushrooms (*Lentinula edodes*) stored at cold storage. *Journal of Integrated Agriculture*, 11, 474–482.

Yoon, M., Jung, K., Lee, K. Y., Jeong, J. Y., Lee, J. W. and Park, H. J. (2014). Synergistic effect of the combined treatment with gamma irradiation and sodium dichloroisocyanurate to control gray mold (*Botrytis cinerea*) on paprika. *Radiation Physics and Chemistry*, 98, 103–108.

Yu, C., Zeng, L., Sheng, K., Chen, F., Zhou, T., Zheng, X. and Yu, T. (2014). γ-Aminobutyric acid induces resistance against *Penicillium expansum* by priming of defence responses in pear fruit. *Food Chemistry*, 159, 29–37.

Yu, J., and Wanga, Y. (2017). The combination of ethoxyquin, 1-methylcyclopropene and ethylene treatments controls superficial scald of 'd'Anjou' pears with recovery of ripening capacity after long-term controlled atmosphere storage. *Postharvest Biology and Technology*, 127, 53–59.

Yun, Z., Gao, H., Liu, P., Liu, S., Luo, T., Jin, S., Xu, Q., Xu, J., Cheng, Y. and Deng, X. (2013). Comparative proteomic and metabolomic profiling of citrus fruit with enhancement of disease resistance by postharvest heat treatment. *BMC Plant Biology*, 13, 44.

Zhang, C., Wang, J., Zhang, J., Hou, C. and Wang, G. (2011). Effects of -aminobutyric acid on control of postharvest blue mould of apple fruit and its possible mechanisms of action. *Postharvest Biology and Technology*, 61, 145–151.

Zhang, J., Jiang, L., Sun, C., Jin, L., Lin, M., Huang, Y., Zheng, X., Yu, T. 2018a. Indole-3-acetic acid inhibits blue mold rot by inducing resistance in pear fruit wounds. *Scientia Horticulturae*, 231, 227–232.

Zhang, D., Xu, X., Zhang, Z., Jiang, G., Feng, L., Duan, X. and Jiang, Y. (2018). 6-Benzylaminopurine improves the quality of harvested litchi fruit. *Postharvest Biology and Technology*, 143, 137–142.

Zhang, D., Xu, X., Zhang, Z, Jiang, G., Feng, L., Duan, X., Jiang, Y. 2018b. 6-Benzylaminopurine improves the quality of harvested litchi fruit. *Postharvest Biology and Technology*, 143, 137–142.

Zhang, J., Jiang, L., Sun, C., Jin, L., Lin, M., Huang, Y., Zheng, X. and Yu, T. (2018). Indole-3-acetic acid inhibits blue mold rot by inducing resistance in pear fruit wounds. *Scientia Horticulturae*, 231, 227–232.

Zhu, F., Chen, J., Xiao, X., Zhang, M., Yun, Z., Zeng, Y., Xu, J., Cheng, Y. and Deng, X. (2016). Salicylic acid treatment reduces the rot of postharvest citrus fruit by inducing the accumulation of H2O2, primary metabolites and lipophilic polymethoxylated flavones. *Food Chemistry*, 207, 68–74.

Betty O Ajibade, Omotola F. Olagunju, Oluwatosin Ademola
Ijabadeniyi

3 Cereals and cereal products

3.1 Introduction

Cereals are one of the major nutrient delivery vehicles all over the world (Serna-Saldivar 2016). They are one of the most popular agricultural products and are regarded as a gift of nature to mankind. Also termed "the staff of life," cereals constitute a major diet staple in many countries around the world. They account for the most part of the diet in the release of nutrients and energy as lots of them contain majorly carbohydrates (Tomić et al., 2019). Cereals including wheat, rice, barley, maize, rye, oats sorghum, millet, buckwheat, teff, amaranth, triticale, and quinoa make up the biggest part of crop production worldwide (FAO 2019). Maize is the most popular cereal in many regions of the world: in Europe, rye, wheat, and barley are popular while millet and sorghum are most known to Africa and India (FAO 2019). Cereals are the world most cultivated plants, globally about 2,342,426 tons were produced in 2007 and about 2,722 million tons in 2018 (FAO 2019). Due to the economic and nutritional importance of cereals to humans, the demand for them and their by-products increases every year. Scientific research focusing on improvement in seed, processing and cultivation methods, increased yield, better storage stability, and so on are constantly being undertaken to improve cereal production (Taylor et al., 2014). The use of biotechnology in the modification of the seeds' genetic makeup offers a possibility in yield increase, disease resistance, and reduced health risks associated with pesticides and insecticides.

Hunger is the inability to obtain necessary nourishment needed for growth and development, and about 821 million people (equivalent to about one in nine persons) face hunger (FAO 2018). The resultant effect is a less active and unhealthy lifestyle. The Food Insecurity Experience Scale also reported that about 770 million people, equivalent to 10% of the world population, suffer a high level of food insecurity based on the level of food availability or level of nutrients present in the available food (Smith et al., 2017). Food security is directly linked to the availability and the available nutrients present in the food. If the available food does not contain the necessary nutrients, food insecurity is presented. Food utilization is related to maximum absorption of nutrients that are present in food product and it must be readily available in order to improve nutrition (FAO 2018).

One of the major tools needed to achieve the implementation of the 2030 agenda is for food production to reach utilization stage, which can combat hunger, malnutrition,

Betty O Ajibade, Oluwatosin Ademola Ijabadeniyi, Durban University of Technology, South Africa
Omotola F. Olagunju, Afe Babalola University, Ekiti State, Nigeria

https://doi.org/10.1515/9783110667462-003

and diseases arising from the consumption of cereals and its by-products (FAO 2018). Cereals utilization has recorded increase since the 1960s with a continued trend. During 2018/19, about 1.7% increase in utilization was recorded. The forecast is also on the increase; thus, processing should be widely improved to feed the teaming world population and to reduce wastage. Continuous increase in world population since the 1960s has put a lot of pressure on food production with the world population reaching 7.5 billion people according to FAO pocketbook (2018).

Cereals contain numerous bioactive compounds which offer different health benefits. However, cereals also contain antinutrients which pose threat to nutrient absorption and digestibility. Tannins and phytates are the most common among the antinutrients identified in cereals grains, and various processing methods are centered on reducing these antinutrients (Devisetti et al., 2014). Adequate processing methods such as dehulling have been standardized to reduce the incidence of tannins in cereals especially in sorghum (Kruger et al., 2012; Taylor et al., 2014). Milling and decortication are the most effective means for phytate reduction in most cereals (Devisetti et al., 2014). This chapter presents recent cereal processing methods and their effects on the final food product.

3.2 Nutritional composition of cereals

Cereals contain high-energy values and make up a good proportion of calorie intake of the world's population (Serna-Saldivar 2016). Typical cereals contain carbohydrates as the major food constituents, and this makes them a good source of energy. Ash, water, and protein are also major constituents. However, the protein quality of cereals is low (Temba et al., 2016), necessitating the application of processing methods such as fermentation and addition of legumes as composites to improve some of the nutritional limitations of cereals. The unavailability of lysine in most cereals has led to protein-energy malnutrition for populations of underdeveloped countries who depend largely on coarse cereals as food sources (Temba et al., 2016).

3.3 Cereal processing

Wheat is one of the widely processed agricultural crops in the world. It is referred to as the "queen of all cereals." This is due to its unique ability to produce various leavened products such as bread and other specialty food products (Serna-Saldivar 2016). This ability has also been attributed to the presence of gluten in wheat. According to reports in May 2019, the annual wheat production was about 730 million tons (FAO 2019). The world-leading producers of wheat include the United States of America, Canada, Australia, Argentina, and some winter regions of Europe and Asia (FAO 2019). Maize and rice are commonly consumed cereals all over the world. Maize production is on a steady rise in Southern Africa and has shifted from its main use as animal feed to a staple in many Africa countries (FAO 2019). Rice is still considered a scarce commodity in some countries due to its cultivation outside the continent, resulting in its importation by several African countries. The most popular cereals that are indigenous to Africa include sorghum, millet, and maize; they are referred to as coarse cereals (FAO 2019).

Cereals utilization as food has increased over the years due to improved processing methods. Traditional methods are still largely used in rural areas due to lack of sophisticated processing equipment, although industrial processing has also largely improved (Temba et al., 2016). Postharvest management of cereals is very significant as it determines the level of consumption among the populace. Similarly, the level of nutrient supply or absorption from cereals largely depends on effective processing methods. Despite a high level of postharvest losses, utilized cereals can benefit from good processing method and overall nutrient intake and absorption (Oghbaei and Prakash 2016).

3.4 Cereal processing methods

Cereal processing comprises primary, secondary, and tertiary processing methods. Processing starts immediately after harvesting from the field (Oghbaei et al., 2016). The preprocessing operations include drying, dehulling/decortication, sorting, cleaning to remove unwanted debris, and storing. All these primary processing operations are important as they usually determine the outcome of other processing steps, and chemically affect the end products (Oghbaei et al., 2016). Secondary processing of cereals includes milling (dry and wet) and dry masa flour processing, also known as nixtamalization (Serna-Saldivar 2016). Nixtamalization is the alkaline/acid cooking of cereals to soften the pericarp, which allows easy dehulling. Processing methods such as milling, fermentation, malting, and cooking (acid or alkaline) affect the composition and bioavailability of many important nutrients in cereals (Serna-Saldivar 2016) (Table 3.1).

3.4.1 Thermal processing

Thermal processing methods include roasting, cooking, popping, high-temperature pressure treatment, microwave heating, autoclaving, and extrusion (Oghbaei et al., 2016; Serna-Saldivar 2016). Thermal processing may cause denaturation of micronutrients, reduction of antinutrients, and improvement in protein and starch bioavailability. Heating improves starch utilization, overheating produces resistant starch, alkali heating lowers lysine content while malting, and fermentation improves protein level and vitamins bioavailability (Oghbaei et al., 2016). Steaming or cooking of buckwheat

Table 3.1: Processing and food use of cereals.

Cereal type	Processing methods	Major food products	Others
Maize	Milling (dry/wet)	Grits, meals and flours, bakery products, batter and breadings, breakfast cereals flakes, puffs, pellets, shredded, snacks second and third generation (puffs, collets, and pellets), brewing adjuncts lager and ale beers, and alcoholic spirits	Animal feeds, ethanol production, Sugar production (syrups), and starches (modified starches)
	Nixtamalization	Fresh and dry masa flour, snacks corn, and tortilla chips	Table tortillas
	Popcorn	Snacks and confectionary products	
	Fermentation	Ogi, Two, Chicha, masa, and Nshima.	
Rice	Milling (dry/wet)	White polished rice, broken kernels and grits, direct home use, breakfast cereals flakes, oven puffs, pellets, shredded, snacks second and third generation (puffs, collets, pellets), starches (syrups and sweeteners), alcoholic spirits and sake.	Flour, snacks, rice noodles, and biofuels
Wheat/rye/ triticale	Decortication/ pounding	Pounded grains, couscous, and bulgur	Animal feeds, biofuel, and bioactive compounds
	Dry-milling	Flour, semolina, yeast leavened bread, pastries, donuts, rolls, chemical leavened cookies, cakes, biscuits, muffins, tortillas, batters and breadings, oriental noodles, pasta products, and couscous.	
	Wet-milling	Starches, gluten, modified starches, and syrups.	
Barley	Malting	Diastatic malt: malt, lager and ale beers, alcoholic spirits, bakery products, nondiastatic: flavorings and syrups.	Functional foods: dietary fiber, and bioactive compounds
	Milling (dry)	Brewing adjuncts: lager and ale beers, alcoholic spirits, refined or whole flour, bakery products composite breads, muffins, and cookies.	

Table 3.1 (continued)

Cereal type	Processing methods	Major food products	Others
Oats	Milling (dry)	Groats, breakfast cereals, rolled oats, granolas, meals and flours, composite cookies and crackers, breakfast cereals flakes, extruded gun-puffed, and composite bread products.	Functional foods: Dietary fiber, bioactive ingredients, brewing
Buckwheat	Milling (dry)	Roasted groats (kasha), flour, buckwheat-enriched bread, buckwheat-enriched cookies, raw groats, flour with different extraction rates, pasta/noodles, tarhana, bread, snacks, and biscuits.	Buckwheat leaves, flowers and hull, tea infusions, and honey
Sorghum or millets	Malting	Diastatic malt, opaque beer, weaning foods, brewing adjuncts, and lager beer.	
	Dry-milling	Grits, meals, and flours, traditional foods: tô, injera, ugali, roti, couscous, and composite bakery products.	Animal feeds, biofuel, ethanol, bioactive compounds, polyphenols, and fortification flour
	Decortication	Decorticated grains, parboiling – parboiled grains.	
	Nixtamalization	Masa: Tortillas, snacks Tortilla Chips.	
	Fermentation	Ogi, Kunu, and tapioca.	

Adapted from Serna-Saldivar (2016), Zhu (2014), Gimenez-Bastida et al. (2015), Gupta et al. (2010), ElMekawy et al. (2013), Nkhata et al. (2018), Serna-Saldivar et al. (1990).

reduces the available bioactive compounds and tannin content. Microwave, roasting, and pressure-steaming treatment lead to reduced antioxidant activity and reduction in phenolics and flavonoids in Tartary buckwheat (Lei et al., 2016). High-pressure treatment offers an alternative to thermal processing in that it prevents the loss of bioactive components and retains antioxidant activity (Serna-Saldivar 2016). High-pressure treatment may be commonly used for cereal processing in the future because they aid in the retention of cereals' nutraceutical capacity. Irradiation and ionization are also now applied in cereals processing to help in the preservation of functional properties of cereals. Irradiation help in flavor preservation, shelf life increase, improvement in sensory quality, and delay microbial spoilage of cereal products (Serna-Saldivar 2016).

3.4.2 Milling

Milling remains one of the ancient methods of cereal processing. Apart from the utilization of whole cereal grains which is becoming popular, milling to obtain flour remains the usual practice in cereal processing. The end use of the cereal is determined by the milling process (Serna-Saldivar 2016). Figure 3.2 gives a flow diagram for the industrial processing and some by-products of cereals into different food products.

3.4.3 Dry milling

Dry milling produces flour which is further used in the production of various cereal products such as brewing adjuncts, bakery products, snacks, breakfast cereals, groats, decorticated or pearled sorghum, and other leavened products such as bread, cake, and fries (Owens 2001). Dry milled maize is used as animal feed and in the production of fuel-ethanol. Flours are produced by milling rice to produce breakfast cereals or cooked for consumption and for making puffed cereals. Rice grits are used as brewing adjuncts (Table 3.1) (Serna-Saldivar 2016). Rice is parboiled, dried, dehulled, decorticated, and then milled to produce the white long grains. Wheat, triticale, and rye are similar in processing operations; they are mostly dry milled into flours for production of various bakery products (Table 3.1) (Serna-Saldivar 2016). They are also dry milled to form different types of composite flours in order to improve specific food properties such as gluten reduction, protein improvement, and flour handling improvement. Barley is dried and dry milled and mostly used as forage and feedstock in some part of Europe (Serna-Saldivar 2016). Oats are used majorly to produce breakfast cereals due to their health benefits (Serna-Saldivar 2016). They are dried and dehulled into groats, meals, and flours. Sorghum and millet are decorticated by abrasion, and then dry milled into groats, meals, and flours (Serna-Saldivar 2016) (Table 3.1). They are also fermented into different beverages such as beer, slurries, and pap. Traditional processing methods are common with sorghum and millet as they are often referred to as "poor man cereal" (Temba et al., 2016). Millet (proso and foxtail) has been found to contain different levels of nutritional properties, phytochemicals, and bioactive compounds based on their milled fractions. The functional properties such as nitrogen solubility, foaming and emulsifying capacity, water and oil absorption capacity, water solubility index, and gelation have all been confirmed to be significantly based on the milled fractions of these millet types. On the other hand, antinutrients such as phenol and phytic acid were also affected by dehulling (Devisetti et al., 2014).

3.4.4 Wet milling

Wet milling involves the production of starch and other modified materials such as gluten meal, germ, and bran from cereals (Serna-Saldivar 2016) (Table 3.1). The starch is converted to various types of sugars such as maltose, maltodextrin, glucose, and high fructose syrup which has found various industrial uses (Serna-Saldivar 2016). Traditional or indigenous wet milling of cereals encourage nutrient enhancement and promote the development of various foods (Serna-Saldivar 2016). Wet milling of wheat, triticale, and rye produce specialty products such as gluten which is used as bakery additives, bioethanol, and starch used in the production of sweeteners. Sprouted and malted barley is processed by wet milling and used in the production of both alcoholic and nonalcoholic beverages (Serna-Saldivar 2016).

3.4.5 Nixtamalization/alkaline/acid cooking

The masa method of maize milling is common in North America, precisely in Mexico where alkaline cooked corn is made into tortillas and corn chips. Millet and sorghum are also processed in a similar way to produce maize-like tortillas and chips (Serna-Saldivar et al., 1990; Serna-Saldivar 2016) (Table 3.1).

3.5 Recent developments in cereal processing

In recent times, cereal processing has received wide attention, with emphasis on fortification and supplementation to improve nutritional constituents of cereal-based food products (Ajibade and Ijabadeniyi 2019). The importance of diet to healthy living has significantly influenced cereal processing, leading to the development of health-promoting cereal-based products. Industrial requirements also influence cereal processing. For example, rheological characteristics of dough are considered to improve handling during bread making (Nkhata et al., 2018). The methods of storage and preservation, consumers' perception of flavor, texture, and acceptability are also considered during cereal processing.

3.5.1 Fortification and supplementation in cereal processing

Due to the deficiency of nutrients in cereals, considerations for the processing methods become important to improve nutrient content, bioavailability, or digestibility (Oghbaei et al., 2016). Since cereals are a staple food in different parts of the world, extensive research into fortification and supplementation of cereal-based products with legumes or other nutrient-dense cereals becomes pertinent. Table 3.2 presents selected research outputs on the nutritional effects of cereal fortification/ supplementation.

3.5.2 Bakery products

Wheat flour may be used in combination with other flours in baking applications. Wheat flour which is low in some essential amino acids such as lysine is usually supplemented/fortified with other flours such as legumes flours, protein concentrates, and milk and milk products (Serna-Saldivar 2016). Some other supplementation focuses on the use of minor but nutrient-dense cereals such as teff, millet, buckwheat, and amaranth. Millet is known for its relative high lysine content and teff (red variety) is known for high iron content. Buckwheat is reportedly rich in lysine and arginine (Gimenez-Bastida and Zielinski 2015). Wheat flour enriched with 15% buckwheat flour has shown high content of free amino acids, sugars, and flavors, and a threefold increase in phenol content (Lin et al., 2009).

Wheat flour was supplemented with pulses known to have a low glycemic index (GI) of 45 ± 4 (e.g., split pea) compared to wheat bread with 101 ± 9 GI. Thermally treated split yellow pea was used to form a composite (20%) with wheat, giving rise to bread with a low GI (Fahmi et al., 2019). Low GI has been attributed to slow digestion of the resistant starch present in food products, thus having a beneficial effect in glucose absorption in the intestines. High carbohydrate foods have been associated with postprandial hyperglycemia which is clinically related to diabetes, hypertension, and cardiovascular diseases (CVD) (Fahmi et al., 2019). Phytochemical-rich and low GI foods have been shown to reduce incidences of noncommunicable diseases (Barrett et al., 2018). Although wheat often loses its unique elastic properties that are essential for baking applications, low gluten content that causes poor dough handling have been resolved by improving dough rheology using enzymes, emulsifiers, and surfactants. When wheat was combined with pearl millet and Bambara groundnut, the rheological characteristics of dough was improved upon after emulsifiers such as sodium stearoyl lactylate, polysorbate 80, diacetyl tartaric acid ester of monoglycerides, and apple pectin had been used as additives. While dough characteristics were improved upon, bread amino acids significantly improved (Ajibade et al., 2019). Gac fruit is known to contain beneficial phytochemicals and bioactive compounds that were used to produce pasta with low starch digestibility and increased dietary fiber (DF) (Chusak et al., 2019). Amaranth, a pseudo-cereal, has been reported to contain a superior protein with various health benefits. These health benefits include stimulation of the immune system, decrease in plasma cholesterol level, reduction in blood glucose level, improved conditions of hypertension and anemia. It also possesses anti-allergic and antioxidant activities. Woldemariam et al. (2019) use

Table 3.2: Improvements/enhancements in cereal processing.

Cereals type	Processing methods	Food product	Types of additives	Positive effect on the final product	Negative effect	Health claims	References
Wheat and red teff	Milling/ baking	Cookies	Soy protein (Okara)	Increased nutritional composition	Antinutritional (tannin and phytate) increased	na	Hawa et al. (2018)
Wheat and buckwheat (Tartary)	Milling/ fermentation/ baking	Bread	buckwheat	Increased free amino acids, flavor, phenolic content, volatile content, and sugar content. Antioxidant activity and rutin level increased. Mineral, fiber and phytic level.	NA	Inhibition of in vitro formation of advanced glycation end-products (AGEs) increase in the plasma antioxidant capacity after consumption	Juan G. B. et al. (2015), Gawlik-Dziki et al. (2009), Vogrinčič et al. (2010), Chlopicka et al. (2012), Szawara-Nowak et al. (2014), Yildiz and Bilgiçli, 2012, Bojňanská et al. (2009)
Wheat and pulses (split yellow pea)	Milling/ fermentation/ baking	Bread	na	High CHO, protein and low GI	na	Low GI bread	Rigard et al. (1998)
Wheat and Gac fruit	Milling/ fermentation/ baking	Bread	Ripe and unripe gac fruit.	Low rapidly digested starch and increased slow digested starch.	na	Low starch digestibility bread	Chusak et al. (2019)

(continued)

Table 3.2 (continued)

Cereals type	Processing methods	Food product	Types of additives	Positive effect on the final product	Negative effect	Health claims	References
Wheat, pearl millet, and Bambara groundnut	Milling/ fermentation/ baking	Bread	Emulsifiers (SSL, PS 80 and DATEM) and pectin	Good dough development, increased protein, and essential amino acids	Dark crusted bread	na	Ajibade and Ijabadeniyi (2019)
Wheat and millet	Milling/ fermentation/ baking	Bread	Protein concentrates from pea, rice, and whey. Transglutaminase.	Increased volume, good crumb strength, improved sensory quality, and loss of bitter after taste.	High protein content reduced the effect of the transglutaminase enzyme.	na	Tomić (2019)
Wheat	Milling/ fermentation/ baking	Bread	Hydrocolloids and probiotic oligosaccharides	Improved sensory perception and low starch digestibility	Low GI bread	Slightly low protein digestibility	Angioloni and Collar (2011)
Wheat	Milling/ fermentation/ baking	Bread	Wheat bran (WB) Locust bean gum (LBG) and granular RS2-type corn resistant starch (RS)	WB, crumb luminosity, increased high-speed mixing time, crumb chroma, and crumb moisture content. Good results in the sensory evaluation.	WB, reduced specific volume, and reduced crumb luminosity (LBG)	na	Almeida et al. (2013)
Teff, barley and amaranth	Milling/ fermentation/ baking	Injera (flatbread)	na	Increased protein, CHO, and mineral content	na	na	Woldemariam et al. (2019)

Raw material	Process	Product	Added ingredient / treatment	Effect			Reference
Teff, sorghum bicolor, and faba bean	Milling/fermentation/baking	Injera (flatbread)	na	Increased proximate composition	na	na	Mihrete and Bultosa (2017)
Wheat and Buckwheat	Sourdough fermentation	Sourdough bread	LAB fermentation: Lactobacillus fermentum AB 15, L. Plantarum AB 16, L. vaginalis AB 17, and L. crispatus AB 19, L. delbrueckii, buckwheat hull hemicelluloses	Reduced phytic content, improved crumb and crust structure, polyphenols and prolonged shelf life. Increased antioxidant capacity and total phenol content.	Decreased network in cells, reduced elasticity, and lower volume with a hard crust.	na	Moroni et al. (2011a), Moroni et al. (2011b), Moroni et al. (2012), Gandhi and Dey (2013)
Wheat (semolina)	Milling/extrusion/baking	Spaghetti	Dietary Fiber: β-glucans (glucagel and barley balance)	Spaghetti with low GI form barley balance β-glucan	na	Low GI spaghetti	Chillio et al. (2016)
Rice (brown) and cornmeal	Milling/extrusion/baking	Pasta	na	Pasta with an improved amino acid score, cooking time, and sensory quality.	na	na	Da Silva et al. (2015)
Wheat (semolina-durum)	Milling/extrusion/baking	Pasta	Dietary fiber (Glucagel, inulin Raftiline HPX, inulin Raftiline GR, and psyllium and oat)	Low glycemic index	Antagonistic effect of different combinations of dietary fiber	Low GI pasta	Foschia et al. (2015)

(continued)

Table 3.2 (continued)

Cereals type	Processing methods	Food product	Types of additives	Positive effect on the final product	Negative effect	Health claims	References
Wheat	Milling/ baking	Muffins	Wheat flour with dietary bra (DF) from oat, wheat, maize, and barley. Cereal bran (CB) from oat, rice, and wheat. Substitution: 10, 20, and 30%	DF addition improved cake volume, texture, sensory characteristics in contrast to CB, and prolonged the shelf-life. Wheat bra and oat bra up to 30%, improved quality characteristics of cupcakes.	CB yielded firm cakes that had low volume, low moisture, compact crumb texture, and low sensory acceptability	na	Lebesi and Tzia (2011)
Wheat	Milling/ baking	Cupcakes	Fibruline Instant, a native inulin (DPw10). Fibruline S20, a highly soluble inulin (DP < 10); Inulin Orafti GR (DP 10), 50, 75, or 100% margarine was removed and replaced with appropriate amounts of inulin and water	product moisture and crumb density increased significantly the replacement of baking fat by inulin – water mixtures significantly affects batter flowability and cohesivity of crumb	Muffin volume decreased.	na	Zahn et al. (2016)

Millet	Milling/ baking	Gluten-free bread	Protein concentrates from pea, rice, and whey. Enzyme: transglutaminase	Improvement in bread texture and crumb network, good sensory acceptability, increased nutritional content, and removal of the bitter taste from millet.	na	na	Tomić, (2019)
Sorghum	Milling/ baking	Gluten-free bread	Pregelatinized cassava starch, egg white, and transglutaminase	Increased crumb firmness and chewiness	Increased incubation time, decreased crumb cohesiveness, chewiness, and resilience	na	Onyango et al. (2010)

Na: Not available

amaranth and barley to improve the nutritional and functional properties of Ethiopian flatbread called Injera. The inclusion of sorghum and faba beans, a legume, produced similar improvements on injera (Mihrete and Bultosa 2017). Improvements in the nutritional composition and consumer acceptability were significant in the flatbread.

Supplementation with DF is also commonly practiced with cereal products (Foschia et al., 2015). DF (soluble and insoluble) has been known to improve the metabolism of cereals and cereal products (Foschia et al., 2013). DF increases the water binding and viscosity alteration, which in turn influences the physiological attenuations such as fat and cholesterol binding, decrease in blood glucose levels, prevention of constipation, and maintaining a healthy colon. Soluble fibers include oligosaccharides, pectins, β-glucans, and galactomannan gums, alginate, psyllium fiber. The insoluble fibers consist of cellulose, hemicellulose, and lignin (Kale et al., 2010; Foschia et al., 2013). Insoluble fibers usually help in smooth bowel movement while soluble fiber help in lowering blood cholesterol and glucose level (Foschia et al., 2015). Addition of DF to nourish cereal-based food products has attracted lots of research work and most common cereals are now benefitting from fiber additions (Foschia et al., 2015) (Table 3.2). The most common fiber that has been added to food in the food industry includes arabinoxylans, β-glucans, and resistant starch (Foschia et al., 2015). Several studies have investigated the effects of adding DF to different cereal-based products such as bread, muffins/cake, pasta, and extruded products (Table 3.2). Addition of DF to wheat flour pasta gave a low GI pasta that can serve as a functional food (Chillo et al., 2011). The addition of bran from buckwheat to pasta led to reduced breakage and improvement in sensory properties. Similarly, improvement in texture, firmness, sensory, and cooking qualities was observed with the addition of amaranth leaf flour (Chillo et al., 2008). With the addition of semolina, the spaghetti produced showed good sensory properties with high antioxidant activity (Aravind et al., 2012; Foschia et al., 2013). Spaghetti made from maize and oat (hydrocolloid addition at 2%) resulted in a product with improved viscoelastic properties, cooking quality, and sensory acceptability (Padalino et al., 2013). Similar improvements were observed in bread, muffins, and extruded products after enrichment with DF (Angioloni and Collar 2011; Almeida et al., 2013) (Table 3.2).

3.6 Whole-grain cereal processing

Recent studies have shown that intake of whole-grain foods and grain fiber protects against type 2 diabetes and CVD (Şanlier et al., 2019). Food-based dietary guidelines from the Department of Health, South Africa, recommend the inclusion of whole cereals in the diet to replace processed cereals (Gani et al., 2012). Dietary Guidelines for Americans also recommends that at least half of cereals consumed should be whole cereals rather than processed cereals. The same recommendation has been adopted in Europe and in other developed countries all over the world and is largely due to the inherent nutrients and health benefits of whole cereal consumption. There is advancement in industrial whole-grain cereal processing and the launch of healthy and nutritious products into the market to meet consumer needs. The American Association of Cereals Chemists International (AACCI, 2017) has defined whole-grain cereals as intact, ground, cracked, or flaked caryopsis, whose principal components: the starchy endosperm, germ, and bran are present in a substantial amount as they exist in the intact caryopsis. Whole-grain cereals contain phytochemicals such as fiber, phytin, lignins, antioxidants, tocotrienols, phytosterols, sphingolipids, fatty acids, vitamins, and minerals (Slavin 2003). All these nutrients promote good health. Processing or refining methods usually reduced the total nutrients and phytonutrients present in cereals. Most vitamins and minerals are lost during refining, although fortification usually helps with this type of nutrients loss.

3.6.1 Health benefits of whole-grain cereals in noncommunicable diseases

Whole-grain cereals have been documented to be beneficial to human health. An array of phytochemicals, DF, and vitamins and minerals are present in the multilayered bran (Micha et al., 2017). The germ contains healthy fats, vitamins, and minerals. The endosperm contains starch and protein which is useful for seed growth. Human health has been shown to be enhanced with these components whether singly or synergistically (Slavin and Jacobs 2001; Slavin 2003; Taylor et al., 2014).

Recent studies have shown that there is reduced risk of contracting noncommunicable diseases such as type 2 diabetes, stroke, certain cancers, cardiovascular disease, and so on. It was found that intake of whole cereal grains reduces incidences of deaths from type 2 diabetes by 17%, 10% of deaths from stroke, 6% of cardiometabolic deaths, and 4% of deaths from coronary heart disease (Micha et al., 2017).

Type 2 diabetes is caused by the inability of the body to metabolize sugars completely in the blood. The condition has been linked to diet and due to its high mortality rate, studies are ongoing to develop an adequate diet that can ameliorate the condition (Liu and Hou 2019). Diets based on whole-grain cereal products have been identified as a potential factor to reduce the incidence of type 2 diabetes. Researchers have found out that the consumption of whole-grain cereals reduces or lowers the risk of developing type 2 diabetes (Meyer et al., 2000; Ye et al., 2012). The risk of developing type 2 diabetes was also reduced when the diet was based on 45 g of whole-grain meal in adults (Chanson-Rolle et al., 2015; Schwingshackl et al., 2017). These studies concluded that whole-grain meals help in sugar metabolism in the intestine by lowering the rate of absorption into the bloodstream.

Postprandial blood glucose response, insulin resistance, and increase in insulin levels have also been reported to be reduced due to the consumption of whole grains (Pereira et al., 2000; Tosh and Chu 2015). Whole grains offer a lower GI, unlike refined grains. The whole cereals also contain an array of nutrients, bioactive compounds, and phytochemicals which can help with glucose and insulin homeostasis, and thus help with type 2 diabetes. Documented reports show that bioactive components present in whole grains function synergistically to protect against CVD (Pereira et al., 2000). A high cholesterol absorption in the blood may trigger heart failures; whole grains, however, contain DF to lower cholesterol absorption in the blood (Slavin 2007). Fermented whole grains also contain resistance starch, DF, plant sterols, and oligosaccharides, which helps in the lowering of serum cholesterol (Brown et al., 1999; Slavin 2007; Durazzo et al., 2014). Consumption of wholegrain foods has been linked to a reduction in the incidence of cancer, a global disease with high mortality rate. Several studies have shown that diet based on whole grain reduces the risk of developing certain cancers (Liu et al., 2019), such as colorectal cancer. This is partly due to the presence of certain anticarcinogenic compounds such as antioxidants, phenolic acids, and phytosterols in whole grains that inhibit cancer cells (McIntosh 2007; Kyrø et al., 2013; Lei et al., 2016).

Obesity is a risk factor for several CVD, as well as type 2 diabetes. Obesity can be prevented through the consumption of whole grains. Similarly, the consumption of whole grains prevents the accumulation of excess fat in the body, helps with lowering the absorption of glucose (lower GI) in the intestine, and reduces the accumulation of sugar in the bloodstream, and gives a feeling of fullness or satiety which helps in weight management (Rigaud et al., 1998; Foerster et al., 2014; Santaliestra-Pasías et al., 2016).

3.6.2 Processing of whole cereal grains

Processing of whole grains, which is similar to the processing of refined grains, generally refines the grains and make them edible. Preprocessing operations carried out during cereal processing include cleaning (removal of unwanted materials), dehulling (removal of outer coat), conditioning (spraying with water), rolling or racking, and toasting to reduce moisture. These operations hydrate the cereal grains, improves the cooking time, starch gelatinization, nutrients bioavailability, chewing, and digestion (Liu et al., 2019).

Milling as a processing method is well utilized for whole grains, and although there are standard methods for milling of refined grains, there are no standard procedures for milling of whole grains (Liu et al., 2019). Milling of cereal grains involves cleaning, tempering, milling, and flour treatment and can be carried out either by single-stream milling or by recombination. Single-stream milling involves milling grains directly between stones or rollers without separation of the grain components while recombination involves the initial separation of grain components into bran, germ, and endosperm. Each of these grain components is treated in different millstreams and then recombined to form proportions of whole grain (Jones et al., 2015; Liu et al., 2019).

Milling operation of whole-grain cereals is very important as it is the basis for most nutritional claims. Milling methods can be classified based on the types of equipment which may include the stone mill, hammer mill, plate mill, and roller mill. The most common equipment is the roller and stone mill method (Jones et al., 2015). The stone mill comprises an upper and lower layer of two stones pressed together to facilitate tearing and shearing of grains, which causes milling of the grains. All the grain components (germ, endosperm, and bran) are still present in the final flour after milling. Stone mill usually imparts earthy flavor to the milled flours, a characteristic that is appreciated by natural food lovers (Liu et al., 2019). However, the level of production by this method (21,600 kg/day/24 h operation) is lower than the demand. The limitation of low productivity by stone mill has been resolved by reducing the weight of the stone on the upper layer of the mill to increase the rotation of the stones and thus increase the rate of milling. Denmark stone miller was designed to improve the output of whole-grain cereals in comparison to the traditional stone mill method. The rotation per minute increased for the Danish stone mill from 20–100 to 300 rpm (Jones et al., 2015; Liu et al., 2019).

Roller mill separates grain components into bran, germ, and endosperm and subjects them to treatments based on the type of grain (Liu et al., 2019). Each of these components follows different streams and then they are recombined based on the original component of the grain. This is usually practiced in the industries because it has a high output and capacity rate (1.4 million kg of flour per day) (Jones et al., 2015). The roller mills also have the advantage of being flexible and can be adjusted based on the type of grain. The possibility of separation of the grain components which are then given specific treatment before recombining makes it easy for manufacturers to stabilize volatile components or even fortify with such nutrients after processing (Jones et al., 2015). While the roller mill offers the advantage of higher capacity production, the loss of sensitive nutrients from the grain limits this process. Decomposition and loss of bioactive compounds, as well as starch degradation, also characterize the roller mill operation. These limitations have not been resolved, even by fortification (Liu et al., 2019). Roller mills are capital intensive as arrays of equipment are needed to enhance the method. Color sorters, washers, scourers, sifters, and purifiers are often required, and this equipment adds a heavy energy utilization to the process. After separation, blending of the different components of grains together can be challenging as bran is light and can float while germs are oily and dense. Adequate blending may mean more equipment and energy requirements. Overall, whole cereals grain processing requires standardization of the process to protect grain integrity and prevent nutrient loss(Liu et al., 2019).

3.7 Gluten-free products

Gluten plays a vital role in bread production. However, the sensitivity of some individuals to wheat gluten which is rapidly increasing around the globe has resulted in consumers' preference for gluten-free products. These gluten-free products are generally characterized by poor quality compared to whole-wheat products. Gluten-free products are produced from cereals that have no gluten such as rice, maize, millet, sorghum, and so on. The products, therefore, require enhancement in terms of dough, crumb, and crust characteristics to meet consumers' expectations. Flours obtained from nonwheat cereals do not possess the ability to form a viscoelastic dough that can retain the gases generated during fermentation and baking. Enhancement can be achieved using enzymes, proteins, starches, cryoprotectants, surfactants, and emulsifiers. In addition, modification of the production process has also improved the technological performance of gluten-free bread (Pineli et al., 2015; Tomić et al., 2019). Recently, combinations of methods of enhancement of gluten-free products have proven effective in the production of gluten-free products. The use of protein concentrates from whey, pea, and rice with the addition of enzymes such as transglutaminase gave millet bread desirable characteristics and eliminated the bitter taste in the product (Tomić et al., 2019). Various studies aimed at the improvement of gluten-free bread using different processing methods and enhancers have been reported (Tiwari and Cummins 2012; Foschia et al., 2013; Gimenez-Bastida et al., 2015; Tomić et al., 2019) (Table 3.2). Examples of such processing methods include starch gelatinization and pressure treatment, and enhancers include sour batter, corn starch, milk protein, enzymes, egg white powder, emulsifiers, gums, and cellulose derivatives (Pineli et al., 2015). Sorghum treated with gelatinized starch, egg white, and transglutaminase enzyme provided a gluten-free bread with good volume, and increased crumb firmness and chewiness. However, the effect of the enzyme on the bread was negative in terms of chewiness after an increased incubation time (Onyango et al., 2011).

3.8 Cereal fermentation

Fermentation is a simple home-based food processing method, commonly practiced in Africa. It aims at food preservation, while also improving the organoleptic properties of the processed food (Achi and Ukwuru 2015). It can be traced back to ancient Egypt where it was employed in the production of bread and beer (Serna-Saldivar 2016). In fermentation, the growth and metabolic activities of microorganisms are used in food preservation. Food fermentation may be aerobic (fungal and alkaline) or anaerobic (alcohol and lactic acid) (Şanlier et al., 2019). Fermentation can also be based on starter cultures, which give the fermented products consistency and allow the development of nutrients and elimination of antinutrients (Osungbaro 2009; Serna-Saldivar 2016).

Fermentation makes food more digestible, improves the sensory attributes, and reduces toxic and ant nutritional factors such as cyanogenic glycosides in foods (Joshi 2016), thereby increasing the bioavailability of nutrients such as vitamin B_{12}. It also improves the protein quality of foods and facilitates the decomposition of enzyme inhibitors and mycotoxins (Achi et al., 2015). Important enzymes and amino acids, active substances, and flavor-enhancing compounds are produced during fermentation (Achi et al., 2015). Further, fermentation can mask undesirable flavors and also replenish intestinal microflora (Joshi 2016).

Cereals generally consist of a starchy endosperm and an embryo containing amino acids, lipids, sugars, minerals, vitamins, and hydrolytic enzymes and antinutrients such as tannins, polyphenols, and enzyme inhibitors, as well as mycotoxins (Achi et al., 2015). Cereals provide nondigestible carbohydrates, which stimulate the growth of probiotics *Lactobacilli* and *Bifidobacteria* (Achi et al., 2015). Cereals also contain β-glucan, a soluble fiber which influences cholesterol and glucose absorption (Beck et al., 2010). Cereal-based fermented food products contain bioactive substances such as dietary and functional fiber (Achi et al., 2015). For example, whole-wheat grain products contain bioactive phytochemicals such as phenolic acids, carotenoids, tocopherols, phytosterols, and lignans. These phytochemicals function as radical scavengers and anticancers, while also playing important roles in cholesterol and diabetes decrease and control (Luthria et al., 2015).

In different parts of Africa, cereals such as maize (*Zea mays*), sorghum (*Sorghum bicolor*), and millet (*Pennisetum americanum*) have been fermented and used in infant complementary feeding and breakfast for adults (Achi et al., 2015). Various factors promote the fermentation of cereals by microbes. These factors include cereal composition, growth capability of the microbes, processing method (Achi et al., 2015), environmental conditions of temperature, humidity, and oxygen availability. Various organisms are associated with cereal fermentation. They include lactic acid bacteria (LAB) such as *Lactobacillus fermentum*, *Lb. brevis*, *Lb. casei*, and *Lb. plantarum*. LAB break down food matrices to produce lactic acid, antimicrobial compounds such as bacteriocin while enhancing safety and shelf life of the product (Achi et al., 2015). Yeasts such as *Saccharomyces cerevisiae* have also been isolated from fermented foods. The types of organisms isolated from any fermentation process are influenced by temperature, pH, inoculum, cereal type, and time (Achi et al., 2015). Table 3.3 presents some indigenous fermented cereals products and the associated microbes.

In recent times, consumers have developed a greater preference for functional foods over regular nutritious foods. Functional foods are nutritious foods which also contain biologically active compounds that can positively impact consumers' health (Achi et al., 2015). The health effects of fermented foods which have been attributed to the synthesis of bioactive peptides during protein degradation by the

Table 3.3: Major cereals-based fermented foods and the associated microorganisms.

Product	Main Cereal	Microorganisms involved	Outcome of fermentation	References
Mawe	Maize (Zea mays)	Lactobacillus fermentum, Lb. cellobiosus, Lb. brevis, Lb. curvatus, Lb. buchneri, Weissella confusa, Candida krusei, Candida kefyr, Candida glabrata, and Saccharomyces cerevisiae	Formation of acidity, flavor, and enhancement of digestibility	Hounhouigan et al., 1994 Hounhouigan et al., 1999
Kenkey	Maize (Zea mays)	Lactobacillus fermentum, C. krusei (Issatchenkia orientalis), and S. cerevisiae Lb. casei, Lb. lactis, Lb. plantarum, Lb. brevis, Lb. acidophilus, Lb. fermentum, and Lb. casei,	Level of available lysine increased from 1.3 in maize kernels to 3.3 g per 16 g nitrogen in ready-to-eat kenkey; flavor compounds (2,3-butanediol, butanoic acid, lactic acid, 3-methylbutanoic acid, octanoic acid, 2-phenylethanol, and propanoic acid) are formed	Jespersen et al., 1994 Nche et al., 1995 Blandino et al., 2003 Olsen et al., 1995, Olasupo et al., 1997
Tchoukout ou	Sorghum (Sorghum bicolor and Sorghum vulgare)	S. cerevisiae and lactic acid bacteria	The solubility of iron in raw sorghum (3% of total Fe) was increased to approx. 20% of total Fe in tchoukoutou protein with a high digestibility	Kayode et al., 2007 Nout, 1987.
Uji	Maize (Zea mays) Sorghum (Sorghum bicolor, Sorghum vulgare), and finger millet	Lactobacillus plantarum, Lb. fermentum, Lb. cellobiosus and Lb. buchneri, Pediococcus acidilactici and Pc. Pentosaceus, and L. mesenteroids	Formation of taste and flavor principles and liquefying of the starch Gel.	Mbugua, 1984, Mbugua, 1991. Lee,1997; Onyango et al., 2003, 2004

Jiu ("daqu")	Sorghum and a the mixture of wheat, barley, and peas	(*Rhizopus, Mucor, Aspergillus* spp.) bacteria (acetic acid bacteria, lactic acid bacteria, bacilli, and yeasts (*Saccharomyces, Candida, Hansenula* spp.)	Source of energy, stimulates the digestive system	Zhang et al., 2007, Wang et al., 2008
Ben-Saalga	Pearl millet (*Pennisetum glaucum*)	The lactic acid bacteria in the natural fermentation include *Lb. fermentum, Lb. plantarum*, and *Pediococcus pentosaceus* as majority organisms	Antinutritional components of pearl millet, e.g., phytate was found to be degraded on average from 0.46 to 0.22 g per 100 g dry matter, i.e., by more than 50% in commercial Ben-Saalga. Increase in dietary uptake of proteins and minerals	Sifer et al., 2005 Mouquet et al., 2008 Lestienne et al., 2005 Tou et al., 2006
Jnard	Finger millet (*Eleusine coracana*)	*Amylomycesrouxii, Rhizopus oryzae, Endomycopsis fibuligera, S. cerevisiae, Enterococcus faecalis, P. pentosaceus*, and others		Tamang et al., 1988 Hesseltine and Ray, 1988
Idli	Rice (*Oryza sativa*), Black gram (*Phaseolus mungo*)	*Leuconostoc mesenteroides, Lb. fermentum, Ent. faecalis, Pc. Dextrinicus* and yeasts especially *Sacch. cerevisiae, Debaryomyces hansenii, Pichia anomala*, and *Trichosporon pullulans, Geotrichum candidum, Torulopsis holmii, Torulopsis candida*, and *Trichosporon pullulans*	The leavening of the batter and flavor formation, starch degradation, and gas formation, as well as the accumulation of vitamin B and free amino acids increase of all essential amino acids and the reduction of antinutrients (such as phytic acid), enzyme inhibitors, and flatus sugars	Nout et al., 2007; Nout, 2009, Blandino et al., 2003 Chavan and Kadam, 1989a; Shortt, 1998. Steinkraus et al., 1983. Soni and Singh et al., 2015; Aidoo et al., 2006; Balasubramanian and Viswanathan, 2007, Farnworth, 2005

(continued)

Table 3.3 (continued)

Product	Main Cereal	Microorganisms involved	Outcome of fermentation	References
Mifen	Rice (*Oryza sativa*)	*Lactobacillus, Leuconostoc, Pediococcus, Streptococcus, Enterococcus* and *Aerococcus* spp. and *the yeasts S. cerevisiae, Candida rugosa, and Candida tropicalis*	Protection against microbial spoilage, as well as the modification of the amorphous region of the starch granules which facilitates their gelatinization during cooking, better eating (chewing) qualities	Lu et al., 2005
Kishk (Fugush)	Wheat (bulgur)	*Lactobacillus plantarum, Lactobacillus casei* and *Lactobacillus brevis, Bacillus subtilis* and yeasts, *S. cerevisiae*	Excellent preservation quality, richer in B vitamins	Beuchat, 1983; Chavan and Kadam, 1989 Tamime and McNulty (1999)
Ogi	Maize (*Zea mays*)	*Lb. plantarum, Lb. fermentum, Leuc. mesenteroides*, and *Sacch. cerevisiae*	na	Odunfa and Adeyele 1985, Adeyemi, 1993, Ijabadeniyi, 2007, Omemu et al., 2007
Boza	Wheat, millet, and rye	*Boza Lb. plantarum, Lb. brevis, Lb. rhamnosus, Lb. fermentum, Leuc. Mesenteroides*, and subsp. *dextranium*	na	Hancioglu and Karapinar 1997; Moncheva et al., 2008; Botes et al., 2007; Todorov et al., 2008.

Adapted from Temba et al. (2016); na, not available.

fermenting bacteria are numerous (Şanlier et al., 2019). They include alleviating symptoms associated with lactose intolerance, increasing immunity, fighting diabetes, obesity and allergies, reducing blood cholesterol levels, and protecting against pathogens (Şanlier et al., 2019).

Cereal fermentation leads to improvement in the texture, nutritional value, and sensory and functional qualities of the product. Nondigestible polysaccharides and oligosaccharides in cereals are reduced during fermentation; group B vitamins are made more available with a significant decrease in antinutrients' concentration such as polyphenols, phytates, and tannins (Şanlier et al., 2019). Enzyme involvement in cereal fermentation causes hydrolysis and solubilization of grain macromolecules, such as proteins and cell wall polysaccharides (Nkhata et al., 2018). Fermentation of cereals has also been linked to the production of probiotics and prebiotics, where LAB are employed. LAB fermentation has been shown to improve nutrients and minerals availability and reduce tannin and phytic acid present in the grains. In addition, some vitamins such as riboflavin, thiamine and niacin, and amino acid lysine have been reported after LAB fermentation of cereals. There was a significant improvement in minerals availability of the final product after pearl millet was fermented with pure cultures of *Lactobacilli* and yeasts cells (Khetarpaul and Chauhan 1990; Sanni et al., 1999).

In the effort to promote probiotics and prebiotics base for consumers, cereals which support the desired growth of LAB can be used to produce defined and standardized fermented food that may have health-promoting properties (Charalampopoulos et al., 2002b). Cereals are also now being developed to encapsulate probiotics in functional foods. Preparation or processing has largely been in laboratories and lots of work is still needed to process fermented cereals into a good probiotic vehicle (Charalampopoulos et al., 2002a). Despite the many benefits of fermentation, some drawbacks have also been identified with this food processing method. These include the proliferation of yeasts and molds which may compromise food safety, loss of vitamins and minerals, and reduction in provitamin A and antioxidant carotenoids (Nkhata et al., 2018). Further studies to overcome these limitations will further promote fermentation as an efficient cereal processing method.

3.9 Conclusion

Cereal science is vital to achieving global food security. Further studies on standardizing cereal processing operations from farm to fork are necessary. The practice of fermentation in the development of cereal-based functional foods offers great potential in contributing to the wellness of consumers.

Acknowledgements: This work is based on research supported in part by the National Research Foundation of South Africa, SA (NRF)/Russia (RFBR) Joint Science and Technology Research Collaboration (Grant Number: 118910).

References

AACC International. 2017. Whole Grains. http://www.aaccnet.org/initiatives/definitions/Pages/WholeGrain.aspx (accessed November 19, 2019).

Adeyemi, I.A. (1993). Making the most of Nigerian ogi. *Food Chain*. 8: 5–6.

Aidoo, K. E., Rob Nout, M. and Sarkar, P. K. 2006. Occurrence and function of yeasts in Asian indigenous fermented foods. *FEMS Yeast Research*, 6 (1): 30-39.

Ajibade, B. O. and Ijabadeniyi, O. A. 2019. Effects of pectin and emulsifiers on the physical and nutritional qualities and consumer acceptability of wheat composite dough and bread. *Journal of Food Science and Technology*, 56 (1): 83-92.

Almeida, E. L., Chang, Y. K. and Steel, C. J. 2013. Dietary fibre sources in bread: Influence on technological quality. *LWT-Food Science and Technology*, 50 (2): 545-553.

Angioloni, A. and Collar, C. 2011. Physicochemical and nutritional properties of reduced-caloric density high-fibre breads. *LWT-Food Science and Technology*, 44 (3): 747-758.

Aravind, N., Sissons, M., Egan, N. and Fellows, C. 2012. Effect of insoluble dietary fibre addition on technological, sensory, and structural properties of durum wheat spaghetti. *Food Chemistry*, 130 (2): 299-309.

Barrett, A. H., Farhadi, N. F. and Smith, T. J. 2018. Slowing starch digestion and inhibiting digestive enzyme activity using plant flavanols/tannins—a review of efficacy and mechanisms. *LWT*, 87: 394-399.

Blandino, A., Al-Aseeria, M.E., Pandiella, S.S., Cantero, D., and Webb, C. (2003). Cereal-based fermented foods and beverages. *Food Research International*. 36:527–543.

Bojňanská, T., Frančáková, H., Chlebo, P. and Vollmannová, A. 2009. Rutin content in buckwheat enriched bread and influence of its consumption on plasma total antioxidant status. *Czech Journal of Food Sciences*, 27 (236.240).

Botes, A., Todorov, S. D., Von Mollendorff, J. W., Botha, A. and Dicks, L. M. 2007. Identification of lactic acid bacteria and yeast from boza. *Process Biochemistry*, 42 (2): 267-270.

Brown, L., Rosner, B., Willett, W. W. and Sacks, F. M. 1999. Cholesterol-lowering effects of dietary fiber: a meta-analysis. *The American Journal of Clinical Nutrition*, 69 (1): 30-42.

Chanson-Rolle, A., Meynier, A., Aubin, F., Lappi, J., Poutanen, K., Vinoy, S. and Braesco, V. 2015. Systematic review and meta-analysis of human studies to support a quantitative recommendation for whole grain intake in relation to type 2 diabetes. *PloS one*, 10 (6): e0131377.

Charalampopoulos, D., Pandiella, S. and Webb, C. 2002. Growth studies of potentially probiotic lactic acid bacteria in cereal-based substrates. *Journal of Applied Microbiology*, 92 (5): 851-859.

Charalampopoulos, D., Wang, R., Pandiella, S. and Webb, C. 2002. Application of cereals and cereal components in functional foods: a review. *International Journal of Food Microbiology*, 79 (1-2): 131-141.

Chavan, J., Kadam, S. and Beuchat, L. R. 1989. Nutritional improvement of cereals by fermentation. *Critical Reviews in Food Science and Nutrition*, 28 (5): 349-400.

Chillo, S., Laverse, J., Falcone, P., Protopapa, A. and Del Nobile, M. A. 2008. Influence of the addition of buckwheat flour and durum wheat bran on spaghetti quality. *Journal of Cereal Science*, 47 (2): 144-152.

Chillo, S., Ranawana, D. and Henry, C. 2011. Effect of two barley β-glucan concentrates on in vitro glycaemic impact and cooking quality of spaghetti. *LWT-Food Science and Technology*, 44 (4): 940-948.

Chlopicka, J., Pasko, P., Gorinstein, S., Jedryas, A. and Zagrodzki, P. 2012. Total phenolic and total flavonoid content, antioxidant activity and sensory evaluation of pseudocereal breads. *LWT-Food Science and Technology*, 46 (2): 548-555.

Chusak, C., Chanbunyawat, P., Chumnumduang, P., Chantarasinlapin, P., Suantawee, T. and Adisakwattana, S. 2019. Effect of gac fruit (Momordica cochinchinensis) powder on in vitro starch digestibility, nutritional quality, textural and sensory characteristics of pasta. *LWT*: 108856.

da Silva, E. M. M., Ascheri, J. L. R. and Ascheri, D. P. R. 2016. Quality assessment of gluten-free pasta prepared with a brown rice and corn meal blend via thermoplastic extrusion. *LWT-Food Science and Technology*, 68: 698-706.

Devisetti, R., Yadahally, S. N. and Bhattacharya, S. 2014. Nutrients and antinutrients in foxtail and proso millet milled fractions: Evaluation of their flour functionality. *LWT-Food Science and Technology*, 59 (2): 889-895.

Durazzo, A., Carcea, M., Adlercreutz, H., Azzini, E., Polito, A., Olivieri, L., Zaccaria, M., Meneghini, C., Maiani, F. and Bausano, G. 2014. Effects of consumption of whole grain foods rich in lignans in healthy postmenopausal women with moderate serum cholesterol: a pilot study. *International Journal of Food Sciences and Nutrition*, 65 (5): 637-645.

ElMekawy, A., Diels, L., De Wever, H. and Pant, D. 2013. Valorization of cereal based biorefinery byproducts: reality and expectations. *Environmental Science and Technology*, 47 (16): 9014-9027.

Fahmi, R., Ryland, D., Sopiwnyk, E. and Aliani, M. 2019. Sensory and Physical Characteristics of Pan Bread Fortified with Thermally Treated Split Yellow Pea (*Pisum sativum L.*) Flour. *Journal of Food Science*.

FAO. 2018. WORLD FOOD AND AGRICULTURE – STATISTICAL POCKET BOOK 2018. Rome. 254 pp. License: CC BY-NC-SA 3.0 IGO. (Accessed: November 28th, 2019).

FAO. 2019 Food Outlook - Biannual Report on Global Food Markets. Rome. License: CC BY-NC-SA 3.0 IGO. (Accessed: November 28th, 2019).

Foerster, J., Maskarinec, G., Reichardt, N., Tett, A., Narbad, A., Blaut, M. and Boeing, H. 2014. The influence of whole grain products and red meat on intestinal microbiota composition in normal weight adults: a randomized crossover intervention trial. *PloS one*, 9 (10): e109606.

Foschia, M., Peressini, D., Sensidoni, A. and Brennan, C. S. 2013. The effects of dietary fibre addition on the quality of common cereal products. *Journal of Cereal Science*, 58 (2): 216-227.

Foschia, M., Peressini, D., Sensidoni, A., Brennan, M. A. and Brennan, C. S. 2015. Synergistic effect of different dietary fibres in pasta on in vitro starch digestion? *Food Chemistry*, 172: 245-250.

Gani, A., Wani, S., Masoodi, F. and Hameed, G. 2012. Whole-grain cereal bioactive compounds and their health benefits: a review. *Journal of Food Process Technology*, 3 (3): 146-156.

Gawlik-Dziki, U., Dziki, D., Baraniak, B. and Lin, R. 2009. The effect of simulated digestion in vitro on bioactivity of wheat bread with Tartary buckwheat flavones addition. *LWT-Food Science and Technology*, 42 (1): 137-143.

Gimenez-Bastida, J. A. and Zielinski, H. 2015. Buckwheat as a functional food and its effects on health. *Journal of Agricultural and Food Chemistry*, 63 (36): 7896-7913.

Gupta, M., Abu-Ghannam, N. and Gallaghar, E. 2010. Barley for brewing: Characteristic changes during malting, brewing and applications of its by-products. *Comprehensive Reviews in Food Science and Food Safety*, 9 (3): 318-328.

Hawa, A., Satheesh, N. and Kumela, D. 2018. Nutritional and anti-nutritional evaluation of cookies prepared from okara, red teff and wheat flours. *International Food Research Journal*, 25 (5).

Hounhouigan, D., Nout, M., Nago, C., Houben, J. and Rombouts, F. 1994. Microbiological changes in mawe during natural fermentation. *World Journal of Microbiology and Biotechnology*, 10 (4): 410-413.

Ijabadeniyi, A. 2007. Microbiological safety of gari, lafun and ogiri in Akure metropolis, Nigeria. *African Journal of Biotechnology*, 6 (22).

Jespersen, L., Halm, M., Kpodo, K. and Jakobsen, M. 1994. Significance of yeasts and moulds occurring in maize dough fermentation for 'kenkey'production. *International Journal of Food Microbiology*, 24 (1-2): 239-248.

Jones, J. M., Adams, J., Harriman, C., Miller, C. and Van der Kamp, J. W. 2015. Nutritional impacts of different whole grain milling techniques: a review of milling practices and existing data. *Cereal Foods World*, 60 (3): 130-139.

Kale, M., Pai, D., Hamaker, B. and Campanella, O. 2010. Incorporation of fibers in foods: a food engineering challenge. In: *Food Engineering Interfaces*. Springer, 69-98.

Khetarpaul, N. and Chauhan, B. 1990. Effect of fermentation by pure cultures of yeasts and lactobacilli on the available carbohydrate content of pearl millet. *Food Chemistry*, 36 (4): 287-293.

Kruger, J., Taylor, J. R. and Oelofse, A. 2012. Effects of reducing phytate content in sorghum through genetic modification and fermentation on in vitro iron availability in whole grain porridges. *Food Chemistry*, 131 (1): 220-224.

Kyrø, C., Skeie, G., Loft, S., Landberg, R., Christensen, J., Lund, E., Nilsson, L. M., Palmqvist, R., Tjønneland, A. and Olsen, A. 2013. Intake of whole grains from different cereal and food sources and incidence of colorectal cancer in the Scandinavian HELGA cohort. *Cancer Causes and Control*, 24 (7): 1363-1374.

Lebesi, D. M. and Tzia, C. 2011. Effect of the addition of different dietary fiber and edible cereal bran sources on the baking and sensory characteristics of cupcakes. *Food and Bioprocess Technology*, 4 (5): 710-722.

Lei, Q., Zheng, H., Bi, J., Wang, X., Jiang, T., Gao, X., Tian, F., Xu, M., Wu, C. and Zhang, L. 2016. Whole grain intake reduces pancreatic cancer risk: a meta-analysis of observational studies. *Medicine*, 95 (9).

Lestienne, I., BESANCon, P., Caporiccio, B., Lullien-Péllerin, V. and Tréche, S. 2005. Iron and zinc in vitro availability in pearl millet flours (Pennisetum glaucum) with varying phytate, tannin, and fiber contents. *Journal of Agricultural and Food Chemistry*, 53 (8): 3240-3247.

Lin, L.-Y., Liu, H.-M., Yu, Y.-W., Lin, S.-D. and Mau, J.-L. 2009. Quality and antioxidant property of buckwheat enhanced wheat bread. *Food Chemistry*, 112 (4): 987-991.

Liu, T. and Hou, G. G. 2019. Trends in Whole Grain Processing Technology and Product Development. *Whole Grains: Processing, Product Development, and Nutritional Aspects*: 257.

Mbugua, S. 1984. Isolation and characterisation of lactic acid bacteria during the traditional fermentation of uji. *East African Agricultural and Forestry Journal*, 50 (1-4): 36-43.

Mbugua, S. 1991. A new approach to uji (an East African sour cereal porridge) processing and its nutritional implications. In: Proceedings of *Development of Indigenous Fermented Foods and Food Technology in Africa. Proceedings from the IFS/UNU Workshop Held at Douala, Cameroon. IFS provisional report*. 288-309.

McIntosh, G. H. 2007. Whole grains and cancer prevention. *Whole Grains and Health*: 69-74.

Meyer, K. A., Kushi, L. H., Jacobs Jr, D. R., Slavin, J., Sellers, T. A. and Folsom, A. R. 2000. Carbohydrates, dietary fiber, and incident type 2 diabetes in older women. *The American Journal of Clinical Nutrition*, 71 (4): 921-930.

Micha, R., Peñalvo, J. L., Cudhea, F., Imamura, F., Rehm, C. D. and Mozaffarian, D. 2017. Association between dietary factors and mortality from heart disease, stroke, and type 2 diabetes in the United States. *Jama*, 317 (9): 912-924.

Mihrete, Y. and Bultosa, G. 2017. The Effect of Blending Ratio of Tef [Eragrostis Tef (Zucc) Trotter], Sorghum (Sorghum bicolor (L.) Moench) and Faba Bean (Vicia faba) and Fermentation Time on Chemical Composition of Injera. *Journal of Nutrition and Food Science*, 7 (583): 2.

Moncheva, P., Chipeva, V., Kujumdzieva, A., Ivanova, I., Dousset, X. and Gocheva, B. 2003. The composition of the microflora of boza, an original Bulgarian beverage. *Biotechnology and Biotechnological Equipment*, 17 (1): 164-168.

Moroni, A. V., Dal Bello, F. and Arendt, E. K. 2009. Sourdough in gluten-free bread-making: an ancient technology to solve a novel issue? *Food Microbiology*, 26 (7): 676-684.

Mouquet-Rivier, C., Icard-Vernière, C., Guyot, J.-P., Hassane Tou, E., Rochette, I. and Trêche, S. 2008. Consumption pattern, biochemical composition and nutritional value of fermented pearl millet gruels in Burkina Faso. *International Journal of Food Sciences and Nutrition*, 59 (7-8): 716-729.

Nche, P., Nout, M. and Rombouts, F. 1995. The effects of processing on the availability of lysine in kenkey, a Ghanaian fermented maize food. *International Journal of Food Sciences and Nutrition*, 46 (3): 241-246.

Nkhata, S. G., Ayua, E., Kamau, E. H. and Shingiro, J. B. 2018. Fermentation and germination improve nutritional value of cereals and legumes through activation of endogenous enzymes. *Food Science and Nutrition*, 6 (8): 2446-2458.

Nout, M. R. 2009. Rich nutrition from the poorest–Cereal fermentations in Africa and Asia. *Food Microbiology*, 26 (7): 685-692.

Nout, M.J.R., Sarkar, P.K., and Beuchat, L.R. (2007). Indigenous fermented foods. In: Food Microbiology: Fundamentals and Frontiers, pp. 817–835. Doyle, M.P. and Beuchat, L.R. Eds., ASM Press, Washington, DC.

Oghbaei, M. and Prakash, J. 2016. Effect of primary processing of cereals and legumes on its nutritional quality: A comprehensive review. *Cogent Food and Agriculture*, 2 (1): 1136015.

Olasupo, N., Olukoya, D. and Odunfa, S. 1997. Identification of Lactobacillus species associated with selected African fermented foods. *Zeitschrift für Naturforschung C*, 52 (1-2): 105-108.

Olsen, A., Halm, M. and Jakobsen, M. 1995. The antimicrobial activity of lactic acid bacteria from fermented maize (kenkey) and their interactions during fermentation. *Journal of Applied Bacteriology*, 79 (5): 506-512.

Onyango, C., Okoth, M. W. and Mbugua, S. K. 2003. The pasting behaviour of lactic-fermented and dried uji (an East African sour porridge). *Journal of the Science of Food and Agriculture*, 83 (14): 1412-1418.

Onyango, C., Mutungi, C., Unbehend, G. and Lindhauer, M. G. 2011. Rheological and textural properties of sorghum-based formulations modified with variable amounts of native or pregelatinised cassava starch. *LWT-Food Science and Technology*, 44 (3): 687-693.

Onyango, C., Mutungi, C., Unbehend, G. and Lindhauer, M. G. 2010. Rheological and baking characteristics of batter and bread prepared from pregelatinised cassava starch and sorghum and modified using microbial transglutaminase. *Journal of Food Engineering*, 97 (4): 465-470.

Osungbaro, T. O. 2009. Physical and nutritive properties of fermented cereal foods. *African Journal of Food Science*, 3 (2): 023-027.

Owens, G. 2001. *Cereals Processing Technology*. CRC Press.

Padalino, L., Mastromatteo, M., Lecce, L., Cozzolino, F. and Del Nobile, M. 2013. Manufacture and characterization of gluten-free spaghetti enriched with vegetable flour. *Journal of Cereal Science*, 57 (3): 333-342.

Pereira, M., Jacobs, D., Pins, J., Raatz, S., Gross, M., Slavin, J. and Seaquist, E. 2000. The effect of whole grains on inflammation and fibrinolysis. In: Proceedings of *Circulation*. Lippincott Williams and Wilkins 530 Walnut St, Philadelphia, PA 19106-3621 USA, 711-711.

Pineli, L. d. L. d. O., Zandonadi, R. P., Botelho, R. B., de Oliveira, V. R. and DE Alencar Figueiredo, L. 2015. The use of sorghum to produce gluten-free breads: A systematic review. *Journal of Advanced Nutrition and Human Metabolism*, 2: e944.

Rigaud, D., Paycha, F., Meulemans, A., Merrouche, M. and Mignon, M. 1998. Effect of psyllium on gastric emptying, hunger feeling and food intake in normal volunteers: a double-blind study. *European Journal of Clinical Nutrition*, 52 (4): 239.

Şanlier, N., Gökcen, B. B. and Sezgin, A. C. 2019. Health benefits of fermented foods. *Critical Reviews in Food Science and Nutrition*, 59 (3): 506-527.

Sanni, A. I., Onilude, A. A. and Ibidapo, O. T. 1999. Biochemical composition of infant weaning food fabricated from fermented blends of cereal and soybean. *Food Chemistry*, 65 (1): 35-39.

Santaliestra-Pasías, A., Garcia-Lacarte, M., Rico, M., Aguilera, C. and Moreno, L. 2016. Effect of two bakery products on short-term food intake and gut-hormones in young adults: a pilot study. *International Journal of Food Science and Nutrition*, 67 (5): 562-570.

Schwingshackl, L., Hoffmann, G., Lampousi, A.-M., Knüppel, S., Iqbal, K., Schwedhelm, C., Bechthold, A., Schlesinger, S. and Boeing, H. 2017. *Food groups and risk of type 2 diabetes mellitus: A systematic review and meta-analysis of prospective studies*: Springer.

Serna-Saldivar, S., Gomez, M. and Rooney, L. 1990. Technology, chemistry, and nutritional value of alkaline-cooked corn products. *Advances in Cereal Science and Technology (USA)*.

Serna-Saldivar, S. O. 2016. *Cereal Grains: properties, processing, and nutritional attributes*. CRC press.

Shortt, C. (1998). Living it up for dinner. *Chemistry and Industry*. 8: 300–303.

Sifer, M., Verniere, C., Galissaires, L., Castro, A., Lopez, G., Wacher, C. and Guyot, J. 2005. DGGE community analysis of lactic acid fermented pearl millet-based infant gruels (ben-saalga, ben-kida) as a tool to characterize relatedness between traditional small-scale production units. In: Proceedings of *8th Symposium on Bacterial Genetics and Ecology*. Lyon, France, 1.

Singh, A. k., Rehal, J., Kaur, A. and Jyot, G. 2015. Enhancement of attributes of cereals by germination and fermentation: a review. *Critical Reviews in Food Science and Nutrition*, 55 (11): 1575-1589.

Slavin, J. 2003. Why whole grains are protective: biological mechanisms. *Proceedings of the Nutrition Society*, 62 (1): 129-134.

Slavin, J. 2007. Whole grains and cardiovascular disease. *Whole grains and health*: 59-68.

Slavin, J. and Jacobs, D. 2001. L Marquart. *Grain processing and nutrition. Critical Reviews in Biotechnology*, 21: 49-66.

Smith, M. D., Rabbitt, M. P. and Coleman-Jensen, A. 2017. Who are the world's food insecure? New evidence from the Food and Agriculture Organization's food insecurity experience scale. *World Development*, 93: 402-412.

Steinkraus, K. H. (1983). Handbook of Indigetrous Fermented Foods. Marcel Dekker, New York.

Szawara-Nowak, D., Koutsidis, G., Wiczkowski, W. and Zieliński, H. 2014. Evaluation of the in vitro inhibitory effects of buckwheat enhanced wheat bread extracts on the formation of advanced glycation end-products (AGEs). *LWT-Food Science and Technology*, 58 (2): 327-334.

Taylor, J. R., Belton, P. S., Beta, T. and Duodu, K. G. 2014. Increasing the utilisation of sorghum, millets and pseudocereals: Developments in the science of their phenolic phytochemicals, biofortification and protein functionality. *Journal of Cereal Science*, 59 (3): 257-275.

Temba, M. C., Njobeh, P. B., Adebo, O. A., Olugbile, A. O. and Kayitesi, E. 2016. The role of compositing cereals with legumes to alleviate protein energy malnutrition in Africa. *International Journal of Food Science and Technology*, 51 (3): 543-554.

Tiwari, U. and Cummins, E. 2012. Dietary exposure assessment of β-glucan in a barley and oat-based bread. *LWT-Food Science and Technology*, 47 (2): 413-420.

Tomić, J., Torbica, A. and Belović, M. 2019. Effect of non-gluten proteins and transglutaminase on dough rheological properties and quality of bread based on millet (*Panicum miliaceum*) flour. *LWT*: 108852.

Tosh, S. M. and Chu, Y. 2015. Systematic review of the effect of processing of whole-grain oat cereals on glycaemic response. *British Journal of Nutrition*, 114 (8): 1256-1262.

Tou, E., Guyot, J.-P., Mouquet-Rivier, C., Rochette, I., Counil, E., Traore, A. and Trèche, S. 2006. Study through surveys and fermentation kinetics of the traditional processing of pearl millet (Pennisetum glaucum) into ben-saalga, a fermented gruel from Burkina Faso. *International Journal of Food Microbiology*, 106 (1): 52-60.

U.S. Department of Health and Human Services and U.S. Department of Agriculture. 2015. 2015–2020 Dietary Guidelines for Americans. 8th edition. http://health.gov/dietaryguidelines/2015/guidelines/ (accessed November 10, 2017).

Vogrincic, M., Timoracka, M., Melichacova, S., Vollmannova, A. and Kreft, I. 2010. Degradation of rutin and polyphenols during the preparation of tartary buckwheat bread. *Journal of Agricultural and Food Chemistry*, 58 (8): 4883-4887.

Wang, H.-Y., Zhang, X.-J., Zhao, L.-P. and Xu, Y. 2008. Analysis and comparison of the bacterial community in fermented grains during the fermentation for two different styles of Chinese liquor. *Journal of Industrial Microbiology and Biotechnology*, 35 (6): 603-609.

Woldemariam, F., Mohammed, A., Fikre Teferra, T. and Gebremedhin, H. 2019. Optimization of amaranths–teff–barley flour blending ratios for better nutritional and sensory acceptability of injera. *Cogent Food and Agriculture*, 5 (1): 1565079.

Ye, E. Q., Chacko, S. A., Chou, E. L., Kugizaki, M. and Liu, S. 2012. Greater whole-grain intake is associated with lower risk of type 2 diabetes, cardiovascular disease, and weight gain. *The Journal of Nutrition*, 142 (7): 1304-1313.

Yıldız, G. 2012. Effects of whole buckwheat flour on physical, chemical, and sensory properties of flat bread, lavaş. *Czech Journal of Food Sciences*, 30 (6): 534-540.

Zahn, S., Pepke, F. and Rohm, H. 2010. Effect of inulin as a fat replacer on texture and sensory properties of muffins. *International Journal of Food Science and Technology*, 45 (12): 2531-2537.

Zhu, F. (2014). Structure, physicochemical properties, and uses of millet starch. *Food Research International*, 64, 200-211.

Tremayne S. Naiker, John J. Mellem

4 Processing operations and effects on the characteristics of legume grains for food system applications

4.1 Introduction

The global population is expected to grow by over a third or 2.30 billion people by 2050. It has been forecasted that almost all growth is to take place in developing countries (FAO, 2017). In addition, consumers will experience difficulty in sourcing their daily nutrient requirements from animal-based products regularly. Thus, consumers will be forced to seek feasible food resources, simultaneously satisfying their recommended daily intake of different nutrients. This presents the developing world with foreseeable challenges in providing poor and undernourished populations with safe, nutritious, and wholesome food resources. Generally, food crops have been viewed to occupy a foundational position in consumer nutrition, where they are seen as protein and energy sources for a large part of global populations (Elhardallou et al., 2015).

Food legumes have been regarded as valuable sources for human nutrition (Bhat and Karim, 2009). The utilization of legume grains by food industries has grown considerably in intermediate forms (e.g., flour, protein concentrates, and isolates) other than whole grains. The techno-functional properties of legume flour are said to be limited, thus continuous efforts are aimed at modifying legume-based materials for providing end products with desirable traits and functional properties. The design of food processing equipment and operations were largely adapted to transform and preserve food resources.

However, more recently it is viewed as an option to produce raw materials with desirable properties whilst preserving their nutritional value, rather than the use of high-grade raw materials in food applications (Bußler et al., 2015). Aguilera (2005) confirmed that improvement to the quality of food-related materials will be largely determined by interventions made at the microscopic level, as many microstructural elements (e.g., fibers, starch, and protein) that critically contribute to their technofunctionality, identity, and quality are below the ~100 μm range.

Tremayne S. Naiker, John J. Mellem, Department of Biotechnology and Food Technology, Durban University of Technology, Durban, South Africa

https://doi.org/10.1515/9783110667462-004

4.2 Water in food materials

Water is omnipresent and influences the quality attributes and physical properties of food materials through various roles. It is well documented for providing an environment during which various sugars, salts, acids, and alternative comparatively tiny hydrophilic molecules are dissolved. Water plays a vital role within the structure and properties of food macromolecules. It interacts with suspended mixture particles and gels and will function as a plasticizer for increasing the elasticity of molecules. Water and environmental conditions (i.e., ionic content and pH) is thought to influence the conformation of enzymes and their functionality, whereby at low water activity most enzymatic reactions are slow.

At a molecular level, water contributes to the flow behavior of dispersions by increasing the space of empty areas during which different molecules can move. In dense systems, molecular chains are largely constrained as a result of entanglements. Water separates the side groups and chains of larger molecules. This allows for easier reptation and on a macroscopic level influences the textural and rheological properties of food materials. Generally, high moisture levels result in higher water activity, higher freezing point, greater flexibility, higher specific heat, greater thermal conductivity, lower viscosity, osmotic pressure, and boiling point. At low moisture levels, several macromolecules may form noncrystalline solid states. In biological materials, more than one dynamic water structure exists and has a considerable influence on water removal as well as biological activity (Damodaran, 2017).

4.2.1 Pretreatment procedures

The objective of soaking grains is to increase the moisture level within the kernel to the required level for ensuring uniform starch gelatinization during cooking. Soaking is known to reduce grain mutilation from osmotic pressure and reduces time for precooking by creating fissures within the kernel. However, extended soaking may result in the loss of water-soluble vitamins, flavor, and mineral elements. Grain quality and costs are largely determined by soaking parameters used and its profound effect on uniformity throughout grain mass during the process of soaking. The purpose of boiling grains is for gelatinizing starch present in the endosperm.

This is an irreversible process carried out by increasing the temperature, whereby starch granules that have absorbed water undergo structural changes from crystalline to an amorphous state. Following starch gelatinization, water enters the interchain space through the disruption of hydrogen bonds between starch chains. For starch gelatinization to occur sufficient moisture and heat transfer around gelatinization temperature are required for grains under treatment (Velupillai, 1993, Ahmad and Noomhorm, 2013).

A variety of processes have been developed for producing value-added grain merchandise. These processes include dry heat, freeze drying, chemical, gamma irradiation, freeze–thaw, gun puffing treatments, and stepwise hydration–cook–dry and soak–cook–dry methods. Freeze–thaw method, freeze drying, and gun puffing are uneconomical as a result of the prices of machine investment. Soak–cook–dry methods combined with dry heat have been recognized for its economy and simplicity (Sabularse et al., 1991, Alfy et al., 2016).

4.2.2 Dehydration

Dehydration or drying refers to the removal of large amounts of water from food materials, thus contributing to preservation and reducing costs related to transportation and storage.

This is accomplished by creating a large difference in water activity between the material and environment. Food preservation against microbiological and chemical forms of deterioration is said to be achieved by reducing the water activity of materials below 0.70–0.80. Previous studies have revealed that there are more than two types of water with different relaxation time constants and rates of rotational motions. Kerr (2013) explained using a traditional model that water exists in bulk and bound phases. During the later phases of the drying process, there is water present that requires more energy to be removed. This water cannot be removed except by freeze drying.

Proteins and polysaccharides containing polar or ionic groups have a greater binding affinity to bound water. Thus, extra energy is required to dissociate from these groups than from other water molecules. In the cell organelles, water is often found compartmentalized, whereby it must move through the cell membrane by diffusion.

In the case of damaged cells, this water is required to move around the cell wall and membrane fragments (Leung, 2017). For spaces present between cells, extracellular water that may be trapped must follow a tortuous path to the material surface. In food materials containing biological tissues, a variety of factors affect the ability of water to diffuse and reach the material surface where it undergoes transformation from a liquid to gaseous state. Hot air dryers, osmotic dryers, and freeze dryers are among the different unit operations available for drying grains. Hot air dryers are operated between 40 and 80°C, whereby drawn air generated by a fan is passed across a bank of heaters.

Continuous dryers have evolved to improve the throughput of products. In tunnel drying, insulated chambers (10–15 m) containing series of trolleys are driven. This provides a semicontinuous movement. In cocurrent model tunnel dryers, the air moves in the same direction as the product. This allows for the moist and coolest product to be exposed to the least humid air. In the initial phases, rapid drying is promoted when the product remains near wet bulb temperature. The product that is most dried at the end of the tunnel is exposed to lower air temperature. Therefore, the product may have undergone fewer quality changes with respect to browning or case hardening (Kerr, 2013).

In countercurrent tunnel drying, air enters in the opposite direction to the product movement. Some of the associated advantages of countercurrent drying are that the lowest moisture product contacts the driest and highest temperature air. This allows for the removal of excess moisture from the respective materials that was hardest to remove. The major physical changes that occur during drying include bed porosity, particle density, shrinkage, and bulk density.

These may vary due to moisture removal, shrinkage of structure, and internal collapse. The extent of these changes also depends on the type of food material, dryer, and conditions employed (Lewicki, 1998).

In terms of food powders, desirable textural properties are related to bulk density, particle size, and ability to disperse and rehydrate in water. Dried food materials often result in large changes to taste profiles but are not always associated with poor quality. These materials are known to retain fat, protein, fiber, and carbohydrates in a denser form compared to their moist precursors. In terms of fruits and vegetables, leaching out of vitamins and minerals may occur during drying preparations and pretreatment procedures. However, minerals are not expected to be destroyed during drying. Vitamins are heat sensitive and are most susceptible to degradation during drying process. Vitamin loss may be dependent on the type of material, drying preparation, process, and parameters employed (Kerr, 2013).

4.3 Microstructure characteristics of legume grains

The fractionation of legume grains by food industries is continuously generating interest for extending the use of its main components (i.e., starch, protein, and dietary fiber). The microstructure of dry legumes (peas, beans, and lentils) is said to be different compared to oilseed grains. This is mainly due to their respective starch and oil contents. In oil-based legumes, the protein bodies are surrounded by a network of lipid bodies whereas in dry legume grains the cotyledon cells are formed by starch granules covered by a protein matrix (Błaszczak et al., 2007). The microstructure of legume cotyledons is known to influence the physical properties of grains and purity of protein isolates.

Acevedo et al. (2017) have evaluated the effects of germination, soaking–cooking (SC), and microwaving treatment on the microstructure of pigeon pea (*Cajanus cajan* L.) legume grains. Soaked grains were germinated using a damp cloth containing sodium hypochlorite for 5 days. SC was conducted in boiling water for 20, 40, and 60 min using a grain to distilled water ratio of 1:10 (m/v). Grains were subjected to microwaving treatment for 10 min at different potencies (50%, 70%, and 100%) in a microwaving oven. All pretreated grains were dried in a hot air oven (55 °C for 24 h) to constant weight.

The cotyledon cells formed by starch granules were identified as the main storage reservoir of legume grains. Most starch granules were found to have retained their shape, size, and smooth surface. The average size of starch granules reported was 24.10 × 17 μm. In raw grains starch granules were found to have a smooth and oval surface. In germinated grains, minimal changes to the cell wall were observed in comparison with its native state. SC (6 h to 20 min) was found to have preserved the structure of starch granules.

However, their surface appeared to be flattened and the protein matrix exhibited contractions because of the heat applied. It was observed that samples produced by SC (6 h to 60 min) resulted in dissemination of compartmentalization between starch granules and the protein matrix. The microwaving treatments did not show significant size variations in starch granules. However, the structure of the protein matrix was affected. This was attributed to the denaturation of protein that was caused by microwaving heating.

These findings were similar to morphological changes observed in our present study that was conducted. The effects of soaking, boiling, and dehydration treatments on the microstructure of *Lablab purpureus* (L.) sweet (hyacinth bean) were determined (Figure 4.1a–c). Pretreated legume grains were dehydrated using a countercurrent tunnel dryer (60–80 °C for 6 h).

Figure 4.1: Scanning electron microscopy of hyacinth bean legume grains: (a) Soak (S), (b) Soak + Boil (S + B), (c) Soak + Boil + Dehydrated. Bar size = 50 μm.

Aguilera et al. (2009) have examined the influence of soaking, cooking, and dehydration treatments on the microstructural characteristics of flour produced from lentil for their effective utilization in various food applications. In this sense, dehydration of food materials is associated with numerous advantages and the adaptation of processing strategies may result in the improved utilization of legume grains. This could further facilitate the development of economical viable products (Vega-Mercado et al., 2001).

Soaked legume material was cooked by boiling for 30 min and dehydrated (75 ± 3°C for 6 h) using an air forced tunnel. The scanning electron micrographs of raw and processed lentil reveal that starch granules were the major storage components. In raw samples they appeared to be spherical–oval (22 μm) in shape. The granules were surrounded by well-defined protein bodies and were characterized by a smooth surface. This observation is in accordance to those reported for hyacinth bean flour produced from soaked legume grains.

In addition, there were drastic changes to shape and size of starch granules in response to various pretreatments (Figure 4.2a–c). In samples produced by soaking, pronounced changes to starch granules were observed.

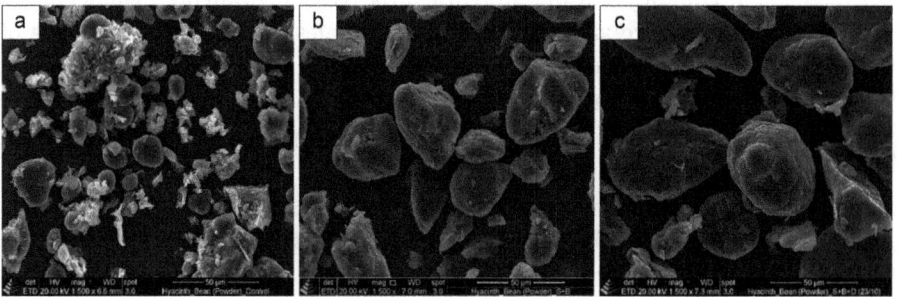

Figure 4.2: SEM of hyacinth bean flour particles: (a) Soak (S), (b) Soak + Boil (S + B), (c) Soak + Boil + Dehydrated. Bar size = 50 μm.

They appeared larger and slightly eroded which was attributed to the water absorption properties of starch. The size of lentil starch granules appeared to increase (121–125 μm) extensively for samples produced by cooking after soaking. Overall, starch granules kept their internal integrity, but their surface was flattened as an effect of heat. Samples produced by dehydration following soaking and cooking resulted in smaller starch granules compared to cooked grains. This was mainly attributed to losses of holding water.

4.4 The physicochemical properties of legume flour

The physicochemical properties of food and its components are functional and chemical which food manufacturers take into consideration when making decisions concerning the quality of ingredients/products. The suitability for using food ingredients in various food systems is largely dependent on their physicochemical properties. The nutritive value of food legumes is their ability to provide nutrients (protein, carbohydrates, vitamins, and minerals) that is digestible in presence/absence of antinutritional compounds. Legume grain (pea, lentil, and bean) varieties are predominantly composed of protein (16.89–34.70%) and starch (54.72–64.60%) on the basis of dry weight (Boye et al., 2010).

Traditional food processing (soaking, decortications, germination, fermentation, and cooking) methods have been found to significantly influence the nutritive values of food legumes. Antinutritional factors (ANF) present in legumes are essential for its further application in various food systems. They have been found to reduce nutrient utilization and/or intake in various legume-based foods. Some of the common ANFs found in food legumes include phytates, tannins, trypsin inhibitors (protease), cyanogenic 12 glycosides, glycosides, flavonoids, oxalates, alkanoids, gossypol, cardiac hemagluttinins (lectins), and coumarins (Shivachi et al., 2012). Soaking, boiling, and germination have been found to increase the utilization of legumes.

Shaahu et al. (2014) examined the influence of processing (decortication, roasting, and boiling in tap water) on the ANFs, proximate, mineral, and amino acid composition of *Lablab purpureus*. All ANFs examined were significantly reduced by the applied processing techniques. Boiling was found to be the best method for the reduction of tannins (37%), alkaloids (33%), oxalates (38%), trypsin inhibitors (100%), and HCN (89%). Soaking legumes in a freshwater solution containing sodium chloride prior to cooking has been found to improve their nutritional quality. Complex sugars, belonging to the raffinose family, have the potential to cause gastric issues (flatulence) if they are not broken down during digestion.

The method of cooking legumes in freshwater containing sodium chloride can result in a tender skin. This was caused by calcium and magnesium ions found in pectin of cell walls replaced by sodium ions (Fabbri and Crosby, 2016). Shivachi et al. (2012) have associated prolonged cooking time to be responsible for the underutilization of legumes in many diets.

This is due to prolonged cooking reducing the nutritive values of legumes, with respect to vitamins and certain amino acids. Furthermore, legumes collected from gene bank accessions have been found to take a longer time to cook in comparison with farmers' collections. Thus, variation in cooking time maybe influenced by several factors including type of water, energy source, genetic characteristics, size, and age of legumes.

Osman (2007) has studied the effects of processing on the chemical composition of *D. lablab* flour produced from legume grains obtained in Saudi Arabia. Grains were prepared by roasting, soaking overnight in water (1:10 m/v), cooking in water for 30 min in a pressure cooker, germination at room temperature for 5 days, and autoclaved for 121 °C for 20 min under 15 lb/in. The moisture content of samples was significantly increased by soaking, germination, and decreased by roasting treatment. Germination was found to further increase the protein content of the prepared flour, whilst other processes used reduced this parameter. The decrease in protein content was attributed to the possibility of leaching out of soluble proteins. Cooking was found to significantly reduce fat content and was attributed to the possibility of the presence of lipolytic enzyme activity which breaks down triacylglycerides to simple fatty acids, sterol esters, and polar lipids.

Mubarak (2005) has associated the dehulling of legume grains prior to soaking to greatly retain mineral elements of mung bean (*Phaseolus aureus*) grains. Germination showed increases in Ca, K, P, Mg, Fe, and Mn. Loss in divalent metal concentrations was attributed to the formation of phytate–cation protein complexes actioned by the binding of proteins. Na, Mg, and Fe concentrations were not significantly reduced by pressure cooking.

Ghavidel and Prakash (2007) reported germination was found to improve protein, thiamin, and in vitro digestibility of iron and calcium for legumes was studied. The concentrations of ANFs phytic acid and tannin were significantly reduced by 47–52% and 43–52%, respectively, compared to the control. Flour produced from dehulled germinated grains improved the overall quality of legumes. This was due to the improvement of nutrient bioavailability and reduction of ANFs. In heterogeneous food systems, functional properties (i.e., gelation, oil absorption, water holding capacity (WHC), swelling, and bulk density) are affected by interactions of major and minor chemical food constituents.

Starch concentrations significantly differ amongst various legumes, whereby it has been reported that raw lentil contains 51–59% (dry matter) of starch. The gelatinization of starch serves as an important property, whereby starch absorbs water resulting in a loss to granular organization (Copeland et al., 2009, Kaur et al., 2010). In legume flour, starch molecules may interact with protein, fiber, and mineral elements. This was found to extensively affect their thermal properties which may include parameters such as heat capacity and gelatinization temperature (Divekar et al., 2017; Ma et al., 2016).

Li and Zhu (2017) studied the influence of starch interactions on the physicochemical properties of whole grain quinoa flour. Pearson correlation analysis revealed significant correlations present amongst starch and properties of quinoa flour. Food lipids and protein were also found to have greatly contributed to the functionality of whole grain quinoa flour. During thermal treatment the structure and physicochemical properties of starch are altered extensively. Thermal treatment of starches has been found to either result in gelatinization or retrogradation of starch granules.

This has been proven to influence functionality and is largely dependent on the type of starch present and degree of modification (Ma et al., 2011). Flours produced from broad bean, black bean, lentil, and chickpea were subjected to heat moisture (HMT) and annealing (ANN) treatments. HMT treated flours resulted in a significant decrease on all thermal parameters investigated compared to ANN treatment.

There were distinct differences observed for the gelatinization temperature range ($T_c - T_o$) between ANN (1.26–2.15 °C broad bean and black bean), HMT (2.41–5.33 °C for black bean and lentil), and native (1.85–5.32°C for chickpea and lentil) flours. The differences observed were attributed to the molecular rearrangement of starch containing flour that may have occurred during HMT and ANN treatments. Thus, heat transfer into starch granules could have been delayed due to increase in starch crystallinity (Chávez-Murillo et al., 2018, de la Rosa-Millán et al., 2017).

Chau and Cheung (1998) compared the functional characteristics of flour produced from *Phaseolus calcaratus*, *Phaseolus angularis*, and *D. lablab* with soybean. The pH values for the respective flour dispersions (10% m/v) ranged between 6.53 and 6.65. Flour produced from soybean legume grains was found significantly denser compared to *P. calcaratus*, *D. lablab*, and *P. angularis*. Benítez et al. (2013) reported similar bulk densities (0.80–0.98 g/mL) for mucuna, dolichos, cowpea, and jack bean grains, whereby flour produced from dolichos and cowpea were found significantly denser compared to jack bean and mucuna flours.

Water and other food components (proteins and polysaccharides) are essential constituents that influence the rheological and textural properties of foods.

Within a protein matrix, proteins can retain water against a gravitational force. This is referred to as WHC which is the sum of physically entrapped, hydrodynamic, and bound water. It is regarded to be more important than water binding capacity in many food applications. Water holding and absorption capacities are important in food applications such as custards, sausage manufacture, and baking of dough. These applications require water to be absorbed without protein dissolution, thus achieving viscous body thickening (Seena and Sridhar, 2005). Other factors that may influence water absorption of food materials include cell wall material and starch content (Damodaran et al., 2008).

Chau and Cheung (1998) found significant differences among the WHC and oil holding capacities (OHCs) for flours produced from *P. angularis*, *P. calcaratus*, *D. lablab*, and soybean. Soybean flour had a greater WHC compared to the other three flours and this was attributed to its higher protein content. The protein content for flours produced from *P. angularis*, *P. calcaratus*, and *D. lablab* ranged between 30.40–32.10%, whereas soybean flour had a protein content of 52.8%. The WHC of *D. lablab* was found comparable to flour produced from faba bean, lentil, and lima bean ranging between 1.04 and1.08 g/g. WHC for flours produced from four nonconventional legumes (*Stizolobium niveum*, *Canavalia ensiformis*, *Vigna unguiculata*, and *Lablab purpureus*) were studied by Benítez et al. (2013).

Differences to capacities observed were attributed to constituents that have high affinity for water molecules such as polar amino acid residues and polysaccharides. OHC of flour is an important property required for flavor retention in many food applications. The basic mechanism for OHC is due to the physical entrapment of oil by capillary action.

Factors that may influence this property include the hydrophobicity, and the different proportions of nonpolar amino acid residues of proteins. OHCs for soybean flour was found significantly greater compared to *P. angularis*, *P. calcaratus*, and *D. lablab* flours.

Lablab had similar OHCs compared to chickpea, cowpea, green gram, and lentil. Emulsion properties of different legume flours are largely influenced by protein specificity, in addition to heating and aging of emulsions having a significant influence on emulsion flocculation stability and coalescence. Protein specificity is thus controlled by interactions involving adsorbed protein molecules. These may include formation of hydrogen, hydrophobic, and disulfide bonds. Emulsion formation are significant in product applications such as formation of frozen desserts, baking, and whiteners (Damodaran et al., 2008).

The influence of pH on the emulsifying activity (EA) of flours produced from *P. calcaratus*, *D. lablab*, and *P. angularis* was compared to soybean flour. At pH 4 minimum emulsion activity (45.80–54.20%) was recorded for flours produced, however at pH 2 (54.00–57.70%) and pH 10 (56.60–62.00%) there was a sizeable increase in activity. Minimum EA observed at pH 4 was attributed to the protein of the flours produced reaching their isoelectric points, which is around pH 4. Similarly, the emulsion stabilities (ESs) of flours produced were found to be pH dependent. Heating of different emulsions at 80 °C did not affect the ES at pH values investigated. The ES profiles of flours produced from *P. calcaratus*, *D. lablab*, and *P. angularis* showed similarities. At pH 4 the ESs (11.40–22.10%) of the mentioned flours were lower compared to soybean flour. In either side of pH 4, an increase in ES was noted mainly attributed to an increase in protein solubility (Chau and Cheung, 1998).

Food foams are formed mechanically or by supersaturation. In addition, the following is required for their formation: conformational change, rearrangement and rapid adsorption at the air–water interface during shear, and viscoelastic film formation through intermolecular interactions. Rapid adsorption at the air–water interface is essential for an increase in the capacity of foams and viscoelastic film formation, respectively. Viscoelastic film formed is said to assist in the stabilization of foam structure as time progresses (Damodaran et al., 2008).

Foam capacities (FCs) for flours were found pH dependent and higher around pH 4 ranging between pH 2 (106–111%) and pH 10 (114–122%). This was attributed to increases in protein flexibility allowing for diffusion to occur more rapidly at the air–water interface resulting in the encapsulation of air particles. Soybean flour had greater FCs and produced more stable foams over time when compared to the other flours studied. This was due to its higher protein content (Chau and Cheung, 1998).

4.5 *In vitro* starch digestibility characteristics

The concept of functional foods in promoting health has been developed during the past two decades. Food legumes and their bioactive properties [sources of resistant starch (RS), saponins, phenolic compounds, dietary fiber, oligosaccharides, antioxidants, and phytochemicals including phytates] have made them unique functional foods. Legumes have a high content of indigestible fibers and low glycemic index (GI) which aid diabetic individuals' glycemic control. In addition, they assist in the prevention of insulin-resistance representing the prodrome to type 2 diabetes.

The inhibition of α-amylase and β-glucosidase activity also known as the hypoglycemic effect has been reported similar for legumes and antidiabetic drugs. In addition, planning diets with the consumption of legumes has been associated with weight management and improvement of dyslipidemia, and attenuates postprandial glycemic response (Bahadoran and Mirmiran, 2015).

In the early 1980s, the GI was defined to serve as a numeric expression for the available carbohydrates in foods (usually containing 50 g available carbohydrate) on blood glucose concentration over a set period. There are three classifications that make up the GI index of foods, low (<55), moderate (56–69), and high (>70). Glycemic index is often expressed relative to a standard usually glucose or white bread. Food-related factors that have been reported to influence the glycemic impact of a food include the nature of carbohydrates available in food, in terms of the nature of starch and its hydrolysis products, physical properties (i.e., hydration properties and particle size), the contents of monosaccharides (galactose, fructose, mannose, and glucose), disaccharides (sucrose, lactose), and oligosaccharides (maltodextrins). In addition to food form and the degree of cooking, other food components such as dietary fiber, fat, protein, phytochemicals, and organic acids may have a significant influence on blood glucose response (Sadler, 2011).

The rate of starch hydrolysis in the digestive tract is an important factor on GI values, for example, amylopectin has been reported to undergo hydrolysis more rapidly in comparison to amylose by pancreatic α-amylase. In terms of starch, there are certain fractions that digest rapidly known as rapidly digestible starch (RDS), slowly digestible starch (SDS), and a fraction that resist digestion known as RS. Other factors that may have a significant influence on starch hydrolysis include the accessibility of digestive enzymes to starch.

This may be further influenced by the physical properties of food (i.e., structure, particle, and size) (Sajilata et al., 2006).

RS refers to the starch fraction that is not hydrolyzed by digestive enzymes in the small intestine. It passes to the large intestine where it is a substrate for bacterial fermentation. It is calculated as the difference between total starch and the sum of both SDS and RDS fractions. RDS and SDS are calculated as the amount of glucose (converted to starch) released between 0, 20, and 120 min of *in vitro* digestion (Singh et al., 2010).

Starch retrogradation prior to gelatinization serves as a mechanism for RS formation and reduction to digestibility, whereby processing techniques may affect both these processes. In terms of legumes, the crystalline structure (type C) is more stable to that found in cereal grains (type A). Thus, processing of cereal grains results in lower RS concentrations compared to legumes. It has been found that starches produced from several legumes subjected to steam treatment had resulted in higher yields of indigestible RS (19–31%, DM basis) compared to raw legumes and were three to five times higher than raw legumes. Furthermore, it was found that dry treatment resulted in higher RS contents compared to wet treatment in several legumes, cereals, and tubers (Sajilata et al., 2006).

Flour produced from *P. angularis*, *P. calcaratus*, and *D. lablab* legume grains subjected to different cooking times were examined by Cheung and Chau (1998), for RS and nonstarch polysaccharides (NSP). The whole legume grains were boiled in tap water for time intervals ranging between 30 and 120 min prior to drying and milling.

Findings revealed that the total dietary fiber contents of flour produced from grains that were processed was significantly higher than raw grains. The maximum percent increase in total dietary fiber were observed between 60 and 120 min of cooking time and ranged between 18.60% and 47.80%. In addition, cooking time was found directly proportional to concentrations of NSP and RS.

Increases in RS concentrations were attributed to the presence of cell-enclosed and retrograded starch formed during cooking. Proteins (exogenous or endogenous) have been found to significantly influence starch digestibility and functional properties in cereals and various food products, whereby exogenous proteins are said to possibly aid starch digestion (Tinus et al., 2012). Protein barrier present around starch granules was confirmed by pronase hydrolysis resulting in a significant increase in *in vitro* starch digestibility. This has made starch readily available for hydrolysis by digestive enzymes (amylase and amyloglucosidase). Endogenous proteins are said to hinder starch digestion as reported in sorghum and pasta. Starch granules in sorghum are reported to be encapsulated by protein bodies making sorghum less digestible (Singh et al., 2010).

Thus, starch digestion could be enhanced or reduced by proteins present in the food matrix. Therefore, understanding starch and protein digestion and nutrient synergy are essential for process and product designs. Piecyk et al. (2019) investigated the effects of processing on in vitro starch digestion of grass pea flour. The analysis of in vitro starch digestibility was carried out on raw flours. The raw flour slurries were heated for 30 min (cooked flour – CF). Cooked and frozen flour (C & FF) was prepared in the first stage as CF, then the samples were cooled for 1 h at room temperature and frozen (–18 °C, 21 days).

Prior to analyses samples were defrosted at room temperature for 2 h. Overall, the flour produced from cooked grass pea grains was found suitable for application in food products for decreasing GI values and to enrich them in RS. Flour samples produced by cooking resulted in a rapid increase in RDS and a decrease in SDS contents.

Following heat treatment in an aqueous environment increases to starch digestibility may be dependent on the denaturation degree of proteins. It has been reported that proteins form layers around starch granules. Flour samples produced by cooking caused a significant decrease in RS content (2.50%) and was attributed to the complete gelatinization of starch.

The cooling period after cooking (30 min) was said to have minimal effects on starch retrogradation. In CSF, higher RS values (12.50%) were found compared to CF. This value is low compared to rapidly digesting (56.00–65.50%), slowly digesting (5.10–9.20%), and RS (29.40–34.80%) contents found in Indian lentil (*Lens culinaris*) cultivars of flour (Kaur et al., 2010).

This was due to the retrogradation of starch during the drying of grains and was confirmed by the content of retrograded starch. The storage of flour after cooking and freezing was found to decrease RDS and cause a small increase in SDS contents. Thus, it was found that the RS content increased only in C & FF compared to CF. Sajilata et al. (2006) related dehydration of amylose rich legumes to high contents of RS. On drying, dispersed polymers of gelatinized starch undergo retrogradation to a semicrystalline form that is highly resistant to digestion by pancreatic α-amylase. Furthermore, when gelatinized starches are cooled/dried, a certain portion can retrograde to a less soluble form that is resistant to acid and amylase action leading to a fall in catalytic efficiency, resulting in low glycemic indices.

Starch hydrolysis was monitored between 0 and 180 min. There was a sharp increase in the content of digested starch in CF followed by an overall decrease in the rate of starch digestion. These differences were reflected in predicted glycemic index (pGI) value and were found significantly higher in CFs. Results conclude that high predicted glycemic values for CFs were largely affected by a low concentration of flour used, relatively long cooking period (30 min), and therefore a high degree of starch disintegration.

4.6 Conclusion

The underutilization of many indigenous crops is mainly attributed to both limited research and marketing. Thus, the application of traditional food processing technologies may still hold considerable potential on influencing the value of leguminous crops through the production of value-added legume-based food products. This may be accomplished through its application in foods that require protein enrichment such as in cereal-based foods and in the production of gluten-free products. In addition, this could help improve the competitiveness of the legume grain sector. Raw material intermediates produced from dehydrated legume grains may hold significance for ingredient application in texture-modified food products that require rehydration for preserving food structure or creating a new one to serve a functional purpose. These products may potentially be recognized for low glycemic responses which are vital for avoiding health illnesses. This presents potential for broadening segmentation of food processing technologies in existing markets for obtaining ingredients and products from legume grains with maximum quality.

References

Acevedo, B. A., Thompson, C. M. B., González Foutel, N. S., Chaves, M. G. & Avanza, M. V. 2017. Effect of different treatments on the microstructure and functional and pasting properties of pigeon pea (*Cajanus cajan* L.), dolichos bean (*Dolichos lablab* L.) and jack bean (*Canavalia ensiformis*) flours from the north-east Argentina. *International Journal of Food Science & Technology*, 52, 222–230.

Aguilera, J. M. (2005). Why food microstructure? *Journal of Food Engineering*, 67, 3–11.

Aguilera, Y., Esteban, R. M., Benítez, V., Mollá, E. and Martín-Cabrejas, M. A. (2009). Starch, functional properties, and microstructural characteristics in chickpea and lentil as affected by thermal processing. *Journal of Agricultural and Food Chemistry*, 57, 10682–10688.

Ahmad, I. and Noomhorm, A. (2013). - Grain process engineering. In: M. Kutz (Ed.), *Handbook of Farm, Dairy and Food Machinery Engineering (Second Edition)* (Chapter 10). San Diego, CA: Academic Press.

Alfy, A., Kiran, B., Jeevitha, G. and Hebbar, H. U. (2016). Recent developments in superheated steam processing of foods – a review. *Critical Reviews in Food Science and Nutrition*, 56, 2191–2208.

Bahadoran, Z. and Mirmiran, P. (2015). Potential properties of legumes as important functional foods for management of type 2 diabetes. A short review. *International Journal of Nutrition and Food Sciences*, 4, 6–9.

Benítez, V., Cantera, S., Aguilera, Y., Mollá, E., Esteban, R. M., Díaz, M. F. and Martín-Cabrejas, M. A. (2013). Impact of germination on starch, dietary fiber and physicochemical properties in non-conventional legumes. *Food Research International*, 50, 64–69.

Bhat, R. and Karim, A. A. (2009). Exploring the nutritional potential of wild and underutilized legumes. *Comprehensive Reviews in Food Science and Food Safety*, 8, 305–331.

Błaszczak, W., Doblado, R., Frias, J., Vidal-Valverde, C., Sadowska, J. and Fornal, J. (2007). Microstructural and biochemical changes in raw and germinated cowpea seeds upon high-pressure treatment. *Food Research International*, 40, 415–423.

Boye, J., Zare, F. and Pletch, A. (2010). Pulse proteins: processing, characterization, functional properties and applications in food and feed. *Food Research International*, 43, 414–431.

Bußler, S., Steins, V., Ehlbeck, J. and Schlüter, O. (2015). Impact of thermal treatment versus cold atmospheric plasma processing on the techno-functional protein properties from Pisum sativum 'Salamanca'. *Journal of Food Engineering*, 167, 166–174.

Chau, C. F. and Cheung, P. C. K. (1998). Functional properties of flours prepared from three Chinese indigenous legume seeds. *Food Chemistry*, 61, 429–433.

Chávez-Murillo, C. E., Veyna-Torres, J. I., Cavazos-Tamez, L. M., de la Rosa-Millán, J. and Serna-Saldívar, S. O. (2018). Physicochemical characteristics, ATR-FTIR molecular interactions and *in vitro* starch and protein digestion of thermally-treated whole pulse flours. *Food Research International*, 105, 371–383.

Cheung, P. C.-K. and Chau, C.-F. (1998). Changes in the dietary fiber (resistant starch and nonstarch polysaccharides) content of cooked flours prepared from three Chinese indigenous legume seeds. *Journal of Agricultural and Food Chemistry*, 46, 262–265.

Copeland, L., Blazek, J., Salman, H. and Tang, M. C. (2009). Form and functionality of starch. *Food Hydrocolloids*, 23, 1527–1534.

Damodaran, S. (2017). Water and ice relations in foods. In: *Fennema's Food Chemistry*. CRC Press.

Damodaran, S., Parkin, K. L. and Fennema, O. R. (2008). *Amino Acids, Peptides and Proteins, Fennema's Food Chemistry*. Boca Raton, FL: CRC Press/Taylor & Francis.

de la Rosa-Millán, J., Orona-Padilla, J. L., Flores-Moreno, V. M. and Serna-Saldívar, S. O. (2017). Physicochemical and digestion characteristics of starch and fiber-rich subfractions from four pulse bagasses. *Cereal Chemistry*, 94, 524–531.

Divekar, M. T., Karunakaran, C., Lahlali, R., Kumar, S., Chelladurai, V., Liu, X., Borondics, F., Shanmugasundaram, S. and Jayas, D. S. (2017). Effect of microwave treatment on the cooking and macronutrient qualities of pulses. *International Journal of Food Properties*, 20, 409–422.

Elhardallou, S. B., Khalid, I. I., Gobouri, A. A. and Abdel-Hafez, S. H. (2015). Amino acid composition of cowpea (*Vigna ungiculata* L. Walp) flour and its protein isolates. *Food and Nutrition Sciences*, 6, 790.

Fabbri, A. D. T. and Crosby, G. A. (2016). A review of the impact of preparation and cooking on the nutritional quality of vegetables and legumes. *International Journal of Gastronomy and Food Science*, 3, 2–11.

FAO. (2017). *The Future of Food and Agriculture – Trends and Challenges*. Rome.

Ghavidel, R. A. and Prakash, J. (2007). The impact of germination and dehulling on nutrients, antinutrients, *in vitro* iron and calcium bioavailability and in vitro starch and protein digestibility of some legume seeds. *LWT – Food Science and Technology*, 40, 1292–1299.

Kaur, M., Sandhu, K. S. and Lim, S.-T. (2010). Microstructure, physicochemical properties and in vitro digestibility of starches from different Indian lentil (*Lens culinaris*) cultivars. *Carbohydrate Polymers*, 79, 349–355.

Kerr, W. L. (2013). Food drying and evaporation processing operations. In: M. Kutz (Ed.). *Handbook of Farm, Dairy and Food Machinery Engineering (Second Edition)* (Chapter 12). San Diego, FL: Academic Press.

Leung, H. K. (2017). Influence of water activity on chemical reactivity. In: *Water Activity*. Routledge.

Lewicki, P. P. (1998). Effect of pre-drying treatment, drying and rehydration on plant tissue properties: a review. *International Journal of Food Properties*, 1, 1–22.

Li, G. and Zhu, F. (2017). Physicochemical properties of quinoa flour as affected by starch interactions. *Food Chemistry*, 221, 1560–1568.

Ma, Z., Boye, J. I., Azarnia, S. and Simpson, B. K. (2016). Volatile flavor profile of Saskatchewan grown pulses as affected by different thermal processing treatments. *International Journal of Food Properties*, 19, 2251–2271.

Ma, Z., Boye, J. I., Simpson, B. K., Prasher, S. O., Monpetit, D. and Malcolmson, L. (2011). Thermal processing effects on the functional properties and microstructure of lentil, chickpea, and pea flours. *Food Research International*, 44, 2534–2544.

Mubarak, A. E. (2005). Nutritional composition and antinutritional factors of mung bean seeds (*Phaseolus aureus*) as affected by some home traditional processes. *Food Chemistry*, 89, 489–495.

Osman, M. A. (2007). Effect of different processing methods, on nutrient composition, antinutrional factors, and in vitro protein digestibility of *Dolichos lablab* bean [*Lablab purpuresus* (L) sweet]. *Pakistan Journal of Nutrition*, 6, 299–303.

Piecyk, M., Worobiej, E., Wołosiak, R., Drużyńska, B. and Ostrowska-Ligęza, E. (2019). Effect of different processes on composition, properties and *in vitro* starch digestibility of grass pea flour. *Journal of Food Measurement and Characterization*, 13, 848–856.

Sabularse, V., Liuzzo, J., Rao, R. and Grodner, R. (1991). Cooking quality of brown rice as influenced by gamma irradiation, variety and storage. *Journal of Food Science*, 56, 96–98.

Sadler, M. (2011). *Food, Glycaemic Response and Health*. Belgium, Brussels: ILSI Europe.

Sajilata, M. G., Singhal, R. S. and Kulkarni, P. R. (2006). Resistant Starch – a review. *Comprehensive Reviews in Food Science and Food Safety*, 5, 1–17.

Seena, S. and Sridhar, K. R. (2005). Physicochemical, functional and cooking properties of under explored legumes, Canavalia of the southwest coast of India. *Food Research International*, 38, 803–814.

Shaahu, D., Carew, S. and Ikurior, S. (2014). Effect of processing on proximate, energy, anti-nutritional factor, amino acid and mineral composition of lablab seeds. *International Journal of Scientific & Technology Research*, 4, 1–4.

Shivachi, A., Kinyua, M., Kiplagat, K., Kimurto, P. and Towett, B. (2012). Cooking time and sensory evaluation of selected Dolichos (*Lablab purpureus*) genotypes. *African Journal of Food Science and Technology*, 3, 155–159.

Singh, J., Dartois, A. and Kaur, L. (2010). Starch digestibility in food matrix: a review. *Trends in Food Science & Technology*, 21, 168–180.

Tinus, T., Damour, M., Van Riel, V. and Sopade, P. A. (2012). Particle size-starch–protein digestibility relationships in cowpea (*Vigna unguiculata*). *Journal of Food Engineering*, 113, 254–264.

Vega-Mercado, H., Marcela Góngora-Nieto, M. and Barbosa-Cánovas, G. V. (2001). Advances in dehydration of foods. *Journal of Food Engineering*, 49, 271–289.

Velupillai, L. (1993). Parboiling rice with microwave energy. *Food Science and Technology-New York-Marcel Dekker-*, 263.

Olugbenga Philip Soladoye

5 Meat and meat products processing

5.1 Introduction

With increasing urbanization, higher income, and population increase, consumers' eating behavior is evolving globally and this consequently drives significant innovations within the meat and poultry industry. Consumers' quality demands have also broadened to include not only qualities related to eating (e.g., flavor and texture), safety, or nutritional values but also credence, convenience, and preference. Considering that consumer demands and quality requirements may vary across world regions, cultures, or even societal trends, it may be imperative to understand the fundamental principles of meat processing in order to satisfy consumer demands regardless of cultural or regional quality requirements.

Meat processing can be grouped into two major categories including primary and secondary processing (Figure 5.1). While the former generally involves animal carcasses or whole meat primals (and may be argued as not actual processing in themselves), the later involves various procedures in further processing. Invariably, the vast array of meat products available in the market would have gone through one or more of these processing procedures prior to market supply. Although this chapter largely focuses on discussing few of these primary and secondary meat processing categories, it is crucial to briefly review the aspects of meat quality attributes important for processing and technological quality of meat.

5.2 Quality aspects in meat and poultry processing

The quality of raw meat is an important aspect of the subsequent product's quality and consequently, consumer acceptability and liking. While nutritional, safety, and sensorial quality attribute are crucial for consumer products acceptability, the technological aspect of quality is also important for industrial delivery of high-quality meat products. This section discusses few quality attributes of meat that may be important in meat processing.

Olugbenga Philip Soladoye, Food Processing Development Centre, Food and Bio Processing Branch, Alberta Agriculture and Forestry, Leduc, AB, Canada

https://doi.org/10.1515/9783110667462-005

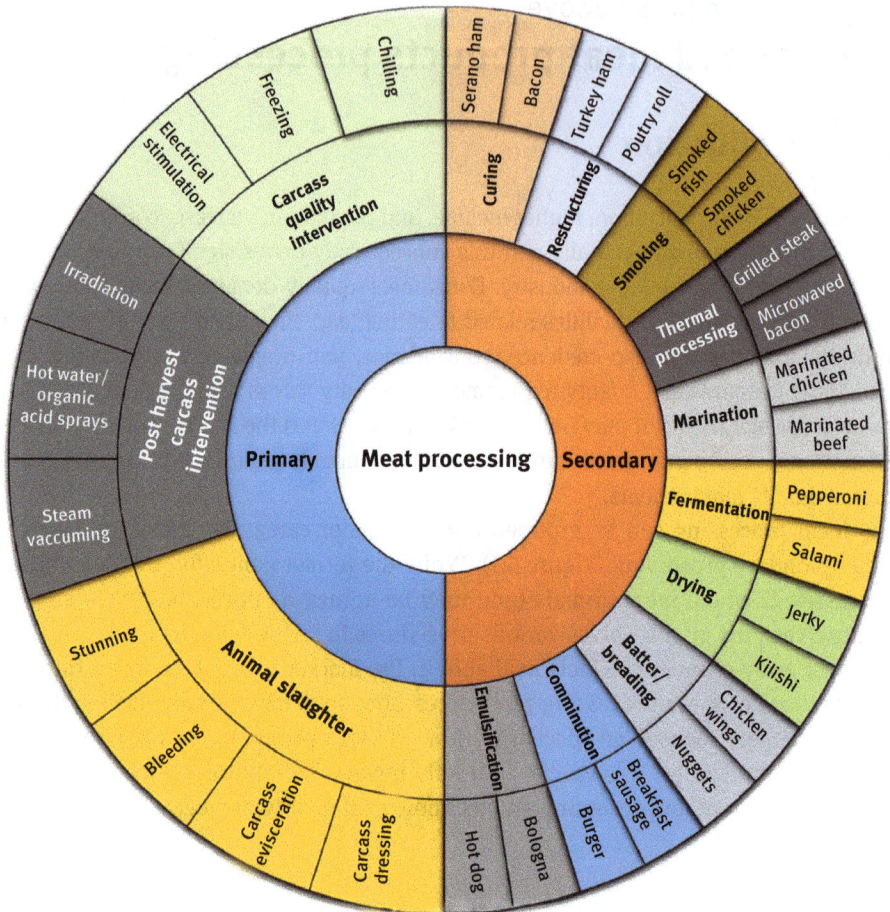

Figure 5.1: Meat processing method categories with corresponding meat products.

5.2.1 Meat composition and nutrients

Meat has been defined as the edible postmortem component derived from live animals (Kauffman, 2012) which may include both the wild (deer, bison, moose, etc.) and domesticated species (cattle, goats, sheep, poultry, hogs, among others) not excluding fishes as well. In fact, the American Meat Science Association defined meat as the skeletal muscle and associated tissues originating from mammalian, avian, reptilian, amphibian, and other aquatic species (Boler and Woerner, 2017). While this definition may vary across regulatory agencies, it will be mainly used in this text. Generally, the skeletal muscle which is about 50–70% of carcass weight of meat animal is made up of whole muscle enveloped in the connective tissue called epimysium. This muscle is further subdivided into fiber bundles separated by perimysium. Inside the bundles are individual muscle fibers/cells encased in endomysium. Within this muscle compartments are also fat layers, other connective tissues, vascular/nerve supply, and large amount of extracellular fluid, mainly water (Kauffman, 2012). Other types of muscles that could be used as meat include cardiac (heart) and smooth (gastrointestinal, respiratory, etc.) muscles (Figure 5.2).

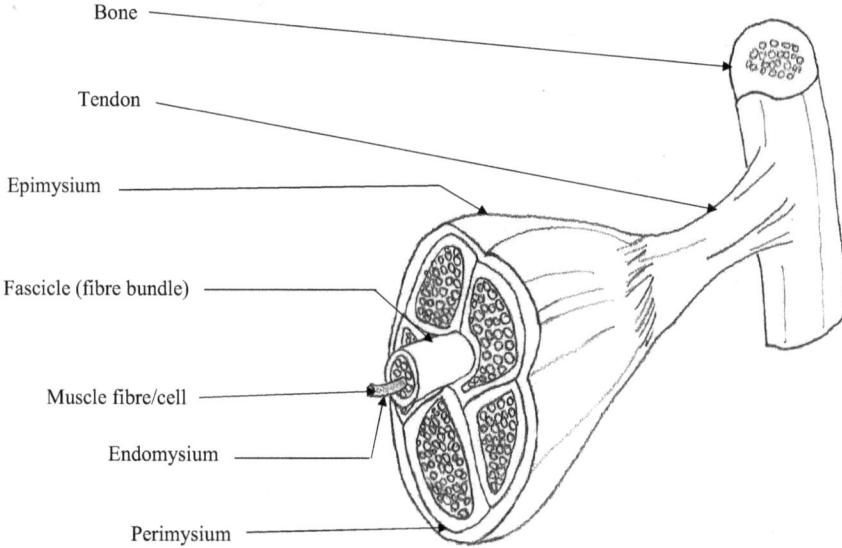

Figure 5.2: Structure of muscle.

Meat composition can be substantially variable depending on animal species, age, production systems and season, feeding regimens as well as specific retail cuts. Red meat can have between 20% and 25% high-quality protein with about 94% digestibility compared to 78% and 86% in beans and whole wheat, respectively (Williams, 2007). Fat content can range from 14% to 19% in beef retail cuts, 8% to 28% in pork, and 1% to 15% in poultry (Pereira and Vicente, 2013). The fatter the animal, the lower the water content in the muscle. For example, beef from matured cattle could have as low as 45% moisture content compared to 75% moisture in veal from a youthful, lean calf. In general, moisture content in lean component of red meat could be between 73% and 75% (Williams, 2007). With carbohydrate content less than 1%, other key nutrients delivered through meat consumption include vitamins (riboflavin, niacin, vitamin B6, pantothenic acid, vitamin B12, and vitamin D), minerals (iron, zinc, phosphorus, selenium, and copper), fatty acids (omega 3, polyunsaturated fatty acid, and conjugated linoleic acid), as well as important bioactive components (including taurine, carnitine, carnosine, anserine, creatinine, and ubiquinone) (Pereira and Vicente, 2013; Williams, 2007). These make meat an important component of a healthy and well-balanced human diet.

5.2.2 Meat color

Color attribute of meat is crucial for consumer's first appeal and subsequent purchase. While fiber and fat composition as well as other pigments such as hemoglobin and cytochrome C may play some role in the color of meat, the quantity and chemical state of myoglobin play the pivotal role (Castigliego et al., 2012). In turn, myoglobin content in meat may be influenced by intrinsic (e.g., animal species, breed, sex, age, and muscle type/function) as well as extrinsic (e.g., animal diet, environmental condition, transportation, and slaughter condition) factors (Castigliego et al., 2012; Lawrie and Ledward, 2014).

Myoglobin is a water-soluble globular protein that consists of a peptide (apoprotein) portion and a heme (prosthetic) group. The heme group, located within the protein's hydrophobic pockets, has a centrally located iron atom that has six coordination sites. Four of these sites keep the iron anchored within the planar heme structure by forming noncovalent bond with pyrrole nitrogen of the heme (Faustman, 2014). The fifth coordination site bonds with the proximal histidine-93 of the apoprotein and the sixth site is available to bind an array of ligands including oxygen, carbon monoxide, nitric oxide (NO), and water (Table 5.1). As such, the redox status of heme iron as well as the ligand bound to the sixth coordination site of heme determines the prevalent color of the meat (Castigliego et al., 2012; Faustman, 2014; Mancini and Hunt, 2005). Table 5.1 shows few forms of myoglobin that can exist in meat. Many other forms may exist by imparting varying color to meat and meat products. Readers are advised to

Table 5.1: Forms of myoglobin and their chemical state in meat color determination.

Form of myoglobin	Ligand at sixth coordination site of heme iron	Redox state of heme iron	Color of meat	Mode of formation	Example
Oxymyoglobin (OMb^{2+})	O_2	Fe^{2+}, ferrous	Bright red	Oxygenation of myoglobin	Bloomed fresh meat
Deoxymyoglobin (Mb^{2+})	No ligand	Fe^{2+}, ferrous	Purplish-red	Freshly cut meat	Vacuum packaged meat
Metmyoglobin (MMb^{3+})	H_2O	Fe^{3+}, ferric	Brown	Oxidation of myoglobin or oxymyoglobin	Stale/old meat

Table 5.1 (continued)

Form of myoglobin	Ligand at sixth coordination site of heme iron	Redox state of heme iron	Color of meat	Mode of formation	Example
Carboxymyoglobin (COMb^{2+})	CO	Fe^{2+}, ferrous	Bright red	Binding of carbon monoxide with myoglobin	Meat in modified atmosphere packaging
Nitrosomyoglobin (nitric oxide myoglobin) (NOMb$^{2+)}$	NO	Fe^{2+}, ferrous	Pink/red	Binding of nitric oxide with myoglobin	Cured meat
Nitrosylhemochrome (denatured NO-heme)	NO	Fe^{2+}, ferrous	Pink/red	Heat denaturation on nitric oxide myoglobin	Cooked cured meat

Table adapted from Faustman (2014) and Castigliego et al. (2012).

refer to various reviews and books that have discussed in detail the chemistry of meat color (Castigliego et al., 2012; Lawrie and Ledward, 2014; Mancini and Hunt, 2005; Suman and Joseph, 2013).

5.2.3 Meat texture

As previously alluded to by Juárez et al. (2012), tenderness and juiciness may be the two crucial components of meat texture since they both contribute to consumer overall experience during mastication. As such, both of these attributes will be discussed together in this section. Historically, meat tenderness has been regarded as the most important organoleptic properties affecting eating quality, consumer satisfaction, and repeated purchase (Bailey, 1972; Egan et al., 2001; Lawrie and Ledward, 2014; Miller et al., 2001). It is also believed to play a significant role in the value of the meat cut. Much recent data is, however, showing that no single palatability trait is most important as failure of any single eating quality trait may result in consumer penalty or rejection of the meat (O'Quinn et al., 2018). While meat tenderness is a function of various complex and interacting factors including preslaughter (genetics, species, muscle type, diet, handling, etc.) and postslaughter (pH, cooling/freezing, electrical stimulation (ES), packaging, cooking etc.) factors, it is also largely influenced by the composition of connective tissue (e.g., collagen and elastin), extent of myofibrillar protein degradation/integrity (actin, myosin, troponin, etc.), as well as the degree of contraction of the sarcomeres (Bhat et al., 2018; Bolumar and Toepfl, 2016). Singh et al. (2019) have added that the organization of perimysium (connective tissue) in muscle tissue contribute to meat background toughness whereas the action of proteolytic enzymes (e.g., μ- and m-calpain) can enhance meat tenderization postmortem. Several other methods have also been explored for meat tenderization including addition of exogenous enzymes (e.g., papain, bromelain) and chemical compounds (e.g., phosphates), mechanical disruption, heat treatment, and shockwaves treatment among others (Tomas Bolumar and Claus, 2017). Moreover, consumer perception of meat tenderness may be further altered by its moisture and fat content which contribute to lubrication during mastication in the mouth (Juárez et al., 2012).

Meat juiciness is a function of both fat and moisture release during chewing (Juárez et al., 2012). It plays a key role in meat texture perception contributing significantly to meat tenderness variability (Winger and Hagyard, 1994). While several objective methods have been used to evaluate meat tenderness (e.g., Warner–Bratzler shear force), the subjectivity of juiciness has made the relationship between the objective and subjective measure of meat texture unpredictable.

5.2.4 Water holding capacity

Water holding capacity (WHC) can be defined as the ability of meat to retain its own water under external influences (Huff-Lonergan and Lonergan, 2005; Swatland, 2002). These water fractions can include bound, entrapped/immobilized, and free water. WHC is an important attribute of meat with significant influence on yield and quality of end products and consequently, with substantial economic implications. In fresh meat, WHC is largely affected by the pH of the tissue as well as the space within and between the muscle cells and fiber bundles. Up to 85% of the water in muscle cell is held in the myofibrils. Numerous factors including genotype, preslaughter management (diet, fasting, lairage period, and stress), stunning method, chilling, ageing, and addition of nonmeat ingredient may influence these intrinsic attributes of meat, subsequently affecting its WHC (Apple and Yancey, 2013; Cheng and Sun, 2008). Several authors have elaborated on the mechanisms of WHC in meat (den Hertog-Meischke et al., 1997; Huff-Lonergan and Lonergan, 2005; Offer and Cousins, 1992; Offer and Trinick, 1983). Typically, the lowest WHC of meat is at the isoelectric point (pI) of its proteins (e.g., myosin $pI = 5.4$). At which point, both the positive and negative charges in the system balance out, resulting in a net zero charge, lowest electrostatic repulsion, and maximum number of bonds between peptide chains (Knipe, 2003). This reduction in intra- and intercellular spaces will result in higher drip, hence, diminished WHC. Denaturation or oxidation of muscle protein could also reduce protein ability to bound water and could result in reduced WHC. Proteolysis of cytoskeletal proteins (e.g., desmin, titin) has been found to enhance WHC in meat due to reduction in shrinkages transmitted through these proteins during rigor (Apple and Yancey, 2013; Huff-Lonergan and Lonergan, 2005).

The rate and extent of pH decline are largely associated with poor WHC and high purge or drip loss. Rapid decline in pH to near or at ultimate pH while the muscle is still warm will lead to protein denaturation and loss of functionality. This may result in pale, soft, and exudative meat with severe purge/drip loss. This condition is usually common to pigs with genetic condition referred to as porcine stress syndrome due to Halothane (HAL) gene. The HAL gene is a result of mutation in ryanodine receptor/calcium channel in the sarcoplasmic reticulum. This condition results in rapid rate of pH decline and poor WHC in pork meat. Other genetic factor that could affect WHC in meat especially pork is referred to as Hampshire effect due to Rendement Napole (RN) gene. This condition results in "acid meat" with lower than normal ultimate pH (typically around 5.3–5.4 compared to normal at 5.5–5.6 for pigs) hence, lower WHC and yield (Greaser and Guo, 2012). RN gene leads to accumulation of glycogen in muscles of pigs which subsequently results in lactic acid accumulation and thus, lower ultimate pH (Castigliego et al., 2012). Management conditions that could lead to stress for animals especially immediately prior slaughter may cause glycogen depletion in muscle, resulting in condition referred to as dark, firm, and dry. This genetic condition is also due to mutation in ryanodine receptor leading to meat with dark color, dry surface, firm texture and high WHC (Greaser and Guo, 2012). WHC of meat is a crucial technological quality important for many meat products' yield and processor's profitability.

5.2.5 Meat flavor

Another important aspect of meat eating quality is its flavor. Meat flavor is a function of an array of both volatile and nonvolatile compounds that develops during cooking from a complex interaction of several precursors derived from lean and fat components mainly through the Maillard reaction, lipids degradation, and the interaction of both (Mottram, 1998). Similar to other meat quality attributes, the flavor of meat relies largely on both preslaughter (breed, diet, age, etc.) and postslaughter (ageing, cooking, etc.) factors that can influence the level of precursors inherent in the muscle meat (Arshad et al., 2018; Brewer, 2007; Jayasena et al., 2013b). With all these reacting compounds and interactions, it is difficult to attribute the perceived meat flavor to a selected group of compounds. An increasing number of chemical compounds are being identified as contributing to the characteristic meat flavor of muscle foods.

Given that the flavor of meat is thermally derived, the lean portion, largely containing water soluble precursors such as free amino acids, peptides, nucleotide among other nitrogenous components, as well as carbohydrates (ribose, ribose-5-phosphate, glucose, and glucose-6-phosphate), has been reported to be responsible for the "meaty flavor" through the Maillard-type reaction (Brewer, 2007; Varavinit et al., 2000). The sulfur and carbonyl containing compounds in the volatile fraction of meat are also considered the key contributors to the "meaty" aroma (Brewer, 2007; Van Ba et al., 2012). Varavinit et al. (2000) have reported that the reaction between cysteine and sugar can lead to the characteristic meat flavors of pork and chicken. For example, 2-methyl-3-furanthiol which is largely believed to be responsible for the characteristic meat flavor in chicken broth is derived from the reaction between pentose sugar (ribose or inosine-5′-monophosphate, IMP) and cysteine or cystine or glutathione or thiamine (Aliani and Farmer, 2005; Jayasena et al., 2013b). The Maillard reaction steps (condensation, rearrangement, and dehydration) as well as the subsequent Strecker degradation of amino acids by dicarbonyl compounds will result in the production of an array of flavor volatiles including furfural, furanone derivatives, hydroxyketone, furans, pyrazine, pyrroles, oxazole, thiophenes, thiazoles, and other heterocyclic compounds which are important classes of flavor compounds in meat (Mottram, 1998; Van Ba et al., 2012).

Another major precursors of flavor in meat are derived from the lipids or water-insoluble components. These lipids could include the intramuscular fat, subcutaneous fat, intermuscular fat, intramyocellular lipids, and structural phospholipids, all of which may contain saturated, unsaturated, or methyl-branched fatty acids (Brewer, 2007). In fact, these lipids components are believed to be mainly responsible for the species-specific differences in flavor (Jayasena et al., 2013a; Perez-Alvarez et al., 2010) and the structural phospholipids are considered the most crucial in this respect due to higher proportion of unsaturated fatty acids. During heating, fatty acids may undergo oxidation and degradation producing intermediate hydroperoxides that later decompose through free radical mechanisms resulting in aliphatic hydrocarbon, aldehydes, alcohols, ketones, esters, lactone, carboxylic acids, and other aromatic hydrocarbons (Jayasena et al., 2013b). Although during prolonged storage, these reactions can cause objectionable rancid off-flavors but in cooked meat, the reactions occur rapidly and provide desirable flavor profiles (Mottram, 1998).

Many other important flavor compounds are also derived from the interaction of the products of Maillard reaction and lipid degradation. Examples of these include 1-heptanethiol in beef, 2-propyl-3-formyldihydro-thiopene in chicken, and 2-hexylthiopene in turkey (Melton, 1999). For more detailed reviews on meat flavors, readers are suggested with other previous literatures (Jayasena et al., 2013a; Jayasena et al., 2013b; Mottram, 1998; Van Ba et al., 2012).

5.3 Meat processing

It can be deduced from the preceding sections that both primary and secondary processing methods could influence the overall quality of meat and meat products and this can be further aggravated by the production/management practices. As mentioned previously, primary meat categories are those applicable to meat animals at or right after slaughter. These processes, which may involve the entire carcass, primal cuts, or the individual retail cuts, do not completely transform the raw meat into a different product but are important for quality of fresh meats as well as further processing. These primary processes have been grouped into animal slaughter, carcass quality intervention, and postharvest carcass intervention (Figure 5.1). The following section briefly discusses some selected primary processing methods.

5.3.1 Primary meat processing

Two operations widely involved in the slaughtering process are stunning/insensibilization and bleeding/exsanguination. Stunning is the process of instantaneously rendering animals insensible and unconscious to pain until there is a complete loss of brain responsiveness due to bleeding out or exsanguination (Velarde et al., 2003). While this is a legal requirement in most western countries, the impact of stunning on meat quality, bone fracture, and ecchymosis in animal carcasses has been a major issue of concern (Channon, Payne, and Warner, 2002). The three main stunning methods widely used in the meat industries are contusion (captive bolt, pneumatic pistol), electrical, and gases (CO_2). These different stunning methods must meet the requirement of easy application, minimum cost, no toxic residue, causing minimum physical damage with fast and prolonged unconsciousness without causing pain (Aguilar-Guggembuhl, 2012; Guerrero-Legarreta and de Lourdes Pérez-Chabela, 2012). Typically, contusion stunning is applied to cattle and sheep, electrical stunning to pigs, sheep, and poultry while gas stunning to pigs and poultry (Aguilar-Guggembuhl, 2012). The underlining principle of contusion stunning is rapid concussion, resulting in trauma to the brain by a blow or skull penetration. Electric stunning occurs in three stages: tonic, clonic, and recovery. The electric flow passing through the brain results in the release of ample neurotransmitters (glutamate and aspartate) which lead to epileptiform seizure and insensibilization (Aguilar-Guggembuhl, 2012; McKinstry and Anil, 2004).

Although other options for gas stunning have been suggested (e.g., argon or mixture of nitrogen and carbon dioxide) (Llonch et al., 2012), CO_2 is still the most common gas stunning method. It works by blocking the animal's neural terminal, reducing the nervous impulses. Meat and carcass qualities are affected differently by the different stunning methods. CO_2-stunned animal may result in less hemorrhages and blemishes (Channon et al., 2002), more tender and possibly better quality meat compared to meat from electric-stunned animals (Vergara et al., 2005). Exsanguination, which is the process of severing blood vessels in the neck of the animal, proceeds following stunning to prevent animal recovering. This is crucial for animal welfare concerns, religious restrictions, and overall meat quality.

Chilling and freezing are another important primary processing procedures with significant effects on meat safety, yield, and eating quality. This is because the rate of carcass heat transfer can have implications on meat microbial growth, WHC, evaporative loss, enzymes' activities, ATP depletion, onset of rigor mortis, and pH decline

(Savell, Mueller, and Baird, 2005; Warriss, 2000). Carcass temperature immediately postslaughter is usually around 37–39 °C and if not lowered quickly, microorganisms may rapidly grow on the carcass resulting in impaired shelf life and safety concerns. Warris (2000) has also shown that the rate of pH decline in beef muscle is minimal at 10 °C. However, the activation of actomyosin ATPase may lead to increases in the rate of pH decline closer to 0 and 37 °C. The depletion of ATP which has repercussion on the onset of rigor mortis is also temperature dependent. The delay in depletion has been reported to be at maximum at about 10–15 ° C. Among the different chilling methods available are spray chilling, rapid chilling (e.g., blast chilling), conventional chilling (batch chilling), delay chilling, and short-duration chilling (Xu, Huang, Huang, Xu, and Zhou, 2012). While rapid chilling or freezing has been found economically viable in the meat industry due to reduced shrink, drip loss, as well as increased product turnover, it has however, been found to result in reduced tenderness due to cold shortening and thaw rigor (Aalhus et al., 2001). Cold shortening ensues when meat is cooled below 10 ° C before the onset of rigor mortis which results in contraction of muscles, hence tougher meat (Warriss, 2000). Similarly, when the rate of cooling is sufficiently high such that muscle freezes before the onset of rigor, during subsequent thawing, the muscle shortens severely, resulting in significant drip loss and toughness. Hence, it is advisable that the temperature of beef carcass for instant does not go lower than 10 °C before the pH has reached 6.0.

An effective way used to prevent the negative effect of rapid chilling in muscle is electrical stimulation (ES). Suitable application of extra low (<100 V), medium (between 100 and 110 V), or high voltage to animal carcass has been found to enable faster chilling without the negative implication on tenderness (Adeyemi and Sazili, 2014). The basis of ES is to fast-tract postmortem glycolysis, with rapid depletion of glycogen, rapid pH fall resulting in the rapid onset of rigor mortis. Aside from the effect of ES in meat tenderization, it has also been reported to improve meat appearance and flavor (Warriss, 2000). Because cold shortening has not been a major issue with pork carcasses due to more rapid rigor onset, ES has been generally applied more to lamb and cattle carcasses.

Carcass irradiation is another primary processing aid that could be employed in the meat industry. Food irradiation is the process whereby food products are exposed to controlled amount of ionizing radiation to control spoilage and enhance safety. This is because controlled doses of ionizing radiation can effectively eliminate pathogenic microorganisms (e.g., *Salmonella*, *Staphylococcus*, *Campylobacter*, *Listeria monocytogenes*, and *Escherichia coli* O157:H7) with minimal impact on sensory, nutritional, and technological qualities (Farkas, 1998). Generally, a dose of 1. 0–10.0 kGy has been shown to be effective in reducing pathogenic levels in food products as well as on animal carcasses without any hazardous health effects (Arthur et al., 2005; Lawrie and Ledward, 2014). In the United States, between 0.3 and 1.0 kGy and less than 3, 4.5, and 7.5 kGy has been approved for the treatment of pork, poultry, fresh meat, and frozen meat, respectively. The three types of ion-izing radiation approved by the Codex General Standard for Irradiated Food com-mercially for food products include high-energy X-rays (from machine source operated at or below 5 MeV), γ-rays (from Cobalt-60; ^{60}Co or Cesium-137; ^{137}CS) and accelerated electrons (from machine source operated at or below 10 MeV) (Badr, 2012). While there is still a huge push back from consumers against this technology due to concerns regarding nutritional loss, chemical modifications, flavor change, unknown risk, and ionizing radiation residues, many countries around the world have approved its application in food. This is because, irradia-tion has several advantages over conventional pasteurization/sterilization/chemi-cal treatments including (1) no potentially harmful residue; (2) minimal heating or processing severity, hence, low impact on nutrient degradation; and (3) can be conveniently used in packaged foods with less subsequent cross-contamination. There has been a growing interest in the application of this technology in the meat industry as it is believed to be generally safe, efficient, environmentally clean, and energy efficient (Farkas, 1998).

Other primary processing methods that may also impact product's quality and value include evisceration, carcass cutout, as well as carcass dressing. Evisceration is the process of separating the visceral organs from the carcass. This is important as it may result in contamination of the carcass if not properly done. The effective-ness and efficiency of carcass cutout and dressing will also have huge economic implications on carcass value and industrial profitability.

5.3.2 Secondary meat processing

The processes that entirely transform whole meat into different meat products are discussed as secondary meat-processing procedures in this text. Some of these pro-cesses are listed in Figure 5.1 with the examples of meat products that could result. The following section will discuss some of these secondary meat-processing methods.

5.3.2.1 Secondary processing with physical structure modification

Structural transformation is a common practice during meat processing. These may include particle size reduction or products reformation. The main purpose of this category of secondary processing may include ingredients reformulation (e.g., aimed at flavor enhancement or mitigating public health concerns), textural modifi-cation, and subsequent value addition. Few of the structural modification processes employed in the meat industry are briefly discussed further.

5.3.2.1.1 Emulsification

Emulsions are two-phase colloidal systems in which one liquid is dispersed (dispersed phase) in another liquid (continuous phase) (Ugalde-Benítez, 2012). Typically, there are two immiscible liquids, one aqueous phase and another oil phase, and the process of transforming these two phases into an emulsion, or reducing oil globules size in previously formed emulsion is called homogenization (McClements, 2015). While this may not fit into the definition of true emulsion, finely chopped meat mixture is conventionally called meat emulsion (Knipe, 2003). Meat emulsion is generally not considered a true emulsion because it consists of solid fat particles dispersed in mixture of water, proteins, connective tissues, and muscle fiber rather than a stable suspension of two liquids (oil and water) and the oil droplets in meat emulsion can also be larger than 100 μm compared to only between 0.1 and 100 μm in true emulsions (McClements, 2015). Owusu-Apenten (2004) has further elucidated that the water, protein, and fat component form the continuous, emulsified, and dispersed phase in meat emulsion, respectively. Chopping is one of the most crucial steps during meat emulsion production. During this process, the raw materials are extensively comminuted in a bowl chopper in order to reduce the size of the particles to attain a stable and homogeneous emulsion (Álvarez et al., 2007). Salt extracted myofibrillar as well as sarcoplasmic proteins enhance the stability of this meat emulsion systems (Ugalde-Benítez, 2012). Most emulsified meat products are distinguishable by their smooth appearance being finely comminuted and they can be either sliceable (e.g., bologna, wieners, mortadella, frankfurters, and meat loaf) or spreadable (e.g., pâté, terrine, roulade, and galantine) products (Ugalde-Benítez, 2012).

5.3.2.1.2 Comminution

While emulsified meat products are finely comminuted, other ground products have particle sizes that range between 0.5 and 2.5 cm (Barbut, 2015). The main purpose of the comminution process is to cut meat down to some required particle sizes and this can be in form of flaking, slicing, or grinding. Comminution is a very common or even a preceding step in many meat products that are discussed in this text. In fact, comminuted meat products have been identified as a significant contributors to consumer fat and salt intake largely due to their prevalence in human diets (Bolger et al., 2017), as well as making it a perfect focus for public health intervention aiming to reduce salt and fat content in consumers' diet. In the United States, half of the beef marketed is in the form of ground beef (Lonergan, Topel, and Marple, 2019). In most cases however, meat cuts used for comminuted products are the less tender yet flavorful cuts (e.g., chuck or rounds) which can benefit from added value through comminution. Mechanically deboned meat or mechanically separated meat (MSM) or mechanically recovered meat are all synonymous terms describing the product obtained from the process of removing meat from flesh-bearing bones following deboning or from poultry carcasses, using mechanical means and resulting in loss of the muscle fiber structure (EFSA, 2013). MSM is sometimes used in the formulation of comminuted meat products. Examples of comminuted meat products include beef burger, meatballs, chicken hamburger, breakfast sausage, pepperoni sticks, and salami among others.

5.3.2.1.3 Restructuring

Restructured meat was developed in a bid to transform lower value cuts and trimmings into products of higher quality and value (Boles, 1999). Restructuring is defined as the utilization of manufacturing steps to create a consumer-ready product which resembles an intact steak, chop, or roast (Sun, 2009). As such, any meat products that are partially or completely dissembled and then reformed into the same or different forms are referred to as restructured meat products (Pearson and Gillett, 1999). Basically, three procedures are used for meat restructuring: (1) chunking and forming, (2) flaking and forming, and (3) tearing and forming (Pearson and Gillett, 1999). Chunking and forming procedure maintains the structure and texture of the original meat cuts, whereas flaking and forming only maintains similar, but not identical texture to intact steaks, roast, or chop (Sun, 2009). The tearing and forming procedure is superior in that it causes less membrane damage (hence, less autooxidation), and maintains the structural integrity and ensures texture that closely resembles that of the intact meat cuts (Sun, 2009). Among factors that can affect the acceptability and quality of restructured meat products are appearance/color, particle size, meat type, comminution type, mixing time, fat content, and added ingredients.

Common steps in meat restructuring process are: particle size reduction, addition of ingredients for binding, mixing of ingredients, forming, and, lastly, allowing time for bind formation (Boles, 1999). The ingredients used may vary depending on whether if hot- or cold-set procedure is employed. The hot setting is a conventional process that employs salt, phosphate, and the mechanical process. In cold-setting procedure, ingredients including Activa[TM], alginate, and Fibrimex[TM] are used. Other ingredients used in meat restructuring include transglutaminase, soy protein, and κ-carrageenan (Sun, 2009). Examples of restructured meat products are turkey ham, poultry roll, cooked duck tenderloin, and turkey luncheon roll (Barbut, 2015).

5.3.2.2 Secondary processing with thermal processing

Another secondary meat-processing method widely used is the thermal processing. While the main objective is the destruction of most pathogenic and spoilage microorganisms as well as the inactivation of enzymes, it also functions to modify the physical and sensorial attributes of meat products (Guerrero-Legarreta and García-Barrientos, 2012). Among several thermal processing techniques used in the meat industry, some are drying, smoking, scalding, frying, roasting, grilling, stewing, pasteurization, and sterilization. Except for meat products that are dry aged for prolonged period of time, thermal treatment is a very pivotal process to ensure safe, palatable, and nutritious products. The impact that each of these cooking methods could have on products' quality, palatability, safety, nutrients availability, and production of chemical compounds with detrimental effect on human health may vary as such, each cooking method should be evaluated individual. Only few will be discussed in this section.

5.3.2.2.1 Drying

Drying is a process of meat preservation whereby moisture is evaporated or sublimed from the meat products by exposure to heat (e.g., solar and microwave) and air stream (Andrés, Barat, Grau, and Fito, 2008). The consequences of this is the reduced water activity (a_w) which impedes microbial growth and biochemical reactions, hence, higher shelf life (Santchurn et al., 2012). While drying can be part of other meat-processing methods especially in the developed countries (e.g., fermentation and smoking), in most developing countries, drying is still solely and widely used traditionally. For example, sun drying, which is among the most ancient meat preservation method, is still widely used in southern countries (Santchurn et al., 2012). Based on the means of heating (convection, radiation, dielectric, or conduction), different types of dryers are available. The convective hot-air dryer seems to be very popular in the food industry functioning in both batch and continuous modes. Sebastian et al. (2005) have also investigated the combination of radiative and convective heat transfers for meat drying.

According to Andrés et al. (2008), the efficiency of drying is sustained only when the heat infiltrates the meat center, forcing moisture to the surface or the region of lower vapor pressure for evaporation. As such, it can be concluded that the moisture evaporation from the meat is a function of the water activity gradient of the forced hot air and the meat. While the typical drying curve will show the induction, constant rate, and the falling rate periods, the moisture sorption isotherm is a sigmoidal curve that describes the equilibrium relationship between water content and the a_w of a product. Other variables that affect drying of meat are air-related (e.g., temperature, velocity, flow characteristics, and humidity) and product-related (size, shape, moisture, structure, etc.) variables (Andrés et al., 2008). Aside from preservation, drying also results in improved sensory properties, harder texture, and decreased mass and volume. Examples of dried meat products include biltong (South Africa), Pastirma (Middle East), Jerky (North America), Kilishi, and Kundi (Nigeria).

5.3.2.2.2 Smoking

Meat smoking involves the use of smoke in combination with drying and/or heating to produce smoked meat products. While historically, smoking was intended to extend meat shelf life and prevent food poisoning, over the years, its purpose has evolved to include enhancement of flavor and other sensory attributes (Sikorski and Sinkiewicz, 2014). The smoke typically comes from smoldering of wood chips or saw dust applied either directly under the meat product or from an external smoke generator (Sikorski and Kolakowski, 2010). The composition of the smoke, which typically varies depending on the species of wood and the smoldering parameters, may include phenols, carbonyls, alcohols, polycyclic aromatic hydrocarbons (PAHs), and acids. These compounds impart color, flavor, as well as antimicrobial and antioxidant properties to meat product. Aside from these, some of these compounds (e.g., PAH) may have some carcinogenic effect on human. Generally, traditional smoking can either be cold or hot smoking. The former usually lasts days or weeks at temperature of about 20–25 °C with relative humidity (RH) of 70–80% while the latter is usually at 75–80 °C with high relative humidity and mostly for previously cooked or blanched products (Andrés et al., 2008). Another form of smoking employed in the meat industry is the electrostatic smoking. In this procedure, electrically charged smoke particles are driven toward the meat products by electrostatic force, creating an electric wind which carries the uncharged components of the aerosol in the same direction (Sikorski and Sinkiewicz, 2014). For purpose of convenience and reduced cost, many processors have employed liquid smoke in meat-processing (Herring and Smith, 2012). Examples of common smoked meat products are bacon, country ham, and pastrami among others.

5.3.2.3 Secondary processing methods intended for sensory and safety enhancement

5.3.2.3.1 Marination

Marination is a value addition technological treatment employed to improve both the sensory and functional attributes of meat, either by immersion, injection, or vacuum tumbling, using aqueous solution containing varying ingredient mixtures (Vlahova-Vangelova, Dragoev, Balev, Assenova, and Amirhanov, 2017). While traditionally, marination is used to extend meat shelf life, it is now widely used to enhance meat flavor, improve tenderness/juiciness, and enhance meat yield (Alvarado and McKee, 2007; Maurizio, Massimiliano, and Claudio, 2010). Based on composition, meat processors may use marinade of either alkaline solution, acid solution, or water–oil emulsion (Vlahova-Vangelova et al., 2017) and this has implications on their functionalities. Salt and phosphates are common ingredients in alkaline marinades. Salt results in transverse expansion of myofibrils due to electrostatic repulsion as well as partial protein solubilization. These promote water uptake and retention in meat (Maurizio et al., 2010). Aside from these, salt enhances meat's flavor and shelf life (Khan et al., 2016). The phosphates help with enhanced water holding due to the shifting of isoelectric point of the muscle protein and protein unfolding (Khan et al., 2016).

Acid marinades (pH < 5) have been found very useful for their antimicrobial qualities and their ability to tenderize meat by protein denaturation. Acid marinades usually contain organic acids including citric, lactic, or acetic acids among other acid blends such propionic, octanoic, and peroxy acids (Smith, 2012). The limitation of acid marinade, however, include negative impacts on color, flavor, WHC, as well as odor (Alvarado and McKee, 2007; Smith, 2012). On the other hand, water–oil emulsion marinades may contain salt, sugar, vegetable oil, vinegar, or citric acid among other supplements (Vlahova-Vangelova et al., 2017). Spices and herbs could also be added into marinade to enhance flavor and aroma while also complementing the meat's shelf stability.

5.3.2.3.2 Curing

The origin of meat curing is lost in ancient history but it is believed to have resulted through the traditional process of the addition of saltpeter to meat for the purpose of preservation (Binkerd and Kolari, 1975). The nitrate in saltpeter was considered an "impurity" which was later discovered helped to fix the red color in the preserved meat (Ramarathnam and Rubin, 1994). More recently, however, meat curing has evolved into a method of flavor and color development through the addition of salt and nitrite/nitrate to meat. Recent research advancement has further revealed that the color fixation of curing salt is mainly due to nitrite that resulted from the bacterial degradation of nitrate (Martin, 2012). As such, nitrite has now been used almost exclusively and approved in most cured meat products. However, nitrate is still used in some dry cured or fermented products that require extended curing or fermentation time. Among the ingredients used in modern day curing are nitrite, salt, sweetener, ascorbates/erythorbates, phosphate, and, where necessary, appropriate seasonings.

The importance of nitrite in cured meat includes the cured color, cured flavor, and aroma as well as the antioxidant and antimicrobial function (Martin, 2012; Gary Anthony Sullivan, 2011). Nitrite produces nitrous acid (HNO_2) and NO. The typical cured meat color is a result of the coordinate–covalent complex of NO with meat myoglobin to form nitrosomyoglobin and when this complex is heated, a stable nitrosylhemochrome is subsequently formed (Castigliego et al., 2012). Due to its possible impact on human health in excess amount, ingoing nitrite or nitrate concentration is regulated in meat products. For instance, United States Department of Agriculture (USDA) established regulatory limits for sodium nitrite at 120 ppm (0.012%), 200 ppm (0.02%), and 156 ppm (0.0156%) in bacon, dry-cured bacon/brine cure or injected products, and frankfurters/cured sausage, respectively (Gary Anthony Sullivan, 2011). Also, in the European Union (EU), the maximum limit of nitrite and nitrate set in traditional cured or dry cured products is 175 and 250 mg/L or mg/kg, respectively (FSAI, 2019).

Among other ingredients used in the curing process is salt (mostly sodium chloride). Salt is crucial in that it drives the curing process by acting as flavor enhancer, serving as antimicrobial and preservative as well as functional ingredient that solubilizes muscle proteins (Ramarathnam and Rubin, 1994). Salt is self-limiting as such, not regulated in meat products. Sweetener, mainly sucrose, dextrose, and corn syrup are also used in the curing process to limit the harshness of salt and provide overall balanced flavor, even through the Maillard reaction (Martin, 2012). While sugar can also function as a preservative, the level required to achieve this function may render the product too sweet and unacceptable to consumers. Being a self-limiting ingredient, its inclusion level is not regulated in meat products. Nitrosamines are carcinogenic compounds formed by the reaction of secondary and tertiary amines with a nitrosating agent. According to Scanlan (1983), nitrosating agent is typically nitrous anhydride produced from nitrite in acidic aqueous solution. To prevent the formation of nitrosamine in meat products, cure accelerators are used in meat curing process. These are reducing agents such as ascorbates and erythorbates (e.g., ascorbic acid, sodium ascorbate, erythorbic acid, and sodium erythorbate) which enhance the rapid reduction of nitrates and nitrites to NO. To this end, it is a requirement by the USDA for injected or brine cured bacon to be produced with the addition of 550 ppm sodium erythorbate or ascorbate when nitrite (not nitrate) is added (Sullivan, 2013). Similarly, the EU also limits the usage of sodium erythorbate or ascorbate in cured and preserved meat products to 500 mg/kg or mg/L (FSAI, 2019). Phosphate is helpful with water retention and hence, juiciness and mouthfeel of cured meat. It is mostly used to reduce shrinkage and purge loss in cured meats through its action on pH elevation and protein solubilization. The maximum allowable level permitted in meat products is 0.5% according to the Canadian Food Inspection Agency (CFIA, 2013). When higher levels are used for phosphates, products could have rubbery texture and soapy flavor (Sullivan, 2013). Although some have been shown to have antimicrobial activity, spices and herbs are generally added for aroma and flavor and from a regulatory standpoint, are self-limiting. Examples of cured meat products are ham, sausages, bacon, prosciutto, pancetta, pepperoni, salami, and chorizo.

5.3.2.3.3 Fermentation

Fermentation of meat is an ancient practice intended for meat and meat product preservation. It is a process where meat is subjected to the action of microorganisms or enzymes resulting in significant biochemical changes and modification to food sensorial attributes including enhanced flavor, aroma, color, palatability, and microbial safety among other desirable attributes (Lücke, 1994). Fermented meats are either inoculated or overgrown with safe to eat microorganisms whose enzymes including amylases, lipases, and proteases hydrolyze proteins, polysaccharides or lipids to produce compounds with desirable flavor and aroma (Singh, Pathak, and Verma, 2012). Traditionally, meat fermentation relied on natural microbiota although in modern processing, pure microbial starter cultures are added. In both cases, lactic acid bacteria (LAB) are the primary microorganism involved in the fermentation. While the LAB results in the acidulation (lowering of pH) of the meat product due to the production of lactic acid from added sugar or reserved glycogen, the added salt and drying process results in reduced water activity (a_w). These hurdles enhance the preservation of these products (Ockerman and Basu, 2008; Petäjä-Kanninen and Puolanne, 2007). Other microbial groups important for the fermentation process include micrococci, enterococci, staphylococci, as well as yeast and mold (Paramithiotis, Drosinos, Sofos, and Nychas, 2010; Yılmaz and Velioğlu, 2009).

Dry-cured hams and sausages are examples of fermented meat products which are largely produced in the Mediterranean countries including Spain, Italy, France, and Portugal due to their favorable climatic conditions. As such, the quality of naturally fermented meat products is a function of the quality of the ingredients and raw materials, the exact conditions of the processing and ripening, and the composition of the microbial population (Yılmaz and Velioğlu, 2009). In other words, both intrinsic and extrinsic factors have to be controlled to achieve the desirable products. For example, the production of dry-cured sausages is carried out in a temperature- and humidity-controlled room where, depending on the microbial population, the first 1–2 days of fermentation is kept at temperature of 25–27 °C and RH of 65–75%. This condition especially favors the rapid growth of LAB responsible for the acidification. This process is followed closely by the drying process that takes between two and three days at temperature of 16–22 °C and RH of 55–65% (Cocolin and Rantsiou, 2012). The final stage of this procedure is the ripening process that ensues for couple of weeks to several months at temperature of 13–15 °C and RH of 75–90% (Cocolin and Rantsiou, 2012). Other factors important for monitoring during fermentation process include pH, water activity, redox potential, light intensity, and wavelength among others (Yılmaz and Velioğlu, 2009). Well-known examples of fermented meat products across the world are Sucuk (Turkey), Salami (Italy, Germany, France, Hungary, Nordic, etc.), Saucisson (French), Kantwurst (Austria), Lup cheong (China), Chorizo (Mexico, Spain), Salchichon (Spain), Fuet (Spain), Pepperoni (Canada, USA), and Summer sausage (USA) (Ockerman and Basu, 2008; Yılmaz and Velioğlu, 2009).

5.3.2.3.4 Batter and breading

There has been a recent surge in the demand for breaded-meat products worldwide. Breading is a thermally processed cereal-based coating, including seasonings and other ingredients, applied to the surface of meat products for enhanced texture, flavor, color, and appearance (Brannan, 2008). Breading also retains moisture, prevents lipid absorption, and preserves the nutrient content of the final products (Jackson et al.,

2009). Several other processes may precede breading during meat processing. These may include predusting and batter application. Predusting is an optional process that involves the application of flour to the wet surface of meat product in order to enhance the adhesion of batter to the meat surface (Mah, 2008). This process, when applied, may also be used to convey seasonings and spices as this layer will be protected during frying and subsequent heat treatments (Barbut, 2015). Afterward, the product is coated with liquid batter which consists of suspension of dry ingredients, namely, flour, starch, protein, and flavoring. The adherence of batter to the meat surface is crucial and the amount of batter that adhered to the surface of food product is called *pick up*. While a direct relationship exists between viscosity of the batter and the thickness of the layer it can form, the fluidity of the batter directly influences the uniformity of the layer it forms (Fiszman and Sanz, 2010). Depending on functions, several types of batter exist including adhesion, cohesion, and tempura batters. Details of these batters are explained in other literatures (Barbut, 2015; Fiszman and Sanz, 2010).

Different examples of breading available are flour, American bread crumbs/home-style, cracker-type crumbs/traditional, and Japanese-styled crumb/panko (Brannan, 2008; Sanz and Salvador, 2012) and these vary in granulation, composition, porosity, and structure (Mah, 2008). Other types of breading include fresh crumbs and mixture of/with seeds and grains (Barbut, 2015). The breading material is usually composed of flour, gum, seasoning, colorant, leavening agent, flavor, and modifying agent among other ingredients that are prepared and baked prior to using for breading. Examples of breaded-meat products include chicken nuggets, chicken breast, chicken drumstick, and breast fillet.

5.4 Future outlook in meat processing

In recent years, consumers more intensely scrutinize their diets and are more interested in the impact of their diets on their health. This is hugely driving the trend toward flexitarian diets, mildly processed foods as well as sustainable food production. As such, given the significant changes that accompany thermal meat processing, consumers are more recently demanding less severely processed meat products. For example, application of smoking in meat processing can impart PAHs that have been identified as carcinogens (Stumpe-Vīksna, Bartkevičs, Kukāre, and Morozovs, 2008). Also, pan-frying can potentially result in more formation of heterocyclic aromatic amines in meat products (Oz, Kaban, and Kaya, 2010; Soladoye et al., 2017). Sous vide is a method of cooking food in heat-stable vacuumized pouches at precisely controlled temperature and time (Baldwin, 2012; del Pulgar et al., 2012). At this condition, the food is cooked at lower temperature (usually lower than 100 °C) and longer times (up to 48 h) (Oz and Zikirov, 2015) exposing it to less severe heat treatment and less production of unhealthy chemical compounds. Another alternative mild processing technology with potential in the meat industry is the ohmic heating. Ohmic (or Joule) heating is a process of heating food by passing electric current through them. This results in uniformly and rapidly heated food that is microbiologically safe and of high quality with less impact on its nutritional and sensorial attributes (Varghese, Pandey, Radhakrishna, and Bawa, 2014; Yildiz-Turp, Sengun, Kendirci, and Icier, 2013).

Another technology that may have potential in meat processing is the high-pressure processing (HPP). HPP allows the decontamination of foods through the application of pressure in the range of 100 and 900 MPa to an already vacuum packaged food placed in a vessel containing a pressure transmitting liquid (water or another aqueous solution) (Campus, 2010). HPP treatment results in microbial inactivation through the disruption of their cellular functions albeit without any impact on covalent bond hence, minimal effects on food chemistry (Muntean et al., 2016). Among the advantages that have been reported for this technology include (1) meat being processed at ambient or lower temperature; (2) no pressure gradient within meat and meat product, hence, uniform treatment of meat irrespective of size and shape; and (3) less degradation of nutrients and overall quality. Although at the moment, this technology has not been practically applied to animal carcasses immediately postslaughter possibly due to its principle of operation and amenability to slaughter line, it has however been applied to raw meats and other meat products (Simonin et al., 2012; Soladoye and Pietrasik, 2018). Furthermore, while this technology may be attractive due to its energy saving, environmentally friendly, and mild processing yet destructive to pathogens nature, its impact on raw meat color may limit its application in raw meats (Jung et al., 2003; Marcos, Kerry, and Mullen, 2010). Depending on the working pressure among other factors, HPP may also contribute to lipid oxidation and drip loss in whole meat products especially during chilled storage (Simonin et al., 2012).

Dielectric heating is another technology that can be explored in the meat industry. It can be considered a volumetric process that allows for efficient temperature rise by eliminating temperature gradient (Xiong, 2017). Although both radio frequency (RF) and microwave (MW) heating are considered dielectric heating, RF operates at intermediate frequencies and has greater penetration power compared to MW and as such shows greater promise than MW in meat processing (Xiong, 2017). With the new generations of consumers and the recent consumer trends, meat industry will have to be more innovative and convincing in their processing and marketing approaches to protect their market share.

5.5 Conclusion

Although several meat-processing methodologies have ancient history and were mostly invented as traditional preservation methods to ensure that meat would be made available for times of scarcity, more recently, however, these processing techniques have evolved into sensory as well as quality-enhancing methods with an accompanied significant value addition to the meat. With the more recent shifting in global consumer trends, the meat industry and researchers have also been making effort to satisfy the ever-changing consumer demands, improving these traditional processes, and exploring some potentially newer options. This chapter has provided a quick overview of some common meat-processing methodologies employed in the meat industry. The inherent meat quality attributes that could affect the successful execution of these processes were also discussed. It is important that meat-processing techniques are robust enough to accommodate the ever-evolving consumer demands in an ever-changing market place.

References

Aalhus, J. L., Janz, J. A. M., Tong, A. K. W., Jones, S. D. M. and Robertson, W. M. (2001). The influence of chilling rate and fat cover on beef quality. *Canadian Journal of Animal Science*, 81(3), 321–330.

Adeyemi, K. D. and Sazili, A. Q. (2014). Efficacy of carcass electrical stimulation in meat quality enhancement: A review. *Asian-Australasian Journal of Animal Sciences*, 27(3), 447.

Aguilar-Guggembuhl, J. (2012). Antemortem handling. In *Handbook of Meat and Meat Processing* (pp. 320–331). CRC Press.

Aliani, M. and Farmer, L. J. (2005). Precursors of chicken flavor. II. Identification of key flavor precursors using sensory methods. *Journal of Agricultural and Food Chemistry*, 53(16), 6455–6462.

Alvarado, C. and McKee, S. (2007). Marination to improve functional properties and safety of poultry meat. *The Journal of Applied Poultry Research*, 16(1), 113–120.

Álvarez, D., Castillo, M., Payne, F. A., Garrido, M. D., Bañón, S. and Xiong, Y. L. (2007). Prediction of meat emulsion stability using reflection photometry. *Journal of Food Engineering*, 82(3), 310–315.

Andrés, A., Barat, J. M., Grau, R. and Fito, P. (2008). Principles of drying and smoking. *Handbook of Fermented Meat and Poultry*, 1, 37–50.

Apple, J. K. and Yancey, J. W. S. (2013). Water-holding capacity of meat. *The Science of Meat Quality* (pp. 119–145). Oxford, UK: John Wiley & Sons, Inc.

Arshad, M. S., Sohaib, M., Ahmad, R. S., Nadeem, M. T., Imran, A., Arshad, M. U., et al. (2018). Ruminant meat flavor influenced by different factors with special reference to fatty acids. *Lipids in Health and Disease*, 17(1), 223.

Arthur, T. M., Wheeler, T. L., Shackelford, S. D., Bosilevac, J. M., Nou, X. and Koohmaraie, M. (2005). Effects of low-dose, low-penetration electron beam irradiation of chilled beef carcass surface cuts on Escherichia coli O157: H7 and meat quality. *Journal of Food Protection*, 68(4), 666–672.

Badr, H. M. (2012). Irradiation of meat. In: *Handbook of Meat and Meat Processing* (pp. 400–425). CRC Press.

Bailey, A. J. (1972). The basis of meat texture. *Journal of the Science of Food and Agriculture*, 23(8), 995–1007.

Baldwin, D. E. (2012). Sous vide cooking: a review. *International Journal of Gastronomy and Food Science*, 1(1), 15–30.

Barbut, S. (2015). *The Science of Poultry and Meat Processing*. Guelph, ON: University of Guelph. ISBN 978-0-88955-626-3. http://download.poultryandmeatprocessing.com/v01/SciPoultryAndMeatProcessing%20-%20Barbut%20-%20v01.pdf.

Bhat, Z. F., Morton, J. D., Mason, S. L. and Bekhit, A. E. D. A. (2018). Applied and emerging methods for meat tenderization: a comparative perspective. *Comprehensive Reviews in Food Science and Food Safety*, 17(4), 841–859.

Binkerd, E. F. and Kolari, O. E. (1975). The history and use of nitrate and nitrite in the curing of meat. *Food and Cosmetics Toxicology*, 13(6), 655–661.

Boler, D. D. and Woerner, D. R. (2017). What is meat? A perspective from the American Meat Science Association. *Animal Frontiers*, 7(4), 8–11.

Boles, J. A. (1999). Meat processing: restructured meats. In: *Paper presented at the Canadian Meat Science Association*. http://www.cmsa-ascv.ca/documents/1999July-99Bolespgs12-14.pdf.

Bolger, Z., Brunton, N. P., Lyng, J. G. and Monahan, F. J. (2017). Comminuted meat products – consumption, composition, and approaches to healthier formulations. *Food Reviews International*, 33(2), 143–166.

Bolumar, T. and Claus, J. (2017). Utilizing shockwaves for meat tenderization. In: *Reference Module in Food Science*: Elsevier.

Bolumar, T. and Toepfl, S. (2016). Application of shockwaves for meat tenderization. In: *Innovative Food Processing Technologies* (pp. 231–258). Elsevier.

Brannan, R. G. (2008). Analysis of texture of boneless, fully fried breaded chicken patties as affected by processing factors. *Journal of Food Quality*, 31(2), 216–231.

Brewer, S. (2007). The chemistry of beef flavour-executive summary. https://www.beefresearch.org/CMDocs/BeefResearch/The%20Chemistry%20of%20Beef%20Flavor.pdf.

Campus, M. (2010). High pressure processing of meat, meat products and seafood. *Food Engineering Reviews*, 2(4), 256–273.

Castigliego, L., Armani, A. and Guidi, A. (2012). Meat color. In: *Handbook of Meat and Meat Processing* (pp. 100–125). CRC Press.

CFIA. (2013). *Annex C: Use of Phosphate Salts and Nitrites in the Preparation of Meat Products*. Canadian Food Inspection Agency.

Channon, H. A., Payne, A. M. and Warner, R. D. (2002). Comparison of CO2 stunning with manual electrical stunning (50 Hz) of pigs on carcass and meat quality. *Meat Science*, 60(1), 63–68.

Cheng, Q. and Sun, D.-W. (2008). Factors affecting the water holding capacity of red meat products: a review of recent research advances. *Critical Reviews in Food Science and Nutrition*, 48(2), 137–159.

Cocolin, L. and Rantsiou, K. (2012). Meat fermentation. In Y. H. Hui (Ed.), *Handbook of Meat and Meat Processing* (pp. 557–572). Boca Raton, FL: CRC Press, Taylor and Francis Group.

del Pulgar, J. S., Gázquez, A. and Ruiz-Carrascal, J. (2012). Physico-chemical, textural and structural characteristics of sous-vide cooked pork cheeks as affected by vacuum, cooking temperature, and cooking time. *Meat Science*, 90(3), 828–835.

den Hertog-Meischke, M. J. A., Van Laack, R. and Smulders, F. J. M. (1997). The water-holding capacity of fresh meat. *Veterinary Quarterly*, 19(4), 175–181.

EFSA. (2013). Scientific opinion on the public health risks related to mechanically separated meat (MSM) derived from poultry and swine. *EFSA Journal*, 11(3), 3137.

Egan, A. F., Ferguson, D. M. and Thompson, J. M. (2001). Consumer sensory requirements for beef and their implications for the Australian beef industry. *Australian Journal of Experimental Agriculture*, 41(7), 855–859.

Farkas, J. (1998). Irradiation as a method for decontaminating food: a review. *International Journal of Food Microbiology*, 44(3), 189–204.

Faustman, C. (2014). Myoglobin chemistry and modifications that influence (color and) color stability. In: *Paper presented at the Reciprocal Meat Conference*.

Fiszman, S. and Sanz, T. (2010). Battering and breading: principles and system development. *Handbook of Poultry Science and Technology, Volume 2: Secondary Processing*, 35–45.

FSAI. (2019). *The Use and Removal of Nitrite in Meat Products*. https://www.fsai.ie/uploadedFiles/Consol_Reg1333_2008.pdf.

Greaser, M. L. and Guo, W. (2012). Postmortem muscle chemistry. In: *Handbook of Meat and Meat Processing* (pp. 82–99). CRC Press.

Guerrero-Legarreta, I. and de Lourdes Pérez-Chabela, M. (2012). Slaughtering operations and equipment. In: *Handbook of Meat and Meat Processing* (pp. 426–433). CRC Press.

Guerrero-Legarreta, I. and García-Barrientos, R. (2012). Thermal technology. In: *Handbook of Meat and Meat Processing* (pp. 542–549). CRC Press.

Herring, J. L. and Smith, B. S. (2012). Meat-smoking technology. In: *Handbook of Meat and Meat Processing* (pp. 566–575). CRC Press.

Huff-Lonergan, E. and Lonergan, S. M. (2005). Mechanisms of water-holding capacity of meat: the role of postmortem biochemical and structural changes. *Meat Science*, 71(1), 194–204.

Jackson, V., Schilling, M. W., Falkenberg, S. M., Schmidt, T. B., Coggins, P. C. and Martin, J. M. (2009). Quality characteristics and storage stability of baked and fried chicken nuggets formulated with wheat and rice flour. *Journal of Food Quality*, 32(6), 760–774.

Jayasena, D. D., Ahn, D. U., Nam, K. C. and Jo, C. (2013a). Factors affecting cooked chicken meat flavour: a review. *World's Poultry Science Journal*, 69(3), 515–526.

Jayasena, D. D., Ahn, D. U., Nam, K. C. and Jo, C. (2013b). Flavour chemistry of chicken meat: a review. *Asian-Australasian Journal of Animal Sciences*, 26(5), 732.

Juárez, M., Aldai, N., López-Campos, Ó., Dugan, M. E. R., Uttaro, B. and Aalhus, J. L. (2012). Beef texture and juiciness. *Handbook of Meat and Meat Processing*, 9, 177–206.

Jung, S., Ghoul, M. and de Lamballerie-Anton, M. (2003). Influence of high pressure on the color and microbial quality of beef meat. *LWT-Food Science and Technology*, 36(6), 625–631.

Kauffman, R. G. (2012). Meat composition. In: *Handbook of Meat and Meat Processing* (pp. 64–81). CRC Press.

Khan, M. I., Lee, H. J., Kim, H.-J., Young, H. I., Lee, H. and Jo, C. (2016). Marination and physicochemical characteristics of vacuum-aged duck breast meat. *Asian-Australasian Journal of Animal Sciences*, 29(11), 1639–1945.

Knipe, L. (2003). Phosphates as meat emulsion stabilizers. *Encyclopedia of Food Sciences and Nutrition*, 2077–2080.

Lawrie, R. A. and Ledward, D. (2014). *Lawrie's Meat Science*. Woodhead Publishing.

Llonch, P., Rodríguez, P., Gispert, M., Dalmau, A., Manteca, X. and Velarde, A. (2012). Stunning pigs with nitrogen and carbon dioxide mixtures: effects on animal welfare and meat quality. *Animal*, 6(4), 668–675.

Lonergan, S. M., Topel, D. G. and Marple, D. N. (2019). Fresh and cured meat processing and preservation. In: S. M. Lonergan, D. G. Topel and D. N. Marple (Eds.), *The Science of Animal Growth and Meat Technology (Second Edition)* (Chapter 13, pp. 205–228). Academic Press.

Lücke, F.-K. (1994). Fermented meat products. *Food Research International*, 27(3), 299–307.

Mah, E. (2008). *Optimization of a Pretreatment to Reduce Oil Absorption in Fully Fried, Battered, and Breaded Chicken Using Whey Protein Isolate as a Postbreading Dip*. Doctoral dissertation. USA: Ohio University.

Mancini, R. A. and Hunt, M. (2005). Current research in meat color. *Meat Science*, 71(1), 100–121.

Marcos, B., Kerry, J. P. and Mullen, A. M. (2010). High pressure induced changes on sarcoplasmic protein fraction and quality indicators. *Meat Science*, 85(1), 115–120.

Martin, J. M. (2012). Meat-curing technology. In *Handbook of Meat and Meat Processing* (pp. 550–565). CRC Press.

Maurizio, B., Massimiliano, P. and Claudio, C. (2010). The use of marination to improve poultry meat quality. *Italian Journal of Animal Science*, 8. DOI: 10.4081/ijas.2009.s2.757

McClements, D. J. (2015). *Food Emulsions: Principles, Practices, and Techniques*. CRC Press.

McKinstry, J. L. and Anil, M. H. (2004). The effect of repeat application of electrical stunning on the welfare of pigs. *Meat Science*, 67(1), 121–128.

Melton, S. L. (1999). Current status of meat flavor. In: *Quality Attributes of Muscle Foods* (pp. 115–133).· Springer.

Miller, M. F., Carr, M. A., Ramsey, C. B., Crockett, K. L. and Hoover, L. C. (2001). Consumer thresholds for establishing the value of beef tenderness. *Journal of Animal Science*, 79(12), 3062–3068.

Mottram, D. S. (1998). Flavour formation in meat and meat products: a review. *Food Chemistry*, 62(4), 415–424.

Muntean, M.-V., Marian, O., Barbieru, V., Cătunescu, G. M., Ranta, O., Drocas, I. and Terhes, S. (2016). High pressure processing in food industry – characteristics and applications. *Agriculture and Agricultural Science Procedia*, 10, 377–383.

Ockerman, H. W. and Basu, L. (2008). Fermented meat products: production and consumption. In: F. Toldra (Ed.), *Handbook of Fermented Meat and Poultry* (pp. 47–68). Oxford, UK: Blackwell Publishing Ltd.

Offer, G. and Cousins, T. (1992). The mechanism of drip production: formation of two compartments of extracellular space in muscle post mortem. *Journal of the Science of Food and Agriculture*, 58(1), 107–116.

Offer, G. and Trinick, J. (1983). On the mechanism of water holding in meat: the swelling and shrinking of myofibrils. *Meat Science*, 8(4), 245–281.

O'Quinn, T. G., Legako, J. F., Brooks, J. C. and Miller, M. F. (2018). Evaluation of the contribution of tenderness, juiciness, and flavor to the overall consumer beef eating experience. *Translational Animal Science*, 2(1), 26–36.

Owusu-Apenten, R. (2004). Testing protein functionality. In: R. Yada (Ed.), *Proteins in Food Processing* (pp. 217–244). New York, NY: Woodland Publishing.

Oz, F., Kaban, G. and Kaya, M. (2010). Heterocyclic aromatic amine contents of beef and lamb chops cooked by different methods to varying levels. *Journal of Animal and Veterinary Advances*, 9, 1436–1440.

Oz, F. and Zikirov, E. (2015). The effects of sous-vide cooking method on the formation of heterocyclic aromatic amines in beef chops. *LWT-Food Science and Technology*, 64(1), 120–125.

Paramithiotis, S., Drosinos, E. H., Sofos, J. N. and Nychas, G.-J. E. (2010). Fermentation: microbiology and biochemistry. *Handbook of Meat Processing*, 185–198.

Pearson, A. M. and Gillett, T. A. (1999). *Processed Meats* (pp. 291–294). Gaithersburg, MD: Aspen Publishers.

Pereira, P. M. d. C. C. and Vicente, A. F. d. R. B. (2013). Meat nutritional composition and nutritive role in the human diet. *Meat Science*, 93(3), 586–592.

Perez-Alvarez, J. A., Sendra-Nadal, E., Sanchez-Zapata, E. J. and Viuda-Martos, M. (2010). Poultry flavour: general aspects and applications. *Handbook of Poultry Science and Technology*, 2, 339–357.

Petäjä-Kanninen, E. and Puolanne, E. (2007). Principles of meat fermentation. *Handbook of Fermented Meat and Poultry*, 31–35.

Ramarathnam, N. and Rubin, L. J. (1994). The flavour of cured meat. In: *Flavor of Meat and Meat Products* (pp. 174–198). Springer.

Santchurn, S. J., Arnaud, E., Zakhia-Rozis, N., & Collignan, A. (2012). Drying: Principles and applications. In *Handbook of meat and meat processing* (pp. 524–541): CRC Press.

Sanz, T. and Salvador, A. (2012). Breading. In: *Handbook of Meat and Meat Processing* (pp. 488–497). CRC Press.

Savell, J. W., Mueller, S. L. and Baird, B. E. (2005). The chilling of carcasses. *Meat Science*, 70(3), 449–459.

Scanlan, R. A. (1983). Formation and occurrence of nitrosamines in food. *Cancer Research*, 43(5 Suppl.), 2435s–2440s.

Sebastian, P., Bruneau, D., Collignan, A. and Rivier, M. (2005). Drying and smoking of meat: heat and mass transfer modeling and experimental analysis. *Journal of Food Engineering*, 70(2), 227–243.

Sikorski, Z. E. and Kolakowski, E. (2010). Smoking. In: F. Toldrá (Ed.), *Handbook of Meat Processing* (pp. 231–245). Blackwell Publishing.

Sikorski, Z. E. and Sinkiewicz, I. (2014). Principles of smoking. *Handbook of Fermented Meat and Poultry*, 39–45.

Simonin, H., Duranton, F. and De Lamballerie, M. (2012). New insights into the high-pressure processing of meat and meat products. *Comprehensive Reviews in Food Science and Food Safety*, 11(3), 285–306.

Singh, P. K., Shrivastava, N. and Ojha, B. K. (2019). Enzymes in the meat industry. In: *Enzymes in Food Biotechnology* (pp. 111–128). Elsevier.

Singh, V. P., Pathak, V. and Verma, A. K. (2012). Fermented meat products: organoleptic qualities and biogenic amines – a review. *American Journal of Food Technology*, 7(5), 278–288.

Smith, B. S. (2012). Marination: ingredient technology. In: *Handbook of Meat and Meat Processing* (pp. 498–513). CRC Press.

Soladoye, O. and Pietrasik, Z. (2018). Utilizing high pressure processing for extended shelf life meat products. In: *Reference Module in Food Science*. Elsevier.

Soladoye, O. P., Shand, P., Dugan, M. E. R., Gariépy, C., Aalhus, J. L., Estévez, M. and Juárez, M. (2017). Influence of cooking methods and storage time on lipid and protein oxidation and heterocyclic aromatic amines production in bacon. *Food Research International*, 99, 660–669.

Stumpe-Vīksna, I., Bartkevičs, V., Kukāre, A. and Morozovs, A. (2008). Polycyclic aromatic hydrocarbons in meat smoked with different types of wood. *Food Chemistry*, 110(3), 794–797.

Sullivan, G. A. (2011). *Naturally Cured Meats: Quality, Safety, and Chemistry*. Doctoral dissertation. Ames: Iowa State University.

Sullivan, G. A. (2013). *A Comparison of Traditional and Alternative Meat Curing Methods* (American Meat Association, Ed., p. 2).

Suman, S. P. and Joseph, P. (2013). Myoglobin chemistry and meat color. *Annual Review of Food Science and Technology*, 4, 79–99.

Sun, X. D. (2009). Utilization of restructuring technology in the production of meat products: a review. *CyTA – Journal of Food*, 7(2), 153–162.

Swatland, H. J. (2002). *On-Line Monitoring of Meat Quality*. Cambridge, UK: CRC Press, Woodhead Pub.

Ugalde-Benítez, V. (2012). Meat emulsions. In: *Handbook of Meat and Meat Processing* (pp. 466–475). CRC Press.

Van Ba, H., Hwang, I., Jeong, D. and Touseef, A. (2012). Principle of meat aroma flavors and future prospect. *Latest Research into Quality Control*, 2, 145–176.

Varavinit, S., Shobsngob, S., Bhidyachakorawat, M. and Suphantharika, M. (2000). Production of meat-like flavor. *Science Asia*, 26(14), 219–224.

Varghese, K. S., Pandey, M. C., Radhakrishna, K. and Bawa, A. S. (2014). Technology, applications and modelling of ohmic heating: a review. *Journal of Food Science and Technology*, 51(10), 2304–2317.

Velarde, A., Gispert, M., Diestre, A. and Manteca, X. (2003). Effect of electrical stunning on meat and carcass quality in lambs. *Meat Science*, 63(1), 35–38.

Vergara, H., Linares, M. B., Berruga, M. I. and Gallego, L. (2005). Meat quality in suckling lambs: effect of pre-slaughter handling. *Meat Science*, 69(3), 473–478.

Vlahova-Vangelova, D. B., Dragoev, S. G., Balev, D. K., Assenova, B. K. and Amirhanov, K. J. (2017). Quality, microstructure, and technological properties of sheep meat marinated in three different ways. *Journal of Food Quality*, 2017.

Warriss, P. D. (2000). *Meat Science. An Introduction Text* (310pp.). Wallingford, UK: CABI Publishing, CAB International.

Williams, P. (2007). Nutritional composition of red meat. *Nutrition & Dietetics*, 64, S113–S119.

Winger, R. J. and Hagyard, C. J. (1994). Juiciness – its importance and some contributing factors. In: *Quality Attributes and Their Measurement in Meat, Poultry and Fish Products* (pp. 94–124). Springer.

Xiong, Y. L. (2017). The storage and preservation of meat: I – Thermal Technologies. In: *Lawrie's Meat Science* (pp. 205–230). Woodhead Publishing.

Xu, Y., Huang, J. C., Huang, M., Xu, B. C. and Zhou, G. H. (2012). The effects of different chilling methods on meat quality and calpain activity of pork muscle longissimus dorsi. *Journal of Food Science*, 77(1), C27–C32.

Yildiz-Turp, G., Sengun, I. Y., Kendirci, P. and Icier, F. (2013). Effect of ohmic treatment on quality characteristic of meat: a review. *Meat Science*, 93(3), 441–448.

Yılmaz, I. and Velioğlu, H. (2009). Fermented meat products. In: I. Yilmaz (Ed.), *Quality of Meat and Meat Products* (pp. 1–16). Indie: Transworld Research Network.

AyoJesutomi O. Abiodun-Solanke

6 Fish and shellfish processing

6.1 Introduction

Fish are vertebrates with backbones. They are cold-blooded animals that live in fresh-water or seawater. They have fins, scales, and gills, and they breathe with their gills (Bene et al., 2015). Shellfish in fisheries is a term for exoskeleton-bearing aquatic invertebrates used as food, which include mollusks, crustaceans, echinoderms, edible species of oysters, clams, mussels, either shucked or in the shell, fresh or frozen, and raw, in-shell scallops (Liverpool-Tasie et al., 2018). They are not fish but because they dwell in water just as fish they are encompassed as fish.

Aquatic organisms have been found to contain superior fat (i.e., very low bad fat and high beneficial fats) and protein (rich in all essential amino acids) and other nutrients that attribute positively to good health (FAO, 2017). Many of these beneficial micronutrients are generally more abundant in aquatic animals and plants than in meats or terrestrial vegetables. Also, fish enzymes have application in many other industries, and a huge number of nutraceuticals from fish have enormous application to human health. A well-balanced diet rich in variety of fish and shellfish contributes immensely to the heart health and proper growth and development in children. Fish and shellfish are recommended to be taken more after certain ages for the superior alternative it provides (FAO, 2018).

Despite these benefits in fish and shellfish and their availability as animal-sourced protein food, their availability is still threatened by high perishability, poor postharvest techniques especially in African countries, and improper education which have resulted in massive losses. Fish, often considered as a "rich food for poor people" (Beveridge, et al., 2010), is widely recognized as "nature's super food" (FAO, 2018). Fish and shellfish are an important part of a healthy diet.

Globally, fish consumption rates are growing faster than the global population growth due to the awareness of the health benefits associated with consuming fish, increased incomes, and rising urbanization (Anderson et al., 2017). There is also increasing demand for fresh fish (particularly in urban or metropolitan areas) and processed fish (predominantly by rural dwellers), which necessitates the improvement in fish processing to enable the delivery of wholesome products to the end users and consumers (Liverpool-Tasie et al., 2018). Smoked fish constitute approximately 80% of widely consumed fish especially in Nigeria while the remaining are consumed as fresh, salted, and sundried or fried (Mafimisebi, 2012); however, in other regions in Africa, the drift is similar but different for the west and developed

AyoJesutomi O. Abiodun-Solanke, Federal College of Fisheries and Marine Technology, Lagos, Nigeria

https://doi.org/10.1515/9783110667462-006

countries. Fresh fish is a more preferred trend in other climes of the globe and this is usually more valued than the processed ones (FAO, 2018).

Fish and shellfish are susceptible to high postharvest losses (e.g., spoilage accounts for up to 10–12 million tons per year), and these fishes are either discarded or sold at relatively low price due to quality deterioration (Kumolu-Johnson and Ndimele, 2011). Postharvest losses are often caused by throwing off by catches, poor processing facilities, predation by animals and insect infestation, and inappropriate packaging leading to damage of the fish product (Yahya, 2011). This means that stakeholders involved in fisheries lose potential income, and less fish is available or consumers are supplied with low-quality fish products (Yahya, 2011).

Fish preservation and processing are necessary processes to tackle postharvest losses. Fish preservation extends the shelf life of fish and fisheries product by applying the principles of chemistry, engineering, and other branches of science in order to improve the quality of the products (FAO, 2018). The term fish processing refers to the preparation of seafood and freshwater fish for human consumption. It is the change of state or form associated with fish and fish products at harvest to the time the final product is delivered to the customer (FAO, 2018).

This aim of this chapter is to assess the status and prospects of fish postharvest handling and processing through review on good handling practices, the composition of fish and shellfish, type of losses, spoilage, current trends in processing, and future prospects especially as it relates to fish and seafood globally.

6.2 The nutritional value of fish and shellfish

Good health and a sense of wellness are important priorities for many people globally today; these are strongly related to diet (Akande and Diei-Ouadi, 2010). This relationship and the effect of diet on the occurrence of illnesses such as heart disease, diabetes, and cancer continue to be active areas of nutrition research by researchers globally. People are generally more deliberate about eating well to reduce the occurrence of terminal diseases and to be better if they have contracted these diseases (FAO, 2017).

Dietary Guidelines for Americans makes recommendations to adequate nutrients within caloric needs, weight management, physical activity, and food safety (Food and Nutrition Board. 2007). This encourages people to limit fat intake to between 20% and 35% of total calories and to consume fats mainly from fish, nuts, and vegetable oils that are good sources of monounsaturated fatty acids and polyunsaturated fatty acids found chiefly in marine fish oils (Food and Nutrition Board. 2007). The guidelines further advice consuming less than 10% of calories from saturated fatty acids such as those found in untrimmed red meat and less than 300 mg/day of cholesterol (Byrd-Bredbenner et al., 2009).

The types of fat in shellfish are beneficial as the proportions of saturated, monounsaturated, and polyunsaturated fats are very healthy. About 15% or even less are calories supplied by shellfish consumption from fat, while in beef and pork more than 40% calories are from fat. Fats from land animals contain high saturated fat perceived as bad fat while shellfish contain higher proportion of polyunsaturated fat. Shellfish also provide high-quality protein with all the dietary essential amino acids for maintenance and growth of the human body. For this reason, shellfish should be considered a low-fat, low-saturated-fat, high-protein food that can be included in a low-fat diet (Food and Nutrition Board, 2007).

Cholesterol is essential in the body because it is used to make important compounds such as bile acids (which help to digest fat in the intestine). Shellfish in general are low in fat and cholesterol though some, for example, clams and scallops contain a high concentration of noncholesterol sterols that are absorbed from the intestine and decrease the absorption of cholesterol (Byrd-Bredbenner et al., 2009) and therefore have a positive effect on health.

Shellfish are rich in several micronutrients that are needed in the body. Examples of these micronutrients are iron (clams have iron in 100 g to meet about 78% of the Dietary Reference Intake for 19- to 50-year-olds), zinc (which is necessary for a healthy diet), copper (it helps to form hemoglobin and collagen), and vitamin B-12 (which helps the body maintain sheathes around nerve fibers among others) (Wardlaw and Smith, 2009).

Water is the main constituent of fish flesh and it accounts for about 80% of the weight of a fresh fish. The average water content contained in a fatty fish is about 70%; a fish sample will contain water content between 30% (though at the extreme) and 90% of their total weight (FAO, 2005).

Protein is next to the water content in ascending order. Protein content in fish muscle is usually between 15% and 20% (Kabahenda et al., 2009). Proteins are chains of chemical units (usually 20 amino acids) linked together to make one long molecule. They are essential in the human diet for the maintenance of good health (FAO, 2005).

Lipids including fats, oils, waxes, as well as fatty acids vary more widely than the water, protein, or mineral content. While the ratio of the highest and the lowest protein or water content is not more than 3:1, the ratio between the highest and lowest fat values is more than 300:1 (FAO, 2005).

The minor components of fish muscle are **carbohydrates**, which are significant in diets with negligible amount (i.e., 1% in light muscle fish and 2% in dark muscle fatty fishes); **vitamins and minerals** include a range of substances that widely differ in character and must be present in the diet only which in minute quantities to promote good health and maintain life itself (FAO, 2005).

6.3 Postharvest losses in fish and shellfish

Postharvest losses of fish and seafood occur in various forms. **Physical loss** refers to fish that are not used or accounted for after harvesting (FAO, 2005). **Quality loss** refers to spoilage or physical damage that leads to quality deterioration. This is the most common losses in many areas (Yahya, 2011). **Economic losses** represent the loss in value due to physical damage such as breakages, which lead to reduction in weight and quantity. The products are usually sold relatively at lower price. **Market force loss** is due to the wide gap that still exists between demand and supply imparting fluctuations in the price of fish. This type of loss is very difficult to measure and sets the ground for quality and physical losses (Mgawe and Bawaye, 2012).

Nutritional loss: This loss happens when fish decomposes and it is rendered unfit or unsafe for human consumption. Likewise, fish that are discarded at sea due to their low economic value are regarded as a nutritive loss and losses due to spoilage (Yahya, 2011; Rahman et al., 2013).

6.4 Fish spoilage

Fish and seafood are highly perishable foods due to their high moisture content and nutrients enabling the growth of microorganisms (FAO, 2018).

A fresh product has its original characteristics intact, while a spoilt product is graded as the changes from absolute freshness to limits of acceptability and then to unacceptability (Akande and Diei-Ouadi, 2010). Spoilage is usually accompanied by changes in physical characteristics, that is, change in color, odor, texture, color of eyes, color of gills, and softness of the muscle (FAO, 2017). Spoilage of fish is caused by a number of factors such as the use of wrong storage temperature, microbial degradation (due to the actions of bacteria, yeast, and molds), the enzymatic reactions, and oxidation of unstable fats in the fish. In addition, the following factors contribute to spoilage of fish (FAO, 2005). Chemical and microbial spoilage are responsible for about 25% losses for fish products annually (Arason et al., 2014). Hence, it is concluded that high water, fat, and protein content, weak muscle, and mishandling result in spoilage of fish.

6.4.1 Fish handling

Handling is a very important aspect of processing as it has so much to say about the quality of the final product. It aimed at preserving the quality of the fish through cold temperature as quickly as possible after harvest. Various aspects of handling have been described further briefly (Emborg et al., 2005).

6.4.1.1 Fish handling onboard/at harvest

Fish handling onboard or at harvest, that is, when on the vessel or being transported to the processing site is very critical for maintenance of high/good-quality product. Working on fish as quickly as possible includes sorting, covering for protection, draining, avoiding bruising, and cooling (using adequate ice), which are cardinal rules in handling of fresh fish during this phase and beyond. These rules help in reducing constraints such as bacteriological (through the invasion of microbes), chemical (through oxidative and enzymatic reactions), and physical (through bruising, tearing, cutting, etc.) processes that cause degradation of fish (FAO, 2018).

6.4.1.2 Fish handling on the land

Handling of fishes on land involves processes such as transportation, preprocessing, and processing to the desired quality. This aspect of handling has similar hazards as the onboard handling to prevent good-quality products. These hazards are directly proportional to the temperature. They are twice as fast at a temperature of 2.5 °C, and at 10 °C they are four times faster (Gram and Huss, 2000).

6.4.1.3 Handling during preprocessing and processing

In this phase of handling, there is grading and sorting according to the type of fish, processing methods, and the intended final use (Arason et al., 2014).

Cleanliness, care (working on fish as quickly as possible, sorting, covering for protection, draining, and avoiding bruising), and cooling (using adequate ice to reach all fishes) are the three cardinal rules to be adhered to in handling of fresh fish to maintain high-quality finished products (FAO, 2005).

6.5 Fish preservation and processing

Fish preservation works with the following principles:
– The growth of microorganisms are prevented or delayed through asepsis; the growth of these microbes are also inhibited through use of cold temperature, high temperature, removal or reduction of water, use of anaerobic environment, use of chemicals or preservatives, and irradiation.
– Delay or prevention of autolysis by the destruction or inactivation of enzymes through processes such as blanching to halt the different chemical reactions takes place, for example, use of antioxidants to reduce oxidation reaction.

– Reduction of infestation by insects or other predators through substances to kill the insects and predators and using storage facilities to eradicate this.

Inhibition of microorganism can be achieved by processes in the first principle above (SSTL, 2009).

Inactivation of microorganisms is achieved by sterilization, pasteurization, irradiation, blanching, cooking, and frying.

Avoidance of recontamination is achieved by packaging, hygienic processing, hygienic storage, aseptic processing, Hazard Analysis Critical Control Points (HACCP), total quality management, risk analysis, and management.

6.5.1 Fish preservation methodology

6.5.1.1 Aseptic techniques

This is defined as a condition in which no living disease-causing microorganisms are present. This can be achieved through modified or controlled packaging not conducive for growth of microorganisms. Modified atmospheric packaging is an advanced food packaging technology that is being used increasingly by the seafood processing industry to address issues associated with maintaining freshness and extending the shelf life of chilled seafood (Emborg et al., 2005). It is also gaining popularity for packaging live seafood (e.g., mussels) (SSTL, 2009).

6.5.1.2 Anaerobic condition

Anaerobic condition is defined as an environment "without oxygen or air." This prevents many disease-causing aerobes from growing in the container. It is believed that the heat treatment the product would have undergone will take care of surviving anaerobes (Leistner, 2000).

6.5.1.3 Temperature control treatment

This is the most commonly used fish preservation method and there are two major types, that is, use of high- and low-temperature storage. High temperature is used to kill and deactivate microorganisms, especially the thermophiles. There are also two processes that are usually used, that is, either high-temperature short time or low-temperature long time. The process to adopt is determined by the microorganisms targeted and the composition of the food (Nuin et al., 2008). For fish that is composed of some heat-labile vitamins, the exposure to heat will be short for the retention of some of these sensitive nutrients. Examples of this method are pasteurization and complete sterilization.

This method especially chilling lowers the activities of microorganisms, thereby reducing chemical and enzymatic reaction. Chilling is usually the temporary method used immediately after harvest. Since spoilage sets in immediately after death, there is a need to bring down the temperature to reduce further irreversible activities to achieve good-quality products. Fish is a highly perishable product urging preservation almost immediately after harvesting. Chilling is the method that produces the least obvious changes; however, storage life is limited depending on fish species, type and season of capture, and of course storage temperature (Jessen et al., 2014). Ice is the most common tool used to achieve chilling.

There are different types of ice to promote efficiency of chilling methods. They are block, flake, plate, tube, liquid/fluid, soft, as well as functional ice, which is an additive-enhanced product currently in field testing that serves to enhance food storage and shelf life by generating lower temperatures and melting slower than the traditional water-based ice. Additional features allow the slow release of antimicrobial solutions that protect raw food by eliminating the build-up of spoilage-causing bacteria (USDA, 2019). In storage achieved through the use of ice, it is not just about the sufficiency of the ice but the effectiveness, which can only be achieved with the use of appropriate ice for the type of fish and shellfish to be preserved. From literature, flakes and smaller pieces of ice have been found to be more effective than big ice blocks (SSTL, 2009). Also there have been wide variations in storage life between the same of fish especially under different conditions. Nonfatty fish have been found to keep longer than fatty fish, while freshwater fish keeps longer than marine fish and tropical fish keeps longer than temperate fish (Miladi et al., 2008).

Freezing, on the other hand though a cold temperature method, is a method of food preservation whereby there is removal of heat to bring down the temperature of fish to below its freezing point causing available water to turn to ice (Silva and Stojanovic, 2013). Since water accounts for 75% of fish weight, this water contains some dissolved substances that are cooled down to a temperature in which there will be same vapor pressure between the solute phase and the solvent phase causing freezing to begin (Silva and Stojanovic, 2013).

There are different freezing methods such as quick and slow freezing through direct and indirect systems. We also have

- *Air freezing*, that is, freezing in still air, air blast, fluidized bed, and rotary drum freezing
- *Indirect contact freezing* such as horizontal plate freezing and vertical plate freezing
- *Freezing by immersion*, that is, brine freezing and brine spray freezing
- *Cryogenic freezing*, that is, freezing through the use of liquid nitrogen, liquid/solid carbon dioxide, and liquid refrigerant (International Institute of Refrigeration [IIR], 2006)

6.5.1.4 Drying or dehydration

Drying is usually done under sunshine or open air. The other term, dehydration, means removal of water through application of artificial heat within controlled conditions (Janjai and Bala, 2012).

6.5.1.4.1 Sun drying

This is the simplest method of fish drying usually ideal for small sized or lean fishes to prevent glut during the on-season and extend the shelf life of these fishes to the off-season. Drying usually takes place when water is removed from the surface and then from the interior of the fish.

6.5.1.4.2 Factors affecting the rate of fish drying
- The size of the fish – smaller fish dry faster than the larger fish
- The fish surface area – exposure of large surface area to heat means increased drying
- The environmental temperature – the temperature is linearly related to the rate of drying, that is, higher temperature means higher drying rate
- The environmental relative humidity – these are inversely related
- The environmental air velocity – more speed of the air means faster drying rate
- The fat content of the fish – fatty fishes will dry longer than lean fishes
- The fish water content – more water content means faster drying

6.5.1.4.3 Solar drying

With this method, the heat of the sun is converged and used for drying fish. This convergence is improved by the use of black surfaces that absorb heat more effectively than light colored surfaces (Akande and Diei-Ouadi, 2010). Cabinet driers and solar tents work with this same principle. Solar panels and batteries are being utilized recently to improve upon the use of solar drying and to continue drying even when there is no sun nowadays; however, the method isn't too accessible for high cost.

6.5.1.4.4 Mechanical drying

This method utilizes mechanical dryers that remove water from fish through thermal energy or heat generated by electricity or other means. It is achieved through burning of fossil fuel or by electricity. It is an expensive method that is not too common in developing economies.

6.5.1.5 Curing

Curing methods are traditional and therefore the oldest methods of fish processing. This is achieved by processes such as salting, pickling, and smoking (Liverpool-Tasie et al., 2018). It is the oldest method of fish preservation and is still widely used in the developing countries where problems such as inadequate processing facilities still exist. It is the cheapest method compared with other fish preservation methods and also very simple making it more acceptable in the fishing communities. Majority of the fish utilized and consumed in Africa and Asia are cured fishes. Many fish processors from these regions export cured fish to the west (Thorarinsdottir et al., 2011).

Smoking is a curing method of fish preservation usually done in a kiln. Smoking employs different preservative means such as brining, impartation of flavors such as aldehydes and ketones, cooking, and then drying. Modern kilns have been introduced to improve upon the traditional smoking method and these facilities are being evaluated for efficiency on a regular basis (Akande and Diei-Ouadi, 2010). The smoke used is generated by wood, charcoal, sawdust, or coconut husk prepared as briquettes or biochar giving varying flavors. Smoking endows fish muscle with flavor and preservative compounds (Alcicek and Atar, 2010). Smoking can be cold, hot, liquid, and electrostatic. Smoking with temperature below 35 °C is referred to as **cold smoking** while the one done with temperature 70–80 °C or even higher is **hot smoking** (Birkerland et al., 2004). Smoking of fish is under serious attack recently for the accumulation of certain carcinogenic compounds from smoke, that is, polycyclic aromatic hydrocarbons (PAH), for example, benzo(a)pyrene is very harmful to human health (Liverpool-Tasie et al., 2018). Liquid smoking with concentrates is performed by dipping the fish in a liquid smoking extract produced by dry distillation of wood followed by concentration (smoke condensate). Smoke condensates are generally produced by the following principles that reduce the formation of PAHs and are often further rinsed with water at 15 °C to reduce condensate with these components (Varlet et al., 2007). The application of smoke condensates on fish was investigated by Muratore et al. (2007), who found that the different smoke flavorings have different effects on taste, usually less salty, depending on the fish species. In electrostatic smoking, the fish is treated with infrared radiation. Muscle is smoked by the creation of a positively charged electrical field while the fish is negatively charged. Electrostatic smoking is fully mechanized, thus entailing savings on labor and production costs. The process renders a higher quality in the final product compared with traditional smoking processes (Arason et al., 2014).

6.5.1.6 Use of preservatives

Using preservatives is another trendy method of fish and shellfish preservation (Arason et al., 2014). The preservatives serve as antimicrobial that prevents or reduces microbial growth. This improves the safety of fish, reduces wastages, reduces postharvest losses, and also extends the shelf life of fish. Synthetic preservatives such as nitrates are used for fish and shellfish products for a long time ago, and traditional preservatives are emerging now. These traditional preservatives such as orange peel, ginger, garlic, and turmeric are replacing synthetic ones used in the past due to some health concerns about the safety of synthetic ones. Also, these trendy organic and plant antimicrobials are readily available, cheap, and are therefore preferred in fishing communities. They have been found to extend the shelf life of smoked fish products for 28, 20, and 18 days for orange peel, ginger, and garlic, respectively (Arason et al., 2014; Abiodun-Solanke, 2019).

Salt is also an ancient preservative known to impart taste and flavor, and dehydrate and inactivate enzymes in marine species (Akande and Diei-Ouadi, 2010). Use of salt has been tested and therefore trusted as a good preservative. It imparts taste and flavor into the fish as well. It is the commonest way of quick preservation done before other methods such as smoking. Sugar, on the other hand, binds free water present in the fish, thereby making this moisture unavailable for microbial activities and reactions. This will further extend the shelf life of the fish. This is usually used to achieve preservation through fermentation, which is not a common preservation method used in Africa (Akande and Diei-Ouadi, 2010). Salt can be used directly as it is being used in deimmobilizing freshwater fishes (e.g., catfish). This is referred to as dry salting. In dry salting, salt is sprinkled or applied to the fish in dry form while for wet salting a solution of salt dissolved in water is used (Abiodun-Solanke, 2019). In wet salting, the fishes are soaked in the salt solution/brine for some minutes. Salt also inactivates enzymes in marine species. Siringan et al. (2006), for example, reported on the autolytic activity of endogenous proteinases, where 25% (w/w) salt reduced this activity by 48%. Also, Sen (2005) found that 10% solution of salt inactivates the enzymatic reaction when shrimp was blanched at 80 °C for just 5 min. Tilapia surimi proteolysis was also reduced by 76% when compared with the control when the product was incubated in 1 M NaCl at 65 °C for 1.5 h (Arasom et al., 2014).

6.5.1.7 Use of acids

Acids act by reducing the pH of the environment thereby making it unconducive for the activities or even survival of microorganisms. Just 20% acids such as vinegar reduce deterioration in most of the fish products. It finds application in fermentation as well as in pickles and sauces (Emborg et al., 2005).

6.5.1.8 Irradiation

In this method, radiation rays and energies are heated up and released on the fish samples. This kills microorganisms at small doses in the fish just like radiation kills cancer cells in chemotherapy. Examples are x-rays, electromagnetic, gamma, and ultraviolet radiations. These are often applied in food preservation though not common in Africa (FAO, 2018).

6.5.1.9 Canning

Canning employs combination of different methods of preservation such as use of high temperature, asepsis, controlled environment, and packaging with the use of heat as the most pronounced. When appropriate containers are used as packaging materials for shellfishes and fishes, reentry of microorganisms becomes very difficult, therefore, preventing spoilage (Alcicek and Atar, 2010). Apart from the canned products being ready-to-eat products, they are safe for consumers since there is proper cooking among the processes involved. Canning is a permanent method of preservation and the products can be stored at room temperature for a very long time (Alcicek and Atar, 2010).

The **important operations** here are preparation of raw material, blanching, filling, medium addition, exhaustion, seaming, retorting, cooling, drying, packaging, and labeling (Alcicek and Atar, 2010).

6.5.1.10 Fermentation

This is one of the oldest preservation technologies that prevents spoilage through the breaking down of sugars into simpler substances, including alcohol, acids, and some other gases. Fermentation improves the physicochemical characteristics and nutritional quality, leading to favorable sensory qualities of the product (Tanasupawat and Visessanguan, 2013). The method is receiving serious attention now more than ever before to produce functional foods with therapeutic values (FAO, 2018). Fermented fish are the products of freshwater and marine fish, shellfish, and crustaceans that are processed with salt to produce simpler substances that help keep the fish for a longer time with more nutritional value products (Tanasupawat and Visessanguan, 2013).

The method is cheap though with changed characteristics and longer shelf life with higher nutritional composition.

6.5.2 Value-added fish products

Valued-added fish is always mentioned in export-oriented fish industry due to the increased realization of valuable foreign exchange from selling these fish products. Value is usually added according to what is required in the different. These range from live fish to shellfish to convenience products that are ready to eat such as fish fingers, fish cakes, fish balls, fish barbeque, breaded shrimp, lobster, oyster, and scallops. Value addition changes the nature and the form of fish products and improves the value when it is to be sold. They are activities that improve the usability, viability, and culinary attributes of the fish products.

6.6 Emerging trends

HACCP is one of the areas of fish processing and preservation gaining attention now (USDA, 2019). The system works by identifying hazards that occur at different points in preservation methods and highlight measures to be taken to control the damage the hazards could have imparted on the product. The preempted expectation for these hazards and the critical control points (CCP) identification are the major systems of HACCP. The process builds a logical system around the control and prevention of hazards through inspectional approach. Once the system is built, the overall quality assurance is directed toward the CCPs and a little away from the final product evaluation which is safer and cheaper. The system is ideal when there are scarce resources. The principle directs more work and resources toward control, which is most useful for the assurance of good-quality product. It encourages remedial actions before problems occur, and builds system around measures that can be monitored easily such as time, temperature, and appearance. It also reduces cost associated with detailed analysis of the final product, predicts potential hazards, and can therefore model equations for future processes and ensures control by familiar personnel (Arason et al., 2014).

6.6.1 Fish packaging

This is another area where improvement is expected in days to come. Appropriate packaging can help achieve fish products that are free of microorganisms of public health importance and other harmful microorganisms. This is irrespective of the storage condition through distribution especially when commercial sterilization was attained during processing. Aseptic techniques have great application in packaging as this will help prevent recontamination or introduction of these microorganisms. Hermitical sealing will follow after with the product closely sterilized (SSTL, 2009). Controlled and modified packaging is also a trend that can be studied adequately to fit in here as controlled environment can be designed to store the product even at ambient temperature for as long as desired. Vacuum packaging was found to extend the shelf life of smoke-dried catfish for 14 days more than the air-packed catfish. Also, the use of opaque or black material with the same thickness and characteristics extended the shelf life of same smoke-dried catfish for 21 days, especially during the dry season from September to February in the year (Abiodun-Solanke, 2019).

6.6.2 Appropriate labeling

Here is another improvement that can help other fish products just like the canned and high-valued products. Having information about the nutritional status, right instruction, storage guidelines in addition to the production date, and shelf life of all processed and fresh fish products will improve the whole fisheries value chains. The information in the labels will guide consumers on appropriate comparison between similar products to make healthy food choices through the provided relevant information (SSTL, 2009).

The seafood industry is increasingly moving toward new product development and adopting innovative processing methods that allow for doing things that were perceived impossible in the past and promoting efficiency with the same tools available. High-pressure processing (HPP) is one of such innovative developments that is trendy and also with potentials to improve efficiency. HPP is a low-temperature pasteurization technique where fish products are subjected to a very high level of hydrostatic pressure (e.g., up to 87,000 psi) for a few seconds to a few minutes. The effect of the HPP treatment is the inactivation of vegetative microorganisms, which extends product's shelf-life, yet there are minimal changes to product's texture, flavor, and nutritional value (USDA, 2019).

Freeze drying is also an emerging trend that involves flash freezing the product followed by water removal via sublimation (i.e., water is removed by going from a frozen state to a gaseous state without passing through the liquid state). Dehydration is done under vacuum with the product being in a solid frozen state (SSTL, 2009). More innovative products can be achieved through freeze drying technology either from traditional or new species (e.g., sea cucumbers). Seafood-based products are now being produced for more refined markets such as nutraceutical, functional foods, and cosmetic and pharmacological uses. But because freeze drying is a complex and expensive form of dehydration, its industrial applications are limited to high-end, high-value products such as functional foods derived from seafood (e.g., collagen powder, shark cartilage, lyprinol, and sea cucumber powder) (USDA, 2019).

In the quest to develop new and innovative products, some seafood processors are using enzyme applications. The application of *meat glue or transglutaminase (TG)* is a relatively new development in seafood processing. TG is an enzyme that occurs naturally in living things such as plants, animals, and bacteria and it speeds up the rate of the chemical bonding between amino acids, thereby giving the ability to bind together protein-containing foods (USDA, 2019). The TG acts as a "glue" to stick pieces of protein together, which can then be formed into various shapes. Innovative seafood products can be created from process trimmings, for example, or novel protein combinations such as lamb and scallops can be produced (USDA, 2019). By using leftover seafood pieces from other processing activities to create novel seafood products, the application of TG can potentially increase product yield and minimize processing waste.

Other applications of enzymes in seafood processing are aimed at converting by-product or waste materials into valuable products (Manuel et al., 2018).

There can be development of pH-driven protein recovery from waste and also the development of value-added food such as surimi from recovered proteins and lipids. Fish and shellfish processing waste will be minced, the pH of the mince adjusted to the pH will result in the solubilization of highest muscle protein, and the mince will thereafter be centrifuged. The top fraction containing lipids will be collected, the middle fraction containing solubilized muscle proteins will also be collected, followed by centrifugation, and then the precipitated proteins will be collected.

6.7 Conclusion

The driving factors for the food industry globally are health, especially through high nutritional value of food products, rising demand for fast/convenient and at the same time healthy foods, and potential growth in the sector. Fish and shellfish can be said to fit positively with these growing factors and so tipped as one of the food products to focus on for unprecedented advances globally (Ravishankar, 2016). There is still a wide gap between this increasing demand and supply for fish products (FAO, 2018), to fill this gap considerably; improved culture is an option to explore. Aquaculture is on the increase globally; however, mariculture which is the culture of marine resources, is still at infancy in many developing economies. To effectively maximize the potentials in this sector, production is advised to improve but much more improvement is required from the processing industry to reduce postharvest losses conservatively put at 40% currently and utilize the anticipated production growth in the sector (FAO, 2018).

Conclusively, this chapter has shown the common practices and gaps in the fisheries sector, while projecting the potentials and future of fish and shellfish processing globally. The sector from this study has huge potential for growth and therefore should be harnessed adequately for improved health, food, and nutritional security of the growing population in the world.

References

Abiodun-Solanke, A. O. (2019). *Performance Evaluation of UI CORAF Smoking Kiln Using the Product Quality of Hot-Smoked Catfish and Tilapia*. Ongoing PhD work. Department of Fisheries and Aquaculture, Faculty of Renewable Natural Resources, University of Ibadan.

Akande, G. and Diei-Ouadi, Y. (2010). *Post-Harvest Losses in Small-Scale Fisheries: Case Studies in Five Sub-Saharan African Countries* (72p.). FAO Fisheries and Aquaculture Technical Paper. No. 550. Rome: FAO.

Alcicek, Z. and Atar, H. N. (2010). The effects of salting on chemical quality of vacuum packed liquid smoked and traditional rainbow trout (Oncorhyncus mykiss) fillets during chilled storage. *Journal of Animal and Veterinary Advances* 9, 2778–2783.

Anderson, J. L., Asche, F. and Garlock, T. (2017). Aquaculture: its role in the future of food. In: *Frontiers of Economics and Globalization* (Volume 17, pp. 159–173). Bradford, UK: Emerald Publishing Limited.

Arason, S., Van Nguyen, M., Thorarinsdottir, K. A. and Thorkelsson, G. (2014). Preservation of fish by curing. In: I. S. Boziaris (Ed.), *Seafood Processing. Technology, Quality and Safety* (pp. 129–151). West Sussex, UK: John Wiley & Sons, Ltd.

Baird-Parker, T. C. (2000). The production of microbiologically safe and stable foods. In: B. M. Lund and T. C. Baird-Parker (Eds.), *The Microbiological Safety and Quality of Food* (pp. 3–18). Gaithersburg, MD: Aspen Publishers Inc. ISBN: 0834213230.

Belton, B. and Thilsted, S. H. (2014). Fisheries in transition: food and nutrition security implications for the global south. *Global Food Security*, 3(1),59–66.

Béné, C., Barange, M., Subasinghe, R., Pinstrup-Andersen, P., Merino, G., Hemre, G. I. and Williams, M. (2015). Feeding 9 billion by 2050 – putting fish back on the menu. *Food Security*, 7, 261–274. DOI: 10.1007/s12571-015-0427-z

Beveridge, M. C. M., Phillips, M. J., Dugan, P. and Brummett, R. (2010). Barriers to aquaculture development as a pathway to poverty alleviation and food security. In: *Advancing the Aquaculture Agenda: Workshop Proceedings* (E. Andrews-Couicha, N. Franz, K. Ravet, C. C. Schmidt and T. Strange, Eds., pp. 345–359), Paris: OECD Publishing.

Birkerland, S., Rørå, A. M. B., Skåra, T. and Bjerkeng, B. (2004). Effects of cold smoking procedures and raw material characteristics on product yield and quality parameters of cold smoked Atlantic salmon (Salmo salar) fillets. *Food Research International* 37, 273–286.

Byrd-Bredbenner, C., Moe, G., Beshgetoor, D. and Berning, J. (2009). *Perspectives in Nutrition (Eighth Edition)* (686pp.). New York, NY: McGraw-Hill.

Cardinal, M., Knockaert, C. and Torrissen, O. (2000). Relation of smoking parameters to yield, colour and sensory quality of Atlantic salmon (Salmo salar). *Food Research International*, 34, 537–550.

Emborg, J., Laursen, B. G. and Dalgaard, P. (2005). Significant histamine formation in tuna (Thunnus albacares) at 2°C: effect of vacuum-and modified atmosphere-packaging on psychrotolerant bacteria. International Journal of Food Microbiology, 101, 263–279. DOI: 10.1016/j.ijfoodmicro.2004.12.001

FAO. (2016). *The State of World Fisheries and Aquaculture. Contributing to Food Security and Nutrition for All* (200pp.). Rome. www.fao.org/3/a-i5798e.pdf. (accessed January 28, 2017).

FAO. (2017). *Strengthening Sector Policies for Better Food Security and Nutrition Results*. Fisheries and Aquaculture Policy Guidance Note 1.

FAO. (2018). *The State of World Fisheries and Aquaculture 2018 – Meeting the Sustainable Development Goals*. Rome. http://www.fao.org/3/i9540en/I9540EN.pdf (accessed September 14, 2018).

FMARD. (2016). *The Agriculture Promotion Policy (2016–2020). Building on the Successes of the ATA, Closing Key Gaps* (59p.). Policy and Strategy Document. http://fmard.gov.ng/wp-content/uploads/2016/03/2016-Nigeria-Agric-Sector-Policy-Roadmap_June-15-2016_Final.pdf (accessed January 14, 2017).

Food and Agriculture Organisation (FAO). (2005). *Post-Harvest Changes in Fish*. Rome, Italy: FAO Fisheries and Aquaculture Department, Food and Agriculture Organization. http://www.fao.org/fishery/topic/12320/en.

Food and Nutrition Board. (2007). *Seafood Choices: Balancing Benefits and Risks*. Washington, DC: Food and Nutrition Board, Institute of Medicine, National Academies Press.

Golden, C. D., Allison, E. H. and Cheung, W. W. L. (2016). Nutrition: fall in fish catch threatens human health. *Nature*, 534, 317–320.

Gram, L. and Huss, H. H. (2000). Fresh and processed fish and shellfish. In: B. M. Lund, A. C. BairdParker and G. W. Gould (Eds.), *The Microbiological Safety and Quality of Foods* (pp. 472–506). London: Chapman and Hall. ISBN: 10: 0834213230.

International Institute of Refrigeration. (2006). Definitions and explanations. In: IIF-IIR (Ed.), *Recommendations for the Processing and Handling of Frozen Foods* (4th ed., pp. 8–33). Paris, France: International Institute of Refrigeration.

Janjai, S. and Bala, B. K. (2012). Solar drying technology. *Food Engineering Reviews* 4, 16–54.

Jessen C, Voolstra C. R and Wild C (2014). In situ effects of simulated overfishing and eutrophication on settlement of benthic coral reef invertebrates in the Central Red Sea.Peer reviewed journal 2:e339; DOI 10.7717/peerj.339

Kabahenda M. K., Omony, P. and Hüsken, S. M. C. (2009). *Post-Harvest Handling of Low Value Fish Products and Threats to Nutritional Quality: A Review of Practices in the Lake Victoria Region*. Project Report, 1975. 15. Fisheries and HIV/AIDS in Africa: Investing in Sustainable Solutions. World Fish Center.

Kader, A. A. (2005). Increasing food availability by reducing postharvest losses of fresh produce. *Acta Horticulturae*, 3, 2169–2176.

Kumolu-Johnson and Ndimele, P. E. (2011). A review on post-harvest losses in artisanal fisheries of some African countries. *Journal of Fisheries and Aquatic Science*, 6, 365–378.

Leistner, L. (2000). Basic aspects of food preservation by hurdle technology. *International. Journal of Food Microbiology*, 55, 181–186. DOI: 10.1016/S0168-1605(00)00161

Liverpool-Tasie, L. S. O., Sanou, A. and Reardon, T. (2018). *Demand for Imported-Frozen Versus Domestic-Traditionally Processed Fish in Africa: Panel Data Evidence from Nigeria. Nigeria Agricultural Policy Project*. Research Paper 115 February 2018. Feed the Future Innovation Lab for Food Security Policy.

Mafimisebi, T. (2012). Comparative analysis of fresh and dried fish consumption in rural and urban households in Ondo State, Nigeria. In: *Visible Possibilities: The Economics of Sustainable Fisheries, Aquaculture and Seafood Trade: Proceedings of the Sixteenth Biennial Conference of the International Institute of Fisheries Economics and Trade, July 16–20, Dar es Salaam, Tanzania. Tanzania Proceedings*. Corvallis, OR: International Institute of Fisheries Economics and Trade (IIFET). http://ir.library.oregonstate.edu/xmlui/bitstream/handle/1957/35117/Mafimisebi.pdf?sequence=4 (accessed December 31, 2017).

Manuel B., Tarub B., Malcolm C.M.B., Kevern I.C., Funge-Smith S and Poulain F (2011). Impacts of climate change on fisheries and aquaculture. Synthesis of current knowledge, adaptation and mitigation options FAO Fisheries and Aquaculture Technical paper. Rome. Italy

Mgawe, Y. and Bawaye, S. (2012) *Report of the Regional Training Workshop on Postharvest Fish Losses in Small-Scale* (p. 53). Ebene, Mauritius: Programme of the Indian Ocean Commission.

Miladi, H., Chaieb, K., Bakhrouf, A., Elmnasser, N. and Ammar, E. (2008). Freezing effects on survival of Listeria monocytogenes in artificially contaminated cold fresh-salmon. *Annals of Microbiology*, 58, 471–476. DOI: 10.1007/BF03175545

Montero, P., Gómez-Guillen, M. C. and Borderías, A. J. (2003). Influence of salmon provenance and smoking process on muscle functional characteristics. *Journal of Food Science*, 68(4), 1155–1160.

Muratore, G., Mazzaglia, A., Lanza, C. M. and Licciardello, F. (2007). Effect of process variables on the quality of swordfish fillets flavored with smoke condensate. *Journal of Food Processing and Preservation*, 31, 167–177.

Nuin, M., Alfaro, B., Cruz, Z., Argarate, N. and George, S. (2008). Modeling spoilage of fresh turbot and evaluation of a Time-Temperature Integrator (TTI) label under fluctuating temperature.

International Journal of Food Microbiology, 127, 193–199. DOI: 10.1016/j. ijfoodmicro.2008.04.010

Rahman, M. S., Khatun, M. B., Hossain, M. N., Hassan, A. A. and Nowsad, K. M. (2013). Present scenario of landing and distribution of fish in Bangladesh. *Pakistan Journal of Biological Sciences*, 16, 148–149.

Ravishankar, C. N. (2016). Recent advances in processing and packaging of fishery products: a review. In: *2nd International Symposium on Aquatic Products Processing and Health ISAPPROSH 2015. Aquatic Procedia*, 7(2016), 201–213. www.sciencedirect.com.

Sen, D.P. (2005). *Advances in Fish Processing Technology* (p. 448). Mumbai, India: Allied Publisher Private Limited. ISBN: 81-7764-655-9.

Silva, J. L. and Stojanovic, J. (2013). *History of Freezing and Chilling*. FST 4/6583 Lecture note. Department of Food Science, Nutrition and Health Promotion.

Siringan, P., Raksakulthai, N. and Yongsawatdigul, J. (2006). Autolytic activity and biochemical characteristics of endogenous proteinases in Indian anchovy (Stolephorus indicus). Food Chemistry, 98, 678–684. DOI: 10.1016/j.foodchem.2005.06.032

Senior Secondary Technology and Living (SSTL). (2009). *Food Preservation Technology*. https://edblog.hkedcity.net/te_tl_e/wpcontent/blogs/1685/uploads/FST/Food%20Booklet% 2010%20eng.pdf.

Tanasupawat, S. and Visessanguan, W. (2013). Fish fermentation. In: I. S. Boziaris (Ed.), *Seafood Processing: Technology, Quality and Safety*. New York, NY: John Wiley & Sons.

Thorarinsdottir, K. A., Arason, S., Sigurgisladittir, S., Valsdottir, T. and Tornberg, E. (2011). Effect of different pre-salting methods on protein aggregation during heavy salting of cod fillets. *Food Chemistry*, 124, 7–17.

United States Department of Agriculture. (2019). *Functional Ice' Shows Food Industry How to Keep Cool and Reduce Loss*. National Institute of Food and Agriculture. https://nifa.usda.gov/.

Varlet, V., Serot, T. and Knockaert, C. (2007). Organoleptic characterization and PAH content of salmon (Salmo salar) fillets smoked according to four industrial smoking techniques. *Journal of the Science of Food and Agriculture*, 87, 847–854.

Wardlaw, G. M. and Smith, A. M. (2009). *Contemporary Nutrition* (750 pp.). New York, NY: McGraw-Hill.

Yongsawatdigul, J., Park, J.W., Virulhakul, P. and Viratchakul, S. (2000). Proteolytic degradation of tropical tilapia surimi. *Journal of Food Science*,65, 129–133. DOI: 10.1111/j.1750-3841.2000.00129.pp.x

Yahva, Y. I. (2011). *Post-Harvest Fish Loss Assessment in Small-Scale* Fisheries, (FAO) Fisheries *and Aquaculture Technical* Paper. *A Guide for the Extension Officer* (p. 559). Rome, Italy.

Part II: **Food microbiology use of microorganisms**

Titilayo Adenike Ajayeoba, Oluwatosin Mary Kaka,
Oluwatosin Akinola Ajibade

7 Microbial food spoilage of selected food and food products

7.1 Introduction

Foods undergo different levels of deterioration after processing and production. The nature of the raw materials and processing conditions possibly determines physical appearance, temperature, atmosphere, water, activity, pH, and microflora that develops during production and storage (Gram et al. 2002). Deterioration may include losses in organoleptic desirability, nutritional value, safety, and aesthetic appeal. Deteriorated foods may be safe to eat, if there are no pathogens or toxins present. They may not cause disease, but differences in texture, smell, taste, or appearance make them to be rejected by consumers. However, spoilage sets in when these products have microbial interactions, and metabolites could lead to food poisoning and intoxication. This is characterized by production of visible microbial growth and production of off-flavor/odor (Siegmund and Pöllinger-Zierler 2006). The implications of consuming contaminated food resulted in various illnesses and diseases. Depending on the type of microorganism, quantity consumed, and immune status, the symptoms could be evident within a few hours to several days. Typical symptoms include diarrhea, vomiting, abdominal cramps, headaches, nausea, dry mouth, and difficulty swallowing and fluke-like symptoms (such as fever, chills, and backache) (Addis and Sisay 2015).

Depending on their shelf life and ease of spoilage, foods can be classified into three main groups: stable foods, semiperishable food, and perishable foods. Throughout production and storage, each and every food product harbors its own unique and distinctive microflora at every stage (Gram et al. 2002). Microbial growth is the main cause of food quality reduction and shelf life deterioration. Microbial contamination may occur naturally via attachment if the food material can be a growing substrate or through improper handling procedures. The existence of a microorganism is an indicator of contamination at any stage of production (Caldera 2014), especially if the presence of such microorganism(s) is not desired. The predominant spoilage of microflora in a food is determined by microbial types, food types, and food environment. Due to chemical, biological, or physical agents, food spoilage may occur (Principles of

Titilayo Adenike Ajayeoba, Department of Microbiology, Faculty of Science, Adeleke University Ede, Nigeria; Department of Biotechnology and Food Technology, Durban University of Technology, Durban, South Africa
Oluwatosin Mary Kaka, Oluwatosin Akinola Ajibade, Department of Biotechnology and Food Technology, Durban University of Technology, Durban, South Africa

https://doi.org/10.1515/9783110667462-007

Food Spoilage). The most important factors influencing microbial growth in food can be either intrinsic [nutrient content, water activity (a_w), pH value, redox capacity, the availability of antimicrobial substances and mechanical barriers to microbial invasion, and environment-related factors in which the foothold occurs] or extrinsic (interactions between food contaminating microorganisms, food composition, packaging and storage conditions of food products, e.g., their ability to use various nutritional sources, withstand stress, and produce growth promoters or inhibitors of other microorganisms; processing factors include treatments such as heating, cooling, and drying, which affect the composition of the food and also the types and numbers of microorganisms that remain in the food after treatment) (Hamad 2012).

7.2 Flour and bakery products

In most of the countries and cultures, bakery products are the main staple foods. Bakery products provide most of the food calories and about half of our protein requirements as a valuable source of nutrients for our diet. In general, the use of whole and organic grains and other natural ingredients has increased the popularity of bakery products (Saranraj and Geetha 2012). The cleaning and grinding processes have little or no direct impact on the level of contamination in wheat; thus, the initial grain quality has a strong influence on the overall safety and quality of the finished products. The majority of microorganisms, including pathogenic and spoilage microorganisms, and mycotoxins originally found on wheat, could therefore be assumed to be present in the milled material (Sabillón and Luis 2014).

Bakery products are spoiled microbially, chemically, and physically because the most common factor of bakery products is a_w. However, the major economic importance of bakery products identified by many researchers is mold spoilage, especially mycotoxigenic fungi (Pitt and Hocking 2009), such as *Aspergillus* and *Penicillium* species, *Bacillus cereus* (Saranraj and Geetha 2012), pathogenic and fecal microorganisms such as *Escherichia coli*, *Salmonella* spp., *Bacillus cereus*, and coliforms (Sabillón and Luis 2014). Although chemical (oxidative and hydrolytic rancidity) and physical spoilage (staling/moisture loss or gain) reduces the shelf life of low and intermediate moisture bakery products, microbial spoilage by bacteria, yeast, and molds is the major issue of high moisture products, that is, food products with $a_w > 0.85$. In addition, several bakery products have also been involved in foodborne diseases caused by *Salmonella* spp., *Listeria monocytogenes*, and *Bacillus cereus*, while *Clostridium botulinum* is a concern in high-humidity bakery products packed under modified atmospheres (Smith et al. 2004). The effect of microbial spoilage on bakery product is shown in Table 7.1. This is greatly influenced by composition of the bakery product (Ijah et al. 2014), a_w (Saranraj and Sivasakthivelan 2016), temperature of the environment of the

Table 7.1: Some microorganisms associated with spoilage of flour products.

Product	Characteristics on food	Associated microorganisms	Point/causes of possible contamination/ spoilage during production	References
Bread/ biscuits/ cake	Ropiness	Bacterial contamination: *B. subtilis, B.licheniformis, B. megaterium, B. cereus, B. pumilus*	Humid environment (e.g., packed bread) and pH values higher than 5.3, storage temperature, handling, packaging, distribution 21–30 °C, pH values 4.0–4.5	Pateras (1999), Saranraj and Geetha (2012), Ravimannan, Sevvel, and Saarutharshan (2016), and Saranraj and Sivasakthivelan (2016)
	Moldiness	Fungi contamination: *Eurotium* spp., *Aspergillus, Penicillium, Cladosporium, Mucor, Rhizopus,*		
	Off-odors from fermentative and filamentous yeast	Yeast contamination: *Saccharomyces cerevisiae, Pichia burtonii*		

baked product (Morassi et al. 2018), baking and cooling procedures of the products, type of packaging material and storage, and relative humidity of the environment (Smith et al. 2004).

7.3 Sugar and confectioneries

Sugar products include sucrose (cane and beet sugar), molasses, syrups, maple sap and sugar, honey, and candy. Confectioneries are sweet, shelf-stable goods such as hard candies, toffee, caramel, fondants, creams, chocolate, sugar confectionery (non-chocolate), liquid sugars, sugar syrups, and honey. Products grouped under the sugar confectionery category include hard candy, soft/gummy, candy, caramel, toffee, licorice, marzipan, creams, jellies, and nougats and pastes (Thompson 2009). These have low a_w, below 0.85, making these products resistant to some bacterial growth/survival in such products (Konkel 2001); however, most of the unit operations during processing can be considered as deterministic factors of product safety (Nascimento and Mondal 2017). In order for spoilage to occur, these products might have been

Table 7.2: Some microorganisms associated with spoilage of confectionery products.

Product	Characteristics on food	Associated microorganisms	Point/causes of possible contamination/spoilage during production	References
Chocolate	Quality deterioration	Bacterial contamination: Salmonella spp., Micrococcus, and Bacillus	Handling, packaging of finished product	Douglas et al. (2000)
	Quality deterioration	Fungi contamination: Penicillium brevicompactum and Eurotium repens	Production environment (factory)	De Clercq et al. (2015)
Honey	Changes in appearance and taste, CO_2, alcohols, nonvolatile organic acid changing in flavor, crystallization, black color	Bacterial contamination: Pseudomonas spp., Bacillus spp., Clostridium spp., coliforms Yeast contamination: Zygosaccharomyces mellis, Z. richteri, Z. nussbaumeri, and Torula mellis	Bee hive, residual level of contamination during production process, handling, type of container, environmental pollution and length of storage	Snowdon and Cliver (1996), Różańska (2011), and Silva et al. (2017)
Candy	Physical and chemical instability of product	Bacterial contamination: Bacillus, Clostridium, Leuconostoc spp., Lactobacillus spp. Fungi contamination: Aspergillus niger, A. sydowi, Penicillium, A. glaucus, and P. expansum Yeast contamination: Saccharomyces spp., Hansenula anomala, Pichia membranefaciens, S. rouxii, S. heterogenicus, S. mellis, and Torulopsis	Raw materials, dusts, floss of crispiness, and changes in water activity, form of packaging, composition of products, handler	Fontana (2006), Thompson (2009), and Rawat (2015)

Product	Effect	Microbial contamination	Factors	References
Sugar (sucrose)	Loss of product recovery during processing, quality deterioration, and product spoilage resulting in some exopolysaccharide-producing microorganism. Interference in crystal formation	Bacterial contamination: *Leuconostoc, Bacillus, Micrococcus, Flavobacterium, Alcaligenes, Xanthomonas, Pseudomonas, Erwinia,* and *Enterobacter* Yeast contamination: *Saccharomyces, Candida,* and *Pichia*	Production, production environment, packaging storage temperature	Hector et al. (2015) and Rawat (2015) Mwambete, Temu and Fazleabbas (2009), Filteau et al. (2012), and Tukur, Muazu, and Mohammed (2012)
Maple syrup/sour syrup/red syrup	Ropy or stingy characteristics Quality and sensorial deterioration/cloudiness, pigmentation	Bacterial contamination: *Salmonella, Vibrio* spp., *Bacillus subtilis, Enterobacter aerogenes, Leuconostoc, Micrococcus roseus, Pseudomonas fluorescens, Alcaligenes,* and *Flavobacterium* Fungal/yeast contamination: *Aspergillus, Penicillium, Mrakia* spp., *Mrakiella* spp., *Guehomyces pullulans*	Processing, production site, storage patterns, sales outlet	
Sap		Bacterial contamination: *Pseudomonas* and *Ralstonia, Staphylococcus, Plantibacter,* and *Bacillus* Fungal/yeast contamination: *Penicillium* spp., *Cryptococcus, Metschnikowia, Candida* and *Aureobasidium* species, *Nadsonia fulvescens, Guehomyces pullulans,* and *Xanthophyllomyces dendrorhous*	Storage temperatures, thawing, type of sap source	Lagacé et al. (2004), and Nikolajeva and Zommere (2018)

contaminated during production or the packaging material damaged, which might in turn trigger the level of a_w. In addition to a_w, pH, processing and storage temperatures, and the presence of preservatives affect the physical, chemical, and microbial stabilities of high-sugar products (Thompson 2009). Spoilage is due to the growth of yeasts and molds, osmophilic/xerophilic, unless the a_w is below 0.61. It is possible to have *Salmonella*. It includes osmophilic/xerophilic yeasts and molds that can ruin the rest of the products.

The processing of chocolate and some other confectioneries includes friction, mixing, conching, tempering, and molding. Depending on the equipment, final shape, and quality, the conching process uses temperatures between 50 and 80 °C for 2–72 h, which is not a reliable step to control heat-resistant bacteria such as *Salmonella*; hence, the quality of the ingredients plays an important role for the safety of the final product (Nascimento and Mondal 2017). Although the processing might have destroyed the presence of toxigenic fungi, ochratoxin A is usually known to contaminate raw products such as cocoa, and the extent of reduction during roasting depends largely on the initial level of contamination.

The process of manufacturing sucrose, which is generally obtained from sugarcane or sugar beet, may be contaminated by soil, insects, workers, and tools, or by the processing environment. The product is exposed to heating conditions to obtain refined sugar, where most vegetative cells are destroyed with the exception of mesophilic and thermophilic spore-forming, like *B. stearothermophilus*, *B. coagulans*, *Desulfotomaculum nigrificans*, and *C. thermosaccharolyticum* (ICMSF 2005; Nascimento and Mondal 2017). In the food industry where sugar is used as an ingredient, these bacteria cause serious problems with spoilage. Additionally, lactic acid bacteria such as *Leuconostoc* and *Lactobacillus* may be introduced into the product after thermal processing through cross-contamination as slime formation in sugar juice has been linked with these microbes (Nascimento and Mondal 2017). *Torulopsis*, *Zygosaccharomyces*, and *Hansenula* are yeasts that are often identified in refined sugar and can cause problems in high-moisture products (Thompson 2009).

Honey is the natural sweet substance produced by secretions or excretions of honey bees. It consists mainly of fructose, glucose, and water. With low a_w, the presence of antimicrobial substances, low protein content, and high viscosity does not favor microbial growth but *Gluconobacter*, *Pseudomonas*, *Micrococcus*, and *Lactobacillus* are the main genus of fermenting honey-isolated bacteria and *Clostridium botulinum* is the only honey-related microorganism of public health concern (Nascimento and Mondal 2017). However, the viability of *C. botulinum* spores in honey is directly related to the storage temperature.

Sugar syrups are known as liquid sugar, starch-based syrup, and maple tree sap. With individual unit operations, each syrup has a different processing history that directly affects the microbial quality of each product. Raw materials and the processing environment are the likely sources of contamination (Nascimento and Mondal 2017). In maple syrup, isolated microorganisms include *Aerobacter*, *Leuconostoc*, *Bacillus* and

Pseudomonas fluorescens group and two subgroups, *Rahnella* spp., *Janthinobacterium* spp., *Leuconostoc mesenteroides*, and *yeast* (*Mrakia* spp., *Mrakiella* spp., and *Guehomyces pullulans*) (Filteau et al. 2012). The effect of microbial spoilage on sugar confectionery products is given in Table 7.2.

7.4 Milk and milk products

Milk and milk products are an important part of human diet and are known as carriers of proteins of higher biological activity, calcium, essential fatty acids, amino acids, fat, water-soluble vitamins, and many bioactive compounds of great importance for several biochemical and physiological functions. These rich nutritional components make it susceptible to contamination by a host of pathogenic microbes that could cause human diseases (Usta and Yilmaz-Ersan 2013). Milk, however, is a highly perishable food that is prone to do the activity of the natural enzyme and contaminating microorganisms in rapid spoilage, thus making the consumption unhealthy (Rawat 2015). The action of microbial metabolism was proposed to catabolize and degrade uncharged milk substrates (such as lactose, proteins, and lipids) to charged electrical conductive species during the milk deterioration process, thus proposing to closely associate the growth of microorganisms and their associated metabolism with the accumulation of charged ionic species in milk (Mabrook and Petty 2002). Bacterial contamination may be introduced from the cow's udder and teats infections or fecal contamination, barn, milk collection materials, various ingredients added to dairy products, or from dairy farm workers and other environmental contamination (Oliver et al. 2009).

Foodborne disease resulting from milk will hinge on the extent to which the act of food safety is controlled by the initial state of raw material, production, pasteurization, processing, and distribution (te Giffel and Wells-Bennik 2010). Storage temperature, pH, packaging, and time greatly influence the survival and growth of microorganisms.

The cooling of raw milk enables control of mesophyll microbiota multiplication, predominantly saccharolytic microorganisms. These microbes are responsible for the acidification and thermal instability of milk proteins, since lactose hydrolysis is a by-product of lactic acid (Ribeiro Júnior et al. 2018). Some cold-tolerant microbes (psychrotrophs) adjust to refrigeration temperatures by producing phospholipids/neutral lipids in order to maintain their fluidity, thus permitting the continual functionality, solute transport, and secretion of extracellular enzymes (de Oliveira et al. 2015). The microbial proteases and lipases are thermostable and may remain active after the removal of the vegetative microorganisms by heat treatments (Oliver et al. 2009; Saranraj and Sivasakthivelan 2016) as a result of thermoresistant exoproteases and lipase production, which may compromise the quality of processed fluid milk and dairy products during storage with organoleptic changes like a characteristic bitter or rancid taste in cheeses or gelation and sedimentation in milk or dairy products (Ribeiro Júnior et al. 2018). Generally, the cultivable psychrotrophic bacteria in milk are represented predominantly by gram-negative genera, which include *Pseudomonas, Achromobacter, Aeromonas, Alcaligenes, Chromobacteriumm,* and *Flavobacterium* spp., and at much lower numbers by gram-positive genera including *Bacillus, Clostridium, Corynebacterium, Streptococcus, Lactobacillus,* and *Microbacterium* spp. (de Oliveira et al. 2015). Among the proteolytic species, more isolated by this study were *Lactococcus lactis, Enterobacter kobei, Serratia ureilytica, Aerococcus urinaeequi,* and *Bacillus licheniformis.* Observed among lipolytics were *E. kobei, L. lactis, A. urinaeequi,* and *Acinetobacter lwoffii.* The isolates *S. ureilytica, E. kobei, Pseudomonas* spp., and *Yersinia enterocolitica* potentially produced alkaline metalloprotease (Ribeiro Júnior et al. 2018). In a recent study, the main zoonotic pathogens identified in raw milk were *Brucella* ssp., mainly *Brucella melitensis, Listeria monocytogenes, Salmonella* spp., *Mycobacterium bovis, Yersinia enterocolitica, Streptococcus pyogenes, Streptococcus agalactiae, Escherichia coli* O157:H7, and *Enterobacter sakazakii* while the new emerging pathogens causing milk food-borne diseases include hepatitis A virus, *Mycobacterium avium* subsp. *paratuberculosis, Streptococcus zooepidemicus, Campylobacter jejuni, Citrobacter freundii, Corynebacterium ulcerans,* and *Cryptosporidium parvum* (Velázquez-Ordoñez et al. 2019).

Various processing techniques such as thermization, low-temperature long-term pasteurization, high-temperature short-term pasteurization, sterilization, ultra-high-temperature, ultraviolet treatment, and microfiltration can be used to treat raw milk to make it safer for human consumption by reducing or eliminating pathogenic micro-organisms. Pasteurization is widely adopted by different techniques but is not intended to sterilize milk, and the microbiological value of pasteurized milk is regulated by the initial raw milk flora, processing conditions, and posttreatment contamination (Sarkar 2015). A wide range of pathogenic microorganism such as *Staphylococcus* spp., *Salmonella* spp., coliforms, and some thermophilic/thermotolerant pathogens such as *Bacillus* spp., *Clostridium botulinum, Campylobacter* spp., and *Yesinia* spp. have been reported. The effect of microbial spoilage on milk products is given in Table 7.3.

Table 7.3: Some microorganisms associated with spoilage of flour products.

Product	Characteristics on food	Associated microorganisms	Point/causes of possible contamination/ spoilage during production	References
Cheese	Quality and flavor deterioration, rancidity, changes in appearance, off-taste	Bacterial contamination: *Yersinia enterocolitica, Campylobacter jejuni, Salmonella typhi, Staphylococcus aureus, Listeria monocytogenes* Fungal contamination: *Mucor, Cladosporium, Fusarium, Penicillium cyclopium, P. viridicatum, Aspergillus flavus,* and *A. ochraceus* Yeast contamination: *Debaryomyces hansenii, Kluyveromyces marxianus, Yarrowia lipolytica, Geotrichum candidum, Saccharomyces cerevisiae, Candida* spp., and *Pishia* spp.	Types/combination of starter cultures, shelf life, type and duration of storage temperature, handlers, packaging, pH, titratable acidity, sugars, and containers	Hayaloglu (2016)
Yoghurt		Spoilage acid-loving bacteria, *Candida kefir* and *Kluyveromyces marxianus, Saccharomyces cerevisiae* and *Saccharomyces bayanus, Byssochlamys, Eurotium,* and *Penicillium*		Mataragas et al. (2011)
Ice-cream		*Salmonella enteritidis, Brucella abortus, Mycobacterium tuberculosis, Salmonella* spp., *Brucella melitensis*		Wallace (1938)

(continued)

Table 7.3 (continued)

Product	Characteristics on food	Associated microorganisms	Point/causes of possible contamination/spoilage during production	References
Kefir	Changes in viscosity	*Escherichia coli* O157:H7 and *Staphylococcus aureus, Leuconostoc, Micrococcus* spp., *Candida kefyr*	Fermentation and pH conditions, geographical location	Hecer, Ulusoy and Kaynarca (2019); KIVANC and YAPICI (2019)
Ryazhenka	Off-odor, other changes in	*Staphylococcus aureus, Salmonella* species, *Listeria* spp., *Listeria monocytogenes*	Storage conditions, starter culture, processing equipment	Okonkwo (2011)
Mozzarella	physicochemical qualities, off-taste	*Klebsiella pneumoniae, Klebsiella oxytoca, Enterobacter aerogenes, Escherichia coli, Staphylococcus* spp., *Salmonella* spp., *Listeria monocytogenes* Fungi/yeast contamination: *Debaryomyces hansenii, Candida lusitaniae, C. parapsilosis, C. catenulate, C. guilliermondii, C. tropicalis, C. parapsilosis, C. lusitaniae, C. catenulata, C. rugosa,* and *C. krusei*	Air, clothes, hands, apparatus and equipment, contact, handlers during the production process and storage	Massa et al. (1992) and Facchin et al. (2013)
Khoa		*Aspergillus niger, Aspergillus flavus, Penicillium citrinum, Rhizopus stolonifer, Aspergillus versicolor, Penicillium frequentans, Mucor* sp., *Aspergillus fumigatus, Aspergillus parasiticus,* and *Fusarium* sp.	Manufacturing and handling practices, tropical climatic factors, and high humidity	Karthikeyan and Pandiyan (2013)
Nunu		E. coli, S. aureus, and Shigella sp.		Okonkwo (2011)
Amasi		Escherichia coli O157:H7		Dlamini and Buys (2009)

7.5 Egg and poultry products

Eggs contain 49% water, 34% lipid, 16% protein, 0.1% carbohydrate, and 1% ash. Egg white contains lysozyme, avidin, conalbumin, and protease inhibitors, hence, their inhibitory effects on microorganisms. In addition, the egg white's high pH (above 9.3) is bactericidal to gram-positive bacteria and yeasts but the yolk's nutrient content and pH (around 6.8) make it an excellent microorganism growth medium. The contents of eggs are usually sterile (Techer, Baron and Jan 2014) and are protected due to the possession of certain structures, including shell, shell membrane, cuticle, lysozyme, conalbumin, avidin, and albumen pH (Wu 2014). Eggshells are not highly susceptible to spoilage microorganisms penetration and development via cracking the eggshell, excessive cleaning, and spoilage of eggs (Erkmen and Bozoglu 2016).

Eggs are of particular interest as they give a moderate calorie source (about 150 kcal/100 g), an excellent quality protein, good culinary flexibility, and low economic price, making eggs available to most citizens. Eggs are also relatively rich in fat-soluble compounds and can be a healthy part of the diet for people of different ages and at various stages of life (Miranda et al. 2015). The eggshell, outer and inner membranes, ordinarily provides resistance to bacterial penetration but microbial penetration is favored by high humidity because the pore size increases during storage and the presence of moisture enhances the entrance of the microorganisms (Erkmen and Bozoglu 2016). However, cracking will increase contamination of egg contents. Gram-negative bacteria, particularly *Salmonella* spp. are most frequently associated with eggs and egg products (Techer, Baron and Jan 2014). Although poultry usually carries *Salmonella* in their intestines, it becomes part of the bacterial flora of the eggshell. Additional sources of *Salmonella* contaminants may include nest, hens, feet, fecal matter, wash water, handling material, human hands, and soiled containers (Shebuski and Freier 2009; Erkmen and Bozoglu 2016).

A variety of bacteria and fungi contaminate eggs and egg products. These include gram-positive bacteria (genera: *Arthrobacter*, *Bacillus*, *Micrococcus*, and *Staphylococcus*); gram-negative bacteria (genera: *Acetobacter*, *Aeromonas*, *Alcaligenes*, *Enterobacter*, *Escherichia*, *Flavobacterium*, *Proteus*, *Pseudomonas*, and *Serratia*), and coliforms and fungi (*Penicillium*, *Alternaria*, *Cladosporium*, *Sporotrichum*, *Thamnidium*, *Botrytis*, *Alternaria*, and *Mucor*) (Shebuski and Freier 2009; Al-Bahry et al. 2012; Erkmen and Bozoglu 2016). If untreated, eggs may also be spoiled. Untreated eggs, depending on humidity and temperature, lose moisture and weight during storage. The egg's white becomes thinner as the egg grows older and the yolk membrane weaker (Shebuski and Freier 2009).

While muscles are sterile in healthy living birds, the gastrointestinal tract, lungs, skin, feathers, and so on harbors various microbiotas. In abattoirs, bacteria cover the surfaces, air (aerosols), and water/liquid that may have contact with the carcass. Therefore, after animal killing, carcasses and cuts can be infected with microbiota from the animal and slaughterhouse climate (Rouger, Tresse and Zagorec

2017). In addition, the different stages that may introduce microbial contamination and spoilage include the use of a water bath (hot or chilled), feather removal, and intestine evacuation.

Poultry products, particularly the poultry meat, are an excellent substrate for the growth of various microorganisms, including those with relatively complex nutritional requirements due to the water content, carbohydrates, amino acids, nucleotides, minerals, and B vitamins (Vihavainen and Björkroth 2010). Although refrigeration extends the time frame for the lag phase and reduces the overall microbial growth and delaying the onset of spoilage, temperature fluctuations may stimulate microbial growth and have a critical impact on quality (Smolander et al. 2004). The kind of wrapping, the formulation and type of the produce, and the quantity and type of initial spoilage bacteria determine the composition of the dominant bacterial population. Among the intrinsic factors of poultry meat, pH and the availability of glucose and other simple sugars affect the development of the spoilage population and the rate of microbial growth and spoilage (Dave and Ghaly 2011). The pH of the meat is highly dependent on the amount of glycogen in the muscle; in breast muscle, the postmortem glycolysis will lead to the accumulation of lactate and a reduction in pH to about 5. 7–5.9. In contrast, the muscles in poultry legs have very low initial glycogen concentration and therefore a pH of 6.2 or above. The pH is also higher in poultry skin, with pH as high as 6.6–7.2. Skin-on poultry cuts and high pH meat and meat products may also support the development of different specific spoilage organisms (SSOs) than those associated with spoilage skinless breast fillets (Vihavainen and Björkroth 2010). Poultry cuts with high pH and low glucose content leads to initiation of amino acid utilization, putrefactive odors, off-odors, and off-flavors, and formation of slime, gas, and purge are typical defects in spoilage of poultry (Vihavainen and Björkroth 2010).

Facultative anaerobic, particularly gram-negative bacteria, develop rapidly on poultry. At low temperature, *Pseudomonas* spp. frequently dominate the spoilage and produce compounds such as ammonia, dimethyl sulfide, and nonvolatile amines. Other spoilage microbes include cold-tolerant members of the family Enterobacteriaceae (*Hafnia* spp., *Serratia* spp., and *Enterobacter* spp.), *Acinetobacter* spp., lactic acid bacteria, and *Brochothrix thermosphacta*.

Although vacuum and modified atmosphere packaging, combined with cold storage, prevent fast-growing aerobic spoilage organisms, packaging of poultry under CO_2-enriched atmospheres favors the growth of psychrotrophic microorganisms with fermentative metabolism, especially genera *Lactobacillus*, *Lactococcus*, *Leuconostoc*, and *Carnobacterium* (Vihavainen et al. 2007).

7.6 Fish and fish products

Fish has a high protein value and lower fat content, which is known as highly valuable meat. In general, fish is the main edible source of omega-3 polyunsaturated fatty acid, specifically docosahexaenoic acid and eicosapentaenoic acid, which are noted both for their anti-inflammatory activity and cardiovascular protective effects (Mei, Ma and Xie 2019). Consumers need fresh fish because it is often linked to health, re-assurance, and superior taste. Due to the endogenous enzyme and rapid microbial growth naturally present in fish or from infection, fresh fish can quickly deteriorate. The spoilage process begins with rigor mortis, the process through which fish loses its flexibility due to stiffening of fish mussels after few hours of its death (Adebowale et al. 2008). It will start within 12 h of their catch in the high ambient temperatures of the tropics and degrade as a result of digestive enzymes and lipases, microbial spoil-age from surface bacteria and oxidation, and as a result, new compounds are respon-sible for the changes in odor, flavor, and texture of the fish meat (Ghaly et al. 2010). During postmortem managing and storage, holding temperature, oxygen, endoge-nous or microbial proteases, moisture may cause harmful changes in fish color, odor, texture, and taste (Sriket et al. 2011; Sriket 2014).

The microflora of a newly caught fish depends on the composition of the seawa-ter, and microorganisms readily found include species of *Pseudomonas*, *Alcaligenes*, *Vibrio*, *Serratia*, and *Micrococcus*, which may result in the production of amines and biogenic amines with unpleasant and unacceptable off-flavors (Pal et al. 2016).

For other seafood such as shell fish (Nordic and tropical shrimps), these are quite prone to rapid deterioration and spoilage because they mostly contain argi-nine, glycine, and low quantity of cysteine and methionine. Melanosis, which is the development of black spots on the cephalothorax, is the main cause of early shrimp spoilage. This method is biochemical, resulting from an enzyme complex called polyphenol oxidase, not actually due to bacterial activity. This produces benzoqui-nones, which interact with amines, amino acids, and oxygen to form the melanin, the major compound responsible for this black coloration. The action of bacteria generates other organoleptic changes (Leroi and Joffraud 2011). After this, activity of *Pseudomonas fragi* may be initiated.

Vacuum packaging is used to prolong the shelf life of fresh and processed fish and to maintain the quality. Nevertheless, to ensure food safety and to guarantee high product quality, strict temperature control and refrigeration storage are required. However, in some of the products some fermenting microorganism has been identified. In both types of rainbow trout products, *Leuconostoc mesenteroides*, *Lactobacillus sakei*, and *Lactobacillus curvatus* strains were detected. *L. mesenteroides* and *L. citreum* were only detected in the cold-smoked product and *Carnobacteria* spp. were only de-tected in the "gravad" product. Furthermore, *Lactobacillus alimentarius* was regarded as the SSO in marinated herring samples (Lyhs 2002).

Fish and fish products are involved in 10–20% of foodborne diseases, and the presence of pathogenic bacteria such as *Staphylococcus aureus*, *Salmonella* spp., pathotypes of *Escherichia coli*, and *Listeria monocytogenes* has been reported in processed fish products like smoked fish (Anihouvi et al. 2019). In addition, differ-ent fungi have been implicated in the contamination and spoilage of dried fish/

smoked fish products. Common fungi identified include *Aspergillus* and *Penicillium* but some mycotoxigenic species such as *A. niger*, *A. fumigatus*, and *A. flavus* have been reported (Prakash et al. 2011; Anihouvi et al. 2019). In dry fish, lipids affect the taste, scent, color, and mouth along with the composition of fatty acids. Lipid oxidation is catalyzed by sun, microorganisms, high temperatures, and enzyme activity. The unsaturated lipids are oxidized to produce hydroperoxides at high temperatures, causing brown and yellow discoloration of the tissue of the fish. More degradation of hydroperoxides leads to aldehydes and ketones developing a heavy rancid taste.

7.7 Meat and meat products

Meat is the animal flesh including muscles, fat, tendons, and ligaments that is basically consumed as food. Meat comprises water, protein and amino acids, minerals, fats and fatty acids, vitamins and other bioactive components, and small amount of carbohydrates depending on the source of meat (Cole and Lawrie 2013). Meat is generally eaten in the cooked, but there are many traditional recipes where meat is eaten raw or is partially cooked. Meat and meat products are usually included in human diet, due to the good-quality nutrients resulting from its high-quality protein, bioavailable minerals, vitamins (Botez et al. 2017), and the diverse forms of presentation. Meat products are an essential source of iron and zinc and also contain considerable levels of phosphorus and potassium, as well as significant amounts of other elements, such as magnesium and selenium, which are important to human health (Bou, Cofrades and Jiménez-Colmenero 2017). The microbial quality and safety of meat products is compromised by system failures or abuses during the production of food animals, the processing and distribution of products, and the preparation for consumption, as well as consumer habits. Despite constant improvements in meat processing, episodes of reported disease and concerns about meat products that act as hazard vehicles are increasing rather than decreasing (Sofos 2014).

A number of factors influence the quality of meat and meat products. Although intrinsic factors such as the species, breed or crossbred, individual genetic aspect, gender, age, and weight at slaughter affect the meat quality, extrinsic factors that are involved previous to slaughter such as management (stress agents)/behavior, diet (type of suckling – dam's milk or milk replacer; type of weaning – suckling or weaned; diets' ingredients – glycerin, dehydrated lucerne, barley straw, and different level concentrates), physical characteristics of ration, chemical characteristics of diet, and additives (hormones and palm oil supplements). And extrinsic factors involved after slaughter such as slaughter and blood loss, freezing, storage, aging, type of conservation, manner, and time of cooking can determine the type of microbial spoilage the meat/meat products undergo (Guerrero et al. 2013).

Preslaughter management of livestock and postslaughter supervision of meat play an important part in deterioration of meat quality. The glycogen content of animal muscles is reduced when the animal is exposed to preslaughter stress which in turn changes the pH of the meat, to higher or lower levels, depending on the production level of lactic acid (Miller 2002). The mechanisms of meat and meat products spoilage including microbial spoilage, lipid oxidation, and autolytic enzymatic start after slaughtering and during processing and storage (Dave and Ghaly 2011).

Cured meat products have been traditionally considered to be safe products from a microbiological point of view due to their physicochemical properties, as a low a_w and, in some products as fermented sausages, a low pH. Although the bacteriostatic and inhibitory effects of salt and nitrates and nitrites have been reported, some of the pathogenic microorganisms, such as *Staphylococcus aureus*, *Listeria monocytogenes*, *Salmonella* spp., *Clostridium botulinum*, and *E. coli*, have been reported to survive in these products (Dourou et al. 2011; Sofos 2014; Holck et al. 2017). The exterior colonization of dry fermented sausages by fungi is almost unavoidable. A safety aspect to the surface growth of molds on fermented sausages is mycotoxin production from the genus *Penicillium*, particularly ochratoxin A, patulin, citrinin, cyclopiazonic acid, and roquefortine (Holck et al. 2017).

7.8 Conclusion

Microorganisms affect a variety of food products, and spoilage occurs across all countries in the world. The intrinsic and extrinsic factors of a particular food product determine the degree of spoilage and the type of microorganisms involved in the spoilage. Although some microorganism contaminates and change the organoleptic characteristics of the food, a larger percentage secrete toxins into the food product. Food handlers and processing plant/equipment are other points/source of contamination that may result in outbreak of foodborne disease, devaluation of food products, and huge economic loss. Creating awareness, synergizing between researchers and industrial producers, development of food-related policies, and policy implementation may show appreciable reduction in microbial contamination.

References

Addis, M. and Sisay, D. (2015). A review on major food borne bacterial illnesses. *Journal of Tropical Diseases*, 3(4), 176–183.

Adebowale, B., Dongo, L., Jayeola, C. and Orisajo, S. (2008). Comparative quality assessment of fish (Clarias gariepinus) smoked with cocoa pod husk and three other different smoking materials. *Journal of Food Technology*, 6(1), 5–8.

Al-Bahry, S., Mahmoud, I., Al-Musharafi, S. and Al-Ali, M. (2012). Penetration of spoilage and food poisoning bacteria into fresh chicken egg: a public health concern. *Global Journal of Bio-Science and Biotechnology*, 1(1), 33–39.

Anihouvi, D., Kpoclou, Y. E., Abdel Massih, M., Iko Afé, O. H., Assogba, M. F., Covo, M., Scippo, M., Hounhouigan, D. J., Anihouvi, V. and Mahillon, J. (2019). Microbiological characteristics of smoked and smoked-dried fish processed in Benin. *Food Science & Nutrition*, 7(5), 1821–1827.

Botez, E., Nistor, O. V., Andronoiu, D. G., Mocanu, G. D. and Ghinea, I. O. (2017). Meat product reformulation: nutritional benefits and effects on human health. *Functional Food: Improve Health Through Adequate Food*, 167.

Bou, R., Cofrades, S. and Jiménez-Colmenero, F. (2017). Fermented meat sausages. In: J. Frias, C. Martinez-Villaluenga and E. Peñas (Eds.), *Fermented Foods in Health and Disease Prevention* (Chapter 10, pp. 203–235). Boston, MA: Academic Press. http://www.sciencedir ect.com/science/article/pii/B9780128023099000108.

Caldera, L. (2014). *Identification and Characterization of Specific Spoilage Organisms (SSOs) in Different Food Matrices*.

Cole, D. and Lawrie, R. A. (2013). *Meat*. Elsevier.

Dave, D. and Ghaly, A. E. (2011). Meat spoilage mechanisms and preservation techniques: a critical review. *American Journal of Agricultural and Biological Sciences*, 6(4), 486–510.

De Clercq, N., Van Coillie, E., Van Pamel, E., De Meulenaer, B., Devlieghere, F. and Vlaemynck, G. (2015). Detection and identification of xerophilic fungi in Belgian chocolate confectionery factories. *Food Microbiology*, 46, 322–328.

de Oliveira, G. B., Favarin, L., Luchese, R. H. and McIntosh, D. (2015). Psychrotrophic bacteria in milk: How much do we really know? *Brazilian Journal of Microbiology: [Publication of the Brazilian Society for Microbiology]*, 46(2), 313–321.

Dlamini, B. C. and Buys, E. M. (2009). Adaptation of Escherichia coli O157: H7 to acid in traditional and commercial goat milk amasi. *Food Microbiology*, 26(1), 58–64.

Douglas, S. A., Gray, M. J., Crandall, A. D. and Boor, K. J. (2000). Characterization of chocolate milk spoilage patterns. *Journal of Food Protection*, 63(4), 516–521.

Dourou, D., Beauchamp, C. S., Yoon, Y., Geornaras, I., Belk, K. E., Smith, G. C., Nychas, G. E. and Sofos, J. N. (2011). Attachment and biofilm formation by Escherichia coli O157: H7 at different temperatures, on various food-contact surfaces encountered in beef processing. *International Journal of Food Microbiology*, 149(3), 262–268.

Erkmen, O. and Bozoglu, T. F. (2016). *Food Microbiology: Principles into Practice*. John Wiley & Sons.

Facchin, S., Barbosa, A. C., Carmo, L. S., Silva, M. C. C., Oliveira, A. L., Morais, P. B. and Rosa, C. A. (2013). Yeasts and hygienic-sanitary microbial indicators in water buffalo mozzarella produced and commercialized in Minas Gerais, Brazil. *Brazilian Journal of Microbiology*, 44, 701–707.

Filteau, M., Lagacé, L., LaPointe, G. and Roy, D. (2012). Maple sap predominant microbial contaminants are correlated with the physicochemical and sensorial properties of maple syrup. *International Journal of Food Microbiology*, 154(1–2), 30–36.

Fontana, A. (2006). Water activity for confectionery quality and shelf-life. *Senior Research Scientist*, 1–20.

Ghaly, A. E., Dave, D., Budge, S. and Brooks, M. (2010). Fish spoilage mechanisms and preservation techniques. *American Journal of Applied Sciences*, 7(7), 859.

Gram, L., Ravn, L., Rasch, M., Bruhn, J. B., Christensen, A. B. and Givskov, M. (2002). Food spoilage – interactions between food spoilage bacteria. *International Journal of Food Microbiology*, 78(1–2), 79–97.

Guerrero, A., Velandia Valero, M., Campo, M. M. and Sañudo, C. (2013). Some factors that affect ruminant meat quality: from the farm to the fork. Review. *Acta Scientiarum. Animal Sciences*, 35(4), 335–347.

Hamad, S. (2012). Factors affecting the growth of microorganisms in food. In: R. Bhat, A. K. A. a. G. P. (Eds.), *Progress in Food Preservation*. Wiley Online Library.

Hayaloglu, A. (2016). *Cheese: Microbiology of Cheese*.

Hecer, C., Ulusoy, B. and Kaynarca, D. (2019). Effect of different fermentation conditions on composition of kefir microbiota. *International Food Research Journal*, 26(2).

Hector, S., Willard, K., Bauer, R., Mulako, I., Slabbert, E., Kossmann, J. and George, G. M. (2015). Diverse exopolysaccharide producing bacteria isolated from milled sugarcane: implications for cane spoilage and sucrose yield. *PloS One*, 10(12), e0145487.

Holck, A., Axelsson, L., McLeod, A., Rode, T. M. and Heir, E. (2017). Health and safety considerations of fermented sausages. *Journal of Food Quality*, 2017, 25.

ICMSF, I. C. o. M. S. f. F. (2005). Microorganisms in foods, 6: microbiological ecology of food commodities. In: *Cocoa, Chocolate, and Confectionery (Second Edition)* (pp. 467–479). New York, NY: Kluwer Academic/Plenum Publishers.

Ijah, U., Auta, H., Aduloju, M. O. and Aransiola, S. A. (2014). Microbiological, nutritional, and sensory quality of bread produced from wheat and potato flour blends. *International Journal of Food Science*, 2014, 6.

Karthikeyan, N. and Pandiyan, C. (2013). Microbial quality of Khoa and Khoa based milk sweets from different sources. *International Food Research Journal*, 20(3).

Kivanc, M. and Yapici, E. (2019). Survival of Escherichia coli O157: H7 and Staphylococcus aureus during the fermentation and storage of kefir. *Food Science and Technology*, 39, 225–230.

Konkel, P. (2001). Confectionery products. *Compendium of Methods for the Microbiological Examination of Foods*, 555–561.

Lagacé, L., Pitre, M., Jacques, M. and Roy, D. (2004). Identification of the bacterial community of maple sap by using amplified ribosomal DNA (rDNA) restriction analysis and rDNA sequencing. *Applied and Environmental Microbiology*, 70(4), 2052–2060.

Leroi, F. and Joffraud, J. (2011). Microbial degradation of seafood. In: *Aquaculture Microbiology and Biotechnology* (pp. 47–72). New Hampshire: Science Pubisher.

Lyhs, U. (2002). *Lactic Acid Bacteria Associated with the Spoilage of Fish Products*.

Mabrook, M. and Petty, M. (2002). Application of electrical admittance measurements to the quality control of milk. *Sensors and Actuators B: Chemical*, 84(2–3), 136–141.

Massa, S., Gardini, F., Sinigaglia, M. and Guerzoni, M. E. (1992). Klebsiella pneumoniae as a spoilage organism in mozzarella cheese. *Journal of Dairy Science*, 75(6), 1411–1414.

Mataragas, M., Dimitriou, V., Skandamis, P. N. and Drosinos, E. H. (2011). Quantifying the spoilage and shelf-life of yoghurt with fruits. *Food Microbiology*, 28(3), 611–616.

Mei, J., Ma, X. and Xie, J. (2019). Review on natural preservatives for extending fish shelf life. *Foods*, 8(10), 490.

Miller, R. (2002). Factors affecting the quality of raw meat. In: *Meat Processing* (pp. 27–63). Elsevier.

Miranda, J. M., Anton, X., Redondo-Valbuena, C., Roca-Saavedra, P., Rodriguez, J. A., Lamas, A., Franco, C. M. and Cepeda, A. (2015). Egg and egg-derived foods: effects on human health and use as functional foods. *Nutrients*, 7(1), 706–729.

Morassi, L. L. P., Bernardi, A. O., Amaral, A. L. P. M., Chaves, R. D., Santos, J. L. P., Copetti, M. V. and Sant'Ana, A. S. (2018). Fungi in cake production chain: occurrence and evaluation of growth potential in different cake formulations during storage. *Food Research International*, 106, 141–148.

Mwambete, K. D., Temu, M. and Fazleabbas, F. S. (2009). Microbiological assessment of commercially available quinine syrup and water for injections in Dar Es Salaam, Tanzania. *Tropical Journal of Pharmaceutical Research*, 8(5).

Nascimento, M. and Mondal, A. (2017). Microbial ecology of confectionary products, honey, sugar, and syrups. *Quantitative Microbiology in Food Processing: Modeling the Microbial Ecology*, 533–546.

Nikolajeva, V. and Zommere, Z. (2018). Changes of physicochemical properties and predominant microbiota during storage of birch sap. *International Food Research Journal*, 25(2).

Okonkwo, O. I. (2011). Microbiological analyses and safety evaluation of nono: a fermented milk product consumed in most parts of Northern Nigeria. *International Journal of Dairy Science*, 6, 181–189.

Oliver, S. P., Boor, K. J., Murphy, S. C. and Murinda, S. E. (2009). *Review Food Safety Hazards Associated with Consumption of Raw Milk*.

Pal, M., Ketema, A., Anberber, M., Mulu, S. and Dutta, Y. (2016). Microbial quality of fish and fish products. *Beverage and Food World*, 43(2), 46–49.

Pateras, I. M. C. (1999). Bread spoilage and staling. *Technology of Breadmaking*, (n), 240–261.

Pitt, J. I. and Hocking, A. D. (2009). *Fungi and Food Spoilage*. Springer.

Prakash, S., Jeyasanta, I., Carol, R. and Patterson, J. (2011). Microbial quality of salted and sun dried sea foods of Tuticorin dry fish market, southeast coast of India. *International Journal of Microbiological Research*, 2(2), 188–195.

Principles of food spoilage. In: *Food Microbiology: Principles into Practice* (pp. 269–278). https://onlinelibrary.wiley.com/doi/abs/10.1002/9781119237860.ch15.

Ravimannan, N., Sevvel, P. and Saarutharshan, S. (2016). Study on fungi associated with spoilage of bread. *International Journal of Advanced Research in Biological Sciences*, 3(4), 165–167.

Rawat, S. (2015). Food spoilage: microorganisms and their prevention. *Asian Journal of Plant Science and Research*, 5(4), 47–56.

Ribeiro Júnior, J. C., de Oliveira, A. M., Silva, F. d. G., Tamanini, R., de Oliveira, A. L. M. and Beloti, V. (2018). The main spoilage-related psychrotrophic bacteria in refrigerated raw milk. *Journal of Dairy Science*, 101(1), 75–83.

Rouger, A., Tresse, O. and Zagorec, M. (2017). Bacterial contaminants of poultry meat: sources, species, and dynamics. *Microorganisms*, 5(3), 50.

Różańska, H. (2011). Microbiological quality of Polish honey. *Bulletin of the Veterinary Institute in Pulawy*, 55, 443–445.

Sabillón, G. and Luis, E. (2014). *Understanding the Factors Affecting Microbiological Quality of Wheat Milled Products: From Wheat Fields to Milling Operations*.

Saranraj, P. and Geetha, M. (2012). Microbial spoilage of bakery products and its control by preservatives. *International Journal of Pharmaceutical & Biological Archives*, 3(1), 38–48.

Saranraj, P. and Sivasakthivelan, P. (2016). Microorganisms involved in spoilage of bread and its control measures. *International Journal of Pharmaceutical and Biological*, 3, 38–48.

Sarkar, S. (2015). Microbiological considerations: pasteurized milk. *International Journal of Dairy Science*, 10(5), 206–218.

Shebuski, J. R. and Freier, T. A. (2009). Microbiological spoilage of eggs and egg products. In: *Compendium of the Microbiological Spoilage of Foods and Beverages* (pp. 121–134). Springer.

Siegmund, B. and Pöllinger-Zierler, B. (2006). Odor thresholds of microbially induced off-flavor compounds in apple juice. *Journal of Agricultural and Food Chemistry*, 54(16), 5984–5989.

Silva, M. S., Rabadzhiev, Y., Eller, M. R., Iliev, I., Ivanova, I. and Santana, W. C. (2017). Microorganisms in honey. *Honey Analysis*, 500.

Smith, J. P., Daifas, D. P., El-Khoury, W., Koukoutsis, J. and El-Khoury, A. (2004). Shelf life and safety concerns of bakery products – a review. *Critical Reviews in Food Science and Nutrition*, 44(1), 19–55.

Smolander, M., Alakomi, H., Ritvanen, T., Vainionpää, J. and Ahvenainen, R. (2004). Monitoring of the quality of modified atmosphere packaged broiler chicken cuts stored in different temperature conditions. A. Time–temperature indicators as quality-indicating tools. *Food Control*, 15(3), 217–229.

Snowdon, J. A. and Cliver, D. O. (1996). Microorganisms in honey. *International Journal of Food Microbiology*, 31(1–3), 1–26.

Sofos, J. N. (2014). Meat and meat products. In: *Food Safety Management* (pp. 119–162). Elsevier.

Sriket, C. (2014). Proteases in fish and shellfish: role on muscle softening and prevention. *International Food Research Journal*, 21(2), 433.

Sriket, C., Benjakul, S., Visessanguan, W. and Kishimura, H. (2011). Collagenolytic serine protease in fresh water prawn (Macrobrachium rosenbergii): characteristics and its impact on muscle during iced storage. *Food Chemistry*, 124(1), 29–35.

te Giffel, M. and Wells-Bennik, M. (2010). Good hygienic practice in milk production and processing. In: *Improving the Safety and Quality of Milk* (pp. 179–193). Elsevier.

Techer, M., Baron, F. and Jan, S. (2014). *Microbial Spoilage of Eggs and Egg Products*.

Thompson, S. (2009). Microbiological spoilage of high-sugar products. In: *Compendium of the Microbiological Spoilage of Foods and Beverages* (pp. 301–324). Springer.

Tukur, M., Muazu, J. and Mohammed, G. (2012). Microbial analysis of brands of multivitamin syrups marketed in Maiduguri, Northeast Nigeria. *Advances in Applied Science Research*, 3(5), 3124–3128.

Usta, B. and Yilmaz-Ersan, L. (2013). Antioxidant enzymes of milk and their biological effects. *Journal of Agricultural Faculty of Uludag University*, 27(2), 123–130.

Velázquez-Ordoñez, V., Valladares-Carranza, B., Tenorio-Borroto, E., Talavera-Rojas, M., Varela-Guerrero, J. A., Acosta-Dibarrat, J., Puigvert, F., Grille, L., Revello, Á. G. and Pareja, L. (2019). Microbial contamination in milk quality and health risk of the consumers of raw milk and dairy products. In: *Nutrition in Health and Disease-Our Challenges Now and Forthcoming Time*. IntechOpen.

Vihavainen, E., Lundström, H., Susiluoto, T., Koort, J., Paulin, L., Auvinen, P. and Björkroth, K. J. (2007). Role of broiler carcasses and processing plant air in contamination of modified-atmosphere-packaged broiler products with psychrotrophic lactic acid bacteria. *Description Applied and Environmental Microbiology*, 73(4), 1136–1145.

Vihavainen, E. J. and Björkroth, J. (2010). Microbial ecology and spoilage of poultry meat and poultry meat products. *Handbook of Poultry Science and Technology, Secondary Processing*, 2, 485–493.

Wallace, G. (1938). The survival of pathogenic microorganisms in ice cream. *Journal of Dairy Science*, 21(1), 35–36.

Wu, J. (2014). Eggs and egg products processing. *Food Processing: Principles and Applications*, 437–455.

Angela Parry-Hanson Kunadu, Emmanuel Addo-Preko,
Nikki Asuming-Bediako

8 Microbiological safety of foods

8.1 Food safety

Food safety refers to practices and conditions that ensure the wholesomeness of food to prevent foodborne illnesses when food is handled and consumed as intended. Food safety encompasses the proper handling, cooking, storing, and preservation of food to protect consumers from foodborne illnesses, which are mostly caused by physical, chemical, and biological hazards in foods. Food exists in diversity that travels across the local, regional, and international borders before ending up on our plate. To assure safety, one requires the understanding of the microbial ecology and chemical constituents of foods, as well as a proactive and systematic approach for preventing, reducing, and eliminating hazards along the food value chain. Implementation of food safety management systems and robust national and international food safety regulations using risk-based standards is required to enhance the food safety.

Among the foodborne hazards that are transmitted through food, microbiological hazards are significant because one-third of developed countries are affected by microbiological foodborne illnesses annually; 70% of diarrheal cases are attributed to consumption of contaminated food, and the burden of disease caused by microbiological agents transmitted through food is very high especially in the developing countries (WHO, 2015). Microorganisms are ubiquitous, and they can contaminate food from multiple sources along the food value chain. A few microorganisms are pathogenic and can cause infections or intoxications when ingested through food. The microorganisms that contaminate food include bacteria, fungi, viruses, and protozoans.

8.2 Foodborne illnesses: statistics, significance, and impact

Food safety is a global issue, and food scares and perceptions of poor hygiene tend to affect a consumer's confidence (Parry-Hanson Kunadu et al., 2019). Apart from microbiological hazards, occurrence of veterinary drug residue, pesticides, and

Angela Parry-Hanson Kunadu, Emmanuel Addo-Preko, Nikki Asuming-Bediako,
University of Ghana, Legon, Accra

https://doi.org/10.1515/9783110667462-008

other environmental pollutants in foods are of grave concern. Foodborne diseases are of global importance, especially in the developing countries, where the diseases have greatly impacted. It affects economic development, social development, and industry. Unsafe food is a significant public health threat; with those at maximum risk being the young, the aged, pregnant women, and immunocompromised, including those suffering from HIV/AIDS, kidney diseases, and cancer. Symptoms of foodborne disease include nausea, vomiting, diarrhea, stomach cramps, paralysis, and neural disorders.

During outbreak situations, medical cost could be very high that puts unnecessary burden on both consumers and the government. In 2015, the World Health Organization (WHO) estimated 31 foodborne hazards, comprising bacteria, viruses, parasites, toxins, and chemicals, responsible for about 600 million cases of foodborne illnesses and 420,000 associated deaths occurring globally each year (WHO, 2015). WHO (2015) also reported that in Africa alone, more than 91 million people suffer from foodborne illnesses, with an estimated 600,000 case fatality of children every year. The United States Centre for Disease Control estimated that each year, 48 million people get sick from a foodborne illness, 128,000 are hospitalized, and 3,000 deaths are reported in the United States. An estimated 600 million people (one in 10) in the world fall ill after consumption of contaminated food. Children under the age of 5 years carry about 40% of foodborne disease burden with 125,000 deaths annually. In addition every year, 220 million children contract diarrheal disease, of which 96,000 deaths are reported. Children under the age of 5 years and people living in the low-income regions of the world suffer the burden of foodborne disease the most.

According to the WHO (2015), the most frequent causes of foodborne illness worldwide are diarrheal disease agents, particularly norovirus, *Campylobacter* spp., and *Salmonella enterica* serovars. Notable microbiological agents that cause death as a result of foodborne diseases include nontyphoidal *Salmonella*, *Salmonella Typhi*, *Taenia solium*, *Escherichia coli* O157:H7, and hepatitis A virus. In the United States, the top five microorganisms that cause illnesses through food are norovirus, *Salmonella*, *Clostridium perfringens*, *Campylobacter*, and *Staphylococcus aureus*. Other pathogens of interest include *Clostridium botulinum*, *E. coli*, and *Listeria monocytogenes*.

Estimates from developing countries are quite difficult to obtain due to underreporting of cases and weak monitoring and surveillance systems. However, it has been reported that 91 million people in Africa fall ill annually as a result of foodborne diseases. In South Africa, for instance, listeriosis outbreak from the consumption of polony during 2017–2018 indicated a cost valuation of US $260 million due to fatalities. There were 204 fatality cases reported, and productivity loses for humans and export value losses for food processers were in excess of US $15 million. In Mali, cholera outbreak in 1984 recorded 1,783 cases, of which 406 deaths were reported. Another outbreak of cholera in Tanzania between September and October 1997 reported 40,000 cases, of which 2,200 people died. This epidemic resulted in an economic loss of US $ 36 million as Europe refused to accept fish imports from the region.

Microbiological contamination of food has been implicated in most food recalls. Its management has become more important due to innovations in the food product development and the emergence of a more discerning and knowledgeable consumer base, who are aware and concerned about the safety of the foods they patronize. There is also a drive toward the use of risk analysis framework to estimate the public health risk and implement evidence-based interventions to minimize the foodborne disease risk among local populations.

8.3 Foodborne intoxications and infections

Foodborne infection is caused by the ingestion of food containing the microorganisms that grow and establish themselves in the human intestinal tracts. The microorganisms continue to multiply in the body and produce toxins that cause infections, for example, salmonellosis. On the other hand, foodborne intoxication is caused by ingesting food containing preformed toxins produced by microorganisms in the food. The reaction is rapid because the toxins are immediately introduced into the body, for example, food poisoning caused by *S. aureus* enterotoxins.

The severity of the foodborne disease is dependent on the type of the microbe or toxin ingested, the amount consumed, and the health status of the individual. Young children, the aged, pregnant women, and the immunocompromised (HIV patients, those undergoing chemotherapy) are more likely to suffer debilitating effects than healthy individuals.

8.4 Causes of foodborne illnesses

Food safety focuses on preventing, eliminating, and reducing hazards to an appropriate level of protection in all foods including raw products, food ingredients, processed and packaged foods, as well as prepared foods throughout the value chain. Foods that are commonly associated with microbiological hazards include poultry, meat, eggs, dairy, vegetables, and fruits. All consumers, regardless of their status, require safe food. Although foodborne hazards may be physical, chemical, or microbiological, it is widely known that microbial hazards present the greatest risk to consumers.

The causes of foodborne illnesses, especially in the developing countries are attributed to poor hygiene, lack of appropriate food safety knowledge, lack of appropriate implementation of food safety and surveillance systems, and rigor of regulatory instruments. A study conducted on food safety knowledge and attitudes and practices of food handlers in food service revealed a general lack of knowledge of appropriate temperature–time combinations for food storage (Parry-Hanson Kunadu et al., 2016). Although food safety attitudes are generally positive, the culture and infrastructure required to support the safety practices in many food establishments are woefully inadequate. Other practices such as the use of untreated animal waste in cultivation of vegetables are common in the developing countries (Grace, 2015). Such practices lead to high prevalence of pathogenic microorganisms on products that are consumed raw or minimally processed.

8.5 Microbiological hazards associated with specific foods

8.5.1 Meat

Meat is a perishable product, which is susceptible to microbial invasion and spoilage because it is an enriched medium that supports the growth of microorganisms (Osei-Asare and Eghan, 2014). Meat gets contaminated by inherent microflora on the animal's hide or during processing and transportation (Ercolini et al., 2006). In developing countries, meat is usually transported to the local markets through unapproved and unhygienic vehicles. In the informal sector, retailing of most meat products occurs in the open under ambient temperatures exposing the beef to flies, bacteria, and other contaminants such as dust (Soyiri et al., 2008). Contamination also occurs during operations such as weighing, processing, cutting, and storage; meat or food preparation in the kitchen environment is also another channel for contamination. The bacteria multiply under the conducive temperature to a level leading to the production of toxins, which result in illnesses such as cholera and typhoid fever. To ensure meat is safe, it is important that the primary producers such as farmers maintain animals that are clean and healthy. It is also important to maintain strict process hygiene and chilled temperature environment for the meat to slow down microbial proliferation. Microbial hazards commonly found in raw meat include bacteria such as *E. coli* O157:H7, *Salmonella*, *Campylobacter*, *L. monocytogenes*, and *Brucella*, and protozoans such as *Toxoplasma gondii*, *Cryptosporidium parvum*, and *Cyclospora cayetanensis*. Factors that are important in ensuring the safety of raw meat include specifications on maximum duration (days) after slaughter, microbiological data on process hygiene, and temperature control during chilling, storage, and transportation (ICMSF, 2011).

8.5.2 Milk and dairy products

Dairy products are rich in both saturated and unsaturated fats, proteins, essential minerals (potassium, magnesium, calcium, and zinc), vitamins A, B, D, E, and lactose. The chemical composition of milk differs with respect to different animal sources from which it is obtained. *Listeria monocytogenes*, diarrheagenic *E. coli*, *S. aureus*, *Mycobacterium bovis*, *Mycobacterium tuberculosis*, *Salmonella*, and *Campylobacter* are common zoonotic pathogens that may be found in ruminants, are shed into milk, and transmitted into other dairy products. Milking from an infected udder (mastitis) leads to contamination of milk. It can also be contaminated during or after milking via dirty teats, feces, unhygienic milking conditions, and processing conditions. Mastitis can be controlled by observing good milking hygiene, use of disinfectant teat dip after milking, and antibiotic infusion at the end of the lactation period. Improving general hygienic conditions in the cattle housing and maintaining milk handling equipment in hygienic conditions are relevant.

8.5.3 Fruits and vegetables

There are numerous health benefits associated with the consumption of fruits and vegetables. However, there is a risk when vegetables and fruits are consumed without proper washing. Consuming contaminated products is detrimental to human health.

E. coli O157:H7, *Salmonella,* and *L. monocytogenes* are common pathogens found in fruits and vegetables. Contamination of fruits and vegetables can result from animals, contact with farm effluents via irrigation and fertilization, environmental hazards, and unhygienic handling and processing. Products such as strawberries, mango, green leafy vegetables, lettuce, and cabbages are possible routes for the transmission of these microbes to humans. Spore-forming microbes such as *C. perfringens* and *C. botulinum* have also been reported in fruits and vegetables and are usually traced to the soils where they were cultivated. Although heat treatments may inactivate certain pathogens, spores from *C. botulinum* or *Bacillus cereus* may germinate in nonacidic products such as processed vegetables. In such situation, chilling is important to control germination and proliferation of microbial agents.

8.5.4 Egg and egg products

Salmonella is the major microbial hazards associated with egg and egg products. Most often, contamination occurs just at the point of lay or handling and storage of eggs. Consumption of cracked, undercooked, or soiled eggs is more likely to lead to infection. The egg could get contaminated from fecal matter while it exits from a bird. Feces from the environment could also contaminate the egg. Cross-contamination during meal preparation is also a medium for transmission and contamination.

8.5.5 Seafood and fish

Finfish and shellfish provide important animal source proteins globally. They are cold-blooded animals whose microbiological quality reflects their aquatic environment. Fish and fish products have been implicated in various foodborne diseases in humans. The etiological agents are viruses, parasites, bacteria, and biogenic amines (ICMSF, 2005). The most important pathogenic bacterium that needs to be monitored in fish and fish products is *Vibrio parahaemolyticus*. *C. perfringens* and *S. aureus* have been isolated from raw and host-smoked fish (Aboagye, 2016), whereas *E. coli* O157:H7, *C. botulinum*, *Plesiomonas shigelloides*, *Aeromonas hydrophila*, *Salmonella*, and *Shigella* have been associated with gastroenteritis through consumption of fish. Moreover, parasites such as *Anisakis* and *Pseudoterranova*; and viruses such as norovirus, rotavirus, and hepatitis A have been reported on fish products (ICMSF, book 6). Illness can arise as a result of eating undercooked seafood that has been contaminated with these microbes. Some of these microbes can be found in water, soil, gills, and the intestinal tract of fishes, oysters, and clams.

8.5.6 Ready-to-eat processed foods

L. monocytogenes is a major pathogen found in ready-to-eat (RTE) processed foods such as ham, salami, fish sauce, canned fish, nuts, and crackers. RTE processed foods can be consumed without additional cooking. Some are shelf-stable, others may require minimal heat treatment, and can be served hot or cold. Safety of RTE food can be compromised before, during, and after the preparation process. Quality raw material, appropriate temperature, water activity, processing environment, and the use of hygienic equipment are some important parameters to consider not to compromise the microbiological safety of RTE foods. For instance, the spores of *C. botulinum* can survive some preservation methods; hence, the use of quality meat and the application of the right temperature–time combinations are important to ensure safety (Table 8.1).

Table 8.1: Microbiological hazards commonly associated with foods.

Food item	Hazard/pathogen
Egg and egg products	*Salmonella*
Milk and dairy products	*Listeria monocytogenes*, verotoxigenic *Escherichia coli*, *Staphylococcus aureus*, *Salmonella* and *Campylobacter*, *Clostridium botulinum*, *Brucella* spp.
Shelf-stable heat-treated foods	*Clostridium* spp., *Bacillus* spp.
Meat products	*Escherichia coli* O157:H7, *Salmonella*, *Campylobacter*
Poultry products	*Escherichia coli* O157:H7, *Salmonella*, *Campylobacter*, *Shigella*
Fish and seafood	*Bacillus cereus*, *Salmonella*, *Aeromonas hydrophilia*
Vegetables and vegetable products	*Salmonella*, verotoxigenic *Escherichia coli*
Fruits and fruit products	*Salmonella*, verotoxigenic *Escherichia coli*, *Aspergillus flavus*
Water	Helminths, protozoa such as *Giardia*, pathogenic bacteria and viruses such as *Shigella*, *Vibrio cholera*, *Escherichia coli*, hepatitis A virus, norovirus
Ready-to-eat processed foods	*Listeria monocytogenes*

8.6 Microbiological assessment of foods

The analysis of food and food environment samples for the presence or concentrations of foodborne pathogens and/or their toxins, spoilage organisms, and microbial indicators is important for ensuring that food is safe and of good microbiological quality. There are standards set to outline the criteria for acceptable limits of target pathogenic microorganisms as well as general concentrations of utility and indicator microorganisms in food. These standards are product dependent and organism specific.

Microbial testing of food is an important aspect in ensuring that food products are safe. Several microbial tests can be performed on food products to determine its wholesomeness or quality. These tests are often influenced by the type of food product, the value chain of the product, the way the product is processed and handled, the chemical composition of the product, and how the product is to be stored and preserved among other factors. Some common examples of microbiological tests include total plate count (TPC), concentrations of Enterobacteriaceae, yeast and molds, and presence of *Salmonella*.

8.6.1 Utility organisms and indicators of microbiological safety

Concentration of utility organisms provides information on the shelf-life, incipient spoilage, and general contamination of a food product. Such information summarizes the use of the product (Skovgaard, 2012). Examples include aerobic plate count (APC) and yeast and molds.

Microbial quality indicators are organisms or their metabolic products whose presence in the food products suggests the presence of pathogenic or spoilage microorganisms. These indicator organisms also indicate the extent of contamination of a product as well as help predict the shelf- life of the product (*Jay and Leossner, 2005*). For example, the presence of *E. coli* in a food product or water suggests the presence of enteropathogens (Goh et al., 2019). Microbial indicators are also useful for verification of process controls and identification of opportunities for improvements in process controls (ICMSF, 2005).

A few criteria must be met to qualify as an indicator organism as listed below (ICMSF, 2005; Jay and Leossner, 2005).

1. They should be present at detectable levels in a given product.
2. They should be easily enumerated and detected, and vividly distinguishable from other organisms that may be present in the product in a short period.
3. Their presence and numbers should have direct negative correlation with the quality of the product.
4. Their presence and concentration should indicate the potential for food spoilage, faulty practice, or faulty process.
5. The results of their presence and concentration should be applicable to process control.
6. Their survival or stability should be similar to or greater than the spoilage organism or pathogen.
7. It should have growth kinetics similar to or faster that the spoilage or pathogenic organism.
8. Their growth should not be affected by other components in the food microbiota.
9. Methods for their analysis should be rapid, inexpensive, reliable, sensitive, validated, and verified.
10. The indicator organism should not present a health risk to the analyst.

8.6.2 Total plate count

TPC is mostly analyzed with general-purpose media. Over the years, plate count agar has been used in the enumeration of the total microbial concentration of a product. Other general-purpose media such as Tryptic Soy Broth, nutrient agar, and Brain Heart Infusion can also be used. TPC is the enumeration of organisms that grow under aerobic conditions between 20 and 45 °C (Smith and Townsend, 2016).

This includes all aerobic bacteria and fungi. This microbial testing method can also be called APC, standard plate count, or total viable count. This method accounts for both pathogens and nonpathogens as the method is not organism specific.

APC is intended to indicate the general microbial concentration of a food product (Jay and Leossner, 2005). In some cases, it gives information about the level of contamination of a product by providing an estimated number of viable cells that can grow under aerobic conditions in the food. In the case of some probiotic and/or fermented foods, TPC is not an appropriate indicator of contamination because they are cultured products (Rusell, 2005).

The TPC is calculated and reported as colony-forming units (CFUs). CFU represents the number of viable cells in a product that can multiply in the culture media to form colonies (Russell, 2005). As much as TPC gives an idea of the microbial concentration of a product, it cannot stand alone as a basis for declaring microbiological quality of a food product. TPC is significant when further studies are conducted on the product to identify the microorganisms that predominate the product.

8.6.3 Enterobacteriaceae

Enterobacteriaceae are a group of bacteria that are tested for in food products to assess the hygienic quality. They are usually considered as hygiene indicators; therefore, they are used to monitor the effectiveness of implemented food safety systems such as good manufacturing practices and good hygienic practices (GHP) (Rigarlsford, 2007). Enterobacteriaceae are a family of gram-negative, rod-shaped bacteria, which are non-spore-forming, facultative anaerobes capable of fermenting glucose (Paterson, 2006). They are also motile, catalase-positive, oxidase-negative, reduce nitrate to nitrite, and produce acid from glucose fermentation (Baylis, Uyttendaele, Joosten, and Davies, 2011). These traits are used to characterize them. Some organisms that belong to this family include *Salmonella*, *E. coli*, *Klebsiella*, *Yersinia*, *Shigella*, *Proteus*, *Enterobacter*, *Serratia*, and *Citrobacter* (Biesta-Peters, Kinders, and de Boer, 2019). They are said to be ubiquitous in nature and can be found in soil, water, and sewage, because of their presence in the intestinal tract (Gould et al., 2019). There are currently 48 genera and 219 species belonging to this family, and these numbers are bound to increase in the future because of the periodic discovery and emergence of new strains (ILSI, 2011). The Enterobacteriaceae family are said to be better indicators of hygiene, as it is a more comprehensive group than total coliforms.

8.6.4 Coliforms

Coliforms are a group of rod-shaped, nonspore-forming gram-negative bacteria that ferment lactose at 35 °C within 24–48 h (Craun, Berger, and Calderon, 1997). They are widely distributed in nature and many are native to the gut of warm-blooded animals and humans. They can be found almost anywhere in the environment, that is, in the soil, water, plants, vegetation, as well as on the skin and in the intestinal tract of warm-blooded animals. Majority of them are harmless (nonpathogenic) although there are a few pathogenic ones (Chambers, 1972).

The presence of coliforms in food products is a useful indicator of hygiene in the product (Deng, Xiao, Xu, and Wang, 2019). They are relatively easy to isolate or identify. Examples of coliforms include *Enterobacter, Citrobacter, Hafnia*, and *Websiella*.

Coliforms are divided into three groups, namely, total coliforms, fecal coliforms, and *E. coli*.

8.6.4.1 Total coliforms

This group consists of closely related harmless bacteria (Divya and Solomon, 2016). They are free living and can be found in environments such as water, vegetation, and soil. Although harmless, their presence in food products is an indication of poor hygiene or loss of process control (Craun et al., 1997). When detected in water samples, they suggest a probable cause of contamination in the water. There are some thermotolerant coliforms belonging to this group that are capable of fermenting lactose at 45 °C.

8.6.4.2 Fecal coliform bacteria

They are a subgroup of the total coliform bacteria that can be found in the intestines and feces of warm-blooded animals (Clarke et al., 2017). Human beings, ruminants, cows, pigs, and dogs among others are common examples of hosts for these group of microorganisms (Divya and Solomon, 2016). *Escherichia coli* is an example of a fecal coliform. It resides in the intestinal tract of the warm-blooded animals and thus, in the fecal matter of these animals (Clarke et al., 2017). Compared to total coliforms, which are generally harmless, fecal coliforms is composed of both pathogenic and nonpathogenic bacteria. Therefore, their detection in food is an indication of fecal contamination (Paludetti et al., 2018).

8.6.4.3 Enterococci

Enterococcus species are gram-positive facultative anaerobes that are morphologically similar in structure to streptococci after gram staining. They are catalase-negative and grow at an optimal temperature of 35 °C (Gira, 2002). This group of bacteria form part of the intestinal flora of humans and other mammals. They are sometimes referred to as "endogenous human flora" (State, 2001). Most of these bacteria have been associated in female genitourinary tract and become opportunistic pathogens when they enter the bloodstream of invalids (Somily et al., 2016). There are over 36 known species belonging to this genus; however, 26 of them have been associated with human infections (Zervos, Chow and Robert, 2010). Some common examples of enterococci species are *Enterococcus gallinarium*, *Enterococcus casseliflavus*, *Enterococcus faecalis*, *Enterococcus faecium*, *Enterococcus hirae*, and *Enterococcus durans*, among others. *Enterococcus faecalis* is the most common human pathogen. Other species that are also associated with human infections are *E. faecium* and *E. durans* (Zervos, Chow and Robert, 2010).

Enterococci have recently emerged as a nosocomial pathogen (Gagetti et al., 2019). They are ubiquitous and thus can be found in soil, water, and in foods. Their presence in foods suggests a certain level of fecal contamination and they render food unsafe as they are known to be pathogens (although they are considered as low-grade pathogens; State, 2001). They withstand adverse environmental conditions and are able to colonize different ecological niches, including diverse foods and food environments (Gira, 2002). *Enterococcus faecalis* and *E. faecium* are associated with infections such as endocarditis, bacteremia, wound infections, and some urinary tract infections. Enterococci also cause food intoxication through the production of biogenic amines (Gira, 2002). There have been increasing occurrences of antibiotic resistance among enterococci species and this has become an issue of public health and safety concern. The most commonly reported resistance strain is the vancomycin-resistant enterococci (Somily et al., 2016).

8.6.5 Bacteriophages

Bacteriophages, popularly known as "Phages," were discovered by Twort and d'Herelle in the early 1900s (Sharma et al., 2017). They are described as the most abundant viral entities on the planet. Bacteriophages literally means "**devourer of bacteria**." The viruses infect the bacterial cells and replicate within the cell, which causes it to lyse (Gorski et al., 2003). Studies have shown that bacteriophages only invade the bacterial cell and not human cells; they are, therefore, used in certain medical therapies to treat infections caused by bacteria. Bacteriophages have two life cycles, the lytic and the lysogenic. The simple difference between these life cycles is that; in the lytic cycle, the host cell is killed after the cycle, whereas in the lysogenic cycle, the host cell is not killed (Duckworth and Gulig, 2002).

Over the years, the use of bacteriophages has been highly discussed as a way to combat antibiotic-resistant bacteria (Díaz-Muñoz and Koskella, 2014). During cheese-making process, some food manufacturers use bacteriophages to eliminate pathogenic bacteria from the cheese (García, Martínez, Obeso, and Rodríguez, 2008). Although adding these viruses to food products can give a certain degree of assurance in terms of microbiological safety, there are issues concerning the use of these viruses in RTE foods. García et al. (2008) argues that till the concerns of having the viruses in RTE foods are resolved, they should be used only in livestock and poultry farming to reduce the occurrence of pathogenic bacteria on and within the animals.

Bacteriophages may seem to be very helpful in eliminating pathogenic bacteria but they can also cause certain nonharmful bacteria to be harmful to humans. Some bacteria species including *E. coli*, *Streptococcus pyogenes* (causes flesh-eating disease), *Vibrio cholerae* (causes cholera), and *Shigella* become harmful when genes that produce toxins are transferred from the bacteriophage to them. These bacteria obtain these disease-causing genes from the bacteriophage in the lysogenic lifecycle, and thus in turn cause various diseases in humans through food poisoning or intoxication (Abedon et al., 2011). Therefore, the presence of bacteriophages in food cannot be totally ascribed with a positive effect, as it could also have a negative effect.

8.7 Microbiological foodborne hazards

Profiles of some microbiological hazards of food safety concern are discussed further.

8.7.1 *Clostridium botulinum*

Clostridium botulinum is a cause of foodborne intoxication. The first outbreak was reported in sausages in the year 1793, in Germany. It is a gram-positive, spore-forming strict anaerobe that grows between 3.3 and 50 °C. However, the optimum temperature for the growth is 37 °C. Its vegetative cells can easily be killed by heat but the spores are very resistant to heat, cold, and low pH used in food preservation. Nonetheless, the time and temperature combination used in canning destroys most of the heat-resistant spores. Symptoms of intoxication include vomiting, vertigo, nausea, and even death by asphyxiation.

8.7.2 *Staphylococcus aureus*

It is a gram-positive facultative anaerobe that can be found on human skin and in many food items. *Staphylococcus aureus* is described as an opportunistic pathogen that infects both humans and animals. *Staphylococcus aureus* produces enterotoxin and is primarily associated with food poisoning. When the toxin is ingested, symptoms such as abdominal cramps, diarrhea, and vomiting are indicated. The toxins of *S. aureus* are heat stable; however, boiling under pressure for 3 h at 121 °C will inactivate the toxins. Also, because it is heat stable, normal cooking does not inactivate toxins to make the food safe.

8.7.3 *Salmonella*

Salmonella is a gram-negative facultative anaerobe. It is nonspore-forming rod with a peritrichous flagella that grows optimally at 37 °C. It is heat sensitive and thorough cooking can be effectively used to control their presence in foods. It can be found in water, milk, shellfish, poultry, and eggs. It is implicated as causing typhoid fever and salmonellosis. Proper chilling, refrigeration, and hygiene are key measures for reducing incidence of salmonellosis. It has about 2,579 serotypes (Lamas et al., 2018). Serotypes are primarily specific variations within the species of the bacteria. In the case of *Salmonella* spp., the determination of these serovars or serotypes is based on the differences in the flagella H antigen, the somatic O antigen, and the phase shift in the H antigen. This was discovered by White and Kauffmann in the early 1900s. Serotyping makes the epidemiology of *Salmonella* infections easy to understand, hence making it possible to track the source of contamination during an outbreak and the severity of the disease (Meneses-Gonzalez, 2010).

The most common vehicles for *Salmonella* infections are animal source products, such as poultry meat and eggs (Greig and Ravel, 2009). Concerns of the safety of poultry meat and its derived products are constantly increasing due to the ability of *Salmonella* to reside in healthy chicken on the farm without any adverse effect on the health of the bird.

GHPs in the farm and slaughterhouses are key to the reduction of cases of zoonosis (van Immerseel et al., 2005). Controlling *Salmonella* among broilers, however, depends on the knowledge of the source of infection. The feed, drinking water, littering material in coops, and the environment both in and out of the broiler house are all considered possible sources of *Salmonella* (FAO/WHO, 2009; Kiilholma, 2007).

8.7.4 *Shigella*

Shigella is a gram-negative facultative anaerobe that forms spores and is rod-like in shape. It is nonmotile and a source of waterborne infections, especially in the tropics. Incubation takes up to 4 days and symptoms are presented by bloody stools and diarrhea. Infection can be prevented by good hygiene, water treatment, and sanitation.

8.7.5 *Vibrio*

Vibrio cholerae is the causative microbe for cholera. Cholera is a highly contagious disease and is characterized by massive acute diarrhea, vomiting, and dehydration with death likely to occur in severe and untreated cases. *Vibrio cholerae* is a gram-negative curved rod that grows well in alkaline medium and is motile with a single polar flagellum. It can easily spread through water, milk, and food. It is often transmitted by water; however, fish or fish products that have been in contact with contaminated water or feces from infected persons also frequently serve as a source of infection (Kam et al., 1995. Dehydration, diarrhea, shock, and even death are some of the symptoms of infection. The most effective therapy is replacement of lost water and electrolytes to correct dehydration and salt depletion.

Vibrio parahaemolyticus causes gastroenteritis and is contracted almost solely from seafood. *Vibrio parahaemolyticus* is known to cause extraintestinal infections in humans. When other foods are involved, they represent cross-contamination from seafood products (Jay et al., 2005).

8.7.6 *Campylobacter*

It is a gram-negative motile microaerophilic organism that has a single polar flagellum. There are various species; however, *Campylobacter jejuni* and *Campylobacter coli* are the most important species for food safety. Optimal growth occurs at 42 °C. *Campylobacter* infection is a major cause of gastroenteritis and symptoms include fever, abdominal pain, headache, and diarrhea. Outbreaks or infections have been related to the consumption of contaminated poultry, meat, and water. It is one of the most commonly diagnosed pathogens known to cause foodborne illness such as nausea, gastrointestinal pain, and diarrhea (Kiilholma, 2007). In extreme and rare cases, it can cause Guillain–Barre syndrome, which is an immunological failure resulting in damage to part of the peripheral nervous system. The primary harborage of *Campylobacter* is said to be in the alimentary canal of mammals and birds. The levels of *Campylobacter* spp. within the guts of these animals can be as high as 10^9 (Henry et al., 2011; Ma et al., 2014). The major route of *Campylobacter* infection is through handling and consumption of *Campylobacter*-contaminated poultry meat, particularly chicken. Broiler chicken is considered one of the most susceptible animals to *Campylobacter* spp. and a common source for human *Campylobacter* spp. infections. As studies have established that *Campylobacter* primarily resides in the gut of bird, the only means the carcass gets contaminated is if there is a leakage of gut or fecal matter onto the meat.

Epidemiological data and resources for the control of *Campylobacter*-related zoonosis is lacking especially in the developing countries. According to the WHO, *Campylobacter* is responsible for 37,600 deaths globally every year. Given the relatively low infective dose of *Campylobacter* spp. (≥500 cells) and the high consumption of poultry, the presence of these organisms in poultry and poultry products presents a public health risk (Vinueza-Burgos et al., 2018).

8.7.7 *Escherichia coli*

Escherichia coli is a gram-negative facultative anaerobic rod that is nonspore form-ing. It is ubiquitous in nature. Most of them are nonpathogenic. *Escherichia coli* is associated with a variety of foods, particularly animal source foods. Pathogenic *E. coli* are grouped into eight based on their properties of virulence, mechanism of pathogenicity, and clinical syndrome (Kai and Aotearoa, 2009; Olsvik et al., 1991). These groups include enteropathogenic *E. coli*, enterohemorrhagic *E. coli*, enteroag-gregative *E. coli*, enterotoxigenic *E. coli*, enteroinvasive *E. coli* (EIEC), diffusely adher-ent *E. coli*, uropathogenic *E. coli*, and neonatal meningitis *E. coli* (NMEC) (Croxen and Finlay, 2010; Kai and Aotearoa, 2009; Saxena et al., 2015). These eight pathovars are further classified into two broad groups, namely diarrheagenic *E. coli*, which consists of the first six pathovars listed above and extraintestinal *E. coli* consisting of the last two pathovars (Croxen and Finlay, 2010).

Following ingestion, these pathovars with the exception of EIEC adhere them-selves to specific host cells with the aid of long appendages called fimbriae or pili (Croxen and Finlay, 2010). These *E. coli* strains overcome the normal functioning of the host cells with the aid of secreted proteins. Each pathovar has a unique way of adhesion, invasion, and colonization of the host cells, which ultimately results in the disease noted for that pathovars or strain (Croxen and Finlay, 2010). *Escherichia coli* grows well at 37 °C and in food; the organism can be destroyed by proper cooking.

8.7.8 *Yersinia enterocolitica*

It is a gram-negative facultative anaerobic nonspore-forming bacterium, which is motile at 20 °C but not at 37 °C. It can grow in refrigerated vacuum-packaged meat as it is a facultative anaerobe and is a psychrotroph. Symptoms upon ingestion in-clude fever, abdominal pain, and diarrhea. Foods that are a source of yersiniosis include chocolate milk, milk powder, tofu, pasteurized milk, and pork products.

8.7.9 *Listeria monocytogenes*

It is a small gram-positive nonspore-forming rod. It grow between 1 and 40 °C. Symptoms of infections include malaise, diarrhea, and mild fever. Infections can also lead to abortions and meningitis in humans. *Listeria* can be transmitted through foods such as poultry, dry sausages, cheese, RTE foods, and vegetables. Because the organism is killed by heat and it is susceptible to sanitizing agents, proper cooking of food and decontamination of food preparation environment are required.

L. monocytogenes causes a highly fatal disease called listeriosis. It is also considered the leading cause of death among foodborne bacterial pathogens, recording very high fatality case ratios (Jay et. al., 2005) and is reported to cause an estimate of 2,500 illnesses and 500 deaths annually in the United States alone (CDC, 2002).

L. monocytogenes is widely distributed in the general environment including fresh water, coastal water, and live fish from these areas. Contamination or recontamination of seafood may also take place during processing (Huss et al., 2000). *Listeria monocytogenes* is also a psychrotroph that cause food spoilage of refrigerated foods and RTE foods.

8.8 Emerging microbiological hazards

New traits of existing pathogens keep emerging. These traits have implications on the ecological niches of the microorganisms, control mechanisms, and their virulence. Factors that contribute to the emergence and reemergence of foodborne pathogens include changes in the agricultural practices and animal production, changing consumer needs and expectations, limited food treatment, the use of food additives and food biotechnology, as well as increase in foreign travel and climate change. Emerging pathogens are usually zoonotic in nature and include parasites, viruses, and bacteria. Some of the features of reemerging microbial hazards are the acquisition of antimicrobial resistance (AMR) traits, resistance to control measure combinations used in food processing, persistence of foodborne pathogens, and acquisition of genes that are more virulent. Examples of these pathogens include non-O157 Shiga toxin-producing *E. coli* serotypes and multidrug-resistant microbes such as *S. aureus*, *Clostridium difficile*, and hepatitis E virus.

8.9 Antimicrobial resistance of microbiological hazards

AMR is becoming a common occurrence in animals and humans and this poses a critical challenge for public health protection. The use, misuse, or overuse of antimicrobial drugs is the major driving force toward resistance. AMR is the ability of a microorganism to stop an antimicrobial (antibiotics, antivirals, and antimalarials) from working against it. When this happens, standard treatments become ineffective in controlling microbial infections; hence, infections persist, spread to others, and cause fatalities. AMR increases the cost of health care with lengthier stays in hospitals, where most intensive care is required.

Currently, AMR is an economic, social, and medical problem. Resistant organisms cause infections that are more difficult to treat; and often, these infections require drugs that are less available and are more expensive. The use of antimicrobials in veterinary practice for therapy and prophylaxis, as well as its use as growth promoters contribute greatly to resistance in animals. When foodborne pathogens become resistant to antimicrobials and are transmitted through food, it increases the burden of foodborne illnesses. Existing data suggest that AMR bacterial infections are common in low- and middle-income countries.

Although AMR can be a natural phenomenon, antimicrobial exposure in health care, agriculture, and the environment have driven its selection. Most bacteria posses multiple mechanisms that act simultaneously to confer resistance (Peterson and Kaur, 2018). There are naturally occurring antimicrobial resistance mechanisms, such as efflux pumps, that have over time become efficient in removing antimicrobials that enter the bacterial cells. Mechanisms that are frequently reported include mutations in the microbial target sites preventing tight binding of antimicrobials, and therefore causing microbial inhibition and sequestration of the antimicrobials by drug-binding proteins that prevent the antibiotic from reaching the target site. Bacteria protect themselves from aminoglycosides (e.g., gentamicin, kanamycin, streptomycin) by producing aminoglycoside-modifying enzymes including acyltransferases, phosphotransferases, adenyltransferases, and nucleotidyltransferases that modify and inactivate the antimicrobials. Others sequester the antimicrobials, degrade them (e.g., β-lactamases), and protect the target DNA by DNA-binding proteins.

The problem of AMR is also compounded by poverty, inadequate health delivery and health care facilities, lack of funds for research into AMR trends, and weak monitoring and surveillance programs. The prudent use of antimicrobials, development of new antibiotics and new vaccines, and improved public health education can contribute to addressing the challenge of AMR.

References

Abedon, S. T., Khul, S., Blasdel, J. and Kutter, M. (2011). Phage treatment of human infections. *Bacteriophage*, 1(2), 66–85.

Aboagye, E. (2016). *Microbial Quality of Fish Along the Tilapia, African Catfish and Sardinella Artisanal Value Chains in Kpong and James Town, Ghana*. MPhil thesis. Accra, Ghana: University of Ghana.

Ashbolt, N. J. (2015). Microbial contamination of drinking water and human health from community water systems. *Current Environmental Health Report*, 2, 95–106.

B. A., & Ayeni, O. (2019). Cost estimation of listeriosis (Listeria monocytogenes) occurrence in South Africa in 2017 and its food safety implications. Food Control 102; 231–239.

Baylis, C., Uyttendaele, M., Joosten, H. and Davies, A. (2011). *ILSI Europe Emerging Microbiological Issues Task Force; The Enterobacteriaceae and Their Significance to the Food Industry.*

Bernabe, K. J., Langendorf, C., Ford, N., Ronat, J. B. and Murphy, R. A. (2017). *International Journal of Antimicrobial Agents*, 50, 629–639.

Biesta-Peters, E. G., Kinders, S. M. and de Boer, E. (2019). Validation by an interlaboratory collaborative trial of EN ISO 21528 – microbiology of the food chain – horizontal methods for the detection and enumeration of Enterobacteriaceae. *International Journal of Food Microbiology*, 288(March 2018), 75–81. https://doi.org/10.1016/j.ijfoodmicro.2018.05.006.

Byarugaba, D. K. (2004). A view on antimicrobial resistance in developing countries and responsible risk factors. *International Journal of Antimicrobial Agents*, 24, 105–110.

CDC. (2002). Listeria outbreak investigation. Retrieved from https://www.cdc.gov/media/pressrel/r021003a.htm

Chambers, M. (1972). Survival of coliform bacteria in natural. 24(5), 805–811.

Clarke, R., Peyton, D., Healy, M. G., Fenton, O. and Cummins, E. (2017). A quantitative microbial risk assessment model for total coliforms and E. coli in surface runoff following application of biosolids to grassland. *Environmental Pollution*, 224, 739–750. https://doi.org/10.1016/j.envpol.2016.12.025.

Craun, G. F., Berger, P. S. and Calderon, R. L. (1997). Coliform bacteria and waterborne disease. 8(3), 96–104. https://doi.org/10.1002/j.1551-8833.1997.tb08197.x.

Croxen, M. A., & Finlay, B. B. (2010). Molecular mechanisms of Escherichia coli pathogenicity. Nature Reviews Microbiology, 8(1), 26–38. https://doi.org/10.1038/nrmicro2265

Deng, Y., Xiao, H., Xu, J. and Wang, H. (2019). Saudi journal of biological sciences prediction model of PSO-BP neural network on coliform amount in special food. *Saudi Journal of Biological Sciences*, (26), 1154–1160. https://doi.org/10.1016/j.sjbs.2019.06.016.

Díaz-Muñoz, S. L. and Koskella, B. (2014). *Bacteria-Phage Interactions in Natural Environments. Advances in Applied Microbiology (First Edition)* (Vol. 89). Elsevier Inc. https://doi.org/10.1016/B978-0-12-800259-9.00004-4.

Divya, A. H. and Solomon, P. A. (2016). Effects of some water quality parameters especially total coliform and fecal coliform in surface water of Chalakudy river. *Procedia Technology*, 24, 631–638. https://doi.org/10.1016/j.protcy.2016.05.151.

dos Santos, C. A. H. and Vieira, R. H. S. F. (2013). Bacteriological hazards and risks associated with seafood consumption in Brazil. *The Revista do Instituto de Medicina Tropical de São Paulo*, 55(4), 219–228.

Duckworth, D. H. and Gulig, P. A. (2002). Bacteriophages: potential treatment for bacterial infections. *BioDrugs*, 16(1), 57–62. https://doi.org/10.2165/00063030-200216010-00006.

Enterobacteriaceae. *World of Microbiology and Immunology*. https://www.encyclopedia.com/science/encyclopedias-almanacs-transcripts-and-maps/enterobacteriaceae (accessed July 17, 2019).

Ercolini, D., Russo, F., Torrieri, E., Masi, P. and Villani, F. (2006). Changes in the spoilage related microbiota of beef during refrigerated storage under different packaging conditions. *Applied Environmental Microbiology*, 72(7), 4663–4671.

FAO/WHO. (2009). Salmonella and Campylobacter in chicken meat: Meeting report. Microbiological Risk Assessment. https://doi.org/10.1016/j.ijfoodmicro.2010.09.004

Food Standards Australia New Zealand. (2009). Public health and safety of eggs and egg products in Australia: Explanatory summary of the Risk Assessment. Australia.

Fung. Y.G. F (2010) Microbial Hazards in foods: Foodborne infections and Intoxications; In Handbook of Meat Processing. Edited by F. Toldra. Blackwell Publishing.

Fung. Y. G. F. (2010). Microbial hazards in foods: foodborne infections and intoxications. In: F. Toldra (Ed.), *Handbook of Meat Processing*. Blackwell Publishing.

Gagetti, P., Bonofiglio, L., García, G., Kaufman, S., Mollerach, M., Vigliarolo, L., et al. (2019). Resistance to β-lactams in enterococci. *Revista Argentina de Microbiología*, 51(2), 179–183. https://doi.org/10.1016/j.ram.2018.01.007.

García, P., Martínez, B., Obeso, J. M. and Rodríguez, A. (2008). Bacteriophages and their application in food safety. *Letters in Applied Microbiology*, 47(6), 479–485. https://doi.org/10.1111/j.1472-765X.2008.02458.x.

Gira, G. (2002). Enterococci from foods. 26, 163–171.

Goh, S. G., Saeidi, N., Gu, X., Vergara, G. G. R., Liang, L., Fang, H. et al. (2019). Occurrence of microbial indicators, pathogenic bacteria and viruses in tropical surface waters subject to contrasting land use. *Water Research*, *150*, 200–215. https://doi.org/10.1016/j.watres.2018.11.058.

Grace, D. (2015). *Food Safety in Developing Countries: An Overview. A Learning Resource for DFID Livelihoods Advisers*. https://www.agrilinks.org/sites/default/files/resource/files/EoD_Learning_Resource_Food%20Safety_updFeb2016-1.pdf (accessed April 6, 2019).

Greig, J. D. and Ravel, A. (2009). Analysis of foodborne outbreak data reported internationally for source attribution. International Journal of Food Microbiology, 130(2), 77–87. https://doi.org/10.1016/j.ijfoodmicro.2008.12.031

Gorski, A., Dabrowska, K., Switala-jele, K., Nowaczyk, M., Weber-dabrowska, B., Boratynski, J. et al. (2003). New insights into the possible role of bacteriophages in host defense and disease. 5, 1–5.

Gould, M., Ginn, A. N., Marriott, D., Norris, R., Sandaradura, I. et al. (2019). Urinary piperacillin/tazobactam pharmacokinetics in-vitro to determine the pharmacodynamic breakpoint for resistant Enterobacteriaceae. *International Journal of Antimicrobial Agents*, (54), 240–244. https://doi.org/10.1016/j.ijantimicag.2019.05.013.

Hafez, M. H. (1999). Poultry meat and food safety: pre and post harvest approaches to reduce foodborne pathogens. *Worlds Poultry Science Journal*, Vol. 55.

Henry, I., Reichardt, J., Denis, M., and Cardinale, E. (2011). Prevalence and risk factors for Campylobacter spp . in chicken broiler flocks in Reunion Island (Indian Ocean). *Preventive Veterinary Medicine*, 100(1), 64–70.

Holmes, A. H., Moore, L. S. P., Sundsfjord, A., Steinbakk, M., Regmi, S., Karkey, A., Guerin, P. J. and Piddock, L. J. V. (2016). Antimicrobials: access and sustainable effectiveness 2: understanding the mechanisms and drivers of antimicrobial resistance. *Lancet*, 387, 176–187.

Huss, H.H., Jørgensen, L.V. and Vogel, B.F. (2000). Control Options for Listeria monocytogenes in Seafoods. International Journal of Food Microbiology 62, 267–74. doi: 10.1016/s0168-1605(00)00347-0.

Hussain, M. A. and Gooneratne, R. (2017). Understanding the fresh produce safety challenges. *Foods*, 6, 23.

ICMSF. (2005). *Microorganisms in Foods 6: Microbial Ecology of Food Commodities (Second Edition)*. New York, NY: Kluwer Academic/Plenum Publishers.

ICMSF. (2011). Part II: Applications of principles to product categories. In: *Microorganisms in Foods 8. Use of Data in Assessing Process Control and Product Acceptance. International Commission on Microbiological Specifications for Foods*. Springer Science + Business Media, LLC.

ILSI. (2011). https://ilsi.eu/publication/the-enterobacteriaceae-and-their-significance-to-the-food-industry/ (accessed July 17, 2019).

Jay, M., Loessner, M.J. and Golden, D.A. 2005. Modern Food Microbiology. 7th Edition. Springer Science + Business Media. USA.

Jay, M. and James, L. J. M. (n.d.). *James M. Jay, Martin J. Loessner – 2005 – Modern Food Microbiology 7th ed.pdf*.

Kai, TM.K. and Aotearoa, A. (2009). Microbiological risk assessment of raw cow milk. Food Standards Australia New Zealand, Australia (2009). Retrieved from: https://www.foodstandards.gov.au/code/proposals/documents/P1007%20PPPS%20for%20raw%20milk%201AR%20SD1%20Cow%20milk%20Risk%20Assessment.pdf

Kam, K. M., C. K. Y. Luey, M. B. Parsons, K. L. F. Cooper, G. B. Nair, M.Alam, M. A. Islam, D. T. L. Cheung, Y. W. Chu, T. Ramamurthy, G. P.Pazhani, S. K. Bhattacharya, H. Watanabe, J. Terajima, E. Arakawa, O.-A.Ratchtrachenchai, S. Huttayananont, E. M. Ribot, P. Gerner-Smidt, and B.Swaminathan.2008. Evaluation and validation of a PulseNet standardizedpulsed-field gel electrophoresis protocol for subtypingVibrio parahaemolyti-cus:an international multicenter collaborative study. Journal of Clinical Microbiology 46: 2766–2773.

Kiilholma, J. (2007). Food-safety concerns in the poultry sector of developing countries. *Food and Agricultural Organisation of the United Nations*, 1–20. Retrieved from http://www.fao.org/ WAICENT/faoINFO/AGRICULT/AGAInfo/home/events/bangkok2007/docs/part2/2_8.pdf

Ma, L., Wang, Y., Shen, J., Zhang, Q., and Wu, C. (2014). International Journal of Food Microbiology Tracking Campylobacter contamination along a broiler chicken production chain from the farm level to retail in China. *International Journal of Food Microbiology*, 181, 77–84. https://doi. org/10.1016/j.ijfoodmicro.2014.04.023

Meneses-Gonzalez, Y. E. (2010). *Identification and Characterization of Salmonella Serotypes Isolated from Pork and Poultry from Commercial Sources*. University of Nebraska-Lincoln. Retrieved from http://digitalcommons.unl.edu/foodscidiss/8

Mensah, P., Mwamakamba, L., Mohamed, C. and Nsue-Milang, D. (2012). Public health and food safety. In: *The WHO African Region African Journal of Food Agriculture, Nutrition and Development*, 12(4).

Nguyen-The, C. (2012). Biological hazards in processed fruits and vegetables-risk factors and impact of processing techniques. *LWT-Food Science and Technology*, 49, 172–177.

Olanya, O. M., Hoshide, A. K., Ijabadeniyi, O. A., Ukuku, D. O., Mukhopadhyay, S., Niemira, B. A. and Ayeni, O. (2019). Cost estimation of listeriosis (*Listeria monocytogenes*) occurrence in South Africa in 2017 and its food safety implications. *Food Control*, 102, 231–239.

Olsvik, Ø., Wasteson, Y., Lund, A. and Hornes, E. (1991). Pathogenic Escherichia coli found in food. *International Journal of Food Microbiology*, 12, 103–113.

Osaili, T. M., Alaboudi, A. R., Al-Quran, H. N. and Al-Nabulsi, A. A. (2018). Decontamination and survival of Enterobacteriaceae on shredded iceberg lettuce during storage. *Food Microbiology*, 73, 129–136. https://doi.org/10.1016/j.fm.2018.01.022.

Osei-Asare, Y. B. and Eghan, M. (2014). Meat consumption in Ghana, evidence from household micro-data. *The Empirical Economics Letters*, 13(2). ISSN 1681 8997.

Paludetti, L. F., Kelly, A. L., Brien, B. O., Jordan, K. and Gleeson, D. (2018). The effect of different precooling rates and cold storage on milk microbiological quality and composition. 1921–1929.

Parry-Hanson Kunadu, A., Aboagye, E., Colecraft, E.K., Otoo, G.E., Adjei, M.Y.B., Acquaah, E., Afrifah-Anane, E. and Amissah, J.G.N., 2019. Low consumption of indigenous fresh dairy products in Ghana attributed to poor hygienic quality. Journal of Food Protection 82, 276–286.

Paterson, D. L. (2006). Resistance in gram-negative bacteria: Enterobacteriaceae. *American Journal of Infection Control*, 34(5 Suppl.), 20–28. https://doi.org/10.1016/j.ajic.2006.05.238.

Peterson, E. and Kaur, P. (2018). Antibiotic Resistance Mechanisms in Bacteria: Relationships Between Resistance Determinants of Antibiotic Producers, Environmental Bacteria, and Clinical Pathogens. Frontiers in Microbiology 9:2928.

Rigarlsford, J. F. (2007). *Microbiological Analysis of Red Meat, Poultry and Eggs: Microbiological Monitoring of Cleaning and disinfection in Food Plants*. UK: Elsevier Science.

Rusell, S. M. (2005). *Rapid Detection and Enumeration of Pathogens on Poultry Meat* (pp. 454–485). USA: Woodhead Publishing. University of Georgia.

Saxena, T., Kaushik, P., & Krishna Mohan, M. (2015). Prevalence of E. coli O157: H7 in water sources: An overview on associated diseases, outbreaks and detection methods. *Diagnostic*

Microbiology and Infectious Disease, 82(3), 249–264. https://doi.org/10.1016/j. diagmicrobio.2015.03.015

Sharma, S., Chatterjee, S., Datta, S., Prasad, R., Dubey, D., Prasad, R. K. and Vairale, M. G. (2017). Bacteriophages and its applications: an overview. *Folia Microbiologica*, 62(1), 17–55. https://doi.org/10.1007/s12223-016-0471-x.

Skovgaard, N. (2012). Microorganisms in foods 8, use of data for assessing process control and product acceptance. *International Journal of Food Microbiology*, 154. http://www.sciencedir ect.com/science/article/pii/S016816051100732X.

Smith, C. F., and Townsend, D. E. (2016). A new medium for determining the total plate count in food. *Journal of Food Protection*, 62(12), 1404–1410. https://doi.org/10.4315/0362-028x-62.12.1404.

Sofos, J. N. (2008). Challenges to meat safety in the twenty-first century. *Meat Science*, 78, 3–13.

Somily, A. M., Al-mohizea, M. M., Absar, M. M., Fatani, A. J., Ridha, A. M., Al-ahdal, M. N. and Al-qahtani, A. A. (2016). Microbial pathogenesis molecular epidemiology of vancomycin resistant enterococci in a tertiary care hospital in Saudi Arabia. *Microbial Pathogenesis*, 97, 79–83. https://doi.org/10.1016/j.micpath.2016.05.019.

Soyiri, I. N., Agbogli, H. K. and Dongdem, J. T. (2008). A pilot microbial assessment of beef sold in the Ashaiman Market, a suburb of Accra. *African Journal of Food, Agriculture, Nutrition and Development*, 8(1), 91–103.

State, H. (2001). Generation of enterococci bacteria in a coastal saltwater marsh and its impact on surf zone water quality. 35(12), 2407–2416.

Stoica, M., Stoean, S. and Alexe, P. (2014). Overview of biological hazards associated with the consumption of the meat products. *Journal of Agroalimentary Processes and Technologies*, 20(2), 192–197.

Tauxe, R. V. (1997). Emerging foodborne diseases: an evolving public health challenge. *Emerging Infectious Diseases*, 3(4), 425–434. https://dx.doi.org/10.3201/eid0304.970403.

Tauxe, R. V., Doyle, M. P., Kuchenmüller, T., Schlundt, J. and Stein, C. E. (2010). Evolving public health approaches to the global challenge of foodborne infections. *International Journal of Food Microbiology*, 139, S16–S26.

Uyttendale, M. and Herman, L. (2015). A review of the microbiological hazards of dairy products made from raw milk. *International Dairy Journal*, 50, 32–44.

van Immerseel, F., Methner, U., Rychlik, I., Nagy, B., Velge, P., Martin, G., Foster, N., Ducatelle, R. and Barrow, P. A. (2005). Vaccination and early protection against non-host-specific *Salmonella* serotypes in poultry: Exploitation of innate immunity and microbial activity. *Epidemiology and Infection*, 133(6), 959–978.

World Health Organisation (2015). WHO Estimates of the Global Burden of Foodborne Diseases, WHO, Geneva.

World Health Organisation+. (2015). *WHO Estimates of the Global Burden of Foodborne Diseases*. Geneva: WHO. https://www.who.int/foodsafety/areas_work/foodborne-diseases/ferg/en/ (accessed August 20, 2019).

Zervos, M. J., Chow, J. W. and Robert, R. (2010). *Infectious Disease and Antimicrobial Agents*. www. antimicrobe.org/new/bo3.asp (accessed July 20, 2019).

Online references

www.cdc.gov/foodsafety/foodborne-germs.html
www.food.unl.edu/food-posioning-foodborne-illness
www.who.int/antimicrobial-resistance/en
www.who.int/news-room/fact-sheets/detail/food-safety

Irina V. Rozhkova, Konstantin V. Moiseenko, Olga A. Glazunova,
Anna V. Begunova, Tatyana V. Fedorova

9 Russia and Commonwealth of Independent States (CIS) domestic fermented milk products

9.1 Introduction

Fermented milk products and, in particular, drinks have a long history. It was assumed that sour-milk drinks were the first products purposefully made from milk by man. Originally, they date back to the dawn of civilization, when humans switched their lifestyle from collection of food (foraging) to its production. Probably, the transition took place in different countries at a different time. Many ancient civilizations with well-developed agriculture (i.e., Sumerians, Babylonians, and Indians) had the technology of milk fermentation. Most likely, the birthplace of sour-milk drinks was the Middle East and Balkans. In Greece and Italy, around third to fourth century BCE, dairy products were prepared from goat's and sheep's milk. The great Homer, in his immortal "Odyssey," described how the hero and his companions found buckets and mugs full of thick sour milk in the cave of Cyclops Polyphemus. The people of India, the Middle East, and the South Caucasus consumed sour-milk drinks on a daily basis. These drinks were prepared from the milk of cow, sheep, or donkey; possessed a great nutritional value; efficiently quenched thirst; and a few were intoxicating. In the distant past, people already learned about the unusual healing properties of fermented milk drinks that could accelerate the treatment of various diseases. Thus, the Indian proverb saying goes as: "Drink sour milk and live long."

Many millennia have passed since the moment the first sour-milk drink was prepared and before the cause of such a transformation of milk was determined. Only in the nineteenth century, the French scientist, Louis Pasteur (1822–1895) discovered that fermentation of milk occurs under the influence of microorganisms. Later, these microorganisms were isolated and classified into different groups (e.g., cocci, coli, bacilli, and sometimes yeast).

Sour-milk products are still widely represented in the diet of many people around the world (Mohamadi Sani, Rahbar, and Sheikhzadeh, 2019; Oberman and Libudzisz, 1998). Currently, there is a clear tendency to expand the assortment of fermented milk products with respect to their taste, texture, and functional properties.

Irina V. Rozhkova, Anna V. Begunova, All-Russian Research Institute of Dairy Industry, Moscow, Russia
Konstantin V. Moiseenko, Olga A. Glazunova, Tatyana V. Fedorova, Research Centre of Biotechnology, Moscow, Russia

https://doi.org/10.1515/9783110667462-009

The nutritional value of fermented milk products is mainly determined by their chemical composition. In the production of fermented milk by the process of lactic acid fermentation (and sometimes other types of microbiological processes), the initial chemical composition of milk (and consequently its digestibility and other functional properties) changes significantly: (1) about 1% of milk sugar (lactose) is converted to lactic acid; (2) casein, the main protein of cow's milk, coagulates in the form of small flakes, which increases its digestability; (3) the content of free amino acids in fermented milk products slightly increases; and (4) the content of vitamins in fermented milk products usually decreases as a result of consumption by lactic acid bacteria; the exception is products in which the development of not only lactic acid bacteria, but also yeast, acetic acid, and propionic acid bacteria occurs.

Unlike other food, along with valuable nutrients, dairy products contain a huge number of living cells of lactic acid microorganisms – up to a billion in each gram. Typically, living cells comprise up to 1–2% of the total mass of the product. Bacterial biomass and its different minor metabolites determine individual health-promoting properties of a particular sour-milk product.

In Russia, fermented milk products began to be widely consumed from the first half of the twentieth century, promoted by the work of I.I. Mechnikov (1845–1916), who first studied their importance in human nutrition. Studying the problem of longevity, Mechnikov developed a theory according to which premature aging is caused by the toxic effect of a putrefactive bacteria in the human intestine and proposed the possibility to fight this condition (Mackowiak, 2013). In his renown work, "The Prolongation of Life: *Optimistic Studies*," he drew attention to the long life expectancy of many residents of Bulgaria. In his opinion, Bulgarians' longevity was caused by the historical tradition of systematic consumption of curdled milk drink – "kiselo mleko." The significant quantity of the lactic acid bacteria and the main metabolic product, lactic acid, found in "kiselo mleko" and other types of sour milk allowed Mechnikov to create his pioneering scientific theory of **Prolongation of Life**; the corner stone of which was use of fermented milk drinks in the prevention of autointoxication of the organism by the putrefactive bacteria.

Due to the national traditions and habits, the assortment of dairy products manufactured in Russia and Commonwealth of Independent States (CIS) countries is extremely diverse (Figure 9.1). These products differ not only in their microflora composition, but also in the technology of their production. The traditional (historically, domestically made) fermented milk products are found in the market in large quantities and have a constant demand from customers. These are classified into three main groups (Figure 9.2, Table 9.1): (1) The products manufactured with *Lactococcus* and the possible addition of *Streptococcus thermophilus* such as **tvorog** (rus: творог /tvorog/) and **smetana** (rus: сметана /smetana/); (2) the products manufactured with *Str. thermophilus* and *Lactobacillus bulgaricus* such as **ryazhenka** (rus: ряженка /rjaženka/) and **varenets** (rus: варенец /varenec/); and

Figure 9.1: The assortment of sour-milk products on the Russian and CIS countries market.

(3) the products manufactured with the cultures of mixed lactic acid and alcohol fermentation: **kefir** (rus: кефир /kəˈfɪər/), **kumis** (rus: кумыс /kumys/), **ayran** (rus: айран /ajran/), **tan** (rus: тан /tan/), **matsoni** (rus: мацони or мацун /matsoni/), and **kurunga** (rus: курунга /kurunga/); although the last one is more specific for the Buryats region. In addition, the fourth group of products recently emerged and started to gain more and more popularity: (4) the products manufactured with probiotic microorganisms – **Acidophilin**, **Bifiton**, **Tonus**, **Biokefir**, and so on.

9.2 Tvorog

Tvorog is a traditional Russian product, the closest analogy of which in the English-speaking countries is curd, quark, or farmer's cheese. Without any exaggeration, products of this kind can be regarded as the first human-made dairy dish. The time and place of their birth are unknown, but their history has already spanned millennia. Mention of curd can already be found in ancient Indian and Chinese treatises, as well as in Roman cookbooks. In the history of traditional Russian cuisine, the **tvorog** was widely known and popular. It was especially recommended for children, pregnant and lactating women, elderly, and sick people. In Russia, for a long time, **tvorog** as well as dishes from it were called "sir" (rus: сыр /syr/). However, after the appearance of hard cheeses, brought in Russia from Europe by Peter the Great (1682–1721), the usage of the word "sir" was gradually dismissed. In modern Russia, the words **tvorog** and sir are separated, the second one being reserved for the different varieties of hard and soft cheeses.

Homofermentative starter

Lactose → Glucose + Galactose → Lactic acid

Heterofermentative starter

Lactose → Glucose + Galactose → Lactic acid + Ethanol + CO_2

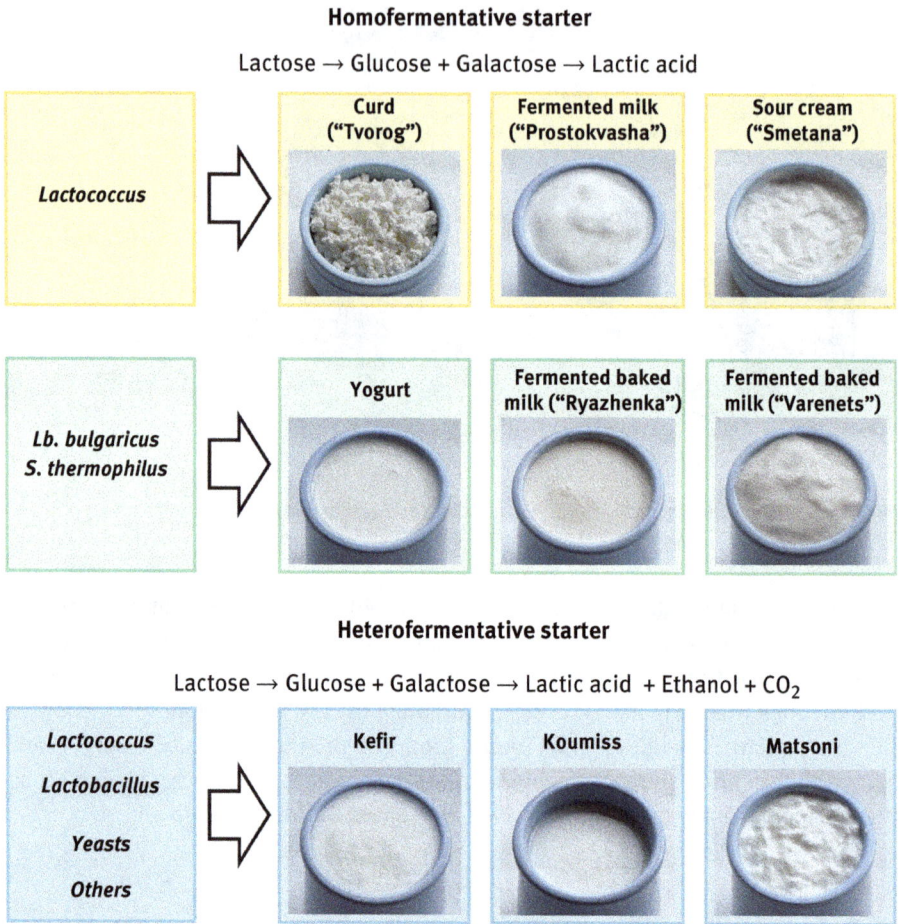

Figure 9.2: The starter types and corresponding domestic products on the Russian and CIS countries market.

In accordance with the "Technical Regulations for Milk and Dairy Products" of the Russian Federation, **tvorog** is a fermented milk product manufactured using starter microorganisms – lactococci or a mixture of lactococci and thermophilic lactic streptococci and acid or acid-rennet coagulation of milk proteins (Figure 9.2). The removal of whey after coagulation can be performed by means of self-pressing, pressing, separation (centrifugation), and/or ultrafiltration. In addition, some amount of milk constituents (before or after fermentation) can be mixed with final product for normalization.

In the industrial production of **tvorog**, the acid-rennet method is used more often. According to this method, the ripening of a milk clot occurs under the influence of lactic acid and rennet; addition of calcium chloride can be used to accelerate the fermentation of milk.

Although **tvorog** can be prepared using just *Lactococcus* culture, usually starter for **tvorog** contains both *Lactococcus* and *Str. thermophilus* (Table 9.1). Inclusion of

Table 9.1: Russia and CIS countries fermented dairy products.

Product	Preparation*	Milk	Associated microorganisms**
Acid milks (homofermentative starters)			
Lactococcus-based starters			
Tvorog	Trad/Com	Cow	*Lc. lactis* subs. *lactis*, *Lc. lactis* subsp. *cremoris*, *Lc. lactis* subsp. *lactis* biovar. *diacetylactis* (*Lc. diacetylactis*)
Cottage cheese	Trad/Com	Cow	
Prostokvasha (sour milk)	Trad/Com	Cow	
Smetana (sour cream)	Trad/Com	Cow	
Lactococcus- and *Str. thermophilus*-based starters			
Tvorog (curd or quark)	Trad/Com	Cow	*Lc. lactis* subs. *lactis*, *Lc. lactis* subsp. *cremoris*, *Lc. lactis* subsp. *lactis* biovar. *diacetylactis* (*Lc. diacetylactis*), *Str. thermophilus*
Cottage cheese	Trad/Com	Cow	
Smetana 10–15% fat (low-fat sour cream)	Trad/Com	Cow	
Str. thermophilus- and *Lb. bulgaricus*-based starters			
Yogurt	Trad/Com	Cow	*Str. thermophilus*, *Lb. bulgaricus*
Ryazhenka (sour-baked milk)	Trad/Com	Cow	
Varenets (sour-baked milk)	Trad/Com	Cow	*Str. thermophilus*
Acid and alcoholic milks (heterofermentative starters)			
Mixed culture-based starters			
Kefir	Trad/Com ("kefir grains")	Cow	**Lactococci:** *Lc. lactis* subsp. *lactis*; **Lactobacilli:** *Lb. kefiranofaciens*, *Lb. casei*, *Lb. hefirgranum*, *Lb. brevis*, *Lb. kefir*, *Lb. parakefir*, *Lb. acidophilus*, *Lb. rhamnosus*; **Leuconostoc:** *Leuc. mesenteroides*, *Leuc. mesenteroides* subsp. *dextranicum*; **Yeast:** *S. cerevisiae*, *K. marxianus* var. *marxianus*, *C. kefir*; **acetic acid bacteria:** *A. aceti*

Table 9.1 (continued)

Product	Preparation*	Milk	Associated microorganisms**
Kumis (Koumiss)	Trad	Mare	**Lactic acid bacteria:** *Lc. lactis* subsp. *lactis* biovar. *diacetylactis*, *Lb. delbrueckii* subsp. *bulgaricus*, *Lb. acidophilus* **Leuconostoc:** *Leuc. mesenteroides*, *Leuc. mesenteroides* subsp. *dextranicum* **Lactose-fermenting yeasts:** *Saccharomyces* spp., *K. marxianus* var. *marxianus*, *C. koumiss* **Nonlactose-fermenting yeast:** *S. cartilaginous*
	Com	Mare Cow	*Lb. bulgaricus*, *Lb. acidophilus*; lactose-fermenting yeast (*K. maxianus*)
Ayran and **Tan**	Trad	Cow	*Lb. lactis* subsp. *lactis*, *Lc. lactis* subsp. *diacetilactis*, *Str. thermophilus*, *Lb. bulgaricus*, *K. lactis*, *S. cerevisiae*, *Geotrichum* spp.
	Com	Cow	*Str. thermophilus*, *Lb. bulgaricus*, lactose-fermenting yeasts
Matsoni (Matzoon)	Trad	Cow Sheep Goat Buffalo	**Lactic acid bacteria:** *Lb. delbruekii* subsp. *lactis*, *Lb. delbrueckii* subsp. *bulgaricus*, *Lb. acidophilus*, *Str. thermophilus*, *Lb. lactis*, *Lc. lactis* subsp. *cremoris*, *Lb. helveticus*, *Lb. paracasei*, etc.; **Leuconostoc:** *Leuc. lactis* **Yeasts:** *K. marxianus*, *S. cerevisiae*, *C. famata*, etc.
	Com	Cow	*Lb. bulgaricus*, *Str. thermophilus*, yeasts
Kurunga	Trad	Cow	**Lactic acid bacteria:** *Lc. lactis* subsp. *lactis*, *Lactobacillus* spp. (*Lb. paracasei*, *Lb. rhamnosus*, *Lb. brevis*, *Lb. buchneri*, *Lb. diolivorans*, *Lb. parabuchneri*, etc.) **Acetic and propionic acid bacteria; lactose- and nonlactose-fermenting yeasts** (predominantly *Torulopsis* spp., *Candida* spp.)
	Com	Cow	*Str. thermophilus*, *Lb. bulgaricus*, lactose-fermenting yeasts

Fermented milks with probiotic microorganisms

Lb. acidophilus-based starters

Product	Preparation*	Milk	Associated microorganisms**
Acidophilin	Com	Cow	*Lb. acidophilus*, *Lc. lactis* subsp. *lactis*, *Lc. lactis* subsp. *lactis* biovar. *diacetylactis*, **kefir** grains
Acidolact	Com	Cow	*Lb. acidophilus*, *Str. thermophilus*
Biolact	Com	Cow	*Lb. acidophilus*

Table 9.1 (continued)

Product	Preparation*	Milk	Associated microorganisms**
Bioyogurt	Com	Cow	*Lb. bulgaricus, Str. thermophilus, Lb. acidophilus*
Bifidobacteria-based starters			
Bifilin and **Bifilin-M**	Com	Cow	*B. adolescentis*
Bifilin-lacto	Com	Cow	*B. adolescentis, Lb. acidophilus, Str. thermophilus*
Bifiton	Com	Cow	*B. adolescentis*
Biokefir	Com	Cow	*B. adolescentis,* **kefir** grains
Biosmetana (bio-sour cream)	Com	Cow	*B. adolescentis, Lb. lactis* subsp. *lactis, Lb. lactis* subsp. *lactis* biovar. *cremoris*
Biorjazenka	Com	Cow	*B. adolescentis, Str. thermophilus, Lb. bulgaricus*
Actifilin	Com	Cow	*B. adolescentis, Lc. lactis* subsp. *lactis* biovar. *cremoris, Lc. lactis* subsp. *lactis* biovar. *diacetylactis, Lb. casei*
Acetic acid- and propionic acid bacteria-based starters			
Tonus	Com	Cow	*Lc. lactis* subsp. *lactis* biovar. *diacetylactis, A. aceti, Pr. freudenreichii* subsp. *shermanii*
Bifidobacteria, acetic and propionic acid bacteria-based starters			
Bifiton and **Bifitonchik**	Com	Cow	*Lc. lactis* subsp. *lactis* biovar. *iacetylactis, Pr. freudenreichii* subsp. *shermanii, A. aceti, B. adolescentis*

*Com: commercial; Trad: traditional.
**A.: *Acetobacter* spp.; *B.: Bifidobacterium* spp.; *C.: Candida* spp.; *K.: Kluyveromyces* spp.; *Lb.*: *Lactobacillus* spp.; *Lc.: Lactococcus* spp.; *Leuc.: Leuconostoc* spp.; *Pr.: Propionibacterium* spp.; *S.*: *Saccharomyces* spp.; *Str.: Streptococcus* spp.

Str. thermophilus provides several advantages such as: improved resistance to the bacteriophage infection; decreased sensitivity to the seasonal changes of milk quality; accelerated curd formation; enhanced water-holding capacity and viscosity of the final product. As *Str. thermophilus* growing together with *Lactococcus* has the ability to develop at the significantly lower temperatures, compared to the optimal for this species, the most favorable temperature for the starters comprising *Lactococcus* and *Str. thermophilus* is determined to be 32 °C.

9.3 Ryazhenka and varenets

History of **ryazhenka** started from immemorial time in Ukraine. The main **ryazhenka** properties are homogeneous dense structure (thick consistency), cream color, and a delicate sweet flavor (Figure 9.2). It can be drunk as it is or used in different recipes. It improves digestion and boosts immunity. Historically, it was made by stewing milk and milk cream at high temperature (without boiling) in special low clay pots until the appearance of a nice cream color. The obtained "stewed milk" was supplemented with small quantity of sour cream and placed into a warm vessel for fermentation. The whole process cycle took 3–5 days.

Nowadays, making **ryazhenka** takes much less time. Initially, milk is poured into a tank and heated up to 50 °C with further steam heating up to 95–99 °C; at this temperature, milk is maintained for 3–5 h until the appearance of a light creamy color. The obtained baked milk is cooled down to 40–42 °C, and *Str. thermophilus*- and *Lb. bulgaricus*-based starter is added. The fermentation process continues for about 4–5 h until the product acidity reaches a definite titratable acidity of 60–65 °T. Subsequently, the product is cooled down, vigorously stirred, and transferred into the dispensing machine.

Currently, analogous to Ukrainian **ryazhenka** product, **varenets** can be found in the Russian market. The industrial technology of **varenets** production is similar to that of **ryazhenka**, with two main differences: (1) the duration of milk baking at 97 ± 2 °C is decreased to 40–80 min and (2) the starter includes only *Str. thermophilus* (Table 9.1).

The stewing processes used in manufacturing of **ryazhenka** and **varenets** initiates the formation of prebiotic lactulose and various antioxidants, which provide both health benefits and enhancement of the shelf-life of the final product. Moreover, a gentle heating process promotes the Maillard reaction, which generates characteristic cream color of the final product (Figure 9.3). In various research works, it has been proved that conjugation of milk proteins with lactose and galactose through Maillard

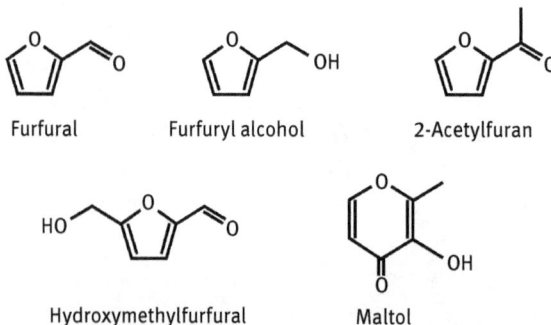

Furfural   Furfuryl alcohol   2-Acetylfuran

Hydroxymethylfurfural   Maltol

Figure 9.3: The possible products of the Maillard reaction.

reaction can be an efficient way to obtain functional food elements with unique dominant prebiotic characteristics (Asgar and Chauhan, 2019). The flavors of **ryazhenka** and **varenets** can be attributed to a large range of compounds including aldehydes, ketones, fatty acids, carboxylic acids, lactones, and alcohols. These flavorings are generated during the heating process by oxidation of lipid, degradation of free fatty acids, and reactions between proteins and lactose (and its various degradation products) (Newton, Fairbanks, Golding, Andrewes, and Gerrard, 2012). Besides being a strong prebiotic, both **ryazhenka** and **varenets** have enormous probiotic properties. The number of cells of thermophilic streptococcus in the final product is 10^7 CFU/mL.

9.4 Kefir

Kefir – a beverage obtained by milk fermentation with a specific type of mesophilic symbiotic culture, "**kefir** grains," originated in the Caucasus mountains (Figure 9.2). The word **kefir** is derived from the Turkish word "Keyif," which means "good feeling." In Russia, the sales volume of **kefir** is two-third of all dairy products available in the market.

The traditional Caucasian method of **kefir** production involves natural fermentation of milk by **kefir** grains. The fermentation process is carried out in special goatskin bags that are regularly shaken to ensure the proper mixing of milk and **kefir** grains. The finished product has high acidity and include variable amount of ethanol and carbon dioxide. Although typically **kefir** is produced from cow's milk, milk of various other species including ewe, goat, and buffalo can also be used.

On a larger scale, **kefir** production is a multistep process (Beshkova, Simova, Simov, Frengova, and Spasov, 2002). Initially, the so-called mother culture is prepared by incubating milk with **kefir** grains (2–3%, v/v). The fermentation is carried out at 20–22 °C for 15–17 h and afterward the mixture is left for ripening at the mentioned temperature for 7–9 h (i.e. the whole process takes up 24 h). Further on, the grains are removed by filtration, the resulting liquid is added to the fresh milk (1–5%, v/v), and the fermentation process is carried out at 20–25 °C for 9–12 h. After the fermentation, the mixture is cooled down to 14–16 °C and kept for 12–15 h to mature (again, the whole process takes up 24 h). **Kefir** grains removed by filtration at the first step are again added to fresh milk. For prolonged storage, **kefir** grains can be lyophilized.

Kefir starter microflora is a specific and complex symbiotic association of bacteria and yeasts, which affects the **kefir** quality (taste and texture) significantly. At minimum five basic functional groups can be determined in the **kefir** starters (Table 9.1) : heterofermentative lactococci, lactobacilli, *Leuconostoc* spp., acetic acid bacteria and yeasts. During the process of **kefir** maturation, the number of acetic acid bacteria increases to 10^3 colony-forming units per gram (CFU/mL) and yeasts to 10^4–10^5 CFU/mL, which provide a strong probiotic potential for **kefir**. For the rest of the microbial association, no substantial change in CFU/mL is observed. Instead, these microorganisms immensely contribute their metabolites to the final product. Lactic acid, bioactive peptides, exopolysaccharides, antibiotics, and numerous bacteriocins are a far from complete list of the bioactive compounds of **kefir** (Chen et al., 2015; Leite et al., 2015).

The main polysaccharide in **kefir** grains and consequently in **kefir** itself is kefiran. Mainly produced by *Lactobacillus kefiranofaciens*, kefiran is a heteropolysaccharide composed of equal proportions of glucose and galactose (Zajsek, Kolar, and Gorsek, 2011). Besides improving the viscosity and viscoelasticity of acidified milk gel, it has been demonstrated that kefiran has number of outstanding properties such as antitumor, antifungal, antibacterial, immunomodulating or epithelium-protecting, anti-inflammatory, healing, and antioxidant (Prado et al., 2015). It is interesting to note that despite *Lb. kefiranofaciens* alone poses an outstanding ability to produce kefiran, the addition of *Saccharomyces* spp. to the culture improves the net kefiran production. This situation is a good illustration of the importance of the symbiosis between the bacteria and yeasts present in **kefir** (Cheirsilp, Shimizu, and Shioya, 2003).

Regular consumption of **kefir** has been associated with improved digestion and tolerance to lactose, antibacterial effect, hypocholesterolemic effect, control of plasma glucose, antihypertensive effect, anti-inflammatory effect, antioxidant activity, anticarcinogenic activity, antiallergenic activity, and healing effects (Koroleva, 1991; Melnikova, Bogdanova, Ponomarev, and Korgov, 2015; Rosa et al., 2017).

9.5 Koumiss (Kumis)

Koumiss is a low alcoholic (1–2%), creamy, and sparkling beverage, with a slight degree of sourness (Figure 9.2). To date, **koumiss** is regarded as the oldest known alcoholic fermented milk beverage. According to Herodotus (484–424 BC), specially prepared mare's milk (that can be identified as **kumis** from further descriptions) was a favorite beverage of the Scythians (Pieszka et al., 2016). **Koumiss** is a traditional beverage in Central Asia. In Bashkortostan, Kazakhstan, Uzbekistan, Kyrgyzstan, and Ukraine, **koumiss** is regarded as a national medicinal product. In Russia, **koumiss** was called "milky wine." The Ipatiev Chronicle of 1182 reports that the Galician prince drank **koumiss** from Batu Khan in 1245, and that Tatars knew this drink long before their attack on Russia during the period 1237–1240.

Traditionally, **koumiss** is a home-made beverage produced by spontaneous fermentation of raw mare's whole milk for 2 or 3 days at room temperature. Compared to cow's milk, mare's milk contains less fat and proteins (1.21% vs 3.61% and 2.14% vs 3.25%, respectively), more lactose (6.37% vs 3.25%), and more vitamin C (1,280–8,100 vs 300–2,300 µg/L). Similar to the human milk, which contains equal amount of casein and whey proteins (α-lactalbumin and β-lactoglobulin), mare's milk with 50–55%

of casein and 45% of whey proteins is an albumin-type milk. In contrast, ruminants milk containing 80% of casein is a casein-type milk (Jastrzębska, Wadas, Daszkiewicz, and Pietrzak-Fiećko, 2017). Although mare's milk is the most frequently used raw material for the preparation of **koumiss**, sometimes it can be replaced by the milk from yak, donkey, camel, sheep, or cow.

Similar to **kefir**, the starter for **koumiss** has a symbiotic origin and contains two distinct microbial groups: (1) lactobacilli that are reported to play a major role in fermentation process affecting aroma, texture, and acidity of the final product; and (2) yeasts, presence of which is crucial for the appropriate production of carbon dioxide and ethanol (Table 9.1).

At the time of Soviets, to organize the industrial production of **koumiss**, the natural starters from mare's milk were analyzed and appropriate consortium of microorganisms together with conditions of fermentation were established. This technology was passed down to the modern days Russia and CIS countries. Industrial **koumiss** starter composed of *Lb. bulgaricus* and lactose-fermenting yeast *Klyvermyces maxianus*. In addition, to intensify **koumiss** medical properties, *Lactobacillus acidophilus* *(Lb. acidophilus)* is usually introduced into the starter. The commercial method of **koumiss** preparation involves pasteurization of mare's milk at 90 °C for 2–3 min. The pasteurized milk subsequently cooled down to 28 °C, inoculated with the starter (10%, v/v), mixed to homogeneity for about 15–20 min, and incubated at 26–28 °C for 5–6 h until the acidity reaches 75–100 °T.

Due to the limited amount of mare's milk, the technology for the production of the **koumiss**-like beverage from a milk of cows has been also developed. When producing **koumiss**-like beverage from cow's milk, the milk composition is adjusted with whey protein concentrate and ascorbic acid to be approximately the same as those of the mare's milk.

Containing enzymes, trace elements, antibiotics, and vitamins A, B1, B2, B12, D, E, and C, **koumiss** helps in treating diseases of cardiovascular, respiratory, gastrointestinal, and urinary systems. It is believed that systematic consumption of **koumiss** stimulates nervous and boosts immune systems. In the early twentieth century, **koumiss** was used as a supporting dietary supplement in the treatment of tuberculosis. Nowadays, **koumiss** is recommended for the people suffering from AIDS, cancer, herpes, and scurvy. Consumption of **koumiss** can substantially alleviate number of psychological disorders such as depression, insomnia, and attention-deficit hypersensitivity disorder.

Currently, **koumiss** is administered in many Central Asian sanatoriums (the best-known are Yumatovo, Gluchovskaya, and Schafranovo) as a supportive treatment. Some sanatoriums even have their own **kumis** production facilities and heards of mares.

9.6 Ayran and tan

Ayran is a popular drink in Iran, Armenia, Syria, Lebanon, Jordan, South India, Iraq, and Cyprus. **Tan** is a Caucasian drink of which homeland is Bulgaria. Essentially, **tan** is an **ayran** with addition of certain amount of water and salt. Both **ayran** and **tan** has a homogeneous thick or moderately dense consistency, undisturbed clot, and pleasant refreshing taste of sour milk.

Microbial consortium of a home-made **ayran** is largely dominated by (Table 9.1): *Lactobacillus lactis* subsp. *lactis*; *Lb. lactis* subsp. *diacetylactis*; *Lactobacillus delbrueckii* subsp. *bulgaricus*; *Str. thermophilus*; and lactose-fermenting and nonfermenting yeast strains, belonging to *Kluyveromyces lactis*, *Saccharomyces cerevisiae*, and *Geotrichum* spp. Recently, high counts of other lactic acid bacteria species, including *Lactobacillus helveticus*, *Lactobacillus fermentum*, and *Lactobacillus paracasei* were also reported (Baruzzi et al., 2016).

Nowadays, commercial manufacture of **ayran** and **tan** have been developed from cow's milk using the starters of mixed-fermenting lactic acid bacteria – *Lb. bulgaricus* and *Str. thermophilus*, and lactose-fermenting yeasts (Table 9.1). A method of the production includes: introduction of starter into normalized, pasteurized, and cooled milk; mixing and fermentation at 30–35 °C (or 34–38 °C) to an acidity of 160–200 °T; and mixing and maturation at 26–28 °C. The pasteurized sodium chloride solution is introduced into the final product after the maturation.

Both **ayran** and **tan** include a number of trace elements that show a beneficial effect on the gastrointestinal tract. Systematic consumption of **ayran** and **tan** promotes lowering of blood cholesterol and removal of heavy metals. It is known that these drinks have a tonic effect, and many people drink them to get rid of a hangover syndrome.

9.7 Matsoni

Matsoni (syn. mazun, matsoon) is a traditional, known from ancient times, Caucasian fermented milk product. With its thick consistency, creamy taste, and pleasant aroma, **matsoni** can be considered analogous to yogurt (Figure 9.2). However, in contrast to yogurt, which is fermented by *Lb. delbrueckii* subsp. *bulgaricus* and *Str. thermophilus*, **matsoni** starters exhibit a large degree of biodiversity, influenced by both geography and type of the fermented milk obtained from cow, sheep, goat, buffalo, or their mixture (Table 9.1) (Bokulich et al., 2015).

Domestic **matsoni** is made from heated or pasteurized milk, cooled to 35–42 °C, and inoculated with a portion (approximately 1%, w/w) of previously finished **matsoni**. Fermentation is carried out at 35–42 °C overnight. Hence, each community maintains an ongoing culture consortium used as a starter for **matsoni**.

It was demonstrated that several core taxa are always present in the most domestic **matsoni** fermentations: *Lactobacillus* spp., *Streptococcus* spp., and *Kluyveromyces marxianus* (Quero et al., 2014). While species of lactobacilli and lactococci present in **matsoni** are similar to those found in other yogurt-type fermentations, the large populations of yeasts are the distinguishing feature of **matsoni**. Besides *K. marxianus*, **matsoni** starters can contain yeasts such as *S. cerevisiae* and *Candida famata*. Producing alcohols, organic acids, and CO_2, and possessing proteolytic and lipolytic activities, yeasts significantly alter the sensory profile of **matsoni** and supplement it with fruity and yeasty flavors. Different populations of yeasts in different fermentations of **matsoni** explains at large the unique characteristics of this product in different regions.

Matsoni is especially widely consumed throughout the Caucasus, where it is considered to have beneficial health effects, particularly for intestinal disorders, and increased longevity.

9.8 Kurunga

Kurunga is a traditional fermented milk product of mixed lactic acid and alcoholic fermentation of cow's milk (the alcohol content usually does not exceed 1%). This beverage is popular in the Northeast Asia among the Buryats, Mongols, Tuvans, and other nationalities. **Kurunga** is a milky liquid foaming product containing small flakes of casein. It has a sour smell and taste.

The method of production of **kurunga** is known from ancient times. For the Mongols and Tuvans, who led a seminomadic lifestyle in the summer, **kurunga** was one of the most important products. Other nationalities, such as Buryats and Khakasses, learned the secret of **kurunga** preparation just from the beginning of eighteenth century.

The consortium of starter of the home-made **kurunga** has a naturally established ratio of microorganisms (CFU/mL): mesophilic lactic acid bacteria -3×10^9; thermophilic lactic acid bacteria -7×10^6; nonlactose-fermenting yeasts -8×10^7; lactose-fermenting yeasts -4×10^5; and acetic and propionic acid bacteria -3×10^4. Lactobacilli of home-made **kurunga** can be attributed to two groups. The first group has similar properties with the *Lb. bulgaricus*, whereas the second has similar properties **with** the *Lb. acidophilus* that produces more acid and can ferment all types of carbohydrates. It is especially interesting that many strains isolated from **kurunga** was characterized by high antimicrobial activity and secreted large amounts of antimicrobial substances (Vodolazov, Dbar, Oleskin, and Stoyanova, 2018).

In the commercial preparation of **kurunga**, according to the Russian Technical Standard, just *Lb. bulgaricus*, *Str. Thermophilus*, and lactose-fermenting yeasts are present in the starter. For successful development, the yeasts present in **kurunga** starter have to use the lactic acid and other metabolites formed by lactic acid bacteria. In turn, in the presence of yeasts, lactic acid bacteria demonstrate faster growth and prolonged fermentation activity. These symbiotic relationships in **kurunga** persist for a long period of time, without any suppression of one group of microorganisms by the other. Consequently, the large scale production of **kurunga** strongly requires the absence of any extraneous microflora in the fermentation process.

9.9 The fermented milk products with probiotic microorganism

Currently, the market of fermented milk products in Russia and CIS countries is constantly expanding. As elsewhere in the world, interest in functional dairy products has increased due to their usefulness and medicinal properties. Many new products are gaining popularity among the people, especially those that poses strong probiotic properties.

"Probiotics" are currently defined by the Food and Agriculture Organization of the United Nations World Health Organization as "live microorganisms that, when administered in adequate amounts, confer a health benefit on the host" (FAO and WHO, 2006). In addition to the well-known bifidobacteria and acidophilic lactic acid bacilli (*Lb. acidophilus*), list of probiotic microorganisms also include various types of lactobacilli, yeasts, acetic acid bacteria, and propionic acid bacteria (De Roos and De Vuyst, 2018; Fazilah, Ariff, Khayat, Rios-Solis, and Halim, 2018; Hatoum, Labrie, and Fliss, 2012) (Table 9.2). In the production of

Table 9.2: Genera of bacteria that are commonly used as a probiotics in fermented dairy product.

Lactobacillus ssp.	*Lb. acidophilus, Lb. amylovarus, Lb. brevis, Lb. bulgaricus, Lb. casei, Lb. cellebiosus, Lb. crispatus, Lb. delbrueckii, Lb. fermentum, Lb. gallinarum, Lb. gasseri, Lb. helveticus, Lb. johnsonii, Lb. plantarum, Lb. reuteri, Lb. rhamnosus, Lb. salivarius*
Bifidobacterium ssp.	*B. adalescentis, B. animalis, B. bifidium, B. breve, B. infantis, B. lactis, B. longum, B. thermophilum*
Propionibacterium ssp.	*Pr. freudereichii* ssp. *shermanii*
Leuconostoc ssp.	*Leuc. mesenteroides*
Acetic acid bacteria	*Acetobacter, Gluconacetobacter,* and *Gluconobacter* ssp.
Yeast	*Saccharomyces cerevisiae, Saccharomyces boulardii, Kluyveromyces lactis*

dairy products, strains of probiotic microorganisms could be used either as a sole starter culture or as an adjunct culture to the standard starters.

It has been demonstrated that systematic consumption of probiotics provides long-standing benefits for the human health (Marco et al., 2017; Sánchez et al., 2017). Inclusion of probiotics into the diet improves intestinal microbiota balance, alleviates lactose intolerance symptoms, decreases food allergy, boosts immune system, reduces risk of colon cancers, and provides antidiabetic, antiobesity, hypo-cholesterolemic, and antihypertensive effects (Ishimwe, Daliri, Lee, Fang, and Du, 2015; Thushara, Gangadaran, Solati, and Moghadasian, 2016; Zoumpopoulou, Pot, Tsakalidou, and Papadimitriou, 2017).

The precise processes by which probiotics exert their effects on the human organism are just starting to be revealed. In the last 10 years, several main mechanisms of probiotics action were described (Figure 9.4) (Benchimol and Mack, 2004; Bermudez-Brito, Plaza-Díaz, Muñoz-Quezada, Gómez-Llorente, and Gil, 2012; Cunninghamrundles, 2000; Ishimwe et al., 2015; Sánchez et al., 2017): (1) With respect to the metabolism of nutrients, probiotics can degrade otherwise indigestible carbon sources (e.g., plant polysaccharides) increasing their bioavailability. Furthermore, they can synthesize various vitamins and digestion-improving enzymes. The most well-known enzymes synthesized by probiotic microflora are β-galactosidase (β-gal) and bile salt hydrolases (BSHs). The β-gal catalyzes the hydrolysis of lactose, alleviating lactose intolerance, whereas BSHs catalyze cleavage of the glycine/taurine moiety from the steroid core, increasing cholesterol excretion. (2) With respect to the suppression of pathogens, probiotics can antagonize pathogenic microflora by producing different antimicrobial metabolites (e.g., bacteriocins and organic acids) and depleting of nutrients available for pathogens. In addition, the presence of probiotics occludes intestinal receptors for pathogen binding and stimulates production of mucin by goblet cells – both of which inhibits pathogens adhesion to gut epithelial surfaces. (3) With respect to the modulation of immune system, probiotics can stimulate the production of cytokines, and activate dendritic cells, macrophages, natural killer cells, and T-lymphocytes. (4) With respect to the functioning of nervous system, probiotics can stimulate the production of a range of neuroactive molecules. One of the best characterized example of which is stimulation of serotonin release by enterochromaffin cells. Recently, a special term "the gut-brain axis" ("microbiome-gut-brain axis") was adopted to refer to the biochemical signaling processes that take place between the gastrointestinal tract and the central nervous system.

Nowadays, in Russia and CIS countries market products with three main types of probiotic microorganisms can be found: the **acidophilic products**, the **bifidobacteria-containing products**, and the **propionibacteria-containing products**.

The **acidophilic products** are prepared either with pure cultures of *Lb. acidophilus* isolated from human intestine or combining it with different lactic acid bacteria-based starters. For example, **acidophilus milk** is traditionally prepared by inoculation of pasteurized-standardized milk with the pure culture of *Lb. acidophilus* and subsequent

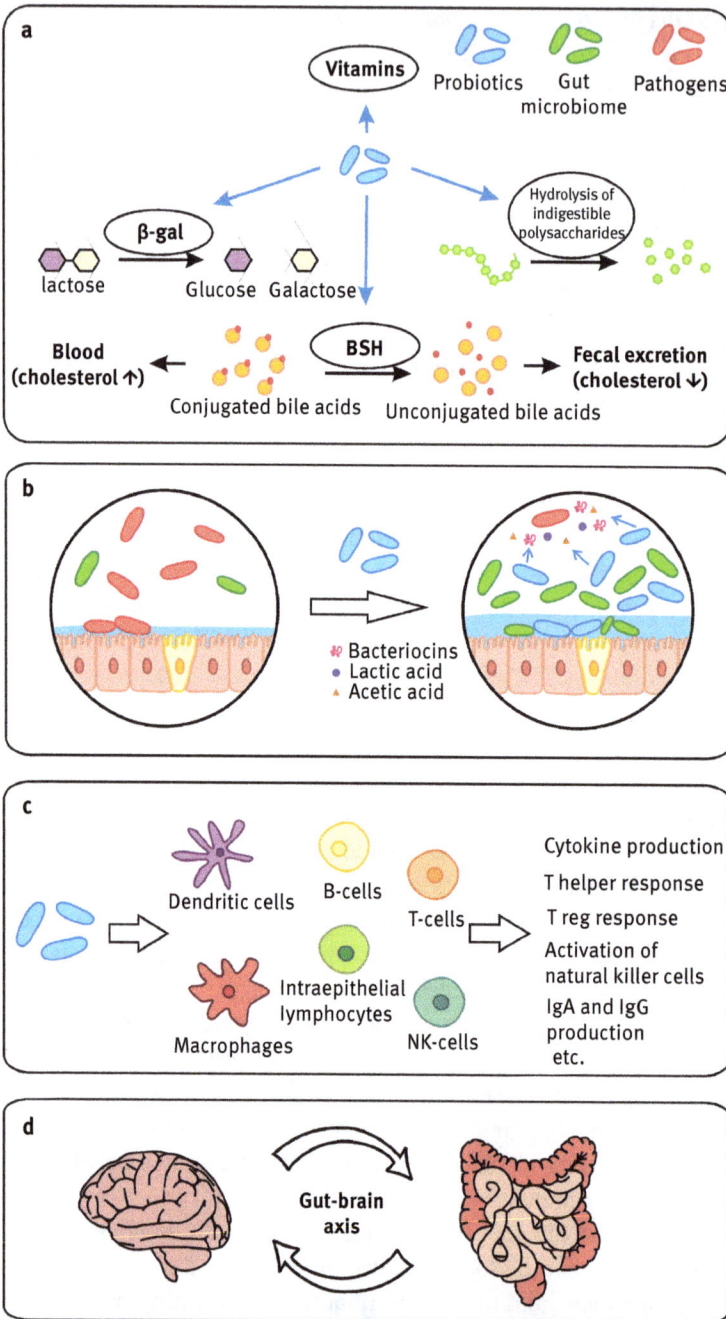

Figure 9.4: The main mechanisms of probiotics action from the perspective of: (a) the metabolism of nutrients; (b) the suppression of pathogens; (c) the modulation of immune system; and (d) the functioning of nervous system.

fermentation at 37 °C for 18–20 h, whereas **acidophilin** is manufactured by inoculation of pasteurized milk with the starter comprising *Lb. acidophilus*, *Lactococcus*, and **kefir** grains. The main peculiarity of *Lb. acidophilus* is their high acid forming capacity and great antagonistic activity toward pathogenic and conditionally pathogenic microorganisms. The **acidophilic products** possess pleasant taste and are widely used as a baby food supplements from the first days of life.

The technology of manufacturing the **bifidobacteria-containing products** can be classified into two groups: (1) fermentation with bifidobacteria monoculture (one-two-three strains composition) and (2) introduction of the precultivated on sterile milk living bifidobacteria or their concentrate into the fermentation performing by lactic acid bacteria.

The first group of the **bifidobacteria-containing products** comprises "Bifilin" (developed on the adapted milk base) and "Bifilin-M" (on whole milk base). Both produced by inoculation of pasteurized or sterilized milk by 7–10% of the starter prepared on sterile milk with growth factors. The fermentation process is performed for 7–10 h, until the final concentration of bifidobacteria reaches 10^8–10^9 CFU/mL. The clinical tests of "**Bifilin**" performed by the Russian State Medical University based on the Russian clinical children hospital showed the efficacy of this food supplement in the treatment of children suffering from insulin-dependent diabetes mellitus.

The second group of **bifidobacteria-containing products** comprises such products as, e.g., "**Biokefir**" and "**Bioryazhenka**." The production cycle of this products is very similar to that of **kefir** and **ryazhenka**, respectively. However, as bifidobacteria grow slower in milk compared to lactic acid bacteria, the amount of the introduced into the starter bifidobacteria should guarantee their required concentration of 10^6 CFU/mL in the final product.

Among the **propionibacteria-containing products**, the most well-known product in the Russian and CIS countries market is "**Tonus**." Its symbiotic starter (comprising *Propionibacterium freudenreichii* subsp. *shermani*, *Lactococcus lactis* subsp. *diacetylactis*, and *Acetobacter aceti*) can ferment milk within 5–6 h. Remarkably, the pure cultures that comprise this starter have a very low growth rates: propionic acid bacteria ferments milk for 5–7 days, lactic acid bacteria for 24–36 h, and acetic acid bacteria for 10–14 days. The "**Tonus**" starter is stable and keeps the initial microorganism's ratio during 25–30 inoculations. The optimal fermentation conditions with this starter are: inoculation of the milk with 5% of starter, cultivation temperature of 30 ± 2 °C, and fermentation time of 6–7 h. The "**Tonus**" supplementation during the treatment of allergy, pancreatic diabetes, and different intestinal diseases of children demonstrated pronounced positive clinical effect.

References

Asgar, S. and Chauhan, M. (2019). Contextualization of traditional dairy products of India by exploring multidimensional benefits of heating. *Trends in Food Science & Technology*, 88, 243–250. https://doi.org/10.1016/j.tifs.2019.03.033.

Baruzzi, F., Quintieri, L., Caputo, L., Cocconcelli, P., Borcakli, M., Owczarek, L., et al. (2016). Improvement of Ayran quality by the selection of autochthonous microbial cultures. *Food Microbiology*, 60, 92–103. https://doi.org/10.1016/j.fm.2016.07.001.

Benchimol, E. I. and Mack, D. R. (2004). Probiotics in relapsing and chronic diarrhea. *Journal of Pediatric Hematology/Oncology*, 26(8), 515–517. https://doi.org/10.1097/01.mph. 0000133291.11443.a1.

Bermudez-Brito, M., Plaza-Díaz, J., Muñoz-Quezada, S., Gómez-Llorente, C. and Gil, A. (2012). Probiotic mechanisms of action. *Annals of Nutrition and Metabolism*, 61(2), 160–174. https://doi.org/10.1159/000342079.

Beshkova, D. M., Simova, E. D., Simov, Z. I., Frengova, G. I. and Spasov, Z. N. (2002). Pure cultures for making kefir. *Food Microbiology*, 19(5), 537–544. https://doi.org/10.1159/000342079.

Bokulich, N. A., Amiranashvili, L., Chitchyan, K., Ghazanchyan, N., Darbinyan, K., Gagelidze, N., et al. (2015). Microbial biogeography of the transnational fermented milk matsoni. *Food Microbiology*, 50, 12–19. https://doi.org/10.1016/j.fm.2015.01.018.

Cheirsilp, B., Shimizu, H. and Shioya, S. (2003). Enhanced kefiran production by mixed culture of Lactobacillus kefiranofaciens and Saccharomyces cerevisiae. *Journal of Biotechnology*, 100(1), 43–53. http://www.ncbi.nlm.nih.gov/pubmed/12413785.

Chen, Z., Shi, J., Yang, X., Nan, B., Liu, Y. and Wang, Z. (2015). Chemical and physical characteristics and antioxidant activities of the exopolysaccharide produced by Tibetan kefir grains during milk fermentation. *International Dairy Journal*, 43, 15–21. https://doi.org/10.1016/j.idairyj.2014.10.004.

Cunninghamrundles, S. (2000). Probiotics and immune response. *The American Journal of Gastroenterology*, 95(1), S22–S25. https://doi.org/10.1016/S0002-9270(99)00813-8.

De Roos, J. and De Vuyst, L. (2018). Acetic acid bacteria in fermented foods and beverages. *Current Opinion in Biotechnology*, 49, 115–119. https://doi.org/10.1016/j.copbio.2017.08.007.

FAO and WHO. (2006). *Probiotics in Food: Health and Nutritional Properties and Guidelines for Evaluation* (Vol. 85). ftp://ftp.fao.org/docrep/fao/009/a0512e/a0512e00.pdf.

Fazilah, N. F., Ariff, A. B., Khayat, M. E., Rios-Solis, L. and Halim, M. (2018). Influence of probiotics, prebiotics, synbiotics and bioactive phytochemicals on the formulation of functional yogurt. *Journal of Functional Foods*, 48, 387–399. https://doi.org/10.1016/j.jff.2018.07.039.

Hatoum, R., Labrie, S. and Fliss, I. (2012). Antimicrobial and probiotic properties of yeasts: from fundamental to novel applications. *Frontiers in Microbiology*, 3. https://doi.org/10.3389/fmicb.2012.00421.

Ishimwe, N., Daliri, E. B., Lee, B. H., Fang, F. and Du, G. (2015). The perspective on cholesterol-lowering mechanisms of probiotics. *Molecular Nutrition & Food Research*, 59(1), 94–105. https://doi.org/10.1002/mnfr.201400548.

Jastrzębska, E., Wadas, E., Daszkiewicz, T. and Pietrzak-Fiećko, R. (2017). Nutritional value and health-promoting properties of mare's milk – a review. *Czech Journal of Animal Science*, 62 (12), 511–518. https://doi.org/10.17221/61/2016-CJAS.

Koroleva, N.S. (1991). Products prepared with lactic acid bacteria and yeasts. *Therapeutic Properties of Fermented Milks*, 159–179.

Leite, A. M. O., Miguel, M. A. L., Peixoto, R. S., Ruas-Madiedo, P., Paschoalin, V. M. F., Mayo, B. and Delgado, S. (2015). Probiotic potential of selected lactic acid bacteria strains isolated from

Brazilian kefir grains. *Journal of Dairy Science*, 98(6), 3622–3632. https://doi.org/10.3168/jds.2014-9265.

Mackowiak, P. A. (2013). Recycling metchnikoff: probiotics, the intestinal microbiome and the quest for long life. *Frontiers in Public Health*, 1. https://doi.org/10.3389/fpubh.2013.00052.

Marco, M. L., Heeney, D., Binda, S., Cifelli, C. J., Cotter, P. D., Foligné, B., et al. (2017). Health benefits of fermented foods: microbiota and beyond. *Current Opinion in Biotechnology*, 44, 94–102. https://doi.org/10.1016/j.copbio.2016.11.010.

Melnikova, E., Bogdanova, E., Ponomarev, A. and Korgov, R. (2015). Preclinical studies of kefir product with reduced allergenicity of βlactoglobulin. *Foods and Raw Materials*, 3(2), 115–121. https://doi.org/10.12737/13128.

Mohamadi Sani, A., Rahbar, M. and Sheikhzadeh, M. (2019). Traditional beverages in different countries: milk-based beverages. In: *Milk-Based Beverages* (pp. 239–272). https://doi.org/10.1016/B978-0-12-815504-2.00007-4.

Newton, A. E., Fairbanks, A. J., Golding, M., Andrewes, P. and Gerrard, J. A. (2012). The role of the Maillard reaction in the formation of flavour compounds in dairy products – not only a deleterious reaction but also a rich source of flavour compounds. *Food & Function*, 3(12), 1231. https://doi.org/10.1039/c2fo30089c.

Oberman, H. and Libudzisz, Z. (1998). Fermented milks. In *Microbiology of Fermented Foods* (pp. 308–350). https://doi.org/10.1007/978-1-4613-0309-1_11.

Pieszka, M., Łuszczyński, J., Zamachowska, M., Augustyn, R., Długosz, B. and Hędrzak, M. (2016). Is mare milk an appropriate food for people? – A review. *Annals of Animal Science*, 16(1), 33–51. https://doi.org/10.1515/aoas-2015-0041.

Prado, M. R., Blandón, L. M., Vandenberghe, L. P. S., Rodrigues, C., Castro, G. R., Thomaz-Soccol, V. and Soccol, C. R. (2015). Milk kefir: composition, microbial cultures, biological activities, and related products. *Frontiers in Microbiology*, 6. https://doi.org/10.3389/fmicb.2015.01177.

Quero, G. M., Fusco, V., Cocconcelli, P. S., Owczarek, L., Borcakli, M., Fontana, C., et al. (2014). Microbiological, physico-chemical, nutritional and sensory characterization of traditional Matsoni: selection and use of autochthonous multiple strain cultures to extend its shelf-life. *Food Microbiology*, 38, 179–191. https://doi.org/10.1016/j.fm.2013.09.004.

Rosa, D. D., Dias, M. M. S., Grześkowiak, Ł. M., Reis, S. A., Conceição, L. L. and Peluzio, M. do C. G. (2017). Milk kefir: nutritional, microbiological and health benefits. *Nutrition Research Reviews*, 30(1), 82–96. https://doi.org/10.1017/S0954422416000275.

Sánchez, B., Delgado, S., Blanco-Míguez, A., Lourenço, A., Gueimonde, M. and Margolles, A. (2017). Probiotics, gut microbiota, and their influence on host health and disease. *Molecular Nutrition & Food Research*, 61(1), 1600240. https://doi.org/10.1002/mnfr.201600240.

Thushara, R. M., Gangadaran, S., Solati, Z. and Moghadasian, M. H. (2016). Cardiovascular benefits of probiotics: a review of experimental and clinical studies. *Food & Function*, 7(2), 632–642. https://doi.org/10.1039/C5FO01190F.

Vodolazov, I. R., Dbar, S. D., Oleskin, A. V. and Stoyanova, L. G. (2018). Exogenous and endogenous neuroactive biogenic amines: studies with Lactococcus lactis subsp. lactis. *Applied Biochemistry and Microbiology*, 54(6), 603–610. https://doi.org/10.1134/S0003683818060157.

Zajšek, K., Kolar, M. and Goršek, A. (2011). Characterisation of the exopolysaccharide kefiran produced by lactic acid bacteria entrapped within natural kefir grains. *International Journal of Dairy Technology*, 64(4), 544–548. https://doi.org/10.1111/j.1471-0307.2011.00704.x.

Zoumpopoulou, G., Pot, B., Tsakalidou, E. and Papadimitriou, K. (2017). Dairy probiotics: beyond the role of promoting gut and immune health. *International Dairy Journal*, 67, 46–60. https://doi.org/10.1016/j.idairyj.2016.09.010.

Abimbola M. Enitan-Folami, Feroz M. Swalaha

10 Application of biotechnology in the food industry

10.1 Introduction

Biotechnology can be defined as any technological application that uses biological systems, living organisms, or derivatives to produce or modify products or processes for specific use (GBO report, 2007). Biotechnology integrates animal, human, and microorganisms that have contributed to every aspect of our lives. It has opened numerous opportunities to different sectors such as food technology, agriculture, animal sciences, cell biology, plant sciences, environmental sciences, and medicine. For example, food technology is not only important but a promising research area that applies biotechnology tools to improve the oldest food processing techniques such as fermentation, enzyme production, as well as conservation of plants (Yu, 2017).

Food processing as an aspect of biotechnology targets the selection and improvement of microbes. The essence is to improve the process of enhancing efficiency, quality, safety, and consistency of bioprocessed products (Pal, 2017). The process uses various biotechnological tools and technologies to transform perishable raw ingredients or inedible raw food materials into more palatable foods, useful shelf stable with long shelf life and potable beverages (Maryam et al., 2017). Through the use of modern biotechnology in the food industry, reduction of food losses has not only made possible, but the efficiency of food quality has also been improved.

Food biotechnology overlaps with genetic engineering in the food industry because genes in food sources like animals, plants, and microorganisms are modified to create new species of animals and plants (Pal, 2017). Similarly, genetic engineering also employs biotechnological tools to convert ingredients in food to other marketable and attractive value-added products that are cost-effective, safe, nutritious, and healthy to support human health. For instance, the use of genetic engineering and fermentation technique has been found useful for the production of "green" products such as antioxidants and preservation of agricultural products. Therefore, food biotechnology aids the improvements of resource utilization rates and reduces environmental pollution through the recycling of food wastes via enzymatic and fermentation technology. Through the use of genetically modified (GM) plants and animals, biotechnological techniques are used to enhance taste, nutrition, shelf

Abimbola M. Enitan-Folami, Feroz M. Swalaha, Department of Biotechnology and Food Technology, Durban University of Technology, Durban, South Africa

https://doi.org/10.1515/9783110667462-010

life, and quality of food products. On the other hand, GM bacteria and yeast are used to produce enzymes to enhance food production most especially through genetic engineering techniques.

10.2 Origin of biotechnology of the food industry

Revolution in the food industry will not be complete without a discussion on the origin and development of biotechnology. Historically, biotechnology as a food engineering process can be traced to three eras, namely (Table 10.1): (i) the ancient biotechnological period, (ii) classical biotechnological period, and (iii) modern biotechnological period. In what follows, each of these phases will be examined.

Ancient biotechnology era occurred prior to the 1800 CE based on common observation (Verma, 2011). Application of biotechnology in sciences and human development is dated back to 7000 BC and 4000 BC, when women inadvertently discover the usefulness of using microbes like bacteria and yeast to create food products such as beer, bread, cheese, yoghurt, and wine production. All of these foods and drinks are made by fermentation through the chemical breakdown of molecules like glucose in the absence of oxygen. Fermentation, from the Latin word *Fevere'* was first defined by Louis Pasteur as *la vie sans l'air* (life without air). The application of biotechnology helps through a fermentation process whereby microorganisms (yeasts, molds, and bacteria) are used to convert sugars into energy and/or use to processing new food products and beverages (FAO, 1998). The oldest fermentation technique was traced back to the Sumerians and Babylonians, and readily explored microbes in the beer, spirits, and wine fermentation (Table 10.1). The Egyptians used yeast to bake leavened bread way back to 4000–3500 BCE (Nair and Prajapati, 2003). The Chinese likewise develop fermentation technique for brewing and cheese production, while the Aztecs used Spirulina algae for baking cake (Habib and Hasan, 2008).

10.2.1 Chronological development of biotechnology

The classical biotechnological science of fermentation can be traced to the identification of microorganisms in 1665 by van Leeuwenhoek and Hooks (Ray and Joshi, 2014). The procedure was applied to food and has been used for several centuries for food production with an increasing success. Using this technological method, Louis Pasteur revoked the "spontaneous generation theory" around 1859 CE through elegantly designed experimentation (Ray and Joshi, 2014). In 1855, there was a discovery of *Escherichia coli* bacterium, which subsequently became a major research development that has opened ways to the production of biotechnological tools (Verma, 2011, Shulman et al., 2007).

Table 10.1: Milestone in the history of fermented foods.

Milestone	Development and location
6000–4000 BCE	Dahi-coagulation sour milk eaten as a food item in India
7000 BC	Cheese production in Iraq following the domestication of animals
6000 BC	Wine making in the Near East, with the Chinese in the village of Jiahu
5000 BC	Nutritional and health value of fermented milk and beverages
4000 BC	Egyptians discovered how to use yeasts to make leavened bread and wine
2000 BCE to 1200 CE	Different types of fermented milks from different regions
1750 BCE	Sumerians fermented barley to beer
1500 BCE	Preparation of meat sausages by ancient Babylonians
500 BCE	Moldy soybean curd as antibiotic in China
300 BCE	Preservation of vegetables by fermentation by the Chinese
500–100 CE	Development of cereal-legume-based fermented foods
1276 CE	First whisky distillery established in Ireland
1500 CE	Fermentation of Sauerkraut and Yoghurt
1851 CE	Louis Pasteur developed pasteurization
1877 CE	Bacterium *lactis* (*Lactococcus lactis*) was shown in fermented milk by John Lister
1881 CE	Published literature on koji and sake brewing
1907 CE	Publication of book *Prolongation of Life* by Eli Metchnikoff describing therapeutic benefits of fermented milks
1900–1930 CE	Application of microbiology to fermentation though the use of defined culture
1928 CE	Discovery of nisin-antagonism of some lactococci to other lactic acid bacteria (LAB) shown by Rogers and Whittier
1970 CE–present	Development of products containing probiotic cultures or friendly intestinal bacteria
1953 CE	Nisin marketed in the UK and since approved for use in over 50 countries
1990 CE–present	Deciphering of genetic code of various LAB isolated from fermented foods
2002 CE	First authoritative list of microorganisms to be used in dairy culture was released by IDF and EFFCA
2012 CE	The list of GRAS Microbial Food Cultures to be used in all types of food fermentation was released by IDF and EFFCA

LAB, lactic acid bacteria; IDF, international dairy federation; EFFCA, European food & feed cultures association; GRAS, generally recognized as safe.
Source: Modified from Nair and Prajapati (2003).

In 1870, the first experimental corn hybrid was produced in the laboratory, while Koch discovered specific organisms of varied morphology and functions causing specific diseases in 1876. In 1877, the role of a sole "bacterium" lactis (*Lactococcus lactis*) in fermented milk was discovered by Sir John Lister (Ray and Joshi, 2014, Santer, 2009). In 1881, the German physician, Robert Kotch, discovered the first solid potato slides medium that supports the growth of bacteria. Therefore, to ascribe a better nomenclature to the application of technology in regard to the conversion of raw materials into more useful products through the use of microorganisms, a Hungarian engineer, Karoly Ereky, coined the term "biotechnology in 1917." He also did this with the aim of providing solutions to societal crisis, such as food and energy shortage.

In 1928, the world's first antibiotics or killer of bacteria "penicillin" was discovered by Alexander Fleming, an antibacterial toxin from mold *Penicillium notatum*. This has been one of the widely used antibiotics for treating various infectious diseases across the world. Another biotechnological discovery was the hybrid corn. This was produced from microbes and that became commercialized in 1933. In 1942, *bacteriophage*, a parasitic virus that reproduces itself, was first identified and characterized. Ever since 1940s, particularly after World War II, there has been an increase in the development of commercial products using molecular techniques. Even in contemporary times, significant discoveries have paved way for modern biotechnology.

A point to note is the fact that there were two key events that resulted in modern biotechnology, namely, (i) the discovery of DNA in 1953 by Watson and Crick; and (ii) recombinant DNA in 1973 by Cohen and Boyer through the transfer of section of DNA from *Escherichia coli* plasmid to another DNA. The latter has further birthed genetic engineering, which is the basis of modern biotechnology. In the 1950s, synthetic antibiotic was created, while 1955 was when an enzyme called DNA polymerase was isolated. In the same year, Dr Jonas Salk also developed the first polio vaccine, which paved the way for the application of the first use of mammalian cells with cell culture techniques for vaccine production.

For the discovery of capacity of genes to regulate enzyme production, Edward Tatum and Joshua Lederberg won a novel award in 1958. According to the Genome News Network (2004), this enzyme was discovered to be the biological catalyst in living organism. In the 1950s, the question on how exactly the information on DNA is being translated into proteins remained obscured. However, in 1960, it had become clear owing to the discovery of messenger RNA (mRNA) by Sydney Brenner, Francis Crick, Francois Jacob, and Jacque Monod. These scientists discovered that genes are activated to make useful proteins through molecular machinery called mRNA, and they have helped in the synthesis of protein using molecular biology (Network, 2004). James D. Watson and Francis H.C. Crick further discovered that the chemical structure of deoxyribonucleic acid (DNA) and the substance that encodes genetic information is double helix in structure (Network, 2004). This forms the basis of molecular, biochemical and chemistry of protein synthesis, as well as reproduction in 1953. With the new discoveries of protein, reverse transcriptase enzyme that induces DNA from an RNA template as well as an enzymatic isolation of DNA, the first automated protein sequencer was launched in 1967. The enzyme was synthesized for the first time in 1969. This concept has been amplified in food processing to facilitate the speed of metabolic reactions in living organisms using enzymatic reaction.

In 1969, the first molecular biology technique on the isolation of a gene from a bacterial chromosome was reported by Jonathan Beckwith (Beckwith, 2002). To this end, a new ground for biotechnology in the modification of microorganisms was set through: (i) the recombination of existing genes or eugenics; (ii) mutation and the production of new genes by direct alteration through genetic engineering; and (iii) control or modification of gene expression. Between the 1970s and 1980s, the field of biotechnology became intertwined with genetics and food industry, leading to the evolution of a DNA-based biotechnology (Kenney, 1998, Wright, 1986 Hughes, 2001). Other major patents in biotechnology and commercialization of molecular biology between 1974 and 1980 were documented in Hughes (2001). Since the early 1980s, there have been different biotechnological tools that have contributed to different areas of food fermentation and GM products for the purpose of increasing food quantity in addition to the quality produced (Zhang et al., 2016). The first genetically engineered (GE) bacteria, plants, and crop were reported in 1980, 1981, and 1997, respectively. Since 1996, food processing and fermentation have experienced new development through the aid of GM crops. This has paved ways to commercialized products and availability of Baker's yeast (*Saccharomyces cerevisiae*) across the world (Lawrence 1988, Goffeau, 1996).

10.2.2 Advances in modern biotechnology for industrial application of food production

Modern biotechnology application is relatively recent and it is one of the growing fields of molecular biology that is changing human ways of life by proffering solutions to numerous challenges confronting the food and agricultural industries. Through domestication and agricultural activities of breeding and selection of plants, new food crops that permit fabrication of more safer, healthier, tastier, nutritional, and edible food items were developed (Ghoshal, 2018). This has enhanced several aspects of modern biotechnological applications in food industries. The contribution to production of potable beverages and foods by converting huge amount of nonedible and perishable food materials into useful and shelf-stable products by employing different operational methods and technologies cannot be underestimated in this regard.

Besides, application of biotechnology has been promoting increase in the production of different targeted consumable food items with specific functionalities naturally, thereby contributing to healthy lifestyle with a reduction in the impact of food waste products on the environment (Pinstrup-Andersen and Cohen, 2000, Gavrilescu and Chisti, 2005). Many aspects of food processing and fermentation through the provision of safe and good-quality edible foods free of hazardous materials have also become enhanced. No doubt, this has been having positive impact on human health and nutrition through the use of improved genes in relation to production of foods and high-quality agricultural produces. This has led to cultivation of plants and animal cells, manufacture of food additives, enzyme engineering, molecular and evolutionary engineering of enzyme molecules, and protein engineering among other biotechnology food-related processes (Pinstrup-Andersen and Cohen, 2000, Gavrilescu and Chisti, 2005, Umeh and Anyanwu, Kirk et al., 2002, Brink et al., 2004, Yang, 2011).

10.3 The role of biotechnology in food industry

There is a wide range of technologies that are being used for food processing in several countries in the world especially in developing countries. Among such technologies are dehydration, dying, freezing, cannin, vacuum packing, sugar crystallization, enzyme production, probiotics, and osmodehydration to mention but a few.

10.3.1 Application of biotechnology in food fermentation

A key aspect of biotechnological process in food production is fermentation. It is an old technique to produce, conserve, or transform food through the use of microorganism or enzymes from raw materials (Maryam et al., 2017) (Table 10.2). Microorganisms are important components of food fermentation and beverages (Table 10.3). Fermentation is a spontaneous process that involves specific microorganisms based on raw materials and environmental conditions such as pH and temperature. Varieties of raw materials used in traditional foods and beverages include fruits and vegetables, milk, plant, leaves, and saps from plants (Table 10.4) (Villarreal-Morales, 2018). The characteristics of fermented foods are determined by microbial inoculant use for fermentation. This plays an important role in the sensory properties such as aroma, taste, shelf life, texture, safety, flavor, constituency, smell, color, and general acceptability (Maryam et al., 2017, Leuchtenberger et al., 2005). Researches have shown that fermentation process and preparation of fermented foods and beverages vary not only among countries, but also from one region of the world to the other (Tables 10.4 and 10.5). There are several steps and sequence of operations involved in food fermentation, namely cleaning of raw materials, size reduction, soaking, and sometimes cooking to complete the process.

Table 10.2: Different classes of fermented foods based on raw materials (Ghoshal, 2018).

Raw material	Product
Cereal-based with/without pulsed fermented foods	Amazake, bread, beer, Choujiu, gamju, kvass, injera, murri, makgeolli, ogi, rejuelac, sikhye, sake, swans, sourdough, rice, grain whisky, malt whisky, wine, dosa, idli, and vodka.
Vegetable, Bamboo Shoot (BS), and unripe fruit-based fermented foods	Gundruk, kimchi, Indian pickle, mixed pickle, sauerkraut, and tursu
Honey-based foods	Mead and metheglin
Legume (pulse)-based foods	Brijing mung bean milk, cheonggukjang, iru, miso, stinky tofu, soysauce, soybean paste, and tempeh
Milk-based fermented foods	Cheese, kefir, kumus (mare milk), shubat (camel milk), cultured milk products like crème fraiche, filjolk, yogurt, smetana, skyr, and quark
Meat-based fermented foods	Chorizo, nemchua, som moo, salami, pepperoni, saucisson
Tea-based	Kombucha and pu-erh tea

Table 10.3: Application of starter culture in fermented foods production using different raw materials around the world.

Product	Raw material	Starter culture	Country
Bread	Grains	Mold (*Aspergillus*) *Saccharomyces cerevisiae*, other yeasts, LAB	International
Wine	Grape juice	Yeast, LAB	
Beer	Cereal	*Yeast*	International
Cheese	Milk	Yeast, mold, LAB (*Lactobacillus lactis*, *Streptococcus thermophilus*, *Lactobacillus bulgaricus*)	
Fermented sausages	Mammalian meat generally pork/beef less often poultry	LAB, catalase-*positive Streptococcus* sp. sometimes yeasts/mold	Southern and Central Europe, USA
Fermented milk	Milk	LAB, *L. thermophilus*, *L. bulgaricus*	International
Pickled vegetables	Cucumber, olives	LAB	
Soy sauce	Soya beans	LAB	

Table 10.3 (continued)

Product	Raw material	Starter culture	Country
Sauerkraut, kimchi	Cabbage	LAB	
Pulque	Juice of agave species	LAB (*Pediococcus parvulus, Lactobacillus brevis, Lactobacillus composti, Lactobacillus plantarum*)	Mexico
Fermented cassava flour (lafun)	Cassava	*Klebsiella oxytoca, Bacillus cereus, Staphylococcus aureus*, and *Clostridium sporogenes*	Nigeria
Wara	Sodom apple plant or pawpaw leaves	*Lactobacillus* sp., *Leuconostoc* sp., *Pediococcus* sp., *Lactococcus* sp. yeast	North Africa
Ogi	Maize, millet, sorghum	*Lactobacillus, Fusarium, Aspergillus niger, Penicillum* sp., *Saccharomyces cerevisiae, Candida mycoderms*	Nigeria, Kenya, West Africa
Tempeh	Soybean	*Rhizopus oligosporus*	Indonesia Sarinam
Gari	Cassava	Yeast, *Lb. plantarum, Streptococcus*	West Africa
Fufu	Cassava	*Bacillus subtilis, Pseudomonas alcaligenes, Lactobacillus plantarum, Corynebacterium manihot, Leuconostoc mesenteroides*, and *Pseudomonas aeruginosa*	West Africa

LAB, lactic acid bacteria.

However, the processing steps differ from one place to another, which in turn determines the biochemical, sensory, and nutritional properties of each product (Table 10.4). It is interesting to note that one-third of all food currently consumed on earth is a fermented food.

As for food biotechnology, it offers a number of benefits due to the traits of microbial culture present during food processing or starter culture used for fermentation (Yu, 2017, Mota de Carvalho, 2018) as follows:

1. An increased shelf-life as compared to raw food material
2. Refined raw materials through the improvement of food quality and yield
3. Offer a high degree of hygienic safety in the food industry
4. Basic technology with low energy consumption during fermentation of foods and beverages
5. In activation of natural harmful or toxins from raw materials (substances) such as goitrogens, mycotoxins, cyanide, oxalic acid, hemagglutinins, glucosinolates, phytic acid, protein inhibitors, and indigestible carbohydrates that can easily degrade (Maryam et al., 2017, Ogunbanwo, 2005)

Table 10.4: Traditional fermented foods and beverages from around the world to improve livelihood.

Source	Distilled beverages	Fermented beverages
Barley	Scotch whisky	Beer, ale
Corn	Bourbon whisky	Corn beer
Rice	Soja (Korea), shochu (Japan)	Sake sonti
Rye	Rye whisky	Rye beer
Wheat	Korn (Germany), wheat whisky	Wheat beer
Apple juice	Apple brandy or Applejack, Calvados	Apfelwein, "hard" cider
Fruit juice apart from apples or pear juice	Cognac (France), brandy, pisco (Peru, Chile), Branntwein (Germany)	Wine mostly of grapes
Pears juice	Pear brandy	Pear cider, perry
Sugar juice or molasses	Rum, cachaca, guaro, aguardiente	Betsa-btesa, basi
Agave juice	Tequila, mezcal	Pulque
Pomace	Trester (Germany), grappa (Italy), marc (France)	Pomace wine
Honey	Distilled mead (honey brandy or mead brandy)	Mead
Plums Juice	Palinca, Tzuica, silvovitz	Plum wine
Milk	Araka	Kumis
Potato and/or grain	Vodka grain or Vodka potato (mostly used in Ukraine)	Potato beer

6. Food biotechnology provides better resource utilization rate and reduces environmental pollution by treating food processing wastes through fermentation and enzymatic engineering
7. Fermented foods and beverages meet the demand of consumers by providing natural and organic foods that influence the intestinal health, thus providing good health benefits (Berni, 2011)

The following are some of the methods employed for fermentation processes in food industry:
1. Spontaneous inoculation of fermentation process: This technique is primarily used in many developing countries. This is because they are largely uncontrolled techniques during fermentation processes. More so, microbes associated with raw materials depend on environmental factors and process conditions. These

microbes serve as natural inoculants in spontaneous fermentation while some of the limitations of this method include low yield of products, varied product quality, inefficiency, and safety issues regarding the hygienic practices during processes.

2. "Appropriate" starter cultures as inoculant for fermentation processes.
3. "Defined starter cultures" as inoculant for fermentation processes.
4. Defined starter cultures developed using diagnostic tools of advanced molecular technologies.
5. GM starter culture. For further review on the appropriate starter and starter cultures use in fermentation processes, see Nissar (2017) and Singh (2015).

10.3.2 Biotechnology tools and enhanced performance of microbial strains used in food fermentation

Large-scale production of fermented foods and beverages is solely dependent on the use of defined starter strains, which have replaced the undefined strain mixture traditionally used for the manufacture of fermented products (Soro-Yao et al., 2014). Therefore, for production of value-added fermented foods and beverages for commercial application, starter cultures development and improvement have been employed in food industry (Table 10.5). A culture used in starting food fermentation is usually referred to as starter culture. Starter cultures are made up of one or more beneficial strains of microorganisms (Pal, 2017, Gilliland, 2018, Holzapfel, 1997). The use of defined bacteria culture helps in controlling the degree of fermentation process.

Table 10.5: Some functional microorganisms used as commercial starters in food fermentation (as compiled by Tamang et al., 2016).

Group	Genera/species	Product/application(s)
Bacteria		
	Acetobacter aceti subsp. *aceti*	Vinegar
	A. pasteurianus subsp. *pasteurianus*	Vinegar, cocoa
	Bacillus acidopulluluticus	Pullulanases (food additive)
	B. coagulans	Cocoa; glucose isomerase (food additive), fermented soybeans
	B. licheniformis	Protease (food additive)
	B. subtilis	Fermented soybeans, protease, glycolipids, riboflavin-B_2 (food additive)
	Bifidobacterium animalis subsp. *lactis*, *B. breve*	Fermented milks with probiotic properties; common in European fermented milks

Table 10.5 (continued)

Group Genera/species	Product/application(s)
Brachybacterium alimentarium	Gruyère and Beaufort cheese
Brevibacterium flavum	Malic acid, glutamic acid, lysine, monosodium glutamate (food additives)
Corynebacterium ammoniagenes	Cheese ripening
Enterobacter aerogenes	Bread fermentation
Enterococcus durans	Cheese and sourdough fermentation
E. faecium	Soybean, dairy, meat, vegetables
Klebsiella pneumoniae subsp. *ozaenae*	*Tempe*; production of vitamin B_{12}
Lactobacillus acetotolerans	Ricotta cheese, vegetables
L. acidophilus	Fermented milks, probiotics, vegetables
L. alimentarius	Fermented sausages, ricotta, meat, fish
L. brevis	Bread fermentation, wine, dairy
L. buchneri	Malolactic fermentation in wine, sourdough
L. casei subsp. *casei*	Dairy starter, cheese ripening, green table olives
L. delbrueckii subsp. *bulgaricus*	Yogurt and other fermented milks, mozarella
L. fermentum	Fermented milks, sourdough, urease (food additive)
L. ghanensis	Cocoa
L. helveticus *L. hilgardii*	Starter for cheese, cheese ripening, vegetables Malolactic fermentation of wine
L. kefiri	Fermented milk (*kefir*), reduction of bitter taste in citrus juice
L. kimchii	*Kimchi*
L. oeni	Wine
L. paracasei subsp. *paracasei*	Cheese fermentation, probiotic cheese, probiotics, wine, meat
L. pentosus	Meat fermentation and biopreservation of meat, green table olives, dairy, fruits, wine
L. plantarum subsp. *plantarum*	Fermentation of vegetables, malolactic fermentation, green table olives, dairy, meat
L. sakei subsp. *sakei*	Fermentation of cheese and meat products, beverages

Table 10.5 (continued)

Group Genera/species	Product/application(s)
L. salivarius subsp. *salivarius*	Cheese fermentation
L. sanfranciscensis	Sourdough
L. versmoldensis	Dry sausages
Lactococcus lactis subsp. *lactis*	Dairy starter, Nisin (protective culture)
L. lactis, *L. mesenteroides* subsp. *cremoris*, *L. mesenteroides* subsp. *dextranicum*, *L. mesenteroides* subsp. *mesenteroides*	Dairy starter
Oenococcus oeni	Malolactic fermentation of wine
Pediococcus acidilactici	Meat fermentation and biopreservation of meat, cheese starter
P. pentosaceus	Meat fermentation and biopreservation of meat
Propionibacterium acidipropionici	Meat fermentation and biopreservation of meat
P. arabinosum	Cheese fermentation, probiotics
P. freudenreichii subsp. *freudenreichii*	Cheese fermentation (Emmental cheese starter)
Streptococcus natalensis	Natamycin (food additive)
S. thermophilus	Yoghurt
Weisella ghanensis	Cocoa
Zymomonas mobilis subsp. *mobilis* Yeasts	Beverages
Candida famata	Fermentation of blue vein cheese and biopreservation of citrus, meat
C. krusei	*Kefir* fermentation, sourdough fermentation
C. guilliermondii	Citric acid (food additive)
Geotrichum candidum	Ripening of soft and semisoft cheeses or fermented milks, meat
Kluyveromyces marxianus	Cheese ripening, lactase (food additive)
Debaryomyces hansenii	Ripening of smear cheeses; meat
S. cerevisiae	Beer, bread, invertase (food additive)
S. pastorianus	Beer
S. florentius	*Kefir* fermentation
S. bayanus	*Kefir* fermentation, juice and wine fermentation

Table 10.5 (continued)

Group	Genera/species	Product/application(s)
	S. cerevisiae subsp. *boulardii*	Used as probiotic culture
	S. unisporus	*Kefir* fermentation
	S. sake	*Sake* fermentation
	Zygosaccharomyces rouxii	Soy sauce
	Schizosaccharomyces pombe	Wine
Filamentous molds		
	Aspergillus flavus	*α-Amylases (food additive)*
	A. oryzae, A. sojae	Soy sauce, beverages; *α*-amylases, amyloglucosidase, lipase (food additives)
	A. niger	Beverages, industrial production of citric acid, amyloglucosidases, pectinase, cellulase, glucose oxidase, protease (food additives)
	P. roqueforti	Blue mold cheeses
	P. notatum *Penicillium camemberti*	Glucose oxidases (food additive) White mold cheeses (camembert type)
	R. oryzae	Soy sauce, *koji*
	Rhizopus oligosporus	*Tempe* fermentation

A variety of microorganisms present in fermented foods especially in the non-standardized starter cultures are mesophilic aerobic bacteria, aerobic bacteria, lactic acid bacteria (LAB) and yeast cells (Ogunbanwo, et al., (2005), Enitan et al., 2011). Most of these organisms are used as starter culture due to their ability to produce one or more metabolites called "the antimicrobial compounds." These metabolites include diacetyl, hydrogen peroxide, lactic acid, and organic acid for the inhibition of undesirable microorganisms and enhancement of desirable organoleptic characteristics of food fermentation (Enitan et al., 2011, Oyedeji et al., 2013, Ogunbanwo, 2005, Adesulu-Dahunsi et al., 2018). For instance, one of the commonly used starter cultures used for food preservation, prevention of pathogens, and spore formation are LAB due to their ability to produce acid and bacteriocins during fermentation process (Ray, 2019, Ren, 2018, Linares, 2017, Mokoena, 2017).

Food fermentation technology is being continually improved with the use of traditional and molecular approaches. In traditional food biotechnology, microbial cultures are improved for utilization in food processing application by improving the quality of microorganism as well as the use of metabolites through mutagenesis, conjugation, and hybridization techniques. These three processes have been the basis of industrial starter culture development in bacteria. Hybridization in the food industry has been used in improving strains that are widely applied in commercial baking beverage and brewing processes (Maryam et al., 2017). Selection of improved strains is based on specific properties such as an improved flavor-producing ability or resistant to bacterial viruses, and this is made possible through hybridization method. On the other hand is the conjugation which is a natural process, whereby genetic materials are transferred among closely related microbial species as a result of physical contact between the donor and the recipient. It involves gene exchange between chromosomal and plasmid-localized cells in similar bacteria species and sometimes there could be transfer between different species (Werckenthin, 2001, El, 2009).

10.3.2.1 Genetic engineering and starter culture for the enhancement of food production

Molecular methods by means of DNA recombination technology and protoplast fusion or cloning as well as plasmid transfer are integral aspects of biotechnology pertaining to the food industry. These tools are used to improve the use of raw materials in food production and to increase the performance of food microorganisms for the production of enzyme, active healthcare, and food ingredients (Yu, 2017, Pariza and Johnson, 2001). Genetically improved starter culture has formed the basis to enhance resistant and defense mechanisms against several enteropathogenic microorganisms (Pal, 2017) and production of new food items and additives with great health important benefits (Gasser and Fraley, 1989).

Lactobacillus bulgaricus, *Streptococcus thermophilus*, *Lueconostoc cremoris*, *Streptococcus lactis*, and *Streptococcus cremoris* are some of the GM bacteria that have been employed during spontaneous secondary fermentation process to produce improved fermented foods, including yogurts, cheese, and butter (Table 10.5). Generally, most GE microorganisms (GMO) are widely used for industrial production with no major safety concerns being raised. For instance, *Lactobacillus* strains are potential probiotics that are employed not only in fermentation of dairy products but also for meat products (Tamang, 2016). Apart from genus *Lactobacillus*, GM *Streptococcus thermophilus* is being used as starter culture for yoghurt production (Lick and Heller, 1998, Lick, 1996). Studies on the modified starter cultures to improve fermented foods such as yoghurts, cheese, and butter among other fermented products have been reported (Kandasamy et al., 2018, Johansen, 2018). Rennet through GM bacteria is being used as starter culture for cheese production (Hartmann, 2018). Some other starter cultures used in the fermentation of cereal, fruit, milk, wine, and vegetables are given in Table 10.5 as retrieved from Tamang (2016).

Yeast is another GMO used for improving the rate of food fermentation (Singh, 2015, Hansen, 2002). Genes in Baker's yeast (*Saccharomyces cerevisiae*) are engineered to enhance bread dough rises by increasing the efficiency of maltose breakdown.

Likewise, the capacity of brewer's yeast to enhance bread production was employed by glucoamylase-encoding gene from yeast for better utilization of carbohydrate in the feedstock. This, in turn, increases the capacity of this yeast to produce high alcohol and full-strength low-carbohydrate beer. The use of GMO in food was first permitted in the United Kingdom in 1990, and most recently modified yeast strains were authorized for wine industry in the Northern America (Buiatti et al., 2013).

10.3.2.2 Genetically modified foods

Recent advancements in agricultural biotechnology and plant sciences offer new opportunities and possibilities to improve quality, yield, and production of economical food crops. Role of modern biotechnology in food production does not only increase food production but raise harvest plant products, improved harvest time, produce better improve raw materials, as well as nutritional values with better flavor. There are GM foods that are produced from microorganisms that have modified genetic materials like DNA in them. This involves the insertion of gene from one organism into a plant. Most of the current GM foods are derived from plants through different ways in order to ensure an adequate supply of foods to meet the demand by growing population in the world (Kamle, 2017).

To add desirable traits to a crop, a foreign genes' transgene encoding the traits must be inserted into the plant cells. Usually, there are two methods used in the transfer of growing genes into the plants. *Agrobacterium tumefaciens*, a plant pathogen that causes crown gall disease in many place species, was first disarmed by removing the T-DNA "tumor-inducing genes" in *Agrobacterium* before inserting it into the plants using genetic engineering techniques (Ghoshal, 2018, Gohlke and Deeken, 2014). The second technique is called "gene gun" biolistic method. The gene gun method involves direct shot of particles carrying foreign DNA into plant chromosomes under high pressure. This method has been applied in the fabrication of many crops especially monocots (Aadel, 2018, Ismagul, 2018, Partier, 2017). Thus, GE foods have become an emerging opportunity that has been carried out in different ways depending on the desired products as highlighted in the following listed products:

1. **Herbicide tolerance plant**: Currently, herbicides and weed killers are being sprayed to kill weeds. This is an expensive process that requires a careful and experienced personnel in order to protect both the crop and the environment from harmful impacts of the herbicides. However, there should be more syntheses of GM crops that can be resistant to herbicides in order to minimize the use of herbicides and reduce the danger of excessive herbicides in the environment. A strain of GM soybean that cannot be damaged by herbicides was synthesized by Monsanto (EFSA GMO Panel, 2013, Bonny, 2008). This application does not only reduce the number of herbicide application but reduces cost and dangers of environmental damages (Maryam et al., 2017, Aris and Leblanc, 2011).

2. **Pesticide resistance:** Damage of crops by insect pests is a great loss to the farmers. Annually, tons of chemical pesticides are being applied to crop fields and this is a cause of concern for consumers due to potential human exposure to health risks from the use of pesticides. Against this background, research and development has given rise to the production of GM foods such as *Bacillus thuringiensis* (Bt) corn. This corn helps in the elimination of the use of chemical pesticides, and reduces costs of the crop production and transportation to the market. It also helps in preventing any health-related risks (Maryam et al., 2017, Aris and Leblanc, 2011). This protects plants against harmful pests and in turn farmers do not necessarily need to spray insecticides/pesticides or reduce their use.

3. **Salinity tolerance/drought tolerance**: Due to increase in the world population, there is a demand for new housing and infrastructure, which has led to an increase in the conversion of available farmlands meant for food production into building of houses (Yeh and Li, 1999, Appiah et al., 2017). Destruction or conversion of farmlands could result in starvation in most countries; hence, farmers are forced to grow their crops in locations that are primarily unsuitable for crop cultivation. To this end, crops are created or synthesized to withstand high salt content in the soil and groundwater and long periods of drought to help people in crop production, especially, in undesirable places (Maryam et al., 2017).

4. **Cold tolerance:** Some crop plants are destroyed by extreme cold; hence, genes from water fish are being inserted into some plants like tobacco, potato, and tomatoes in order for them to withstand cold weather (Prasad et al., 2018). Also, GE soybean are processed to withstand applications of different herbicides through the expression of double-mutated maize 5-enolpyruvyl shikimate-3-phosphate synthase and phosphinothricin acetyltransferase enzymes which have made soybean to be tolerant to glyphosate and glufosinate, respectively (Maryam et al., 2017, Herman, 2018, Kocak, 2019, Carden et al., 2019, Cole and Erdahl, 2018). Likewise, genes that confer resistance to insects or disease and oleic acid have been encoded in soybean (Kocak, 2019).

5. **Delayed ripening tomato:** This is the first GM food crops produced in the developed countries. These tomatoes spend more days before ripening with better flavor and longer shelf life (Maryam et al., 2017). Pandey (2015) associated alteration in expression pattern of ripening specific genes such as diamine putrescine (Put) and polyamines, and spermidine (Spd) and spermine (Spm) to a delay of on-vine ripening and prolonged shelf life over untransformed tomato cultivars. These transgenic fruits could not only express the pattern of ripening specific genes but they were also fortified with important nutraceuticals including antioxidants, ascorbate, and lycopene in order to enhance the quality of tomato (Pandey, 2015). Other studies on delayed ripening of tomato fruits had been reported as a possible solution to postharvest losses (Gupta, 2019, Torrigiani, 2012, Javanmardi et al., 2013).

6. **Biotechnological-based maize:** Maize is one of the widely used grains in the world that is fit for use in different purposes in the food and agroallied industries. It is a raw material for processing cooking oil, livestock feedstock, biofuel, and food additives. Biotechnological maize is a GM crop that has been engineered to express desirable traits. Transgenic maize contains one or more proteins from *Bt* strains with pesticide-resistant and herbicide-tolerant strains. These species allow better flexibility in using certain herbicides to control weeds that can damage crops, while the latter is a maize species with built-in insecticidal protein from a naturally occurring soil microorganism that gives a plant long seasonal protection against corn borer. This protein serves as organic insect control and reduces toxin contamination that may arise due to fungal attack (Aris and Leblanc, 2011). Presently, transgenic corn, cotton, soybean, and brinjal expressing insecticidal proteins against target insect pests are being cultivated in one or more countries (ISAAA, 2018).

7. Production of biotechnology insect-resistant (IR) and virus-resistant potatoes (Alyokhin, 2008). IR potato contains protein that provides clients with built-in protection from the Colorado potato beetle which is beneficial to farmers, consumers, and the environment (Alyokhin, 2008). Virus-resistant potato was modified to resist both potato virus Y and potato leaf roll virus. These potato varieties are protected from viruses through biotechnology.

8. Canola is another rapeseed with genetic variation as modified by Canadian plant breeders due to its nutritional properties and low level of saturated fatty acids (Ronald and Mcwilliams, 2010). Canola varies from herbicide tolerant to high-laureate canola. High-laureate canola contains high levels of laureate, and oil produced from the novel canola cultivars is similar to coconut oil. These canola oil species are now considered to be source of good raw materials for industry in the production of coffee whiteners, frosting, icings, chocolate, candy coating, and whipped toppings. Additionally, cosmetic industries are also enjoying the benefits of this novel high-laureate canola species.

9. Additionally, biotechnology has allowed scientists to produce tasty and better fruits such as eggplants, cherries, seedless watermelon, tomatoes, and pepper. Elimination of seeds from these fruits results in more soluble sugar content that enhances the sweetness in them (Haroon and Ghazanfar, 2016, Falk, 2002). Other GM food crops that have been produced in this respect include herbicide-tolerant and IR rice (Aris and Leblanc, 2011) and cotton (ISAAA, 2018, Paul Singh and Heldman, 2013), herbicide-resistant sugar beet (Singh, 2008), herbicide-tolerant alfalfa (Aris and Leblanc, 2011), and virus-resistant biotechnology yellow crookneck squash that can resist both Zucchini yellow mosaic virus and watermelon mosaic virus (Gonsalves, 2004).

10.3.3 Biotechnological improvement of nutritional values

Today, there is an increasing improvement in the nutritional value, flavor, and texture of raw materials because not all food items have the essential components (essential nutrients, amino acid, and metabolites) needed by the body. However, with advances in biotechnology and biofortification of foods through fermentation procedures and recombination, DNA technology is gaining increasing attention in the food industry in order to avoid the deficiency of these essential components in food materials (Pal, 2017). Designer functional foods are becoming an important part of human life, due to their role in disease prevention and improved human health (Cashman and Hayes, 2017). There are many designer foods that are fortified with health-enhancing ingredients like vitamin, essential minerals, and amino acid–rich foods (Garg, 2018, Muchenje et al., 2018). Among the next-generation GM plants are the engineered food products with an elevated nutritional molecules like amino acid, omega-3 fatty acid, and vitamins.

Several researchers are working on the introduction of micronutrients like iron, zinc, lipids, and vitamins into the most staple foods (Oladosu, 2019). One of such staple foods in many countries of the world is rice, of which parental rice cultivar is deficient in vitamin A and this makes it unfit as staple food. The first pro-vitamin-rich transgenic rice was synthesized by incorporating Psy and CrtI genes from daffodils and bacteria (Sun, 2008). New iron-rich rice was produced by introducing soybean ferritin cDNA into rice plants (Goto, 1999). In 2002, fungal (*Aspergillus niger*), phytase, cDNA inserted in rice with an increased phytic acid degradation was produced (Lucca et al., 2002). Cho (2016) found that transgenic drought-tolerant rice Agb0103 is an improved resistance type. This resistance rice has similar nutritional characteristics with parental rice cultivar in terms of lipid profile, vitamins, proximate components, amino acids, and antinutrients.

Another is the designer omega-3 fatty acid egg that started in 1934 by Cruickshank. This egg showed better stability of polyunsaturated fatty acids during storage and cooking with high availability of nutrients like vitamins, selenium, vitamin E, carotenoids, and antioxidants. Thus, eating such eggs improves omega-3 and antioxidants in people who consume them. Other designer eggs that are enriched and enhanced with conjugated linoleic acid (CLA) (Raes, 2002), β-carotene, and vitamins have been reported (Jiang et al., 1994).

Some studies have also demonstrated that designer milk produced through biotechnological application could contain healthier fatty acids such as omega fats and CLA, casein, improved amino acid profile, less lactose, more protein that is free from β-lactoglobin (β-LG) (Rajasekaran and Kalaivani, 2013). According to the published results by Rajasekara and Kalaivani (2013), elimination of β-LG from bovine could reduce children allergies to cow milk (Sabikhi, 2007). Other fortified foods include selenium (Se)-enriched beef, chicken, and pork. These are produced by feeding farm animals and poultry with organic Se diet, which indirectly help in producing nutrient-rich eggs (Pal, 2017).

10.3.4 Biotechnological applications to improve food yield

Increase in world population by year 2050 is estimated to be 9 billion. This will lead to the demand for more food production due to increase in population. Hence, more yield will be required from using the same land mass. Thus, biotechnology has proven to be of great benefit with potential to achieve higher food yield in order to fight against starvation, hunger, malnutrition, and diseases (Ramón et al., 2008). Its application will not only produce better and high-quality products, but increase the quantity of food production, raise health standard, and lower mortality rates. Dairy products such as modified cattle or cow, mice, milk, egg, fish, sheep, and rabbit are examples of transgenic bioengineered animal and food products. These are fabricated with the aid of biotechnology techniques like (i) transfer of nuclear materials, (ii) microinjection, (iii) viral vector infection, and (iv) transfer of embryonic stem cell (Ghoshal, 2018, Ramón et al., 2008, Cima, 2016).

Fish is an important protein source consumed by most people on earth. However, overfishing has put the oceans under stress to the point that future viability is under threat. However, GM fish have considerable potential to increase the yield of fish farms, because they are produced with an increased growth rate. GM salmon fish with disease resistance capacity, improved environmental tolerance, as well as increased growth rates than its non-GM farm-raised counterparts have been reported (Cima, 2016, Forabosco, 2013). Also, tilapia growth rate was found to be 33% more than the wild type of tilapia and reduces farmer's production cost [United States Food and Drug Administration (US FDA)]. It has been proven that the consumption of such fish is safe and healthy. Likewise, transgenics swine with reduced fat content was fabricated through artificial insemination by inserting recombinant bovine growth hormone into pig genome. This is to reduce the fat content in pigs, so that consumers can get more benefits by eating healthy pork with lower unsaturated fatty acid (Pal, 2017).

Likewise, maize with high essential amino acid (methionine) was synthesized by adding a single *E. coli* gene into the maize chromosomes (Planta et al. 2017, Rutgers Today). In 2011, herbicide-resistant GM corn was grown in 14 countries, and 26 of the ratings were also raised for importation into the European Union (EU) by 2012 (European Commission held and consumers, 2012). Cassava is another example that serves as main source of calorie for millions of people around the world. Protein content in the wild type is low, and this could lead to an unbalanced diet for people who lack access to varieties of foods. However, with the help of genetic engineering, amino acid and protein contents of cassava were modified to increase nutritional value of this type of staple food.

Some developed countries like the USA and EU have commercialized the release of GM food crops to increase food yield. According to the Global Status of Commercialized Biotech/GM Crops (ISAAA, 2018), the USA has accomplished the first planted biotech crop area followed by Brazil, Argentina, Canada, India, ten other Latin American countries, and nine Asia and Pacific countries in 2018 (Table 10.6). About 191.7 million hectares of biotech crops were grown by 26 countries – five industrial countries and 21 developing countries in the world (Figure 10.1). Developing countries grew 54% of the

Table 10.6: Global area of biotech crops in 2018 by country in million hectares (ISAAA, 2018).

Rank	Country	Biotech crops	Area (million hectares)
1	USA *	Maize, soybeans, cotton, canola, sugar beets, alfalfa, papaya, squash, potatoes, apples	75.0
2	Brazil*	Soybeans, maize, cotton, sugarcane	51.3
3	Argentina*	Soybeans, maize, cotton	23.9
4	Canada	Maize, canola, sugar beets, alfalfa, apple	12.7
5	India	Cotton	11.6
6	Paraguay	Maize, soybeans, cotton	3.8
7	China	Cotton, papaya	2.9
8	Pakistan	Cotton	2.8
9	South Africa	Maize, soybeans, cotton	2.7
10	Uruguay	Maize, soybeans	1.3
11	Bolivia	Soybeans	1.3
12	Australia	Cotton, canola	0.8
13	Philippines	Maize	0.6
14	Myanmar	Cotton	0.3
15	Sudan	Cotton	0.2
16	Mexico	Cotton	0.2
17	Spain	Maize	0.1
18	Colombia	Cotton, maize	0.1
19	Vietnam	Maize	<0.1
20	Honduras	Maize	<0.1
21	Chile	Maize, soybeans, canola	<0.1
22	Portugal	Maize	<0.1

Table 10.6 (continued)

Rank	Country	Biotech crops	Area (million hectares)
23	Bangladesh	Brinjal/eggplant	<0.1
24	Costa Rica	Cotton, Soybeans	<0.1
25	Indonesia	Sugarcane	<0.1
26	eSwatini	Cotton	<0.1
	Total		191.7

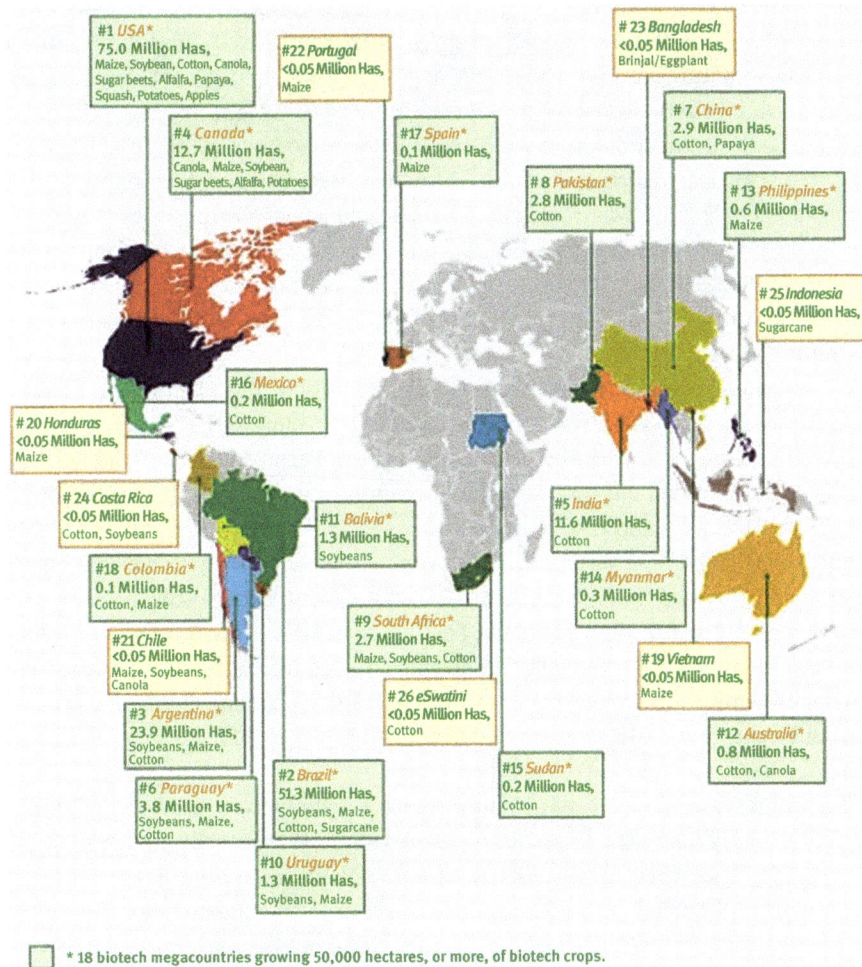

Figure 10.1: Major countries that are involve in biotechnological modified food cultivation and production in the world. Source: ISAAA: Brief 54 (ISAAA,2018).

global biotech crop area compared to 46% for industrial countries (Table 10.6). Additional 44 countries (18 plus 26 EU countries) imported biotech crops for processing, feed, and food. Thus, a total of 70 countries in total have adopted biotech crops worldwide ISAAA Brief 54 (ISAAA, 2018). The four major biotech crops – soybeans, maize, cotton, and canola – were the most adopted biotech crops by the 26 countries. For example, soybeans at 95.9 million hectares attained 50% of the global biotech crop adoption, a 2% increase from 2017. This was followed by maize (58.9 million hectares), cotton (24.9 million hectares), and canola (10.1 million hectares).

Based on the International Service for the Acquisition of Agri-biotech Applications (ISAAA): 2018 (Brief 54) (ISAAA, 2018) and FAO (2017), the global biotech crop area for individual crops was 29% of canola, 30% of maize, 76% of cotton, and 78% of soybeans in 2018 (ISAAA, 2018). The average biotech crop adoption rate in the top five biotech crop-growing countries increased in 2018 to reach close to saturation, with the USA at 93.3% (average for soybeans, maize, and canola adoption), Brazil 93%, Argentina ~100%, Canada 92.5%, and India 95% (ISAAA, 2018). In addition to the 2018, 93.3% adoption rate by the USA for planting biotech soybeans, maize, and cotton, 75 million hectares of land, covering 39% of the global biotech area was recorded. The biotech crops planted were: soybeans cover 34.08 million hectares, maize accounted for 33.17 million hectares, cotton 5.06 million hectares, canola 900,000 hectares, sugar beets 491,000 hectares, alfalfa 1.26 million hectares, and some 1,000 hectares of papaya, squash, potatoes, and apples (ISAAA, 2018).

Till today, African continent remains the region with the biggest potential to reap the benefits associated with modern agricultural biotechnology for food production when compared with the global application. For instance, in South Africa, trials have been carried out on GM potatoes, canola, sugarcane, apple, eucalyptus, sugar beets, tomato, and sweet potato (Kruger Park Times, GM foods in SA) (Haroon and Ghazanfar, 2016). In 2004, Biowatch South Africa estimated that about 35% of maize grown in South Africa was GM, while about 80% of grown cotton was *Bt* cotton with an insect-tolerant cotton. Lastly, about 30% of soybean was GM. It is interesting to know that in the last two decades, the continent recorded impressive growth with South Africa planting 2.7 million hectares of biotech crops (maize, soybean, and cotton) in 2018, and ranked among the top 10 biotech crop countries (ISAAA, 2018).

It is interesting to know that Nigeria added a new biotech crop to the global biotech basket by being the first country in the world to openly approve the cultivation of GM Bt cowpea that is pest resistant. This cowpea only needs just two sprays instead of current varieties that need eight sprays of pesticides. The kingdom of eSwatini (formerly Swaziland) has started commercial planting of IR Bt cotton with an initial launch of 250 hectares but in 2018, a total of 3.14 million hectares have been planted. This makes the country the latest and third African country to plant biotech cotton apart from South Africa (2.7 million hectares) and Sudan (243,000 hectares IR cotton), the third state to have planted biotech crops in Africa, while also currently growing commercial IR Bt cotton, one of the biotech crops. Four other African countries Ethiopia, Kenya, Nigeria, and Malawi have also granted approvals for planting biotech cotton as proof that Africa is ready for biotech crop adoption (ISAAA, 2018).

Generally, GE foods do not only increase productivity but lower the cost of production, increase and improve harvesting through plant cloning, and reduce the work load of farmers. Consequently, the final consumers also benefit by eating healthy food. Stacked traits with IR and herbicide tolerance increased by 4% and covered 42% of the global area – a testimony to farmers' adherence to smart agriculture and food production with reduced insecticide use has been made possible by biotechnology. In addition, these food products have been tested and found to be digestible with no toxic or allergic effects on human health (Adenle, 2011). Approval and commercialization of biotech crop in different areas of the world have provided new biotech crops and traits in order to target problems related to climate change and the emergence of new pests and diseases as well as increase in food production to alleviate hunger. In contrast to regulations by other countries, US FDA approval of the consumption of GM foods (or crops) that are available at the moment do not need special regulations because they are not inherently dangerous. And they are as safe as the traditionally bred non-GM cultivar and do not need to be labeled as "GMO" if the novel food was not materially different from its conventional counterpart (AgBioResearch, 2018). Up to 70% of processed foods on the shelf in the USA contain GE ingredients ranging from crackers to condiments, and soup products as compared to 0.5% consumption rate in the Europe (ISAAA, 2018, Kathage et al., 2016).

10.3.5 Biotechnology for industrial enzyme production

Enzymes are biological catalysts used by humans for decades to facilitate metabolic reactions of living organisms. They are proteins used for production, processing, preservation, and as raw ingredients in different food industries (Torres-León, 2018). They are used for production of different food items such as brewing, baking, mayonnaise production, cheese production, and dairy products (Pal, 2017, Torres-León, 2018). In the past, enzymes were isolated from plants but today, they are either in the form of microorganisms or in plants and animal-based sources. Microorganisms such as bacteria, yeast, and fungi are important sources of diverse commercial enzymes that are available in the market depending on the substrate used and the environmental condition of the fermentation process (Ghoshal, 2018). Some strains of microorganisms are GM to increase their capacity in enzyme production at an optimum condition (Table 10.7).

Table 10.7: Enzymes produced from genetically modified microorganisms using gene technology complied from Ghoshal (2018).

Enzymes	Microorganisms	Application
Chymosin	*Aspergillus niger*	Mayonnaise
Phytase		Dairy
Lipase		Baking, cheese flavor development
Aspartic proteinase	*Aspergillus oryzae*	
Esterase-lipase		
Glucose oxidase		Starch
Lipase		Cheddar cheese production
Laccase		
α-Amylase	*Bacillus licheniformis*	Brewing, baking, starch liquefaction, clarification of fruit juice, improve bread quality, rice cakes
Pullulanase		Baking, dairy, distilling, fish, meat, starch, vegetable
α-Acetolactate decarboxylase	*Bacillus subtilis*	Brewing
α-Amylase		Baking, brewing, starch liquefaction
Maltogenic amylase		Fats, oils
Pullulanase		
Chymosin	*Escherichia coli K-12*	Cheese
Xylanase	*Fusarium venenatum*	Brewing, baking, starch and digestive aid for only α-amylase
Chymosin	*Kluyveromyces marxianus* var. *lactis*	Cheese production
α-Amylase	*Pseudomonas fluorescens*	
Pectin lyase	*Trichoderma reesei*	Brewing, baking, distilling, starch, vegetable
α-Glucanotransferase	LAB	Brewing
Microbial rennet	*Mucor miehei*	Baking, starch, cheese
Protease	*Aspergillus usamii*	Starch, brewing, meat tenderizer, milk coagulation, improvement of bread quality

Table 10.7 (continued)

Enzymes	Microorganisms	Application
Glucoamylase	*Aspergillus awamori, Aspergillus niger, Rhizopus oryzae*	High fructose and high glucose syrups, beer production, improvement of bread quality
β-Galactosidase (lactase)	LAB	Lactose intolerance reduction in people Prebiotic food ingredients

Some strains of microorganisms are GM to increase their capacity in enzyme production at an optimum condition. In most cases, modified genes of a GM microorganism that produces an enzyme may come from another microorganisms of different kingdom. Many industrial enzymes like glucoamylase, lipase, α-amylase, pectinase, antibiotics, amino acids, lactic acid, nucleic acid, and polysaccharide are bio-based chemicals that are produced using GM starter cultures (Table 10.7). For instance, an enzyme found in stomach of calves, one of the DNA coding for chymosin that causes milk to curdle or coagulates during cheese fermentation, was cloned into bacteria (*Escherichia coli*), yeast (*Kluyveromyces lactis*), and mold (*Aspergillus niger*). Modified *E. coli* is currently being used in Thailand for production of lysine, so as to get higher yield in less time. In another case, aspartyl protease from *Mucor miehei* is commonly used as a substitute for chymosin in cheese production due to its properties (Silveira, 2005, Thakur et al., 1990). Currently, chymosin produced via this recombinant DNA is produced commercially and is widely used in the production of cheese. Many microorganisms (*Rhizomucor miehei, Rhizomucor pusillus, Aspergillus oryzae, Endothia parasitica*, and *Irpex lactis*) are extensively used in the production of protein as a substitute for calf rennet that are used by cheese manufacturers (Pal, 2017, Sun, 2014, Hayaloglu et al., 2014).

It is worth noting that problem of lactose intolerance during consumption of milk and dairy products has been solved through the production and incorporation of β-galactosidase or β- galactosidase-producing organisms into different dairy products. Commercial production of β- galactosidase from *Aspergillus oryzae, A. niger* ATCC 9142, LAB, and *Kluyveromyces lactis* NRRL Y-8279 developed by researchers has made it possible to produce these enzymes at lower cost (Kazemi et al., 2016, Dagbagli and Goksungur, 2008, Enitan, 2008). Likewise, various animal and microbial lipases producing organisms such as *Serratia marcescens, Pseudomonas aeruginosa, Bacillus subtilis*, and *Staphylococcus aureus* have been made available for the production of various cheese flavors with lower bitterness and strong rancidity. In the same vein, combination of lipases, peptidases, and proteinases have also been found to give better cheese flavor with low levels of bitterness (Adrio and Demain, 2014). Other applications of fungal in food biotechnology are illustrated in Figure 10.2 (Hyde, 2019). Most of the GM enzymes used in food industry are purified enzymes that are free of cells

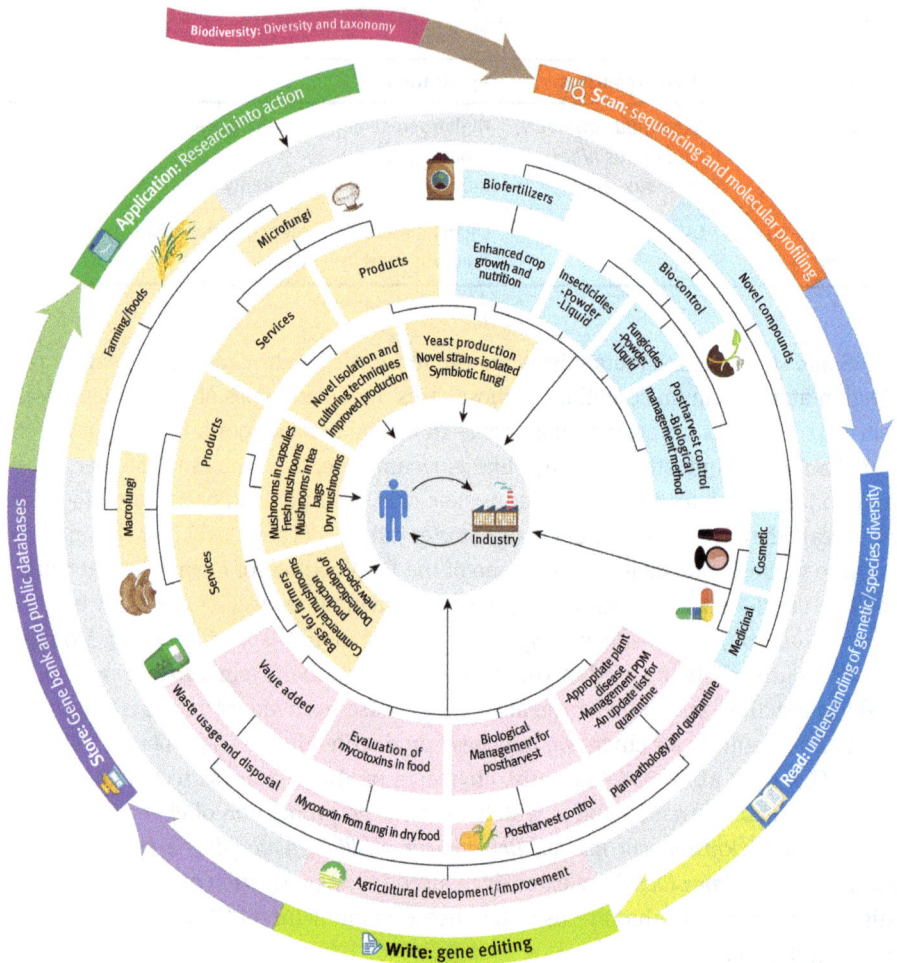

Figure 10.2: Potential beneficial uses of fungi in food biotechnology (source: Hyde, 2019).

from the organisms and do not contain any macromolecules such as DNA. The safety of enzyme produced through GM microorganisms has been evaluated by the joint effort of the FAO/WHO experts Committee on food additives (JECFA) (EFSA GMO Panel, 2010). Presently, a great number of enzymes being used in the USA for production in the food industries are more than 50% of GM carbohydrase, protease, and alpha amylase due to their characteristics (Pal, 2017, Hyde, 2019, Raveendran, 2018, Singh, 2016).

10.3.6 Application of biotechnology to improve shelf life of food

The shelf life of fermented foods and beverages are extended by bacteria due to the production of secondary metabolites. For example, LAB including the genera *Lactobacillus, Lactococcus, Pediococcus*, and *Streptococcus* produce metabolites like bacteriocins (nisin, pediocin PA-1, Table 10.8), hydrogen peroxide, and lactic acid that have been incorporated into food as natural preservatives (Enitan, 2008, Kaškonienė, 2017). The use of LAB as starter culture or their metabolites is considered safe for food production (Pal, 2017, Enitan, 2008, Patel and Prajapat, 2013). Nisin-producing LAB as a starter culture is also useful for fermentation of milk. Biotechnological techniques to improve the storage time of tomatoes have also been developed and patented (Zhao, 2007). This is carried out by reducing the enzymes that soften ripe tomato fruit, thereby causing early deterioration of tomato at postharvest period or during storage (Osorio, 2019, Masoodi, 2018, Deltsidis et al., 2018). Regardless, long shelf life for ripe tomato has been developed while margarine and canned foods have received approval across the world (Monteiro, 2018, Monteiro, 2019).

Table 10.8: Classes and properties of bacteriocins from lactic acid bacteria (Mokoena, 2017).

Class	Typical producing species	Properties	Examples
1	*Lactobacillus lactis* subsp. *lactis*	Contain unique amino acids, i.e., lanthionine and methyl lanthionine <5 kDa	Lactocin, nisin, mersacidin
IIa	*Leuconostoc gelidum*	Heat-stable, nonmodified, cationic, hydrophobic peptides contain a double-glycine leader peptide, pediocin-like peptides: <10 kDa	Pediocin PA1, sakicin A, leucocin A
IIb	*Enterococcus faecium*	Require synergy of two complementary peptides, mostly cationic peptides	Lactococcin G, plantarum A, enterocin X
IIc	*Lactobacillus acidophilus*	Affect membrane permeability and cell wall formation	Acidocin B, enterocin P, reuterin 6
III	*Lactobacillus helveticus*	Heat-labile, large molecular mass peptides, <30 kDa	Lysostaphin, enterolysin A, helveticin J

10.3.7 Production of food flavors and sweeteners through biotechnology

Other benefits of biotechnology include the production of food ingredients, amino acids, flavors, and sweeteners (Table 10.9). Currently, microorganisms-produced flavor is now competing with the traditional agricultural-produced sources with more than 100 commercial aroma chemicals derived using biotechnological applications (Ghoshal, 2018). DNA recombinant is being used to enhance efficiency of microorganisms in the production of nonnutritive sweeteners such as aspartame, thaumatin, and flavor-enhanced glutamate. This is done through the help of high-performance strains of *E. coli* and *Corynebacterium glutamicum* during fermentation of sugar sources like glucose, sucrose, and molasses.

Table 10.9: Some food additives and processing aids produced by genetically modified microorganisms.

Categories of food additives/product	Food additives	Application
Enzymes	Rennet	Cheese production
	Pullulanase	Lite beer
	Xylanase	Beer, bread, and juice production
	Isomerase, amylase	High-fructose corn syrup production
	Proteases	Meat tenderizer
Amino acids	Aspartic acid and phenylalanine	Sweetener production ingredient
	Lysine, methionine, and tryptophan	Nutritional supplement
Organic acids	Acetic acid, benzoic, citric acid, and probionic acid	Acidulant and food preservatives
Single-cell protein	Peptide, amino acid	Animal and human food supplement
Flavors and pigments	Vanillin and monascin	Coloring and flavoring agents
Low-calorie products	Modified fatty acid triglycerides	Cooking oil and food additives
	Aspartame, monellin, thaumatin	Nonnutritive sweeteners
Microbial polysaccharides	Xanthan gum	Stabilizers, gelling agents, and thickeners

Sources: retrieved from Ghoshal (2018) and Maryam et al. (2017).

10.3.8 Social and safety concerns of GM foods

Food technology has proven to be an improved area to increase crops, feeds, and food processing due to the ability to express foreign genes through transgenic technologies. This has opened ways for producing quality and large quantity of food and pharmaceutical products. Notwithstanding, there are still a multitude of concerns about the social, health, and environmental impacts of GM crops. This is because GM foods remain controversial among food experts in some countries, especially in Europe (Fu, 2019, EFSA GMO Panel, 2010). For example, many European environmental organizations, NGO, and public interest groups have from time to time actively protested against GM food for years. Part of their concern is that the introduced genes in GMO foods could cause new food allergies and allergic reactions in humans (Fu, 2019, EFSA GMO Panel, 2010). Also, it is believed that it is capable of serving as potential contaminant to environmental and food chains (EFSA GMO Panel, 2010).

The aforementioned concerns have made scientists to conduct further research on the potential benefits and risks of GMO foods to humans and the environment as compared to traditional agricultural and food-processing practices (Kathage et al., 2016, Ali and Rahut, 2018, Amir, 2014, ISAAA, 2017, Smale, 2017). This is to provide more awareness on the merits and demerits of GM foods and products.

10.3.8.1 Benefits of biotechnology in food production

1. Increased food production: It is now possible to produce crops that are drought resistant and IR, and pesticide tolerant in order to increase the yield of food production to feed growing population in the world.
2. Herbicide- and pesticide-tolerant crops made it possible to reduce the cost and application of these chemicals to farmland. This also helps the consumers to pay less, since the cost of food production is lower.
3. High nutrient-rich healthy foods or crops are made possible due to biotechnological application. This application makes it healthy for consumers to get nutrient-rich crops that parent crops do not possess at lower price, for example, GM golden rice has extra vitamin A and iron that saves people from blindness.
4. Due to biotechnology, shelf life of food and beverages, as well as crops can be extended. Making food to stay fresh for a longer period of time and this is not only beneficial for the producers, but to the sellers and consumers.

10.3.8.2 Risks of biotechnology in food production

There are two basic potential risks that GM foods could pose. This includes the health and environmental impacts as earlier mentioned. The following are other few examples of the risks associated with biotechnological method of food production:

1. Ecosystem contamination: One of the concerns is that there is a possibility of gene transfer to other crops or organisms through horizontal gene transfer when transgenic organisms are exposed to natural environment resulting in the spread of transgene everywhere. This cool resorts in contamination of ecosystem.

2. Allergic reaction: Another concern raised against GM food is that it contains foreign genes that could cause allergic reaction or hypersensitivity to the consumers as reported by local laboratory (Adenle, 2011, Fu, 2019, Goodman et al., 2013). Similarly, Jia (2005) reported the potential of foreign protein OVA in allergic reactions such as increase in histamine level and drop in systolic blood pressure. As further reviewed by Haroon and Ghazanfar, cases of allergic reactions by animal to Cry9 foreign protein from soil bacteria *Bt* have been reported.

3. Yet-to-be-discovered health hazards: Despite the fact that the use of GM foods is gradually gaining ground as a possible solution to the problems of hunger and malnutrition in the world with high productivity, however, there are still some health-related risks that are yet to be discovered that could arise in years to come.

4. Economic risks: Another concern has to do with economic risk of high cost of GM seeds for crop production. It has been argued that this will hamper the ability of poor farmers to purchase such seeds and prevent them from benefiting from biotechnological improvements, hence affect food industries in getting less expensive and highly nutritious raw materials for food production.

10.3.8.3 Safety concerns related to consumption of biotechnologically produced food products

A number of food and health organizations as well as scientific evidences support the consumption of GM foods. Government agencies such as the Institute of Food Technologists and American Medical Association support the use of GM foods. However, there are still some areas that should be improved and this includes labeling of GM foods, because it is one of the weak areas in the field of food biotechnology. It is necessary to properly brand GM foods as such so as to allow people to make their own choice, which could lead to successful commercialization of GM foods. This is so because people around the world value transparency. Thus, proper and positive labeling system could help in galvanizing public trust. Therefore, an effective universal standard labeling in food biotechnology should be prioritized and developed.

10.3.9 Modern biotechnological diagnostics to ensure food safety

The food industry is a field where biotechnological application continues to receive a boost to improve quality and quantity of food crops from varieties of raw materials around the world. To this end, many of the classical food microbiological techniques are culture based with microorganisms grown on agar plates and identification through biochemical characterization. These conventional methods are tedious, slow, and labor-intensive while some microorganisms are not easily detected. This is because about 99% of the microorganisms are considered non-culturable (Gokal, 2016). Consequently, different molecular techniques and advanced metagenomic methods that are culture independent have been developed and used to study and detect microorganisms in different fermented foods (Figure 10.3) (Cocolin, 2013, Sarethy et al., 2014).

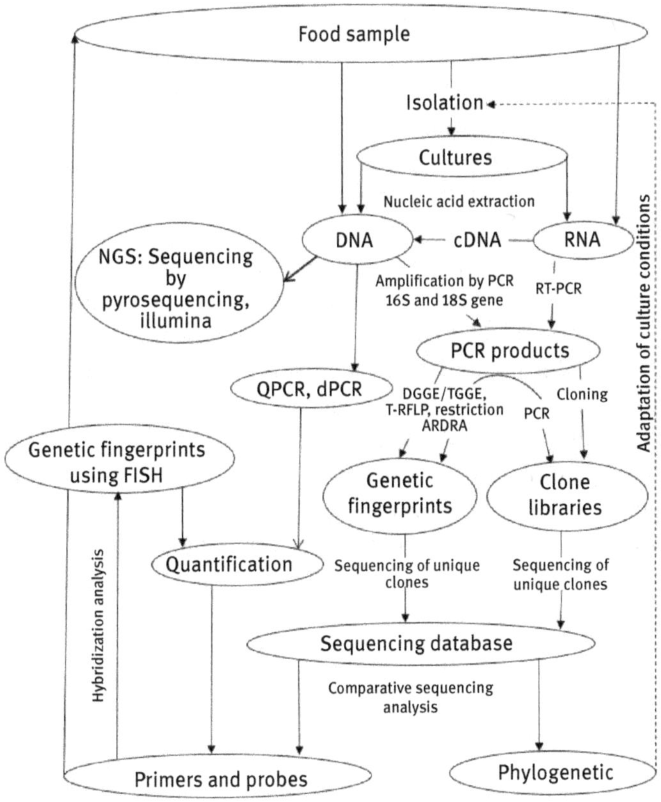

Figure 10.3: Biotechnological techniques employed in studying microbial communities in food samples to ensure safety.

In recent times, genetic diagnostic and identification of microorganisms in food samples have gained more attention due to sensitivity, speed, and specificity of microbial testing (Table 10.10) and as such culture-independent molecular techniques are based on the isolation of total DNA or RNA of microbial population ecological niche with subsequent application of metabolic genes or marker genes. It is expected that such genes must be present in the target organisms and have conserved regions. These methods use primers in encoding the 16S rRNA Gene (V3 and V4 regions) and primers for different phylogenetic markers such as Tu elongation factor, *RecA* protein, RNA polymerase β-subunit (rpoβ) 26S rRNA gene (D1/D2 regions), and other genes (Villarreal-Morales, 2018).

The essential biotechnological culture-independent techniques that have been employed in food sciences include (Figure 10.3) cell culture, DNA fingerprinting and recombinant DNA/RNA technology, and amplified fragment length polymorphism (terminal restriction fragment length polymorphism, T-RFLP). Others are ribosomal intergenic spacer analysis, length heterogeneity polymerase chain reaction (PCR), T-RFLP, amplified ribosomal DNA restriction analysis, single-stranded conformational polymorphism, as well as denaturing gradient gel electrophoresis or temperature gradient gel electrophoresis and fluorescence *in situ* hybridization (Rocha, 2019, Dias and Rmsuk, 2018). The methods also include random amplified polymorphic DNA (Pal, 2017, Kamle, 2017, Krasznai, 2018). The most recent application of molecular technology in food industry is the use of metagenomic tools through the next-generation sequencing technologies such as pyrosequencing and HiSeq or MiSeq Illumina sequencing (Gokal, 2016).

These techniques have been used for the detection of microbial contamination in food, diversity of microbial communities, and dynamics in (i) raw materials use for fermentation; (ii) fermentation process, and (iii) final fermented products. Applications of these molecular techniques in food processing are also significant to the development of newly improved products, tracing genetic or foreign materials in foods and maintenance of food safety for general consumption. Food biotechnology does not only entail reduction of food losses. Rather, it provides ways of improving the quality and safety of processed food products through the use of knowledge in the fields of bioengineering, molecular biology, and genetic engineering.

Table 10.10: Culture-independent biotechnology methods employed for monitoring and safety of fermented foods and beverages.

Countries	Raw materials	Beverages	Reported microorganisms	Methods	Genes	References
South Africa	Apple, grape, pear, mango, and aloe vera	Fruit juice	*Alicyclobacillus acidoterrestris* and *Alicyclobacillus pomorum*	PCR-DGGE	16S rRNA	Duvenage et al. (2007)
Republic of Congo	Maize	Poto poto	*Lactobacillus gasseri*, *Enterococcus* sp., *Escherichia coli*, *Lactobacillus plantarum*/*paraplantarum*, *Lactobacillus acidophilus*, *Lactobacillus delbrueckii*, *Bacillus* sp., *Lactobacillus reuteri*, and *Lactobacillus casei*	TGGE	16S rRNA	Abriouel (2006)
Burkina Faso	Millet	Dégué dough	*L. gasseri*, *Enterococcus* sp., *E. coli*, *Lactobacillus fermentum*, *Lactobacillus brevis*, and *L. casei*	TGGE	16S rRNA	Abriouel (2006)
China	Milk	Commercial yoghurts	*L. delbrueckii* and *S. thermophilus*	PCR-DGGE	16S rRNA	Ma et al. (2009)
China	*Camellia sinensis* leaves	Xiaguan Tuo tea	*Aspergillus niger* and *Blastobotrys adeninivorans*, *Debaryomyces hansenii*, *Penicillium*, *Rhizopus*, *Trichoderma*	FISH, Illumina MiSeq	RNA, 16S rRNA V4 region	Li (2018)
Ghana, Nigeria, Benin, Burkina Faso, Uganda, Kenya, Ethiopia and South Africa	Cassava, millet, sorghum, maize, milk	Ogi, Motoho, Mawe, Boule d'akassa, Nono, wara, fura, Obusera, cassava dough	Lactobacillales (phylum Firmicutes) such as *Lactobacillus fermentum*, *Weissella* and *Streptococcus*, *Acetobacter*, pathogens found include *Escherichia* or *Clostridium*, *Zymomonas*	HiSeq. Illumina	16S rRNA V4 region	Diaz, (2019)

(continued)

Table 10.10 (continued)

Countries	Raw materials	Beverages	Reported microorganisms	Methods	Genes	References
Nigeria	soybean	Soy-daddawa	Bacteria: *Bacillus tequilensis, Bacillus thuringiensis, Bacillus pumilus, Brevibacterium massiliensis, Nosocomiicoccus ampullae,* and *Atopostipes suicloacalis.* Fungi: *Alternaria lini, Aspergillus brasiliensis, Candida rugose, Candida tropicalis, Cladosporium cladosporioides, Cladosporium oxysporum, Dokmaia* sp., *Issatchenkia orientalis, Kodamaea ohmeri, Lecythophora hoffmannii, Phoma* sp., *Pichia kluyveri, Pichia rabulensis, Saccharomyces cerevisiae,* and *Starmerella* sp. *Rhizopus delemar* was the only Zygomycota detected. Pathogens also detected.	PCR-DGGE	V3–V5 region of the 16 S rRNA	Ezeokoli (2016)
China	Cooked rice, fresh fish	Yucha	*Lactobacillus, Lactococcus, Enterococcus, Vibrio, Weissella, Pediococcus, Enterobacter, Salinivibrio, Acinetobacter, Macrococcus, Kluyvera,* and *Clostridium*	Illumina MiSeq, q-PCR	16S rRNA gene V4 region	Zhang (2016)
Mexico (SLP)	Agave salmiana	Mezcal	*Candida ethanolica, Kluyveromyces marxianus, Saccharomyces cerevisiae, Clavispora lusitaniae, Torulaspora delbrueckii, Pichia kluyveri, Saccharomyces exiguus,* and *Zygosaccharomyces bailii*	PCR-DGGE	26S rRNA	Valdez (2011)

Zambia	Maize	Mabisi, Chibwantu and Munkoyo	*Lactobacillus* and *Weisella* genera dominated Chibwantu products; Mabisi dominated by members of the *Lactobacillus, Lactococcus,* and *Streptococcus*; Munkoyo dominated by members of the *Weisella* and *Lactobacillus* genera. Few *Acinetobacter, Chryseobacterium, Acetobacter,* and *Gluconobacter* (negative genera)	ARDRA	16S rRNA	Schoustra (2013)
USA	Rhizome and roots of kava (Piper methysticum)	Kava	*Weissella soli, Lactobacillus* spp., and *Lactococcus lactis*	PCR-DGGE	16S rRNA	Dong (2011)
Mongolia	raw milk of mares or camels	Airag and Tarag	*Lactobacillus helveticus* and *Lactobacillus kefiranofaciens, Lactobacillus (L.) delbrueckii* ssp. *bulgaricus, L. helveticus, L. fermentum, L. delbrueckii* ssp. *Lactis,* and *Lactococcus lactis* ssp. *lactis*	RAPD PCR	16S rRNA	Takeda (2011)

PCR-DGGE, polymerase chain reaction denaturing gradient gel electrophoresis; TGGE, temperature gradient gel electrophoresis; FISH, fluorescence *in situ* hybridization; ARDRA, amplified ribosomal DNA restriction analysis; RAPD, random amplified polymorphic DNA.

10.4 Conclusion

Thus far, it is arguable that the application of biotechnological techniques in food industries across the world has made incontrovertible contributions to human advancement. Even though there exists evidence of shortcomings in some areas where the knowledge in the field of biotechnology has been applied, it is worth noting that benefits far outweigh any known limitations. For instance, with the aid of modern biotechnology in food processing such as enzyme production, fermentation, and high nutritional foods and beverages, contemporary issues of hunger, malnutrition, poverty, and diseases are being creatively addressed. As noted earlier in the preceding paragraphs, production of GM foods through food biotechnology to enhance nutritional values, yields, and shelf life has provided much-needed assistance in meeting the demand for self-sufficiency in food production of many people, especially in the Third World countries.

Notwithstanding the foregoing, it is necessary that every GM foods produced by any manufacturer or food industries must be properly labeled. Awareness on the potential pros and cons of the GM foods should be improved such that people can make their choice. Seminars and debates should be conducted to create more awareness on the benefits of biotech foods. In as much as biotechnological application helps in increasing food production, experts in the field should intensify their research output for more valuable discoveries and avoidance of any potential risk of biotechnologically produced foods. Finally, it is suggested that collaboration between food industries and scientists in both research institutes and universities should be strengthened in order to improve or produce more nutritional foods through the application of biotechnological proficiencies.

Acknowledgments: The authors are grateful to our academic editor Mr Folami Ola-Oluwa at the Department of Public Management and Economics, Urban Futures Centre, Durban University of Technology, Durban, South Africa.

References

Aadel, H., et al. (2018). Agrobacterium-mediated transformation of mature embryo tissues of bread wheat (Triticum aestivum L.) genotypes. *Cereal Research Communications*, 46(1), 10–20.

Abriouel, H., et al. (2006). Culture-independent analysis of the microbial composition of the African traditional fermented foods poto poto and dégué by using three different DNA extraction methods. *International Journal of Food Microbiology*, 111(3), 228–233.

Adenle, A. A. (2011). Response to issues on GM agriculture in Africa: are transgenic crops safe? *BMC Research Notes*, 4(1), 388.

Adesulu-Dahunsi, A., Jeyaram, K. and Sanni, A. (2018). Probiotic and technological properties of exopolysaccharide producing lactic acid bacteria isolated from cereal-based Nigerian fermented food products. *Food Control*, 92, 225–231.

Adrio, J. L. and Demain, A. L. (2014). Microbial enzymes: tools for biotechnological processes. *Biomolecules*, 4(1), 117–139.

AgBioResearch, M. S. U. (2018). *Are GMOs Safe?* https://www.canr.msu.edu/news/are-gmos-safe (accessed December 9, 2019).

Ali, A. and Rahut, D. B. (2018). Farmers willingness to grow GM food and cash crops: empirical evidence from Pakistan. *GM Crops & Food*, 9(4), 199–210.

Alyokhin, A., et al. (2008). Colorado potato beetle resistance to insecticides. *American Journal of Potato Research*, 85(6), 395–413.

Amir, P. (2014). GM crops: no gain for small farmers. *Appropriate Technology*, 41(3), 44.

Appiah, D. O., Asante, F. and Nketia, B. A. (2017). Perceived agricultural land use decisions in a peri-urban district, Ghana. *Journal of Agricultural and Crop Research*, 5(1), 1–10.

Aris, A. and Leblanc, S. (2011). Maternal and fetal exposure to pesticides associated to genetically modified foods in Eastern Townships of Quebec, Canada. *Reproductive Toxicology*, 31(4), 528–533.

Beckwith, J. R. (2002). *Making Genes, Making Waves: A Social Activist in Science*. Harvard University Press. ISBN 978-0-674-00928-8.

Berni, R. C., et al. (2011). Saccharomyces boulardii: a summary of the evidence for gastroenterology clinical practice in adults and children. *European Review for Medical and Pharmacological Sciences*, 15(7), 809–822.

Brink, J., McKelvey, M. and Smith, K. (2004). *Conceptualizing and Measuring Modern Biotechnology*.

Bonny, Sylvie. (2008). Genetically modified glyphosate-tolerant soybean in the USA: adoption factors, impacts and prospects. A review. Agronomy for Sustainable Development, 28 (1), 21–32.

Buiatti, M., Christou, P. and Pastore, G. (2013). The application of GMOs in agriculture and in food production for a better nutrition: two different scientific points of view. *Genes & Nutrition*, 8(3), 255.

Carden, B. A., Kalvig, A. B. and Kyle, D. (2019). Soybean variety 5PSUP67. Google Patents.

Cashman, K. D. and Hayes, A. (2017). Red meat's role in addressing 'nutrients of public health concern'. *Meat Science*, 132, 196–203.

Cho, K., et al. (2016). RNA interference-mediated simultaneous suppression of seed storage proteins in rice grains. *Frontiers in Plant Science*, 7, 1624.

Cima, G. (2016). GE salmon gains FDA approval. *Journal of the American Veterinary Medical Association*, 248(1), 25.

Cocolin, L., et al. (2013). Culture independent methods to assess the diversity and dynamics of microbiota during food fermentation. *International Journal of Food Microbiology*, 167(1), 29–43.

Cole, C. B. and Erdahl, B. S. (2018). Soybean cultivar CL1564482. Google Patents.

Dagbagli, S. and Goksungur, Y. (2008). Optimization of b-galactosidase production using Kluyveromyces lactis NRRL Y-8279 by response surface methodology. *Electronic Journal of Biotechnology* 11(4), 11–12.

Deltsidis, A. I., Sims, C. A. and Brecht, J. K. (2018). Ripening recovery and sensory quality of pink tomatoes stored in controlled atmosphere at chilling or nonchilling temperatures to extend shelf life. *HortScience*, 53(8), 1186–1190.

Dias, P. and Rmsuk, R. (2018). Fluorescence in situ hybridization (FISH) in food pathogen detection. *International Journal of Molecular Biology: Open Access*, 3(3), 141–147.

Diaz, M., et al. (2019). Comparison of the microbial composition of African fermented foods using amplicon sequencing. *Scientific Reports*, 9(1), 13863.

Dong, J., et al. (2011). PCR-DGGE analysis of bacterial community dynamics in kava beverages during refrigeration. *Letters in Applied Microbiology*, 53(1), 30–34.

Duvenage, W., Gouws, P. A. and Witthuhn, R. C. (2007). PCR-based DGGE identification of bacteria present in pasteurised South African fruit juices. *South African Journal of Enology and Viticulture*, 28(1), 56–60.

El, Mohamed Salahel-Din AM Abd. (2009). Production of Volatile Secondary Metabolites in Plant Tissue Cultures. UK: The University of Manchester.

Enitan, A., Adeyemo, J. and Ogunbanwo, S. (2011). Influence of growth conditions and nutritional requirements on the production of hydrogen peroxide by lactic acid bacteria. *African Journal of Microbiology Research*, 5(15), 2059–2066.

Enitan, A. M. (2008). *Production of Hydrogen Peroxide and of β-Galactosidase by Lactic Acid Bacteria Isolated from Raw and Fermented Milk*. Master of Science in Microbiology Dissertation submitted to the Department of Botany and Microbiology, Faculty of Science. Dissertation submitted to the Department of Botany and Microbiology. Ibadan, Nigeria: Faculty of Science, University of Ibadan.

EFSA GMO Panel (EFSA Panel on Genetically Modified Organisms) (2010). Draft Scientific Opinion on the assessment of allergenicity of GM plants and microorganisms and derived food and feed. EFSA Journal, 8(7):1700. doi: http://doi.org/10.2903/j.efsa.2010.1700. Available online: www.efsa.europa.eu.

EFSA GMO Panel (EFSA Panel on Genetically Modified Organisms) (2013). Scientific Opinion on application EFSA-GMO-NL-2011-93 for the placing on the market of the herbicide-tolerant genetically modified soybean MON 87708 for food and feed uses, import and processing under Regulation (EC) No 1829/2003 from Monsanto. European Food Safety Authority (EFSA) Journal, 11 (10), 3355. doi: http://10.2903/j.efsa.2013.3355 www.efsa.europa.eu/efsajournal.

Ezeokoli, O. T., et al. (2016). PCR-denaturing gradient gel electrophoresis analysis of microbial community in soy-daddawa, a Nigerian fermented soybean (Glycine max (L.) Merr.) condiment. *International Journal of Food Microbiology*, 220, 58–62.

Falk, M. C., et al. (2002). Food biotechnology: benefits and concerns. *The Journal of Nutrition*, 132(6), 1384–1390.

FAO. (1998). *Fermented Fruits and Vegetables: A Global Perspective*, by M. Battcock & S. Azam-Ali. Rome: FAO Agricultural Services Bulletin N. 134.

Forabosco, F., et al. (2013). Genetically modified farm animals and fish in agriculture: a review. *Livestock Science*, 153(1–3), 1–9.

Fu, L., et al. (2019). Allergenicity evaluation of food proteins. In: *Food Allergy*. Singapore: Springer.

Garg, M., et al. (2018). Biofortified crops generated by breeding, agronomy, and transgenic approaches are improving lives of millions of people around the world. *Frontiers in Nutrition*, 5, 12.

Gasser, C. S. and Fraley, R. T. (1989). Genetically engineering plants for crop improvement. *Science*, 244(4910), 1293–1299.

Gavrilescu, M. and Chisti, Y. (2005). Biotechnology – a sustainable alternative for chemical industry. *Biotechnology Advances*, 23(7–8), 471–499.

Ghoshal, G. (2018). Biotechnology in food processing and preservation: an overview. In: *Advances in Biotechnology for Food Industry* (pp. 27–54). Elsevier.

Gilliland, S. E. (2018). *Bacterial Starter Cultures for Food*. ISBN 9781315890968, Taylor & Francis. CRC Press.

Global Biodiversity Outlook (GBO) report (2007) "Convention on Biological Diversity, Chapter 2. https://www.cbd.int/gbo1/summary.shtmL"

Goffeau, A., et al. (1996). Life with 6000 genes. *Science*, 274(5287), 546–567.

Gohlke, J. and Deeken, R. (2014). Plant responses to Agrobacterium tumefaciens and crown gall development. *Frontiers in Plant Science*, 5, 155.

Gokal, J., et al. (2016). Molecular characterization and quantification of microbial communities in wastewater treatment systems. In: P. Shukla (Ed.). *Microbial Biotechnology* (pp. 73–128). CRC Press.

Gonsalves, D. (2004). *Transgenic Papaya in Hawaii and Beyond*.

Goodman, R. E., Panda, R. and Ariyarathna, H. (2013). Evaluation of endogenous allergens for the safety evaluation of genetically engineered food crops: review of potential risks, test methods, examples and relevance. *Journal of Agricultural and Food Chemistry* 61(35), 8317–8332.

Goto, F., et al. (1999). Iron fortification of rice seed by the soybean ferritin gene. *Nature Biotechnology*, 17(3), 282.

Gupta, A., et al. (2019). Improvement of post-harvest fruit characteristics in tomato by fruit-specific over-expression of oat arginine decarboxylase gene. *Plant Growth Regulation*, 88(1), 61–71.

Habib, M. A. and Hasan, M. (2008). *A Review on Culture, Production and Use of Spirulina as Food for Humans and Feed for Domestic Animals and Fish*.

Hansen, E. (2002). Commercial bacterial starter cultures for fermented foods of the future. *International Journal of Food Microbiology*, 78, 119–131.

Haroon, F. and Ghazanfar, M. (2016). Applications of food biotechnology. *Journal of Ecosystem and Ecography*, 6(215), 2.

Haroon, F. and Ghazanfar, M. Applications of food biotechnology. *Journal of Ecosystem and Ecography*, 6, 215.

Hartmann, K., et al. (2018). Swiss-type cheeses (propionic acid cheeses). In: *Global Cheesemaking Technology* (pp. 336–348). Wiley Online Library.

Hayaloglu, A., Karatekin, B. and Gurkan, H. (2014). Thermal stability of chymosin or microbial coagulant in the manufacture of Malatya, a Halloumi type cheese: proteolysis, microstructure and functional properties. *International Dairy Journal*, 38(2), 136–144.

Herman, R. A., et al. (2018). Food and feed safety of DAS-44406-6 herbicide-tolerant soybean. *Regulatory Toxicology and Pharmacology*, 94, 70–74.

Holzapfel, W. (1997). Use of starter cultures in fermentation on a household scale. *Food Control*, 8(5–6), 241–258.

Hughes, S. S. (2001). Making dollars out of DNA: the first major patent in biotechnology and the commercialization of molecular biology, 1974-1980. *Isis*, 92(3), 541–575.

Hyde, K. D., et al. (2019). The amazing potential of fungi: 50 ways we can exploit fungi industrially. *Fungal Diversity*, 97(1), 1–136.

International Service for the Acquisition of Agri-biotech Applications (ISAAA) (2017). Global Status of Commercialized Biotech/GM Crops in 2017: Biotech Crop Adoption Surges as Economic Benefits Accumulate in 22 years. ISAAA Brief.

ISAAA, International Service for the Acquisition of Agri-biotech Applications. (2018). Biotech crops continue to help meet the challenges of increased population and climate change. In: *Brief 54: Global Status of Commercialized Biotech/GM Crops: SAAA Brief 54-2018: Executive Summary*. https://www.isaaa.org/resources/publications/briefs/54/executivesummary/default.asp.

Ismagul, A., et al. (2018). A biolistic method for high-throughput production of transgenic wheat plants with single gene insertions. *BMC Plant Biology*, 18(1), 135.

Javanmardi, J., Rahemi, M. and Nasirzadeh, M. (2013). Post-storage quality and physiological responses of tomato fruits treated with polyamines. *Advances in Horticultural Science*, 173–181.

Jia, X.-D., et al. (2005). Studies on BN rats model to determine the potential allergenicity of proteins from genetically modified foods. *World Journal of Gastroenterology*, 11(34), 5381.

Jiang, Y., McGeachin, R. and Bailey, C. (1994). α-Tocopherol, β-carotene, and retinol enrichment of chicken eggs. *Poultry Science*, 73(7), 1137–1143.

Johansen, E. (2018). *Use of Natural Selection and Evolution to Develop New Starter Cultures for Fermented Foods.*

Joint, FAO, and WHO Expert Committee on Food Additives. (2006). *Safety evaluation of certain food additives.*

Kamle, M., et al. (2017). Current perspectives on genetically modified crops and detection methods. *3 Biotech*, 7(3), 219.

Kandasamy, S., Kavitake, D. and Shetty, P. H. (2018). Lactic acid bacteria and yeasts as starter cultures for fermented foods and their role in commercialization of fermented foods. In: *Innovations in Technologies for Fermented Food and Beverage Industries* (pp. 25–52). Springer.

Kaškonienė, V., et al. (2017). Current state of purification, isolation and analysis of bacteriocins produced by lactic acid bacteria. *Applied Microbiology and Biotechnology*, 101(4), 1323–1335.

Kathage, J., Rodríguez-Cerezo, E. and Gómez-Barbero, M. (2016). *Providing a Framework for the Analysis of the Cultivation of Genetically Modified Crops: The First Reference Document of the European GMO Socio-Economics Bureau.*

Kazemi, S., Khayati, G. and Faezi-Ghasemi, M. (2016). β-Galactosidase production by Aspergillus niger ATCC 9142 using inexpensive substrates in solid-state fermentation: optimization by orthogonal arrays design. *Iranian Biomedical Journal*, 20(5), 287.

Kenney, M. (1998). Biotechnology and the creation of a new economic space. *Private Science: Biotechnology and the Rise of the Molecular Sciences* (pp. 131–143).

Kirk, O., Borchert, T. V. and Fuglsang, C. C. (2002). Industrial enzyme applications. *Current Opinion in Biotechnology*, 13(4), 345–351.

Kocak, K. J. (2019). Soybean variety 5PWAV53. Google Patents.

Krasznai, D. J., et al. (2018). Compositional analysis of lignocellulosic biomass: conventional methodologies and future outlook. *Critical Reviews in Biotechnology*, 38(2), 199–217.

Lawrence, R. H. (1988). New applications of biotechnology in the food industry. In: *Biotechnology and the Food Supply: Proceedings of a Symposium*. Washington, DC: National Academies Press; National Research Council (US) Commission on Life Sciences. https://www.ncbi.nlm.nih.gov/books/NBK235032/ (accessed December 9, 2019).

Leuchtenberger, W., Huthmacher, K. and Drauz, K. (2005). Biotechnological production of amino acids and derivatives: current status and prospects. *Applied Microbiology and Biotechnology*, 69(1), 1–8.

Li, H., et al. (2018). Microbial diversity and component variation in Xiaguan Tuo Tea during pile fermentation. *PloS One*, 13(2), e0190318.

Lick, S., et al. (1996). Rapid identification of Streptococcus thermophilus by primer-specific PCR amplification based on its lacZ gene. *Systematic and Applied Microbiology*, 19(1), 74–77.

Lick, S. and Heller, K. (1998). Quantitation by PCR of yoghurt starters in a model yoghurt produced with a genetically modified Streptococcus thermophilus. *Milchwissenschaft*, 53, 671–675.

Linares, D. M., et al. (2017). Lactic acid bacteria and bifidobacteria with potential to design natural biofunctional health-promoting dairy foods. *Frontiers in Microbiology*, 8, 846.

Lucca, P., Hurrell, R. and Potrykus, I. (2002). Fighting iron deficiency anemia with iron-rich rice. *Journal of the American College of Nutrition*, 21(Suppl. 3), 184S–190S.

Ma, J., Kong, J. and Ji, M. (2009). Detection of the lactic acid bacteria in commercial yoghurts by PCR-denaturing gradient gel electrophoresis. *Chinese Journal of Applied and Environmental Biology*, 15(4), 534–539.

Maryam, B. M., Datsugwai, M. S. S. and Shehu, I. (2017). The role of biotechnology in food production and processing. *Industrial Engineering*, 1, 24–35.

Masoodi, K. Z., et al. (2018). Genetic modification in fruits and vegetables for improved nutritional quality and extended shelf life. In: *Preharvest Modulation of Postharvest Fruit and Vegetable Quality* (pp. 359–379). Elsevier.

Mokoena, M. (2017). Lactic acid bacteria and their bacteriocins: classification, biosynthesis and applications against uropathogens: a mini-review. *Molecules*, 22, 1255.

Mokoena, M. P. (2017). Lactic acid bacteria and their bacteriocins: classification, biosynthesis and applications against uropathogens: a mini-review. *Molecules*, 22(8), 1255.

Monteiro, C. A., et al. (2018). The UN Decade of Nutrition, the NOVA food classification and the trouble with ultra-processing. *Public Health Nutrition*, 21(1), 5–17.

Monteiro, C. A., et al. (2019). Ultra-processed foods: what they are and how to identify them. *Public Health Nutrition*, 22(5), 936–941.

Mota de Carvalho, N., et al. (2018). Fermented foods and beverages in human diet and their influence on gut microbiota and health. *Fermentation*, 4 (4), 90.

Muchenje, V., Mukumbo, F. E. and Njisane, Y. Z. (2018). Meat in a sustainable food system. *South African Journal of Animal Science*, 48, 818–828.

Nair, B. M. and Prajapati, J. B. (2003). The history of fermented foods. In: *Handbook of Fermented Functional Foods* (pp. 17–42). CRC Press.

Network, G. N. (2004). *Genetics and Genomics Timeline*. http://www.genomenewsnetwork.org/resources/timeline/1960_mRNA.php.

Nissar, J., et al. (2017). *Applications of Biotechnology in Food Technology*.

Ogunbanwo, S. T., et al. (2005). Influence of lactic acid bacteria on fungal growth and aflatoxin production in Ogi, an indigenous fermented food. *Advances in Food*, 27(4), 189–194.

Oladosu, Y., et al. (2019). Drought resistance in rice from conventional to molecular breeding: a review. *International Journal of Molecular Sciences*, 20(14), 3519.

Osorio, S., et al. (2019). Genetic and metabolic effects of ripening mutations and vine detachment on tomato fruit quality. *Plant Biotechnology Journal*, 2019.

Oyedeji, O., Ogunbanwo, S. T. and Onilude, A. A. (2013). Predominant lactic acid bacteria involved in the traditional fermentation of fufu and ogi, two Nigerian fermented food products. *Food and Nutrition Sciences*, 4(11), 40.

Pal, M., et al. (2017). A review of biotechnological applications in food processing of animal origin. *American Journal of Food Technology*, 5(4), 143–148.

Pandey, R., et al. (2015). Over-expression of mouse ornithine decarboxylase gene under the control of fruit-specific promoter enhances fruit quality in tomato. *Plant Molecular Biology*, 87(3), 249–260.

Pariza, M. W. and Johnson, E. A. (2001). Evaluating the safety of microbial enzyme preparations used in food processing: update for a new century. *Regulatory Toxicology and Pharmacology*, 33(2), 173–186.

Partier, A., et al. (2017). Molecular and FISH analyses of a 53-kbp intact DNA fragment inserted by biolistics in wheat (Triticum aestivum L.) genome. *Plant Cell Reports*, 36(10), 1547–1559.

Patel, A. and Prajapat, J. (2013). Food and health applications of exopolysaccharides produced by lactic acid bacteria. *Advances in Dairy Research*, p. 1–8.

Paul Singh, R. and Heldman, D. R. (2013). *Introduction to Food Engineering (Fifth Edition)* (p. 1). Academic Press. ISBN 0123985307.

Pinstrup-Andersen, P. and Cohen, M. (2000). Modern biotechnology for food and agriculture: risks and opportunities for the poor. In: *Agricultural Biotechnology and the Poor* (pp. 159–172). Washington, DC: Consultative Group on International Agricultural Research.

Prasad, R., Gill, S. S. and Tuteja, N. (2018). *New and Future Developments in Microbial Biotechnology and Bioengineering: Crop Improvement Through Microbial Biotechnology*. Elsevier.

Raes, K., et al. (2002). The deposition of conjugated linoleic acids in eggs of laying hens fed diets varying in fat level and fatty acid profile. *The Journal of Nutrition*, 132(2), 182–189.

Rajasekaran, A. and Kalaivani, M. (2013). Designer foods and their benefits: a review. *Journal of Food Science and Technology*, 50(1), 1–16.

Ramón, D., Diamante, A. and Calvo, M. D. (2008). Food biotechnology and education. *Electronic Journal of Biotechnology*, 11(5), 1–2.

Raveendran, S., et al. (2018). Applications of microbial enzymes in food industry. *Food Technology and Biotechnology*, 56(1), 16–30.

Ray, B. (2019). *Food Biopreservatives of Microbial Origin*. CRC Press.

Ray, R. and Joshi, V. (2014). *Fermented Foods: Past, Present and Future*.

Ray, R. C. and Joshi, V. (2014). Fermented foods: past, present and future. In: *Microorganisms and Fermentation of Traditional Foods* (pp. 1–36).

Ren, D., et al. (2018). Antimicrobial characteristics of lactic acid bacteria isolated from homemade fermented foods. *BioMed Research International*, 2018.

Rocha, R., et al. (2019). Development and application of peptide nucleic acid fluorescence in situ Hybridization for the specific detection of Listeria monocytogenes. *Food Microbiology*, 80, 1–8.

Ronald, P. C. and Mcwilliams, J. E. (2010). Genetically engineered distortions. *New York Times*, May 14.

Sabikhi, L. (2007). Designer milk. *Advances in Food and Nutrition Research*, 53, 161–198.

Santer, M. (2009). Joseph Lister: first use of a bacterium as a 'model organism' to illustrate the cause of infectious disease of humans. *Notes and Records of the Royal Society*, 64. http://doi.org/10.1098/rsnr.2009.0029.

Sarethy, I. P., Pan, S. and Danquah, M. K. (2014). Modern taxonomy for microbial diversity. In: *Biodiversity – The Dynamic Balance of the Planet*. IntechOpen.

Schoustra, S. E., et al. (2013). *Microbial community structure of three traditional zambian fermented products: mabisi, chibwantu and munkoyo. PloS One*, 8(5), e63948.

Shulman, S. T., Friedmann, H. C. and Sims, R. H. (2007). Theodor Escherich: the first pediatric infectious diseases physician? *Clinical Infectious Diseases*, 45(8), 1025–1029.

Silveira, G. G. d., et al. (2005). Microbial rennet produced by Mucor miehei in solid-state and submerged fermentation. *Brazilian Archives of Biology and Technology*, 48(6), 931–937.

Singh, B., et al. (2015). Textbook of animal biotechnology. In: Frontiers in Diary Biotechnology. New Delhi, India: The Energy and Resources Institutes (TERI).

Singh, P., et al. (2008). Functional and edible uses of. soy protein products. *Comprehensive Reviews in Food Science and Food Safety*, 7(1),14–28.

Singh, R., et al. (2016). Microbial enzymes: industrial progress in twenty-first century. *3 Biotech*, 6(2), 174.

Smale, M. (2017). GMOs and poverty: the relationship between improved seeds and rural transformations. *Canadian Journal of Development Studies/Revue canadienne d'études du développement*, 38(1), 139–148.

Soro-Yao, A. A., et al. (2014). The use of lactic acid bacteria starter cultures during the processing of fermented cereal-based foods in West Africa: a review. *Tropical Life Sciences Research*, 25(2), 81.

Sun, Q., et al. (2014). Purification and characterization of a chymosin from Rhizopus microsporus var. rhizopodiformis. *Applied Biochemistry and Biotechnology*, 174(1), 174–185.

Sun, S. S. (2008). Application of agricultural biotechnology to improve food nutrition and healthcare products. *Asia Pacific Journal of Clinical Nutrition*, 17.

Takeda, S., et al. (2011). The investigation of probiotic potential of lactic acid bacteria isolated from traditional Mongolian dairy products. *Animal Science Journal*, 82(4), 571–579.

Tamang, J. P., et al. (2016). Functional properties of microorganisms in fermented foods. *Frontiers in Microbiology*, 7, 578.

Thakur, M., Karanth, N. and Nand, K. (1990). Production of fungal rennet byMucor miehei using solid state fermentation. *Applied Microbiology and Biotechnology*, 32(4), 409–413.

Torres-León, C., et al. (2018). Food waste and byproducts: an opportunity to minimize malnutrition and hunger in developing countries. *Frontiers in Sustainable Food Systems* 2, 52.

Torrigiani, P., et al. (2012). Spermidine application to young developing peach fruits leads to a slowing down of ripening by impairing ripening-related ethylene and auxin metabolism and signaling. *Physiologia Plantarum*, 146(1), 86–98.

Umeh, O. and Anyanwu, C. *Breeding Genetically Modified Organisms: Their Benefits, Issues and Regulations in Nigeria.*

Valdez, A. V., et al. (2011). Yeast communities associated with artisanal mezcal fermentations from Agave salmiana. *Antonie Van Leeuwenhoek*, 100(4), 497–506.

Verma, A. S., et al. (2011). Biotechnology in the realm of history. *Journal of Pharmacy and Bioallied Sciences*, 3(3), 321.

Villarreal-Morales, S. L. et al. (2018). *Metagenomics of Traditional Beverages*, in *Advances in Biotechnology for Food Industry* (pp. 301–326). Elsevier.

Werckenthin, C., et al. (2001). Antimicrobial resistance in staphylococci from animals with particular reference to bovine Staphylococcus aureus, porcine Staphylococcus hyicus, and canine Staphylococcus intermedius. *Veterinary Research*, 32(3–4), 341–362.

Wright, S. (1986). Recombinant DNA technology and its social transformation, 1972–1982. *Osiris*, 2, 303–360.

Yang, S.-T. (2011). *Bioprocessing for Value-Added Products from Renewable Resources: New Technologies and Applications.* Elsevier.

Yeh, A. G.-O. and Li, X. (1999). Economic development and agricultural land loss in the Pearl River Delta, China. *Habitat International*, 23(3), 373–390.

Yu, P. (2017). Food biotechnology – an important and promising research field. *Journal of Food Biotechnology Research*, 1(1), 1.

Zhang, C., Wohlhueter, R. and Zhang, H. (2016). Genetically modified foods: a critical review of their promise and problems. *Food Science and Human Wellness*, 5(3), 116–123.

Zhang, J., et al. (2016). Metagenomic approach reveals microbial diversity and predictive microbial metabolic pathways in Yucha, a traditional Li fermented food. *Scientific Reports*, 6, 32524.

Zhao, J. (2007). Nutraceuticals, nutritional therapy, phytonutrients, and phytotherapy for improvement of human health: a perspective on plant biotechnology application. *Recent Patents on Biotechnology*, 1(1), 75–97.

Part III: **Food chemistry: analysis and nutrition**

Aribisala Jamiu Olaseni, Madende Moses, Oyedeji Amusa Mariam
Oyefunke, Sabiu Saheed

11 Dietary nutrients, antinutritional factors, and valorization of food waste

11.1 Introduction

Human nutrition refers to the supply of nutrients to cells in the body through food consumption (Newton et al., 1979). These nutrients are used by the body tissues and provide energy for the biological processes taking place in the body. Seven major nutrient groups can be obtained from foods. These include: **macronutrients** in the form of proteins, fats, carbohydrates, fibers, and water; and **micronutrients** such as vitamins and minerals. Extremely important role of a balanced diet and hence nutrition in human health has been appreciated since ancient times (Latham, 1998). A variety of naturally occurring foods such as fruits and vegetables have been shown to possess functional properties that aid in stalling disease progression, inhibition of pathophysiological mechanisms, and suppression of infection (Hartmann and Meisel, 2007).

With the advent of new technologies in the detection of human diseases and an in-depth understanding of the core causes of these ailments, a wide variety of drugs have been developed to target certain pathogenic molecules participating in the disease processes (Kannan et al., 2011). Considering that drugs often treat the symptoms rather than cure diseases, a more affordable alternative is to identify and consume natural foods as well as active food components with proven therapeutic properties such as anticancer, antianemic, antidiabetic, and antimicrobial (Reedy et al., 2014). Consequently, the study of human nutrition has evolved to be interdisciplinary involving physiology, biochemistry, molecular biology, psychology, anthropology, public health, economics, and political science, with the ultimate goals being to promote optimal health, prevent nutritional deficiency diseases, and reduce the risk of chronic nutritional illnesses (Wilk, 2012). The goal also encompasses management of nutritional problems through valorization of food wastes, as

Aribisala Jamiu Olaseni, Department of Biotechnology and Food Technology, Durban University of Technology, Durban, South Africa; Department of Microbiology, Federal University of Technology, Akure, Nigeria
Madende Moses, The Food BioSciences Department, Teagasc Food Research Centre, Ashtown, Dublin 15, Ireland
Oyedeji Amusa Mariam Oyefunke, Department of Botany and Plant Biotechnology, University of Johannesburg, Auckland Park APK, South Africa
Sabiu Saheed, Department of Biotechnology and Food Technology, Durban University of Technology, Durban, South Africa

https://doi.org/10.1515/9783110667462-011

this could be envisaged to minimize malnutrition, hunger, and the accompanying ill environmental impacts of food wastes.

A variety of methods have been developed for the analysis of micro- and macro-nutrients in foods. Some methods are simple, rapid, and less time consuming, whereas others could be complex, more accurate, and may require special equipment (Cruz et al., 2016). Moreover, the suitability of each method is governed by factors such as the physical state of the sample, coexistence of the subject nutrient with other nutrients, and the purpose of analysis among other factors. This chapter highlights the significance and methods of determination of the major dietary nutrients and antinutritional factors, as well as the potential of food waste valorization as a sustainable alternative to reducing malnutrition, hunger, and ill health.

11.2 Significance of dietary nutrients and methods of determination in food samples

11.2.1 Proteins

Most functional proteins comprise 20 standard amino acids that are peptide bond linked to form a variety of conformations, which gives each protein a unique structure and function (Kannan et al., 2011). Nine of these amino acids (valine, histidine methionine, isoleucine, lysine, threonine, leucine, tryptophan, and phenylalanine) cannot be synthesized by the body in sufficient quantities, hence they are considered essential, whereas the remaining 11 (proline, arginine, aspartic acid, tyrosine, cysteine, glutamic acid, glycine, asparagine, serine, glutamine, and alanine) are nonessential. The name **protein** originated from the Greek word "proteios," meaning primary. Dietary proteins are hydrolyzed by proteases into amino acids or smaller peptides, which are assimilated in the small intestine lumen. As a result, the nutritional value of dietary proteins is determined by their content, digestibility coefficients, and relative proportions of amino acids (Tomé, 2013). The nutritional role of proteins cannot be overstated. Proteins provide essential compounds such as nitrogen and sulfur, which are building blocks for the synthesis of low-molecular-weight substances such as dopamine, creatinine, serotonin, nitric oxide, RNA, glutathione, and DNA, which have extensive physiological functions Wu et al., 2014b.

Sources of good-quality proteins include eggs, fish, dairy products, and meat, which are of animal origin. Apart from soybeans, plant origin foods are generally poor-quality protein sources. The average daily protein requirement for an adult weighing 90 kg is 75 g, assuming a score of 1.0 for the protein digestibility corrected amino acid of all protein sources, conforming to previous studies on nitrogen balance as reported by the World Health Organization (WHO, 2007). Table 11.1 shows the average content of protein in selected foods. Excess consumption of proteins leads to its

Table 11.1: Average protein content in a variety of foods.

Food group	Serving (g)	Food	Protein (g)
Meat	85	Cooked lean beef	28
	85	Cooked chicken	26
	150	1 cup cooked haddock fish	36
Legumes	172	1 cup cooked soya beans	29
	256	1 cup cooked red kidney beans	13
	196	1 cup cooked split peas	16
Dairy	245	1 cup full-fat milk	8
	28	Cheddar cheese	7
Starchy foods and cereals	185	White rice, cooked (1 cup)	15
	25	1 slice whole wheat bread	3
Vegetables and fruits	180	1 cup cooked spinach	5

Adapted from Schonfeldt and Hall (2012).

break down into keto acids and nitrogen. The keto acids are converted to carbohydrates or fat, whereas nitrogen is integrated into urea and excreted as urine.

Another very important role of proteins in human health and diseases is their nutraceutical effects. Many *in vitro* or *in vivo* released peptides from animal or plant proteins have bioactive properties and play a role in regulation, beyond that of nutrition (Hartmann and Meisel, 2007). A variety of bioactive proteins can be generated from animal or plant protein sources using different technologies such as fermentation. Some bioactive peptides extracted from soy (Bylund et al., 2000), milk (Håkansson et al., 1995), and mushroom (Ng et al., 2003) have been shown to have anticancer effects. Similarly, several peptides from different sources have been identified as possessing antihypertensive action through angiotensin-converting enzyme inhibition (Meisel, 1997). Many peptides have also been shown to have a cholesterol-lowering effect (Potter, 1995), antioxidant activity (Takenaka et al., 2003), antimicrobial activity (Ha and Zemel, 2003), and antimutagenic activity (Parodi, 2007). As a result, many commercial products are either available or currently under development by food companies that exploit the potential of bioactive peptides derived from foods with health claims that have been scientifically proven.

Having established that proteins are such an important constituent of the human diet, it is paramount to know the protein content using reliable analytical methods. Many methods exist for evaluating the protein content of different foods and indeed for the determination of their bioactive potential for marketing as functional foods

(Mæhre et al., 2018). The commonly reported protein content of foods is dependent on the method used for the analysis, which presents a problem when making a direct comparison between studies. Measurement of the protein content of foods is made complex by the presence of a heterogenic mixture of proteins with other nutrients such as lipids and carbohydrates; hence, the protein content of foods can be determined either directly or indirectly. Direct methods calculate the protein content based on the analysis of the total amino acid residues, whereas indirect methods correlate protein content to the nitrogen content or chemical reactions with a variety of functional groups within proteins (Wilson and Walker, 2002). The choice of the preferred analytical method is thought to be influenced by factors such as using established analytical procedures in laboratories, lack of analytical infrastructure, or high economic costs associated with certain methods (Mariotti et al., 2008).

Before determining the protein content of foods, a protein extraction step is undertaken. This step typically involves extracting proteins using a salt or alkaline extraction method (Maehre et al., 2016) or the Good's buffer extraction method (Alhamdani et al., 2010). After extraction, direct protein determination by total amino acid analysis can be done. During this stage, an internal standard (norleucine) is usually introduced and samples are chemically hydrolyzed using concentrated acids (hydrochloric acid). After hydrolysis, the hydrolysate is dried and analyzed chromatographically. The protein content is calculated as the sum of the individual amino acid residues.

A few indirect methods of protein determination in foods exist. These include the Kjeldahl (Kjeldahl, 1883), Lowry (Hartree, 1972), and Bradford (Bradford, 1976) methods. The Kjeldahl method involves hydrolyzing the sample with sulfuric acid using copper as the catalyst. The hydrolysate is then neutralized and titrated before determining the total nitrogen using a conversion factor. The Bradford and Lowry assays typically target the reaction of the functional group of the proteins with chemicals in the assay solution followed by a color change, which can be measured calorimetrically.

Maehre et al. (2018) undertook a study to document the varying protein content in several common foods with different matrix compositions due to the choice of analytical and extraction methods. The study showed that amino acid determination was the most accurate way of determining the protein content of the food. Spectrophotometric protein determination methods, such as the Bradford and Lowry assays, are often affected by interfering substances and could thus overestimate the protein content. Similarly, correlating the protein content of foods with the nitrogen content overestimates the protein content.

11.2.2 Carbohydrates

Dietary carbohydrates include molecules that can be metabolically transformed directly into glucose or alternatively undergo oxidation into pyruvate, including some sugar alcohols such as sorbitol (Ludwig et al., 2018). The main role of carbohydrates in the diet is to provide energy for all the physiological and biochemical processes taking place in the body. They can be grouped into monosaccharides (sugars), disaccharides, oligosaccharides, and polysaccharides (starch). Sugars are either intrinsic (found naturally in fruits and milk) or extrinsic (added to foods during processing and preparation; Slavin and Carlson, 2014). Sugars can also be used in food preservation and for functional properties such as viscosity, texture, and browning capacity. Starches are often found in vegetables, legumes, and grains, and comprises many glucose units linked by glycosidic bonds. Most starches can be broken down into sugars and used by the body for energy. Fibers consist of many sugar units bonded together with complex bond formation making then resistant to breakdown by the human small intestine enzymes. As a result, fibers enter the large intestines relatively intact and can be fermented by the microflora in the colon or it can pass through the large intestine and increase the stool weight by binding water.

Carbohydrates are the only macronutrients with no established minimum requirement although the recommended daily allowance (RDA) has been set at 130 g for adults and children aged 1 year or older (Trumbo et al., 2002). Carbohydrates are the only food constituent that directly increases blood glucose. However, the intake of certain types of carbohydrates has been associated with a risk of chronic diseases (Ley et al., 2014). In view of this, two empirical metrics have been introduced to rank foods according to effects on blood glucose; (1) **glycemic index** (GI), which compares foods based on a standardized amount of available carbohydrate and (2) **glycemic load**, which applies to the GI multiplied by the amount of carbohydrate in a typical serving (Bao et al., 2011). The carbohydrate content of a variety of representative foods is presented in Table 11.2.

Table 11.2: Carbohydrate content of selected food samples.

Food	Serving (g)	Carbohydrate (g)
Rice, jasmine, boiled	120	32
Oatmeal	250	26
Potato, boiled	150	20
Pasta (white/brown boiled)	120	31
Bread (white/brown)	40	19
Fruit juice	250 mL	24
Boiled legumes	150	22
Temperate fruit	120	14
Milk	250 mL	12
Nuts	30	7

Adapted from US Department of Agriculture, Agricultural Research Service (2009).

A variety of quantitative and qualitative analytical methods have been developed over the years to measure the total concentration and type of carbohydrates present in foods. Carbohydrate analysis is important for several reasons. **Qualitative analysis** makes certain that ingredient labels present accurate compositional information, whereas **quantitative analysis** makes certain that added components are listed in their correct quantities on ingredient labels (Heimo and Günther, 1998). The carbohydrate content of a food can be determined directly or indirectly by calculating the percentage remaining after all the other components have been measured. Before the concentration of carbohydrates is determined in foods, extraction steps can be undertaken, which often involve boiling a defatted sample in 80% alcohol followed by filtration, clarification, or ion-exchange chromatography (AOAC, 2016).

The type and concentration of carbohydrates in foods can be determined by chromatographic techniques such as thin-layer chromatography, gas chromatography (GC) (Bradbury, 1990), and high-performance liquid chromatography (HPLC) (Bonn, 1985). Moreover, HPLC and GC are commonly used in association with nuclear magnetic resonance (NMR) or mass spectrometry (MS) so that the chemical structure of the molecules can be identified (Kazmaier et al., 1998). Electrophoretic methods can also be used to analyze carbohydrates after a derivatization step using chemicals such as borates. Chemical methods of monosaccharide and oligosaccharide determination in foods take advantage of the presence of reducing groups in these carbohydrates. These chemical methods include titration (Lane–Eynon method), gravimetric (Munson and Walker method), and calorimetric (Anthrone method) methods (Dubois et al., 1951). Enzymatic methods, such as D-glucose/D-fructose and maltose/sucrose, can also be used to determine the carbohydrate content of foods. These methods often require very little sample preparation and are rapid, highly specific, and sensitive to low concentrations of carbohydrates. Carbohydrate concentration in food can also be determined by physical methods that rely on the physicochemical properties of carbohydrates in foods. These methods include polarimetry, refractive index, infrared, and density methods (Hodge and Hofreiter, 1962; Letzelter et al., 1995; Caprita et al., 2014). More sophisticated methods for carbohydrate analysis involve immunoassays. The major advantages of immunoassays are that they are extremely sensitive, specific, easy to use, and rapid. However, their expensive nature is usually a deterrent for use (Ibrahim, 1986).

Starch, the most common digestible polysaccharide found in foods, consists of the linear amylose and the extensively branched amylopectin (Wu et al., 2014). The starch content of foods cannot be determined directly. Thus, extraction and solubilization method is usually the first step. Following extraction, several methods such as enzymatic, iodine, and physical can be used to determine the total starch concentration (Wu et al., 2014a). Some starch is resistant to extraction or breaking down, which may be a disadvantage in this method.

Dietary fiber, which is also a component of carbohydrates, is defined as plant polysaccharides that are indigestible by humans, including lignin (Burton-Freeman,

2000). Adequate dietary fiber intake is beneficial to good health as there is evidence that it helps protect against conditions such as cardiovascular diseases, constipation, and colon cancer. It consists mainly of cellulose, hemicellulose, pectin, hydrocolloids, and lignin. Most of the techniques that have been developed for fiber analysis in foods mimic the processes that occur in the human digestive system (Prosky et al., 1988). Before fiber analysis is conducted, procedures such as lipid, starch, and protein removal, as well as selective precipitation of fibers are performed. Gravimetrical methods of fiber determination include the crude fiber method and the total, insoluble, and soluble fiber methods. The crude fiber method relies on acid hydrolysis and is, therefore, less accurate as it does not account for hemicellulose and pectins that require an additional alkali for hydrolysis. The total, soluble, and insoluble fiber method is commonly used in the food industry to determine the fiber content of many foods (AOAC, 2016). The sample is fractionated by selective precipitation followed by mass determination of the fraction of interest. The method also uses a variety of enzymes to isolate the fiber fraction of interest. However, this method tends to overestimate the fiber content of foods containing high concentrations of simple sugars. Chemical methods of determining the fiber content of foods, such as the Englyst–Cummings method, are based on calculating the sum of all nonstarch monosaccharides plus the remaining lignin once all the digestible carbohydrates have been removed (Englyst, 1989). Chemical methods are useful in determining the total, soluble, and insoluble fiber contents of food; however, they do not account for the lignin content, which cannot be acid hydrolyzed. As most foods contain very small quantities of lignin, this method could be used but when foods contain large quantities of lignin, a gravimetric method could be more useful.

11.2.3 Fat

Lipids can be defined as a group of substances that, in general, are sparingly soluble in water but are soluble in ether, chloroform, or other organic solvents (Fahy et al., 2011). They mostly consist of triacylglycerols, which are either referred to as fat (solid at room temperature) or lipids (liquid at room temperature) and as a result, the terms lipids, fats, and oils are often used interchangeably. To differentiate lipids in foods, they are usually classified into simple lipids (fats and waxes), compound lipids (phospholipids, cerebrosides, and sphingolipids), and derived lipids (fatty acids, fat-soluble vitamins, long-chain alcohols, sterols, and hydrocarbons) (Sud et al., 2012). Fats provide the highest source of energy for the body; generally, 9 kcal of energy is released per gram of fat consumed (Rolls, 2009). In comparison, proteins or carbohydrates provide only 4 kcal energy per gram consumed (Blatt et al., 2011).

Fats also play very important roles as structural components of the cell membrane, carrier of fat-soluble vitamins, and other biological functions such as immunity, blood clotting, inflammation, and wound healing (van Itallie, 1957; Hubler and Kennedy 2016). Dietary fats also play several critical roles in the normal functioning of the central nervous system in the brain and associated disorders. For example, there is some evidence that supplementation with n-3 polyunsaturated fatty acids (eicosapentaenoic acid and docosahexaenoic acid) in the diet could be useful for the treatment of conditions such as Huntington's, depression, antisocial behavior, bipolar disorder, aggression, cognitive decline, schizophrenia, age-related macular degeneration, and Alzheimer's (Crawford et al., 1976). According to the Food and Agriculture Organization (FAO), the recommendations for fat and fatty acids intake for adults are as follows, 20–35% total fat, 10% saturated fatty acids, and less than 1% transfatty acids of the daily energy intake (FAO, 2008). Table 11.3 shows the fat content of selected foods.

Table 11.3: % Fat content of selected foods.

Food	% Fat (wet weight basis)
Rice, white, raw	0.7
Sorghum	3.3
Macaroni, dry, enriched	1.5
Milk, reduced fat, fluid, 2%	2
Cheddar cheese	33.1
Yogurt, plain, whole milk	3.2
Lard, shortening, oils	100
Butter with salt	81.1
Mayonnaise, soybean oil, with salt	79.4
Apples, raw with skin	0.2
Avocados, raw	14.7
Sweet corn, yellow, raw	1.2
Soybeans, mature seeds, raw	19.9
Separable lean and fat beef	5.0
Chicken breast	1.2
Pork, fresh, loin, whole, raw	12.6
Pork, fresh, loin, whole, raw	2.3
Almonds, dried, unblanched, dry roasted	52.8
Whole fresh raw eggs	10.0

Adapted from Nielsen (2010).

Despite the health benefits of fat intake toward human health and diseases, many conditions such as obesity, coronary heart diseases, and even certain types of cancers have been attributed to the intake of specific fat types (Das, 2006). Saturated fatty acids have been shown to increase the serum cholesterol levels and therefore increase the risk of coronary heart diseases (Suresh and Das, 2006). Intake of transfats has also been attributed to a reduction of the favorable high-density lipoprotein and increase of low-density lipoprotein leading to inflammation, which is associated with heart disease, stroke, diabetes, as well as insulin insensitivity (Mozaffarian et al., 2006). Monounsaturated fatty acids have been shown to reduce the levels of oxidized low-density lipoprotein, which may lead to atherosclerosis (Parthasarathy et al., 2010). Intake of polyunsaturated fatty acids such as *n*-6 and *n*-3, which are mostly plant derived in the correct ratios, has been shown to have significant health benefits (Das, 2008).

The analysis of fat composition or content of different foods requires knowledge of the chemistry and structure as well as the occurrence of the principal lipid classes and their different constituents. Before the actual analysis is conducted, several preanalysis steps (drying, particle size reduction, and acid hydrolysis) are carried out to remove water, reduce particle size, and separate the lipids from bound carbohydrates and proteins (AOCS, 2009). Methods for fat extraction in foods may include the use of a solvent or eliminate the use of a solvent. Continuous solvent extraction methods, such as the Goldfish method, give faster and more efficient extraction compared to semicontinuous extraction methods (AOAC International, 2016). However, their disadvantage is that it may cause incomplete extraction due to channeling. Semicontinuous solvent extraction methods such as the Soxhlet method do not cause channeling but take longer to complete (Li et al., 2014). Discontinuous solvent extraction methods such as the Mojonnier method do not require drying of the sample before analysis and are suitable for analysis of the fat content of dairy products such as milk (Lunder, 1971).

Methanol and chloroform combination has been commonly used to extract lipids. Various methanol–chloroform extraction procedures are often rapid, well suited to low-fat containing samples, and can be used to generate lipid samples suitable for subsequent fatty acid compositional analysis such as GC-MS (Bligh and Dyer, 1959). However, chloroform and methanol are highly toxic compounds; thus, the extraction must be done in well-ventilated areas. Total fat analysis for nutrition labeling can also be conducted by GC (Golay et al., 2007). Fats are usually extracted from foods by hydrolytic methods. Following the extraction, an internal standard is added before transesterification is done and fats are quantitatively measured by capillary GC against the internal standard. Nonsolvent wet extraction methods for fat analysis in foods also exist. Methods such as the Babcock and Gerber method have been applied for fat analysis in milk and are best suited for wet samples (Robertson and Black, 1949). Similar to protein analysis in foods, instruments such as the NMR, infrared, and Foss-Let are also available for fat determination of specific foods (Cronin and McKenzie, 1990). These methods are very simple, reproducible, fast, and useful for quality control. However, they are only available for fat determination of specific foods, require correlation to an internal standard, and are expensive to run and maintain.

11.2.4 Vitamins

Vitamins are defined as relatively low-molecular-weight organic compounds that are required by humans in small quantities for normal metabolism. It does not provide energy. Plants and microorganisms mainly synthesize vitamins, whereas animals cannot synthesize sufficient quantities; thus, animals obtain most of their vitamin supply from the diet (Mora et al., 2008). Vitamins are grouped into fat-soluble (vitamins A, D, E, and K) and water-soluble vitamins (vitamins C and B-complex). Vitamins and their metabolites are essential for driving a variety of physiological processes such as blood coagulants (vitamin K), hormones (vitamins A and D), coenzymes (vitamins B_1, B_2, B_3, B_5, B_6, and B_7), and antioxidants (vitamins E and C), as regulators of tissue growth and differentiation (vitamin B_9 and B_{12}), in embryonic development as well as calcium metabolism (Uribe et al., 2017). In addition, vitamins have been shown to play an important role in both innate and adaptive immune responses (Bhalla et al., 1986) and in neurological functioning and central metabolism (Kennedy, 2016). No single food contains all the vitamins. Most foods except for sucrose, refined grains, and alcoholic beverages, provide variable quantity and a number of vitamins. The RDA, selected food sources and deficiency syndromes of vitamins are presented in Table 11.4.

Table 11.4: Vitamins – food sources, deficiency syndromes, RDA.

Vitamin	RDA	Food sources	Deficiency syndrome
Vitamin A (retinol)	800 µg/day	Liver, dairy products, meat, fish, butter, leafy vegetables, and egg yolk	Xerophthalmia, night blindness, keratinization of the corneal epithelium, dry mucous membranes
Vitamin D (cholecalciferol)	15 µg/day	Fish liver oils, egg yolk, fortified dairy products, fatty fish, and fortified cereals	Rickets (in children), osteomalacia (in adults), and osteoporosis
Vitamin E (α-tocopherol)	15 mg/day	Vegetable oils, fruits, unprocessed cereal grains, vegetables, meats, and nuts	Peripheral neuropathy, spinocerebellar ataxia, and pigmentary retinopathy.
Vitamin K (phylloquinone)	90–120 µg/day	Green vegetables, cabbage, and margarine	Hemorrhages
Vitamin B_1 (thiamine)	1.2 mg/day	Enriched, bread, mixed foods whose main ingredient is grain, cereals, liver, pork, potatoes, and eggs	Beri beri, Wernicke-Korsakoff syndrome, polyneuritis, heart failure, anorexia, and gastric atony
Vitamin B_2 (riboflavin)	1.2 mg/day	Organ meats, milk, bread products and fortified cereals	Oral-ocular-genital syndrome
Vitamin B_3 (niacin)	15 mg/day	Meat, fortified cereals poultry, whole grain breads and bread products, fish, and mushrooms	Pellagra (dermatitis, dementia and diarrhea)

Table 11.4 (continued)

Vitamin	RDA	Food sources	Deficiency syndrome
Vitamin B$_5$ (pantothenic acid)	5 mg/day	Chicken, potatoes, cereals, tomato, beef, products, liver, kidney, yeast, oats, egg yolk, and whole grains	Hypertension, gastrointestinal disturbances, muscular cramps, hypersensitivity, neurological disorders
Vitamin B$_6$ (pyridoxine)	1.3 mg/day	Fortified soy-based meat substitutes, fortified cereals, organ meats, and bananas	Neuropathy (paresthesia) Epileptiform convulsions in infants. Hypochromic anemia, seborrheic dermatitis, and glossitis
Vitamin B$_7$ (biotin)	30 μg/day	Liver, pork, vegetables, and egg yolk	Dermatitis, conjunctivitis, alopecia, and abnormalities of the CNS (depression, hallucinations and paresthesia)
Vitamin B$_9$ (folic acid)	400 μg/day	Enriched cereal grains, enriched and whole-grain bread, liver, dark leafy vegetables, and nuts	Macrocytic anemia
Vitamin B$_{12}$ (cobalamin)	2.4 μg/day	Fortified cereals, meat, fish, and poultry	Hematological (macrocytic anemia), paresthesia
Vitamin C (ascorbic acid)	80 mg/day	Citrus fruits, tomatoes, potatoes, cauliflower, broccoli, Brussel sprouts strawberries, spinach, and cabbage	Scurvy. Sjögren syndrome, gum inflammation, dyspnea, edema, and fatigue. Bone abnormalities, hemorrhagic symptoms, and anemia

RDA: recommended daily allowance.
Adapted from Uribe et al. (2017).

Precise vitamin analysis of food is important for determining animal and human nutritional requirements (Pegg et al., 2010). There are a variety of methods that have been developed over the years for the analysis of vitamins in foods. Because of their relatively low quantity in foods, the vitamin content is often expressed as mg or μg per food serving. However, vitamins can also be expressed as international units, United States Pharmacopeia units, and % daily value. Methods for measuring vitamin content of foods are divided into mainly three groups: (1) bioassays involving humans and animals; (2) microbiological assays making use of bacteria, protozoan organisms, and yeast; (3) and physicochemical assays such as fluorometric, spectrophotometric, chromatographic, immunological, enzymatic, and radiometric (Pegg et al., 2010). Correct sampling and subsampling, as well as the preparation of a homogeneous sample, are very important aspects of vitamin analysis.

Vitamin analysis in foods often begins with an extraction step using heat or extraction compounds such as acid, alkali, solvents, and enzymes specific for each vitamin (Zhang et al., 2018). Following extraction, a specific assay method is used to determine the vitamin content. Bioassays are currently only used for the determination of a limited number of vitamins such as A, B_{12}, and D (Newton et al., 1979). Bioassays are limited to animals because they usually involve sacrificing the test organisms. Microbiological assays are only applicable to the analysis of water-soluble vitamins, but these methods are very specific and very sensitive for each vitamin (Tanner and Barnett, 1986). The growth of test organisms can be measured in terms of optical density, respiration, acid production, or gravimetry. However, these methods can be time-consuming and laborious although the use of microtiter plates results in saving time. Physicochemical methods such as HPLC methods are relatively simple, accurate, and precise and therefore, much more favored. HPLC methods also allow multianalyte procedures that save on time (Giorgi et al., 2012). Moreover, when the HPLC method is used in conjunction with MS, this method becomes very powerful and accurate in determining the content of most vitamins with relative ease.

11.2.5 Ash

Ash refers to the inorganic residue such as minerals remaining after either ignition or complete oxidation of organic matter in a food compound (Ismail, 2017). The amount and composition of ash in food is dependent on the nature of the food incinerated as well as the method used for ashing. Ash in food is composed of a variety of minerals such as calcium and iron that are involved in many structural and regulatory functions in the body (Prashanth et al., 2015). Minerals required by the body in smaller quantities are called **trace elements** (iron, zinc, copper, etc.), whereas those required in larger quantities are known as **macrominerals** (calcium, sodium, magnesium, etc.). Table 11.5 gives the functional attributes, deficiency syndromes, food sources, and the reference daily intake (RDI) of selected minerals of health significance.

Many methods have been developed for ash analysis in foods (Liu, 2019). Determining the ash content may be crucial for proximate analysis, for nutritional evaluation and labeling, as well as for the analysis of specific elements. Certain foods are higher in mineral content than others; thus, the determination of their ash content becomes important (Wolf and Hamly, 1984). Before ashing, samples are usually dried and ground or milled. For specific element analysis, contamination by the milling or grinding tool due to friction should be avoided. Two major types of ashing are dry and wet ashing (Harris and Marshall, 2017). These methods can both be accomplished by conventional means or using microwave systems.

Dry ashing is typically incineration in a muffle furnace at high temperatures typically, 525 °C or higher. Its major advantages are that it is a safe method that requires no blank subtraction or added reagents. Also, a large number of samples can be analyzed at the same time and the resultant ash is suitable for use in other

Table 11.5: Role of minerals in human health, deficiency syndromes, food sources, and the daily requirement.

Minerals	Functional attributes	Deficiency syndrome	Food sources	RDI
Calcium	Involved in bone and teeth formation and maintenance, muscle contraction and relaxation, nerve functioning, blood clotting, blood pressure regulation, immune system health	Osteopenia, osteoporosis, eczema, psoriasis, extreme fatigue, muscle aches, cramps, spasms, dental changes, cataracts	Milk, milk products, fish with bones broccoli, mustard, legumes	1 g
Magnesium	Incorporated into bones, required for protein synthesis, muscle contraction, nerve transmission, immune system	Muscle spasm, depression, confusion, agitation, hallucinations, weakness, neuromuscular irritability (tremor), athetoid movements, and convulsive seizures	Nuts, seeds, legumes, leafy, vegetables, seafood, "hard" drinking water	0.3–0.42 g
Potassium	Involved in proper fluid balance, nerve transmission, and muscle contraction	Weakness and fatigue, muscle cramps, muscle aches and stiffness, heart palpitations, breathing difficulties, digestive symptoms	Meat, milk, fresh fruits, and vegetables, whole grains, legumes	3.5–4.7 g
Sodium	Involved in proper fluid balance, nerve transmission, and muscle contraction	Fatigue, headache, nausea, vomiting, muscle cramps or spasms, confusion, irritability	Table salt, soy sauce, processed foods, milk, bread, vegetables	2.3 g
Phosphorous	Healthy bones and teeth, found in every cell, involved in hemostasis balance	Bone diseases such as rickets, osteomalacia, and osteoporosis	Meat, fish, poultry, eggs, milk, processed foods	700 mg
Chloride	Proper fluid balance and stomach acid	Hypochloremia	Table salt, processed foods, milk, meat, vegetables	3.6 g

Table 11.5 (continued)

Minerals	Functional attributes	Deficiency syndrome	Food sources	RDI
Sulfur	Component of protein molecules, synthesis of connective tissue	Acne, arthritis, brittle nails and hair, convulsions, depression, memory loss, gastrointestinal issues, rashes, slow wound healing, chronic fatigue	Meat, fish, eggs, milk, nuts, legumes	910 mg
Iron	Facilitates oxygen transport by Hb, boosts immune defenses, involved in the synthesis of collagen and elastin for skin, hair, and nails	Anemia, microcytic hypochromic RBCs, tiredness, achlorhydria, Plummer–Vinson syndrome, atrophy of epithelium, impaired attention, irritability, and lowered memory	Grains, liver, nuts, poultry, seafoods, eggs, shellfish, fish, legumes	1–2 mg
Zinc	Cell proliferation, differentiation, and metabolic activity, normal spermatogenesis and maturation, genomic integrity of sperm, normal organogenesis, proper functioning of neurotransmitters, proper development of thymus, proper epithelialization in wound healing, taste sensation, and secretion of pancreas and gastric enzymes	Growth retardation, alopecia, dermatitis, immunological dysfunction, psychological disturbances, gonadal atrophy, faulty spermatogenesis, congenital malformation, keratogenesis, taste disorders, and delayed wound healing	Meat, shellfish, legumes, seeds, nuts, dairy products, eggs, whole grain cereals	15–20 mg
Copper	Play roles in normal enzyme functioning, Hb synthesis, connective tissue metabolism, bone development, synthesis of tryptophan, and iron transport	Hypochromic anemia, neutropenia, hypopigmentation of hair and skin, abnormal bone formation with skeletal fragility and osteoporosis, joint pain, lowered immunity, vascular abnormalities, and steely hair	Liver, shellfish, dried fruits, milk and milk products	2–5 mg

Table 11.5 (continued)

Minerals	Functional attributes	Deficiency syndrome	Food sources	RDI
Iodine	A constituent of thyroid hormones, proper functioning of the parathyroid glands, promotes general growth and development, aid in metabolism	Thyroid malfunction, extreme fatigue, slowing of both physical and mental processes, weight gain, facial puffiness, constipation, and lethargy	Dairy products, iodized salt, eggs, seaweed, kidney beans	0.15 mg

Hb: hemoglobin; RBCs: red blood cells; RDI: reference daily intake.
Adapted from Prashanth et al. (2015).

analyses (Harris and Marshall, 2017). However, this method requires longer periods to complete and uses expensive equipment. In addition, volatile elements such as copper, zinc, and nickel are lost during analysis. **Wet ashing** is a preparatory technique for specific mineral analysis and metallic poisons. Advantages of wet ashing are that minerals usually stay in solution and the lower temperature used to ensure that there is little-to-no loss from volatilization and sample oxidation is limited. However, this method requires constant attention, uses corrosive reagents, and only a few samples can be analyzed at a time (Harris and Marshall, 2017). The advent of microwave technology in food analyses has greatly reduced the time required for both analytical methods.

11.2.6 Water

Water comprises about 75% of the body weight in infants and 55% in adults. It is essential for processes such as cellular homeostasis and life (Wang et al., 1999). It is a major constituent of cells, tissues, and organs (Hoffmann et al., 2009). Water represents a multifunctional and critical nutrient whose absence has lethal consequences. It acts as a building material in all the cells, as a solvent, reaction medium, reactant, and reaction product in many biochemical processes. Water is also important in cellular homeostasis as a carrier of nutrients and waste and plays a very crucial role in thermoregulation. It is also a major component of lubricants and shock- absorbing compounds in the body (Jéquier and Constant, 2010).

Dehydration or lack of water in the body can cause dry mouth, sunken eyes, poor skin turgor, cold hands and feet, weak and rapid pulse, rapid and shallow breathing, confusion, exhaustion, and coma. Moreover, dehydration is also related to several chronic diseases such as urolithiasis, hyperglycemia, ketoacidosis, hypertension, stroke, dental disease, gallstone, bladder and colon cancer, as well as urinary tract infections (Popkin et al., 2010). Apart from drinking water, an additional supply of water to the body is through food intake and oxidation of macronutrients. The RDI of water is between 2.0 and 3.7 L for both adult men and women (Krutzen et al., 1995). Table 11.6 shows the amount of water in some food samples.

Table 11.6: The average water content of selected food samples.

Food	% of water
Water	100
Fat-free milk, cantaloupe, watermelon, lettuce, cabbage, spinach, and pickles	90–99
Fruit juice, apples, oranges, carrots, pears, pineapple	80–89
Bananas, avocados, cottage cheese, ricotta cheese, potato (baked), shrimp	70–79
Pasta, legumes, ice cream, chicken breast	60–69
Ground beef, feta cheese, tenderloin steak (cooked)	50–59
Cheddar cheese, bagels, bread	30–39
Butter, margarine, raisins	10–19
Oils, sugars	0

Adapted from Popkin et al. (2010).

Water or moisture exists in food as free water, adsorbed water, or water of hydration. The methods used to determine its content in foods are dependent on the form in which it exists. Water analysis methods are grouped into **oven-drying methods** (forced draft oven, vacuum oven, microwave, infrared, etc.; Lewis et al., 2013), distillation procedures (direct and reflux distillations), **chemical methods** (Karl Fischer titration; Jones, 1980), and **physical methods** (dielectric technique, hydrometer, refractometry, infrared spectroscopy, freeze-drying; Huang et al., 2008). Oven-drying methods usually involve the removal of moisture from the sample, followed by the determination of weight of the remaining solids to calculate the moisture content. Although negligible, these methods may result in the loss of nonwater volatiles during drying. Distillation methods also involve separating the moisture from the solids, followed by a direct volume quantitation (Bradley, 2010). Chemical methods involve measuring the moisture content of food by direct correlation with the volume of titrant used. The dielectric method depends on the electrical properties of water, whereas hydrometric methods determine moisture content through its relationship to gravity (Kim et al., 2006). The basis of refractive index and near-infrared methods is refraction characteristics of water and absorption wavelength, respectively (Vera et al., 2019).

The nature of the food sample determines the suitable method of moisture analysis. For example, vacuum oven drying at reduced temperatures is suitable for foods that undergo chemical reactions at high temperatures (AOAC International, 2016). On the other hand, distillation techniques can be used to minimize volatilization and decomposition. Titration methods are more suitable for foods of very low moisture content or high in fats and sugars, whereas a pycnometer, hydrometer, and refractometer are more suitable for liquid samples with limited constituents (Bradley, 2010). For faster drying in quality control, oven drying is much more preferred although it could be less accurate. Physical methods are usually rapid, but they require reference to another less empirical method, and chemical methods are very specific to certain foods (Bradley, 2010).

11.3 Antinutritional factors

11.3.1 Occurrence and significance

Antinutritional factors or antinutrients are compounds that inhibit the utilization and absorption of nutrients from the digestive system and hence reduce nutrient intake and digestion, and may produce other adverse effects (Popova and Mihaylova, 2019). They decrease the body's ability to absorb essential nutrients and are mostly present in food of plant origin by normal metabolism and mechanisms, which exerts contrary to optimum nutrition (Akande et al., 2010). The antinutrients occur mainly in grains, legumes, beans, nuts, coffee, wine, and tea. They are also present in roots, fruits and leaves of certain plants, and vegetables including eggplant, tomatoes, and peppers (Popova and Mihaylova, 2019). Among other biological activities and inhibition of optimal utilization of nutrients, they are also used by plants for self-defense (Gemede and Ratta, 2014). Although they have been reported to be deleterious to health, they could also evidently be beneficial to human and animal health, if eaten moderately (Ugwu and Oranye, 2006; Gemede and Ratta, 2014; Popova and Mihaylova 2019). Huge amount of antinutrients in the body can result in rashes, nausea, headaches, bloating, and nutritional deficiencies (Essack et al., 2017; Popova and Mihaylova, 2019).

Most of these antinutrients, which are secondary metabolites, potentiate harmful biological responses and at the same time are implicated as beneficial in nutrition and as pharmacologically active agents (Oakenfull and Sidhu, 1989; Soetan 2008). Some of the antinutrients are highlighted as follows.

11.3.1.1 Protease inhibitors

The protease inhibitors are found in all cells and tissues of raw cereals and legumes, mostly soybeans. They bind to their target proteins reversibly or permanently. Their activities are associated with pancreatic hypertrophy, growth inhibition, and poor food utilization (Adeyemo and Onilude, 2015). Logsdon and Ji (2013) reported high secretion of digestive enzymes by the pancreas due to high level of protease inhibitors in an *in vivo* experiment. Protease inhibitors inhibit pepsin, trypsin, and other proteases in the gut, hence inhibiting digestion and absorption of proteins and amino acids.

11.3.1.2 Oxalate

Oxalates are salts (soluble or insoluble) formed from oxalic acid and are mostly found in leafy vegetables or synthesized in the body (Akwaowo et al., 2000). Foods rich in oxalates include broccoli, cauliflower, radishes, spinach, beets, chocolate, rhubarb, black peppernuts, berries (blue berries, black berries), and beans (Mamboleo, 2015). Oxalic acid forms strong bonds with different minerals including calcium, potassium, sodium, and magnesium, which result in the formation of oxalate salts. Sodium and potassium oxalates are soluble, whereas calcium oxalates are essentially insoluble and have harmful effects on human nutrition and health as they can solidify in the kidneys or urinary tract forming kidney stones when highly accumulated and the acid is expelled in the urine (Nachbar et al., 2000; Gemede and Ratta, 2014; Olawoye and Gbadamosi, 2017). Oxalates also bind with nutrients in the gastrointestinal tract, making them inaccessible to the body. Nutritional deficits as well as plain irritation to the lining of the gut are possible when food containing high quantities of oxalate is eaten (Gemede and Ratta, 2014).

11.3.1.3 Saponin

Some saponins are toxic, while others are beneficial to human health. They are toxic in high concentrations and known as antinutrient components, due to their adverse effects such as ingrowth loss, ability to decrease nutrient absorption by inhibiting enzyme activity, and protein digestibility and their ability to bind with nutrients such as zinc (Gemede and Ratta, 2014; Parca et al., 2018; Popova and Mihaylova, 2019). They have been reported to be the causative agent of leaky gut because of their ability to inhibit the uptake of minerals and vitamins in the gut. Despite these effects, their activity such as hypocholesterolemic, immunostimulatory, and anticarcinogenic agents cannot be overemphasized (Popova and Mihaylova, 2019).

11.3.1.4 Phytates

Phytates occur as phytin or phytate salt in nuts, seeds, legumes, and grains. Their occurrence may affect bioavailability of minerals, functionality, solubility, and digestibility of proteins and carbohydrates (Sakamoto et al., 2014). Phosphorus stored as phytate can be unlocked by the digestive enzyme, phytase. When phytase is absent, phytate binds and inhibits the absorption of magnesium, iron, zinc, and calcium (Akond et al., 2011). This results in reduced bioavailability of minerals as a result of increased insoluble salts that are inefficiently used by the gastrointestinal tract. They can hinder digestive enzymes such as trypsin, pepsin, and amylase (Kumar et al., 2010). Mineral deficiency caused by high level of consumption of phytate-containing foods depends on the other foods jointly consumed. Associated intake of phytate is a cause of concern in areas of the world where cereal proteins are mostly a dietary factor (Mueller, 2001).

11.3.1.5 Tannins

These are astringent, bitter group of antioxidant polyphenolic compound found in foods and beverages. They precipitate proteins and various organic compounds such as alkaloids and amino acids (Redden et al., 2005). Tannins inhibit the uptake of protein or digestive enzymes and increase fecal nitrogen, hence decreasing protein digestibility. They hinder the activities of lipase, amylase, trypsin, chemotrypsin; reduce the protein value of foods; and affect dietary iron uptake (Felix and Mello, 2000). Microbial enzyme activities and intestinal digestion may be depressed due to high tannin concentration in a diet (Aletor, 2005). Tannins interfere with the digestion of various nutrients, hence preventing the body from absorbing valuable bioavailable substances (Ertop and Bektaş, 2018).

11.3.1.6 Lectins

Lectins, also known as phytohemagglutinins, are glycoproteins found in wheat, beans, quinoa, peas, and some certain oil seeds (Peumans et al., 2001; Akande et al., 2010). They are characterized by their ability to bind with carbohydrates, especially animal cell carbohydrates and their ability to bind directly to the intestinal mucosa (Akande et al., 2010; Popova and Mihaylova, 2019). They travel all over the body, binding with the enterocytes, hence hindering the uptake and transportation of nutrients during digestion, which causes epithelial lesions within the intestine (Akande et al., 2010). They can also cause acne, inflammation, migraines, joint pains, and leaky gut syndrome when consumed in large quantity (Himansha and Sarathi, 2012).

11.3.1.7 Cyanides

Cyanides and specifically **hydrogen cyanide** (HCN) are produced from cyanogenic glycosides (phytoanticipins) in plants when consumed (Gemede and Ratta, 2014). It interferes with some amino acids and absorption of supplementary nutrients, leading to depressed growth as well as interference with many enzyme systems. They cause neuropathy and intense toxicity, which could culminate into death (Röös et al., 2018; Mihrete., 2019).

Despite the nutrient inhibitory effect of antinutrients, some were reported to have noteworthy health benefits. For instance, tannins and saponins were reported to exhibit antiviral, antibacterial, and antiparasitic activities (Akiyama et al., 2001; Lü et al. 2004; Kolodziej and Kiderlen 2005). Saponins are reported to prevent osteoporosis, maintain liver functions, as well as platelet agglutination (Kao et al., 2008). Moreover, when consumed in low amount, phytate, lectins, saponins, and protease inhibitors have been reported to decrease cancer risk, plasma cholesterol or blood glucose, and triacylglycerol (Popova and Mihaylova 2019). Hence, antinutrients are appreciable tools for the management of diseases as long as the dosage intake is taken into consideration.

11.3.2 Ways of reducing antinutrients in food samples

The quality of food can be improved by removing undesirable components such as antinutrients. Techniques including cooking, fermentation, sprouting, soaking, and gamma radiation could be used in ameliorating or removing antinutrients from the food samples (Bains et al., 2014; Gupta and Nagar, 2014; Popova and Mihaylova, 2019).

11.3.2.1 Cooking

Antinutrients such as phytic acid, oxalic acid, protease inhibitors, and tannins were reported to be reduced in legumes and vegetables when subjected to heat during cooking mostly at temperatures less than the boiling point for 15 min (Chai and Liebman, 2005; Onwuka 2006; Fernando et al., 2012; Udousoro and Akpan, 2014). Although the cooking time required is influenced by the type of antinutrient, food plant, and the cooking method, lengthier cooking time enhances greater reductions of antinutrients.

11.3.2.2 Fermentation

This is an ancient method used in preserving food. Controlled fermentation is usually applied in food production, as accidentally fermented food is regarded as spoilt. Several researchers have reported the reduction of phytic acid, polyphenol, hydro cyanide, oxalate, protease inhibitor, and tannin in foods such as soyabean, cowpea, and sorghum using this method (Babalola and Giwa 2012; Ojokoh et al., 2013; Adeyemo et al., 2016; Ovaldez-González et al., 2018).

11.3.2.3 Sprouting

Sprouting, also known as germination, is a stage in the phase of plants' life, where they develop from seeds. Sprouting increases the availability of nutrients in grains, legumes, and seeds as well as reduction of antinutrients (Soetan, 2008). This method has been effectively reported in the reduction of phytate, protease inhibitor, amylase inhibitor, and tannins in various types of grains, legumes, amaranth seeds, and flax seeds (Vidal-Valverde et al., 2002; Dikshit and Ghadle, 2003; Kanensi et al., 2011; Kajla et al., 2017).

11.3.2.4 Soaking

This is an easy process of removing soluble antinutrients from food samples. As most of the antinutrients in foods are found in the skin, they tend to dissolve easily when soaked in water as many of the antinutrients are water soluble (Fernandes, 2010). Soaking has been reported to reduce antinutrients such as tannins phytate, calcium oxalate, protease inhibitors, and lectins in grains, nuts, seeds, legumes, and leafy vegetables (Agume et al., 2017; Devi et al., 2018; Shi et al., 2017). Specifically, while overnight soaking of legumes reduces antinutrients to an appreciable level, the effect of soaking depends on the type of legumes involved.

11.3.2.5 Gamma radiation

Gamma radiation at low doses were effectively reported in decreasing the level of antinutrients such as phytic acid, trypsin inhibitor, oligosaccharides, and tannins (Al-Kaisey et al., 2003; Osman et al., 2014). It could also be applied to minimize antinutrients of millet grains during postharvest (Mahmoud et al., 2015).

Although, each of the methods discussed could be highly effective in eliminating or reducing the level of antinutritional factors in food samples, a combination of various methods could be highly efficient and results in total degradation of antinutrients. While Valencia et al. (1999) reported the combination of soaking, sprouting, and fermentation in the reduction of phytate by 98% in quinoa, a 98–100% elimination of lectins, tannins, and protease inhibitors in pigeon peas by soaking and boiling was reported by Onwuka (2006).

11.4 Valorization of food wastes

Annually, nearly 1.3 billion tons of food is wasted directly after harvesting or discarded as a result of bad quality or been left over after cooking (Mirabella et al., 2014). Three billion people suffering from hunger across the globe can be fed by this quantity of waste foods (Thi et al., 2015). Roughly 1.4 billion hectares or nearly 30% agricultural land worldwide is used to cultivate food that is ultimately wasted (Mirabella et al., 2014). Food loss and wastage happen worldwide at different points from farm to fork, and this depend on the region and available technology that supports farming, food storage and processing. About 40% food loses are recorded to be prominent immediately after harvesting and during processing in developing countries according to FAO of the United Nations. Meanwhile, food wastage, as opposed to food losses, is more frequent in developed countries, where over 40% of food waste take place at stores and homes after purchase (Mirabella et al., 2014). Food such as milk and eggs, roots and tubers, cereals, fruits and vegetables, oilseed and pulses, and meat are all implicated in food losses. A recent study by FAO revealed that food is lost and wasted at every production processes from agricultural production to consumer levels (Mirabella et al., 2014). However, irrespective of where the food is lost or wasted, waste generation and its byproducts are having negative effect on the economic, environmental, and social sectors. From the economic viewpoint, the unfavorable influence is as a result of the costs involved in taking care of solid waste in landfills. Moreover, the management of huge quantity of various degradable materials also poses a challenge (Milica et al., 2018). From the environment point of view, food waste and loss promote greenhouse gas emissions (Mirabella et al., 2014). Majority of these biomaterials are not used and end up in municipal landfills where they pose serious environmental challenge as a result of microbial degradation and leachate production (Semih et al., 2015). In the social sector, food waste impacts the ethical and moral dimensions within the general concept of global food security as billions of people across the globe are hunger-stricken (Milica et al., 2018).

Frequently, strategies such as landspreading and landfilling, anaerobic digestion, incineration, animal feeding, and composting for dealing with the food wastes are elementary and offer a low economic and environmental value (Mirabella et al., 2014). However, food wastes comprise various compounds with a broad range of potential commercial applications, which make them exploitable raw material for valorization.

11.4. 1 Selected food wastes and their nutritional values

11.4. 1. 1 Mango

Mango (*Mangifera indica* L.) belongs to the family Anacardiaceae and bears fruits that ripen in 3–4 months after flowering. These fruits change in color from dark green to light green on maturity. In most cases, reaping of the fruit usually begins after few fruits drop during the early hours of the day. Harvested fruits are kept in gunny bags under shade. Whole fruits are categorized into two to three ranks corresponding to color, shape, and size and are collected in harvest boxes, whereas unripe, underdeveloped, bruised, and contaminated fruits are classified as wastes. Mango has a large percentage of bioactive compounds such as polyphenolic compound, lupeol, β-carotene, and ascorbic acid. A high significant amount of byproducts such as peels (13–16%) (Serna et al., 2015) and seeds (9.5–25%) are generated during the processing of mangos (Torres-Leon et al., 2017). Approximately 60% by weight of the fruit is estimated to be discarded after processing (O'Shea et al., 2012). Mango seed accounts for 58–80% carbohydrates, 6–13% protein with appreciable contents of essential amino acids, and lipids. The lipid content is made up of 6–16% oleic and stearic acids (Siaka, 2014). The protein content of the seeds is rich in majority of the essential amino acids at higher quantity more than those indicated by the FAO as quality protein. Despite report on the antinutritional properties of mango seeds by Sandhu and Lim (2008), these antinutrients can be reduced to improve the nutritional composition of the seed. Mango seed flour has been reported by Arogba (1997) to have the prospect of being used in food formulation for adults in several homes and could also be preferable in the preparation of infant foods. On the other hand, mango peels have large amount of soluble fiber ranging between 51.2% and 78.4% (Serna-Cock et al., 2016). The fibers are considered both vital additive needed for the preparation of functional foods and indispensable constituent of the human diet (Juarez Garcia et al., 2006).

11.4.1.2 Pineapple

Pineapple (*Ananas comosus*) is a tropical fruit that generates high byproducts during processing. These byproducts are often discarded as low-value product and comprise largely of peel and pomace, which account for approximately 30–35% of the whole fruit (Varzakas et al., 2016). The fact that it is difficult to recognize when a pineapple is ready to be reaped have continually resulted in the harvest of unripe pineapple that are later discarded as waste (Varzakas et al., 2016). For fresh consumption, fruits are reaped at "one- on two-eyes ripe," i.e., the bottom one or two eyes have turned to be colored. However, the cultivar and the location of the market determine which harvesting index is to be used. Fruits are usually harvested with a long knife locally referred to as "parang" and gathered in a basket that is normally carried on the back of the harvester (Nakthong et al., 2017). During harvesting, when the peduncle is cut, nearly 3–5 cm of it is left attached to the fruit. As injury to the fruit can affect the quality and shortens the storage life, it is essential to harvest and handle the fruit with care. Any unhealthy or abnormal fruits are rejected in the field. Damages caused as a result of poor handling, abrasion, or dropping, result in localized areas of softening and development of secondary microbial infection (Nakthong et al., 2017). Although the pineapple pomace is rich in total fiber (45.22%), its soluble dietary fiber contents have been reportedly low. Roughly 76% of pineapple byproduct (peel and heart) is fiber, of which soluble fraction comprises 0.8%, whereas the insoluble fraction is 0.2% (Martinez et al., 2012). It has been reported that pineapple bagasse has the potential to improve dietary fiber, which is nutritionally poor (Mabel et al., 2014). Pineapple stem is an agricultural waste that has great potential to serve as an alternative source of starch. The extracted starch has clearly unique properties compared to commercial corn, cassava, and rice starches (Nakthong et al., 2017).

11.4.1.3 Banana

Banana belongs to the Musaceae family and is normally harvested and carefully carried to a conveyor system, which transports the bananas very gently to the packing station. However, during harvesting and transportation, mechanical damage usually occurs and the affected fruits are usually sorted out later as waste (Babbar et al., 2011). Also, mechanical damages can occur during bananas commercialization resulting into quality loss and encouraging pathogen invasion (Memon et al., 2008). Only 12% of the total weight of the plants represent the fruit, which is the only edible portion of the plant; hence, creating a huge amount of agro-industrial waste such as peel, which is commonly used in industrial processes to manufacture new products (dried pulps, chips, wine, jams, sauces, and beer). Of the whole banana fruit, banana peels account for 30–40% of the total weight (Boudhrioua et al., 2003; Zuluaga et al., 2007; Happi Emaga et al., 2008; Bankar et al., 2010; Babbar et al., 2011). A huge amount of banana peel is accumulated and discarded like garbage after banana processing, as it is considered as a fruit waste. These residues are of critical pollution problem in the environment, as it is dispersed in the planting area or burnt. As a result of the rise in the number of banana processing industries, there is a great increase in the amount of banana wastes that is accumulation; hence, the development of appropriate strategies for reuse is necessary (Happi Emaga et al., 2007; Memon et al., 2008; Xu et al., 2015). Banana peel has a huge amount of phytochemicals such as carotenoids (β-carotene, α-carotene, and xanthophylls), flavonoids (anthocyanins, delphinidines, cyanidines), and catecholamines (dopamine and L-dopa) with antioxidant properties (Happi Emaga et al.,

2007; Babbar et al., 2011; Pelissari et al., 2014). Furthermore, banana peel also has micronutrients and minerals such as phosphorus, potassium, calcium, iron, sodium, magnesium, and zinc in appreciable amount. Other compounds of importance in banana peel are the essential amino acids and vitamins (Happi Emaga et al., 2007; Mohapatra et al., 2010; Babbar et al., 2011).

Banana peel contains a huge amount of dietary fiber, which accounts for approximately 50% dry mass of the peel and large amount of this fiber are insoluble dietary fiber. (Happi Emaga et al., 2007, 2008; Pelissari et al., 2014). Due to the health benefits of dietary fibers, it has gained more importance in recent years and as a result of this, the market is concentrating on the production of novel products with great levels of dietary fiber. Juarez Garcia et al. (2006) have reported that dietary fibers are implicated in the prevention and treatment of several diseases, mostly in the digestive tract. This is so because dietary fiber creates an environment and conditions that promote the growth of the beneficial intestinal flora (Juarez Garcia et al., 2006). Making use of fibers from food wastes and byproducts fortifies the final product with added value and at the same time contribute to the reduction of environmental pollution (Bilba et al., 2007).

11.4.1.4 Tomato

Tomato cultivation has a very essential social importance because a significant portion of the economically active populace is involved in tomato cultivation directly or indirectly, making it one of the most notable crops globally (Colle et al., 2010). The use of wooden crates or woven baskets by farmers during harvesting of tomatoes often cause injury to the fruit and most of the time, the delay in transporting tomatoes to the market as a result of the poor road infrastructure often increases postharvest losses (Sulaiman et al., 2014). The use of unsuitable methods of transportation to market such as motorbikes that vibrate and oscillate, during transportation often result in mechanical damage to the fruits; moreover, the lack of facilities where tomatoes can be kept at the recommended cooling temperature to minimize spoilage, has also contributed to the huge loss of tomatoes recorded. In addition, instead of staggering their planting, most farmers tend to produce tomatoes all at the same time, thus resulting in lack of readily available markets for selling of their produces and hence increasing postharvest losses (Sulaiman et al., 2014). Seed and skin are the major residues obtained from tomato processing. Nevertheless, whole fruits might also be rejected because of damage (Verrijssen et al., 2015). These wastes usually end up giving rise to environmental problems such as bad odors and pest promotion if not processed properly (Borel et al., 2015). In addition, this type of waste occurs usually after the end of the tomato harvest, meaning that just in a short period of time, a large quantity of waste is piled up (Colle et al., 2010). Tomato residues are rich in bioactive molecules, especially carotenoids, such as lycopene and β-carotene, which confers not only beneficial health properties due to their high antioxidant content but also high nutritional value (Borel et al., 2015). Moreover, they contain sugars, proteins, fiber, seed oil, and waxes (Colle et al., 2010). The health benefits of β-carotene (despite its low quantity in tomato) in disease prevention and as precursor of vitamin A have been reported (Martí et al., 2016; Wang et al., 2014). Vitamin A is an important micronutrient in diet, which enhances immunity and healthy development (Verrijssen et al., 2015). Lycopene, which represents approximately 80–90% of tomato, is the main pigment responsible for the red color of the fruit (Borel et al., 2015; Basuny, 2012).

11.4.1.5 Fish

Due to aquaculture and other fishery activities, fishes and fishery products are an essential source of food and thus a source of income in most economies of the world. About 50–70% remains from fish mostly comprise of head, viscera, skin, bones, or carcass, depending on their species and processing approaches from which only about 30% is intended to be reused (Wong et al., 2016). Based on this, efforts are presently being channeled to revalorize this category of byproducts to be used as a replacement for protein in animal diets and to formulate functional products. In 2013, there was a consistent rise in annual per capita consumption of fish by 18.8 kg in the developing countries (FAO, 2016b); however, the quantity of captured fishery remains quiet static since 1980s. Food security is being guaranteed through fishes and fishery to many countries as skills like aquaculture help in accomplishing this aim. One of the major expenditures in aquaculture is the formulation of feeds. This involves more than 50% of the total cost for fish culture industry (Malaweera and Wijesundara, 2014). The entire fish or their byproducts constitute portion where fish meal can be sought for. This portion contains a high nutritional value comprising approximately 70% protein, 10% minerals, 9% fat, and 8% of other components (Olsen and Hasan, 2012; Younis et al., 2017). Utilization of an alternative technology such as food valorization is a solution toward providing nutritionally balanced diets for livestock and improves the environmental management (Malaweera and Wijesundara, 2014). Aquatic byproducts are commonly preferable for the production of new products, which guarantee an improved quality of life made possible by the provision of bioactive substances such as protein hydrolysates and other products, which could be used as antihypertensive, antimicrobial, and antioxidants (Harnedy and FitzGerald, 2012). The existence of peptide chains, 2–20 amino acids, is encoded into protein sequences and discharged through various means, which include enzymatic hydrolysis, and account for the properties of these products (Korhonen and Pihlanto, 2006).

11.4.1.6 Others

Other significant wastes finding pharmacological and biotechnological applications include shells from palm fruits and groundnuts (Sadaa et al., 2013); peels from potato, yam, cassava, and orange (Mahmood et al., 1998; Kongkiattikajorn and Sornvoraweat, 2011; Mohammad et al., 2014; Tawakalitu et al., 2017); silk and cob from maize (Nangole et al., 1983; Sabiu et al., 2017); apple and orange seeds (Xiuzhu et al., 2007; Mohammad et al., 2014); rice husk (Humayatul et al., 2015); and so on. The beneficial use of these food wastes is mainly due to their nutritional and bioactive constituents.

11.4.2 Strategies in valorization of food wastes

Frequently used strategies for food wastes treatment are elementary and offer a low environmental and economic value. Food waste, however, comprises several compounds with a diverse range of probable commercial application, which makes them suitable raw materials for valorization (Mirabella et al., 2014). Strategies such as anaerobic digestion, animal feeding, incineration, composting, landfilling, and landspreading are some of the commonly used methods for dealing with food waste (Mirabella et al., 2014).

Furthermore, to achieve a circular economy in any industrial sector, two main strategies that aim at finding the most sustainable method to manage the remaining waste and reducing the amount of waste generated are needed. Food waste valorization, which is the process of converting food waste into more valuable products, is a suitable approach in the management of waste materials. This method gives rise to the production of a wide range of products, thereby boosting the sustainability of food production and helping to curb the negative effect, which other strategies have on the environment (Venkata et al., 2014). They have been shown to be an important feedstock that can be exploited to obtain many useful chemicals. Moreover, food manufacturing industries have already identified a good prospect in applying industrial symbiosis in valorizing food wastes (Mirabella et al., 2014). Therefore, food waste valorization has a greater potential to provide economic, environmental, and social benefits; many countries are already promoting strategies for food waste valorization. By using waste as resources, these strategies give support to the development of a circular economy in the food sector by closing the circle; an approach that is being encouraged by recent policies in Europe. Food waste valorization is not only beneficial to food manufacturing companies, in terms of the increase of economic value of the material itself, but also because of the fact that these materials are usually available in huge amounts.

11.4.3 Techniques in food waste recovery

11.4.3.1 Macroscopic pretreatment

The first step in the recovery of valuable compounds from food wastes involves a stage of macroscopic pretreatment, which modifies the waste material by preparing the food matrix to a form that is suitable for the later stages of the recovery process (Capson-Tojo et al., 2018; Lia et al., 2015). The step involves the moderation of enzyme activity, prevention and control of any microbial growth, as well as the modification of the phase content and properties such as the water, solid, and fats content (Capson-Tojo et al., 2018). Operations such as freeze drying, thermal or vacuum concentration, mechanical pressing, centrifugation, reduction of particle size, and microfiltration can be used for conventional macroscopic pretreatment (Misra et al., 2015). Recently, several innovative techniques have been investigated, improved, and modified from other fields for sterilization, drying, enzyme inactivation, and enhancing mass transfer in food and biomaterials. Other upcoming techniques of note includes radiofrequency drying, high-pressure processing, foam-mat drying, cold plasma technology, and electro-osmotic drying (Misra et al., 2015).

11.4.3.2 Macro- and micromolecules separation

Macro- and micromolecules separation is the second step in the recovery of compounds from food wastes. Several technologies such as isoelectric solubilization/precipitation, alcohol precipitation, ultrafiltration, and extrusion, have been implemented for this purpose (Chiranjib et al., 2015). For example, in the presence of an antisolvent agent, solutes are precipitated even if they are present at a very low concentrations within the substrate. In addition, the solubility of proteins or lipids might be reduced when pH value turns to their isoelectric point, causing their precipitation (Chiranjib et al., 2015). On the other hand, the separation of macro- and micromolecules based on their size is mostly done using ultrafiltration, which is a pressure-driven membrane separation technology (Krishnamurthy et al., 2015). Extrusion is a process of using shear energy and heating to physically or chemically modify the physicochemical properties of micromolecules (sugars, vitamins, polyphenols, or minerals) and macromolecules (polysaccharides, dietary fibers, or proteins). Following modification of the physical and chemical properties, macro- and micromolecules in biological substances are nowadays separated using emerging techniques such as ultrasound-assisted crystallization, colloidal gas aphrons, and pressurized microwave extraction (Krishnamurthy et al., 2015). Colloidal gas aphrons are surfactant-stabilized microbubbles that have been used for the selective separation of both macromolecules (e.g., protein) and micromolecules (e.g., polyphenols; Krishnamurthy et al., 2015). Furthermore, ultrasound-assisted crystallization has been used for the segregation of proteins, a process considered in terms of nucleation and crystal growth. In addition, an extraction technique based on microwaves has been recently combined with pressure to separate macromolecules such as pectin effectively from food waste (Krishnamurthy et al., 2015).

11.4.3.3 Extraction

Solvent extraction and steam distillation are the conventional extraction technologies adopted for the recovery of naturally occurring extracts from both plant and animal sources for ages (Milica et al., 2018). The utilization of these extracts has mainly been as food additives and medicines. Other effective and environment-friendly approaches such as enzyme-, ultrasound-, and microwave-supported extraction, and supercritical fluid extraction were recently developed (Juliana et al., 2015). In response to the possibility of the civilized society to reduce pollution, recovery of bioactive compounds from food wastes is another strategy that improves the values of such remains and simultaneously reduces their environmental impacts (Juliana et al., 2015). Conventional extraction approaches include a typically high temperature treatment (>100 °C) accompanied with a consequential risk of thermal deformation or alteration of the target molecules (Juliana et al., 2015). However, these approaches are extremely arduous and demands quite huge quantities of solvents (Milica et al., 2018). Nevertheless, the use of environment-friendly technologies has led researchers and the food industry to develop new alternative methods that can extract beneficial compounds from diverse sources and from food wastes of several origins (Francisco et al., 2015). These evolving technologies include ultrasound-assisted extraction, laser ablation, pulsed electric fields, high-voltage electrical discharge, and membrane-assisted extraction. The current approaches align with the theories of green chemistry and sustainability within the food industry (Francisco et al., 2015).

11.4.3.4 Purification and isolation

Isolation and purification of compounds isolated from waste being the fourth step of the **five-stage universal recovery process**, produce pure end products, which are cost-effective, more compact, and environmentally friendly (Lorenzo et al., 2015). Adsorption, electrodialysis, chromatography-based techniques, and nanofiltration are recognized and consolidated operations for the isolation and purification of valued biomolecules extracted from several organic residues are performed (Lorenzo et al., 2015). Emerging techniques for purification and isolation, such as aqueous two-phase system, magnetic fishing, and ion-exchange membrane chromatography, have recently been administered (Arijit et al., 2015). Based on their selective affinity, solutes are separated in the presence of magnetic particles and recovered by suitable buffer. The ion-exchange membrane chromatography is a method in which biomolecules (mostly protein) are separated based on the ionic charge (adsorption and elusion by pH of buffer), whereas the aqueous two-phase system is a liquid/liquid partitioning technique, comprising two types of phase-forming components (Arijit et al., 2015).

11.4.3.5 Product formation

The quality and technological aspects of a given food matrix are important factors that need to be considered during the application of micro- and macronutrients obtained from food wastes as additives in food products development. The quality aspect includes any possible interaction that might occur during processing and storage of the product, which could alter the general quality and stability of the product (Paola and Adem, 2015). Conventional methods such as microencapsulation and emulsification have been used for a long time as the final step of downstream processing in an attempt to avoid any reduction of food product functionality and stability (Paola and Adem, 2015). However, it is important to have the maximum protection of sensitive ingredients, along with targeted delivery systems in the right place at the right time during the formulation of products that are rich in nutraceuticals and bioactive components (Paola and Adem, 2015). Conventional methods have several drawbacks in this regard, whereas alternative innovative product development techniques such as nanoencapsulation, which is one of the most commonly used innovative methods could be applied at this stage of the universal recovery process (Seid et al., 2015). Nanoencapsulation has several advantages that include eradication of inconsistencies, protection of the sensitive bioactive food ingredients from unfavorable environmental conditions, solubilization, and masking of unpleasant taste and odor (Seid et al., 2015). Furthermore, nanocrystals created especially from food waste materials are a promising functional ingredient for both encapsulation purposes and development of biodegradable composite films. In addition, the pulsed fluidized bed agglomeration technique is among the most latest techniques for the encapsulation of bioactive compounds (Seid et al., 2015).

11.5 Conclusion

The locution "all for the love of nutrients," is an embodiment of the significance of balanced diets and nutrition in human health, and this has been appreciated since ancient times. Despite that adequate nutrition is central to well being and overall health, the human nutritional science is still in its infancy. Since the discovery of the essential nutrient requirements of human body, many mechanisms of the interactions of nutrients with genes, hormones, enzymes, and other food-based chemicals for maintenance of health, growth, and development have been unraveled.

Human health goes beyond the absence of disease. It also involves emotional, intellectual, spiritual, and social well-being. To achieve and maintain that, optimum nutrition is highly recommendable and is the mainstream of disease prevention. Foods consumed or ingested and their bioavailable nutrients remain the most essential factors influencing human health and well-being. Huge strides in alleviating mortality and morbidity caused by debilitating diseases such as cancer, diabetes, coronary heart diseases, and stroke can be achieved through nutritional education and knowledge application. Moreover, to manage the major nutritional challenges in this modern age, robust sources of composite nutrients are required, and this could be achieved from food wastes and byproducts that are vital due to the presence of macro- and micronutrients in adequate quantities. Globally and specifically in the developing countries, where hunger and malnutrition exist substantially, there is a huge potential of valorization of food wastes. The utilization of residues and byproducts recovered from food wastes to formulate unique food products automatically benefits the masses as well as their use as a raw material for animal feeds. Such creation of value-added products not only supports the infrastructural development, food packaging, and processing industries by contributing to waste reduction and improved economy through financial gains, but also aids in achievement of the Sustainable Development Goals-2 related to hunger. Equally important is the awareness of antinutrients that hinders the effective metabolism, absorption, and bioavailability of nutrients in foods and value-added products from food wastes. Upholding strategies to annihilating these antinutritional factors and ensuring adequately balanced diet are germane to sound health and overall well-being.

References

Adeyemo SM, Onilude AA, Olugbogi DO. Reduction of Anti-nutritional factors of sorghum by lactic acid bacteria isolated from Abacha - an African fermented staple. Front Sci 2016; 6(1): 25–30.

Adeyemo SM, Onilude AA. Enzymatic reduction of anti-nutritional factors in fermenting soybeans by lactobacillus plantarum isolates from fermenting cereals. Niger Food J 2015; 31(2): 84–0.

Agume AS, Njintang NY, Mbofung CM. Effect of soaking and roasting on the physicochemical and pasting properties of soybean flour. Foods 2017; 6(2): 12.

Akande, K.E., Doma, U.D., Agu, H.O. and Adamu, H.M., 2010. Major antinutrients found in plant protein sources: their effect on nutrition. Pakistan Journal of Nutrition, 9(8), pp.827–832.

Akiyama H, Fujii K, Yamasaki O, Oono T, Iwatsuki K. Antibacterial action of several tannins against Staphylococcus aureus. J Antimicrob Chemother 2001; 48(4): 487–91.

Akwaowo EU, Ndon BA, Etuk EU. Minerals and antinutrients in fluted pumpkin (Telfairia occidentalis Hook f.). Food Chem 2000; 70(2): 235–40.

Aletor VA. (2005). Anti-nutritional factors as nature's paradox in food and nutrition securities. Inaugural lecture series 15, delivered at The Federal University of Technology, Akure (FUTA).

Al-Kaisey MT, Alwan AKH, Mohammad MH, Saeed AH. Effect of gamma irradiation on antinutritional factors in broad bean. Radiat Phys Chem 2003; 67(3): 493–6.

Alhamdani, M. S. S., Schroder, C., Werner, J., Giese, N., Bauer, A. and Hoheisel, J. D. (2010). Single-step procedure for the isolation of proteins at near-native conditions from mammalian tissue for proteomic analysis on antibody microarrays. *Journal of Proteome Research*, 9(2), 963–971. https://doi.org/10.1021/pr900844q.

AOAC International. (2016). Official methods of analysis of AOAC International. In: G. W. J. Latimer (Ed.), *Association of Official Analysis Chemists International (Twentieth Edition)*. https://doi.org/10.3109/15563657608988149.

AOCS. (2009). *Official Methods and Recommended Practices of the AOCS*. Urbana, IL: AOCS.

Arijit, N., Ooi C. W., Sangita, B. and Chiranjib, B. (2015). *Emerging Purification and Isolation* (Chapter 12, pp. 273–292). Chania, Greece: Galanakis Laboratories, Academic Press.

Arogba, S. (1997). Physical, chemical and functional properties of nigerian mango (Mangifera indica) kernel and its processed flour. *Journal of Agricultural and Food Chemistry*, 73, 321–328.

Babalola RO, Giwa OE. Effect of fermentation on nutritional and anti-nutritional properties of fermenting Soybeans and the antagonistic effect of the fermenting organism on selected pathogens. Int Res JMicrobiol 2012; 3(10): 333–8.

Babbar, N., Oberoi, H. S., Uppal, D. S. and Patil, R. T. (2011). Total phenolic content and antioxidant capacity of extracts obtained from six important fruit residues. *Food Research International*, 44, 391–396.

Bains K, Uppal V, Kaur H. Optimization of germination time and heat treatments for enhanced availability of minerals from leguminous sprouts. J Food Sci Technol 2014; 51(5): 1016–20.

Bankar, A., Joshi, B., Ravi, A. and Zinjarde, S. (2010). Banana peel extract mediated novel route for the synthesis of silver nanoparticles. *Colloids and Surfaces A: Physicochemical and Engineering Aspects*, 368, 58–63.

Bao, J., Atkinson, F., Petocz, P., Willett, W. C. and Brand-Miller, J. C. (2011). Prediction of postprandial glycemia and insulinemia in lean, young, healthy adults: glycemic load compared with carbohydrate content alone. *The American Journal of Clinical Nutrition*, 93(5), 984–996. https://doi.org/10.3945/ajcn.110.005033.

Basuny, A. M. M. (2012). The anti-atherogenic effects of lycopene. In: S. Frank and G. Kostner (Eds.), *Lipoproteins- Role in Health and Diseases* (pp. 489–506). Rijeka: In Tech.

Bhalla, A. K., Amento, E. P. and Krane, S. M. (1986). Differential effects of 1,25-dihydroxyvitamin D3 on human lymphocytes and monocyte/macrophages: inhibition of interleukin-2 and augmentation of interleukin-1 production. *Cellular Immunology*, 98(2), 311–322. https://doi.org/10.1016/0008-8749(86)90291-1.

Bilba, K., Bilba, K., Arsene, M. and Ouensanga, A. (2007). Study of banana and coconut fibers-botanical composition, thermal degradation and textural observations. *Bioresource Technology*, 98, 58–68.

Blatt, A. D., Roe, L. S. and Rolls, B. J. (2011). Increasing the protein content of meals and its effect on daily energy intake. *Journal of the American Dietetic Association*, 111(2), 290–294. https://doi.org/10.1016/j.jada.2010.10.047.

Bligh, E. G. and Dyer, W. J. (1959). A rapid method of total lipid extraction and purification. *Canadian Journal of Biochemistry and Physiology*, 37(8),911–917. https://doi.org/10.1139/o59-099.

Bonn, G. (1985). High-performance liquid chromatography of carbohydrates, alcohols and diethylene glycol on ion-exchange resins. *Journal of Chromatography A*, 350, 381–387. https://doi.org/https://doi.org/10.1016/S0021-9673(01)93543-5.

Borel, P., Desmarchelier, C., Nowicki, M. and Bott, R. (2015). Lycopene bioavailability is associated with a combination of genetic variants. *Free Radical Biology and Medicine*, 83, 238–244.

Boudhrioua, N., Giampaoli, P. and Bonazzi, C. (2003). Changes in aromatic components of banana during ripening and air-drying. *LWT – Food Science and Technology*, 36, 633–642.

Bradbury, A. G. W. (1990). Gas chromatography of carbohydrates in food. In M. Gordon (Ed.), *Principles and Applications of Gas Chromatography in Food Analysis* (pp. 111–144). https://doi.org/10.1007/978-1-4613-0681-8_4.

Bradford, M. M. (1976). A rapid and sensitive method for the quantitation of microgram quantities of protein utilizing the principle of protein-dye binding. *Analytical Biochemistry*, 72, 248–254. https://doi.org/10.1006/abio.1976.9999.

Bradley, R. L. (2010). Moisture and total solids analysis. In: S. S. Nielsen (Ed.), *Food Analysis (Second Edition)* (pp. 85–104). https://doi.org/10.1007/978-1-4419-1478-1_6.

Burton-Freeman, B. (2000). Dietary fiber and energy regulation. *The Journal of Nutrition*, 130(2), 272S–275S. https://doi.org/10.1093/jn/130.2.272S.

Bylund, A., Zhang, J. X., Bergh, A., Damber, J. E., Widmark, A., Johansson, A., et al. (2000). Rye bran and soy protein delay growth and increase apoptosis of human LNCaP prostate adenocarcinoma in nude mice. *The Prostate*, 42(4),304–314. https://doi.org/10.1002/(sici)1097-0045(20000301)42:4<304::aid-pros8>3.0.co;2-z.

Caprita, R., Caprita, A. and Cretescu, I. (2014). Determination of lactose concentration in milk serum by refractometry and polarimetry. *Scientific Papers: Animal Science and Biotechnologies*, 47(1), 158–161.

Capson-Tojo, G., Ruiz, D., Rouez, M., Crest, M., Steyer, J. P., Bernet, N., Delgenès, J.-P. and Escudié, R. (2018). Comparison of different strategies for stabilizing food waste anaerobic digestion. In: *The 2nd International Conference on Anaerobic Digestion Technology Sustainable Alternative Bioenergy for a Stable Life, June 4–7, Chiang Mai, Thailand.*

Chai, W. and Liebman, M., 2005. Effect of different cooking methods on vegetable oxalate content. Journal of agricultural and food chemistry, 53(8), pp.3027–3030.

Chiranjib, B., Arijit, N., Alfredo, C., Reza, T. and Sudip, C. (2015). *Conventional Macro- and Micromolecules Separation* (Chapter 5, pp. 105–126). Chania, Greece: Galanakis Laboratories, Academic Press.

Colle, I. J., Lemmens, L., Tolesa, G. N., Van Buggenhout, S., De Vleeschouwer, K. and Van Loey, A. M. (2010). Lycopene degradation and isomerization kinetics during thermal processing of an olive oil/tomato emulsion. *Journal of Agricultural and Food Chemistry*, 58, 12784–12789.

Crawford, M. A., Casperd, N. M. and Sinclair, A. J. (1976). The long chain metabolites of linoleic avid linolenic acids in liver and brain in herbivores and carnivores. *Comparative Biochemistry and Physiology. B, Comparative Biochemistry*, 54(3), 395–401. https://doi.org/10.1016/0305-0491(76)90264-9.

Cronin, D. A. and McKenzie, K. (1990). A rapid method for the determination of fat in foodstuffs by infrared spectrometry. *Food Chemistry*, 35(1), 39–49. https://doi.org/https://doi.org/10.1016/0308-8146(90)90129-R.

Cruz, R. M. S., Khmelinskii, I. and Vieira, M. (2016). Methods in food analysis. In R. M. S. Cruz, I. Khmelinskii and M. Vieira (Eds.), *Methods in Food Analysis (First Edition)*. https://doi.org/https://doi.org/10.1201/b16964.

Das, U. N. (2006). Essential fatty acids – a review. *Current Pharmaceutical Biotechnology*, 7, 467–482. https://doi.org/http://dx.doi.org/10.2174/138920106779116856.

Das, U. N. (2008). Can essential fatty acids reduce the burden of disease(s)? *Lipids in Health and Disease*, 7(1), 9. https://doi.org/10.1186/1476-511X-7-9.

Devi, R., Chaudhary, C., Jain, V., Saxena, A.K. and Chawla, S., 2018. Effect of soaking on anti-nutritional factors in the sun-dried seeds of hybrid pigeon pea to enhance their nutrients bioavailability. Journal of Pharmacognosy and Phytochemistry, 7(2), pp.675–680.

Dikshit M, Ghadle M. Effect of sprouting on nutrients, antinutrients and in vitro digestibility of the MACS-13 soybean variety. Plant Foods Hum Nutr 2003; 58(3): 1–11.

Dubois, M., Gilles, K., Hamilton, J. K., Rebers, P. A. and Smith, F. (1951). A colorimetric method for the determination of sugars. *Nature*, 168(4265), 167. https://doi.org/10.1038/168167a0.

Englyst, H. (1989). Classification and measurement of plant polysaccharides. *Animal Feed Science and Technology*, 23(1), 27–42. https://doi.org/https://doi.org/10.1016/0377-8401(89)90087-4.

Ertop, M.H. and Bektaş, M., 2018. Enhancement of bioavailable micronutrients and reduction of antinutrients in foods with some processes. Food and health, 4(3), pp.159–165.

Essack H, Odhav B, Mellem JJ. Screening of traditional South African leafy vegetable for selected anti-nutrient factors before and after processing. Food Sci Technol 2017; 3: 462–1.

Fahy, E., Cotter, D., Sud, M. and Subramaniam, S. (2011). Lipid classification, structures and tools. *Biochimica et Biophysica Acta*, 1811(11), 637–647. https://doi.org/10.1016/j.bbalip.2011.06.009.

FAO. (2008). Fats and fatty acids in human nutrition. In: *Proceedings of the Joint FAO/WHO Expert Consultation, November 10–14,Geneva, Switzerland* (Vol. 91). https://doi.org/10.1159/000228993.

FAO. (2016a). *Reducing Food Losses and Waste in Sub-Saharan Africa*. http://www.fao.org/africa/news/en/Food%20waste/Reducing%20food%20losses%20and%waste%20in%20sub-Saharan%20Africa.html (accessed May 2, 2017).

FAO. (2016b). *The State of World Fisheries and Aquaculture 2016. Contributing to Food Security and Nutrition for All*. Rome: FAO.

Felix, J.P., and Mello, D. (2000). Farm Animal Metabolism and Nutrition. United Kingdom: CABI.

Fernando R, Pinto P, Pathmeswaran A. Goitrogenic food and prevalence of goitre in Sri Lanka. J Food Sci 2012; 41: 1076–81.

Fernandes, A. C., Nishida, W. and da Costa Proença, R. P. (2010). Influence of soaking on the nutritional quality of common beans (*Phaseolus vulgaris* L.) cooked with or without the soaking water: a review. *International Journal of Food Science & Technology*, 45(11), 2209–2218. https://doi.org/10.1111/j.1365-2621.2010.02395.

Francisco, J. B., Eduardo, P., Mladen, B., Ivan, N. P., Dimitar, A. D., Violaine, A. D., Marwen, M. and Isabelle S. (2015). *Emerging Extraction* (Chapter 11, pp. 249–272). Chania, Greece: Galanakis Laboratories, Academic Press.

Gemede, H.F. and Ratta, N., 2014. Antinutritional factors in plant foods: Potential health benefits and adverse effects. International Journal of Nutrition and Food Sciences, 3(4), pp.284–289.

Giorgi, M. G., Howland, K., Martin, C. and Bonner, A. B. (2012). A novel HPLC method for the concurrent analysis and quantitation of seven water-soluble vitamins in biological fluids (plasma and urine): a validation study and application. *The Scientific World Journal*, 2012, 359721. https://doi.org/10.1100/2012/359721.

Golay, P.-A., Dionisi, F., Hug, B., Giuffrida, F. and Destaillats, F. (2007). Direct quantification of fatty acids in dairy powders with special emphasis on trans fatty acid content. *Food Chemistry*, 101(3), 1115–1120. https://doi.org/https://doi.org/10.1016/j.foodchem.2006.03.011.

Gupta V, Nagar R. Minerals and antinutrients profile of rabadi after different traditional preparation methods. J Food Sci Technol 2014; 51 (8): 1617–21.

Ha, E. and Zemel, M. B. (2003). Functional properties of whey, whey components, and essential amino acids: mechanisms underlying health benefits for active people (review). *The Journal of Nutritional Biochemistry*, 14(5), 251–258. https://doi.org/10.1016/s0955-2863(03)00030-5.

Håkansson, A., Zhivotovsky, B., Orrenius, S., Sabharwal, H. and Svanborg, C. (1995). Apoptosis induced by a human milk protein. *Proceedings of the National Academy of Sciences of the United States of America*, 92(17), 8064–8068. https://doi.org/10.1073/pnas.92.17.8064.

Happi Emaga, T, Rado HA, Bernard W, Jean TT and Michel, P. (2007). Effects of the stage of maturation and varieties on the chemical composition of banana and plantain peels, Food Chemistry, 103(2): 590–60.

Happi Emaga, T., Robert, C., Ronkart, S. N., Wathelet, B. and Paquot, M. (2008). Dietary fibre components and pectin chemical features of peels during ripening in banana and plantain varieties. *Bioresource Technology* 99, 4346–4354.

Harnedy, P. A. and FitzGerald, R. J. (2012). Bioactive peptides from marine processing waste and shellfish: a review. *Journal of Functional Foods*, 4, 6–24.

Harris, G. and Marshall, M. (2017). Ash analysis. In: S. S. Nielsen (Ed.), *Food Analysis (Fifth Edition)* (pp. 287–297). https://doi.org/10.1007/978-3-319-45776-5_16.

Hartmann, R. and Meisel, H. (2007). Food-derived peptides with biological activity: from research to food applications. *Current Opinion in Biotechnology*, 18(2), 163–169. https://doi.org/ https://doi.org/10.1016/j.copbio.2007.01.013.

Hartree, E. F. (1972). Determination of protein: a modification of the Lowry method that gives a linear photometric response. *Analytical Biochemistry*, 48(2), 422–427. https://doi.org/ 10.1016/0003-2697(72)90094-2.

Heimo, S. and Günther, B. (1998). *Analytical Chemistry of Carbohydrates (Sixth Edition)*. Stuttgart: Thieme Medical Publishers.

Himansha, S. and Sarathi, S.P., 2012. Insight of Lectins–A review. Int'l J. Scientific and Eng. Research, 3(4), pp.1–9.

Hodge, J.E. and Hofreiter, B.T. (1962) Determination of reducing sugars and carbohydrates. In: Whistler, R.L. and Wolfrom, M.L., Eds., Methods in Carbohydrate Chemistry, Academic Press, New York, 380–394.

Hoffmann, E. K., Lambert, I. H. and Pedersen, S. F. (2009). Physiology of cell volume regulation in vertebrates. *Physiological Reviews*, 89(1), 193–277. https://doi.org/10.1152/ physrev.00037.2007.

Huang, H., Yu, H., Xu, H. and Ying, Y. (2008). Near infrared spectroscopy for on/in-line monitoring of quality in foods and beverages: A review. *Journal of Food Engineering*, 87(3), 303–313. https://doi.org/https://doi.org/10.1016/j.jfoodeng.2007.12.022.

Hubler, M. J. and Kennedy, A. J. (2016). Role of lipids in the metabolism and activation of immune cells. *The Journal of Nutritional Biochemistry*, 34, 1–7. https://doi.org/10.1016/j. jnutbio.2015.11.002.

Humayatul, U., Dadang, A. S., Mary, S. and Abdul Wahid, W. (2015). Analysis of chemical composition of rice husk used as absorber plates sea water into clean water. *ARPN Journal of Engineering and Applied Sciences*, 10(14), 6046–6050.

Ibrahim, G. (1986). A review of immunoassays and their application to salmonellae detection in foods. *Journal of Food Protection*, 49(4), 299–310.

Ismail, B. P. (2017). Ash content determination. In: S. S. Nielsen (Ed.), *Food Analysis (Fifth Edition)* (pp. 117–119). https://doi.org/10.1007/978-3-319-44127-6_11.

Jéquier, E. and Constant, F. (2010). Water as an essential nutrient: the physiological basis of hydration. *European Journal of Clinical Nutrition*, 64(2), 115–123. https://doi.org/10.1038/ ejcn.2009.111.

Jones, F. E. (1980). Determination of moisture in grain by automatic Karl Fischer titration. In: F. T. Corbin (Ed.), *World Soybean Research Conference II: Abstracts* (pp. 56, 57). Boulder, CO: Westview Press.

Juarez Garcia, E., Agama Acevedo, E., Sáyago Ayerdi, S. G., Rodríguez Ambriz, S. L. and Bello Pérez, L. A. (2006). Composition, digestibility and application in breadmaking of banana flour. *Plant Foods for Human Nutrition*, 61, 131–137.

Juliana, M., Prado, R. V., Isabel, C. N., Debien, M. A., de Almeida, M., Lia, N., Gerschenson, H., Bogegowda, S. and Smain, C. (2015). *Conventional Extraction* (Chapter 6, pp. 127–148). Chania, Greece: Galanakis Laboratories, Academic Press.

Kajla PS, Sharma A, Sood DR. Effect of germination on proximate principles, minerals and antinutrients of flaxseeds. Asian J Dairy Food Res 2017; 36(1): 52–7

Kanensi OJ, Ochola S, Gikonyo NK, Makokha A. Optimization of the period of steeping and germination for amaranth grain. J Agric FoodTech 2011; 1(6): 101–5.

Kao TH, Huang SC, Inbaraj BS, Chen BH. Determination of flavonoids and saponins in Gynostemma pentaphyllum (Thunb.) Makino by liquid chromatography-mass spectrometry. Anal Chim Acta 2008; 626(2): 200–11.

Kannan, A., Hettiarachchy, N. and Marshall, M. (2011). Food proteins and peptides as bioactive agents. In: N. Hettiarachchy (Ed.), *Bioactive Food Proteins and Peptides: Application in Human Health (First Edition)* (pp. 1–28). https://doi.org/10.1201/b11217-2.

Kazmaier, T., Roth, S., Zapp, J., Handing, M. and Kuhn, R. (1998). Quantitative analysis of malto-oligosaccharides by MALDI-TOF mass spectrometry, capillary electrophoresis and anion exchange chromatography. *Fresenius' Journal of Analytical Chemistry*, 361(5), 473–478. https://doi.org/10.1007/s002160050928.

Kennedy, D. O. (2016). B vitamins and the brain: mechanisms, dose and efficacy – a review. *Nutrients*, 8(2), 68. https://doi.org/10.3390/nu8020068.

Kim, K. B., Park, S. G., Kim, J. Y., Kim, J. H., Lee, C. J., Kim, M. S. and Choi, M. Y. (2006). Measurement of moisture content in powdered food using microwave free-space transmission technique. *Key Engineering Materials*, 321–323, 1196–1200. https://doi.org/10.4028/www.scientific.net/KEM.321-323.1196.

Kjeldahl, J. (1883). Neue Methode zur Bestimmung des Stickstoffs in organischen Körpern. *Zeitschrift Für Analytische Chemie*, 22(1), 366–382. https://doi.org/10.1007/BF01338151.

Kongkiattikajorn, J. and Sornvoraweat, B. (2011). Comparative study of bioethanol production from cassava peels by monoculture and co-culture of yeast. *Kasetsart Journal*, 45, 268–274.

Korhonen, H. and Pihlanto, A. (2006). Bioactive peptides: production and functionality. *International Dairy Journal*, 16, 945–960.

Kolodziej H, Kiderlen AF. Antileishmanial activity and immune modulatory effects of tannins and related compounds on Leishmanian parasitised RAW 264.7 cells. Phytochemistry 2005; 66(17): 2056–71.

Krishnamurthy, N. P., Giorgia, S., Paula, J., Misra, P. and Cullen, J. (2015). *Emerging Macro- and Micromolecules Separation* (Chapter 10, pp. 227–248). Chania, Greece: Galanakis Laboratories, Academic Press.

Krutzen, E., Back, S., Nilsson-Ehle, I., Nilsson-Ehle P., Armstrong, L. E., Hubbard, R. W., et al. (1995). Electrolytes and water nutrient. *Medicine and Science in Sports and Exercise*, 29(2), 276A. https://doi.org/10.3945/ajcn.115.129858.

Kumar, V., Sinha, A.K., Makkar, H.P. and Becker, K., 2010. Dietary roles of phytate and phytase in human nutrition: A review. Food Chemistry, 120(4), pp.945–959.

Latham, M. C. (1998). Human nutrition in the developing world. In: *FAO* (Vol. 24). https://doi.org/10.2307/2808139.

Letzelter, N. S., Wilson, R. H., Jones, A. D. and Sinnaeve, G. (1995). Quantitative determination of the composition of individual pea seeds by Fourier transform infrared photoacoustic spectroscopy. *Journal of the Science of Food and Agriculture*, 67(2), 239–245. https://doi.org/10.1002/jsfa.2740670215.

Lewis, M., Trabelsi, S., Nelson, S., Tollner, E. and Haidekker, M. (2013). An automated approach to peanut drying with real-time microwave monitoring of in-shell kernel moisture content.

Applied Engineering in Agriculture, 29(4), 583–593. https://doi.org/https://doi.org/http://dx.doi.org/10.13031/aea.29.9929.

Ley, S. H., Hamdy, O., Mohan, V. and Hu, F. B. (2014). Prevention and management of type 2 diabetes: dietary components and nutritional strategies. *Lancet (London, England)*, 383 (9933), 1999–2007. https://doi.org/10.1016/S0140-6736(14)60613-9.

Li, Y., Ghasemi Naghdi, F., Garg, S., Adarme-Vega, T. C., Thurecht, K. J., Ghafor, W. A., et al. (2014). A comparative study: the impact of different lipid extraction methods on current microalgal lipid research. *Microbial Cell Factories*, 13, 14. https://doi.org/10.1186/1475-2859-13-14.

Lia, N., Gerschenson, Q. D. and Alfredo, C. (2015). *Conventional Macroscopic Pretreatment* (Chapter 4). Chania, Greece: Galanakis Laboratories, Academic Press.

Liu, K. (2019). Effects of sample size, dry ashing temperature and duration on determination of ash content in algae and other biomass. *Algal Research*, 40, 101486. https://doi.org/https://doi.org/10.1016/j.algal.2019.101486.

Liener, I.E., (2003). Phytohemagglutinins: Their nutritional significance. J. Agric. Food Chem., 22: 17.

LIU, K. and MARKAKIS, P., 1987. Effect of maturity and processing on the trypsin inhibitor and oligosaccharides of soybeans. Journal of Food Science, 52(1), pp.222–223.

Logsdon CD, Ji B. The role of protein synthesis and digestive enzymes in acinar cell injury. Nat Rev.

Lorenzo, B., Dario, F., Herminia, D., Elena, F., Francisco, A. R. and Silvia, A. B. (2015). *Conventional Purification and Isolation* (pp. 149–172). Chania, Greece: Galanakis Laboratories, Academic Press.

Ludwig, D. S., Hu, F. B., Tappy, L. and Brand-Miller, J. (2018). Dietary carbohydrates: role of quality and quantity in chronic disease. *British Medical Journal (Clinical Research Edition)*, 361, k2340. https://doi.org/10.1136/bmj.k2340.

Lunder, T. L. (1971). Simplified procedure for determining fat and total solids by Mojonnier method. *Journal of Dairy Science*, 54(5), 737–739. https://doi.org/https://doi.org/10.3168/jds.S0022-0302(71)85917-9

Mabel, M., Guidolin, S., Brazaca, C., Tadeu, C., Ratnayake, W. S. and Flores, R. A. (2014). Characterisation and potential application of pineapple pomace in an extruded product for fibre enhancement. *Food Chemistry*, 163, 23–30.

Mæhre, H. K., Dalheim, L., Edvinsen, G. K., Elvevoll, E. O. and Jensen, I.-J. (2018). Protein determination-method matters. *Foods (Basel, Switzerland)*, 7(1), 5. https://doi.org/10.3390/foods7010005.

Maehre, H. K., Jensen, I.-J. and Eilertsen, K.-E. (2016). Enzymatic pre-treatment increases the protein bioaccessibility and extractability in dulse (Palmaria palmata). *Marine Drugs*, 14(11). https://doi.org/10.3390/md14110196.

Mahmood, A. U., Greenman J. and Scragg A. H. (1998). Orange and potato peel extracts: analysis and use as *Bacillus* substrates for the production of extracellular enzymes in continuous culture. *Enzyme and Microbial Technology*, 22(2), 130–137.

Malaweera, B. O. and Wijesundara, W. M. N. M. (2014). Use of seafood processing by-products in the animal feed industry. In: S.-K. Kim (Ed.), *Seafood Processing by-Products* (pp. 315–339). New York, NY: Springer.

Mamboleo, T. F. (2015). *Nutrients and Antinutritional Factors at Different Maturity Stages of Selected Indigenous African Green Leafy Vegetables*. Doctoral dissertation. Sokoine University of Agriculture.

Mariotti, F., Tome, D. and Mirand, P. P. (2008). Converting nitrogen into protein--beyond 6.25 and Jones' factors. *Critical Reviews in Food Science and Nutrition*, 48(2), 177–184. https://doi.org/10.1080/10408390701279749.

Martí, R., Roselló, S. and Cebolla-Cornejo, J. (2016). Tomato as a source of carotenoids and polyphenols targeted to cancer prevention. *Cancers*, 8, 1–28.

Martínez, R., Torres, P., Meneses, M. A., Figueroa, J. G., Pérez-álvarez, J. A. and Viuda-martos, M. (2012). Chemical, technological and in vitro antioxidant properties of mango, guava, pineapple and passion fruit dietary fibre concentrate. *Food Chemistry*, 135, 1520–1526.

Meisel, H. (1997). Biochemical properties of bioactive peptides derived from milk proteins: potential nutraceuticals for food and pharmaceutical applications. *Livestock Production Science*, 50(1),125–138. https://doi.org/https://doi.org/10.1016/S0301-6226(97)00083-3.

Memon, J. R., Memon, S. Q., Bhanger, M. I., Memon, G. Z., El-turki, A. and Allen, G. C. (2008). Characterization of banana peel by scanning electron microscopy and FT-IR spectroscopy and its use for cadmium removal. *Colloids and Surfaces B: Biointerfaces*, 66, 260–265.

Mihrete, Y. (2019). Review on mineral malabsorption and reducing technologies. *International Journal of Neurologic Physical Therapy*, 5(1), 25–30. https://doi.org/https://doi.org/10.11648/j.ijnpt.20190501.15.

Milica, P., Aleksandra, M. and Brijesh, T. (2018). Eco-innovative technologies for extraction of proteins for human consumption from renewable protein sources of plant origin. *Trends in Food Science and Technology*, 75, 93–104.

Mirabella, N., Castellani, V. and Sala, S. (2014). Current options for the valorization of food manufacturing waste: a review. *Journal of Cleaner Production*, 65, 28–41.

Misra, N. N., Patrick, J., Cullen, F. J., Ching, L. H., Henry, J., Julia, S., Attila, K. and Hiroshi, Y. (2015). *Emerging Macroscopic Pretreatment* (Chapter 9, pp. 197–225). Chania, Greece: Galanakis Laboratories, Academic Press.

Mohammad, R., Issa, M., Shahrokh, N., Mahdi, J. and Leila, S. (2014). Physicochemical characteristics of citrus seed oils. *Journal of Lipids*, 3(1), 56–74.

Mohapatra, D., Mishra, S. and Sutar, N. (2010). Banana and its byproduct utilisation: an overview. *Journal of Scientific and Industrial Research*, 69, 323–329.

Mora, J. R., Iwata, M. and von Andrian, U. H. (2008). Vitamin effects on the immune system: vitamins A and D take centre stage. *Nature Reviews Immunology*, 8(9), 685–698. https://doi.org/10.1038/nri2378.

Mozaffarian, D., Katan, M. B., Ascherio, A., Stampfer, M. J. and Willett, W. C. (2006). Trans fatty acids and cardiovascular disease. *New England Journal of Medicine*, 354(15), 1601–1613. https://doi.org/10.1056/NEJMra054035.

Nakthong, N., Wongsagonsup, R. and Amornsakchai, T. (2017). Industrial crops & products characteristics and potential utilizations of starch from pineapple stemwaste. *Industrial Crops and Products*, 105, 74–82.

Nangole, F. N., Kayongo-Male, H. and Said, A. N. (1983). Chemical composition, digestibility and feeding value of maize cobs. *Animal Feed Science and Technology*, 9(2), 121–130.

Newton, G. L., Tucker, R. E., Mitchell, G. E. J., Schelling, G. T., Alderson, N. E. and Knight, W. M. (1979). Bioassay of vitamin A in liquid supplements. *International Journal for Vitamin and Nutrition Research. Internationale Zeitschrift Fur Vitamin-Und Ernahrungsforschung. Journal International de Vitaminologie et de Nutrition*, 49(3), 240–245.

Ng, T. B., Lam, Y. W. and Wang, H. (2003). Calcaelin, a new protein with translation-inhibiting, antiproliferative and antimitogenic activities from the mosaic puffball mushroom Calvatia caelata. *Planta Medica*, 69(3), 212–217. https://doi.org/10.1055/s-2003-38492.

Nielsen, S. S. (2010). Fat characterization. In: S. S. Nielsen (Ed.), *Food Analysis (Second Edition)* (pp. 251, 252). https://doi.org/10.1038/1841347a0.

O'Shea, N., Arendt, E. and Gallagher, E. (2012). Dietary fibre and phytochemical characteristics of fruit and vegetable by-products and their recent applications as novel ingredients in food products. *Innovative Food Science and Emerging Technologies*, 16, 1–10.

Olsen, R. L. and Hasan, M. R. (2012). A limited supply of fishmeal: impact on future increases in global aquaculture production. *Trends in Food Science & Technology*, 27, 120–128.

Paola, P. and Adem, G. (2015). *Conventional Product Formation* (Chapter 8, pp. 173–193). Chania, Greece: Galanakis Laboratories, Academic Press.

Parodi, P. W. (2007). A role for milk proteins and their peptides in cancer prevention. *Current Pharmaceutical Design*, 13(8),813–828. https://doi.org/10.2174/138161207780363059.

Parthasarathy, S., Raghavamenon, A., Garelnabi, M. O. and Santanam, N. (2010). Oxidized low-density lipoprotein. *Methods in Molecular Biology (Clifton, N.J.)*, 610, 403–417. https://doi.org/10.1007/978-1-60327-029-8_24.

Pegg, R. B., Landen, W. O. and Eitenmiller, R. (2010). Food analysis. In: S. Nielsen (Ed.), *Food Analysis (Fourth Edition)* (p. 602). https://doi.org/10.1038/1841347a0.

Pelissari, F. M., Sobral, P. J. D. A. and Menegalli, F. C. (2014). Isolation and characterization of cellulose nanofibers from banana peels. *Cellulose*, 21, 417–432.

Popkin, B. M., D'Anci, K. E. and Rosenberg, I. H. (2010). Water, hydration, and health. *Nutrition Reviews*, 68(8), 439–458. https://doi.org/10.1111/j.1753-4887.2010.00304.x.

Potter, S. M. (1995). Overview of proposed mechanisms for the hypocholesterolemic effect of soy. *The Journal of Nutrition*, 125(3 Suppl.), 606S–611S. https://doi.org/10.1093/jn/125.3_Suppl.606S.

Prashanth, L., Kattapagari, K., Chitturi, R., Baddam, V. and Prasad, L. (2015). A review on role of essential trace elements in health and disease. *Journal of Dr. NTR University of Health Sciences*, 4(2), 75–85. https://doi.org/10.4103/2277-8632.158577.

Prosky, L., Asp, N. G., Schweizer, T. F., DeVries, J. W. and Furda, I. (1988). Determination of insoluble, soluble, and total dietary fiber in foods and food products: interlaboratory study. *Journal – Association of Official Analytical Chemists*, 71(5), 1017–1023. http://europepmc.org/abstract/MED/2853153.

Reedy, J., Krebs-Smith, S. M., Miller, P. E., Liese, A. D., Kahle, L. L., Park, Y. and Subar, A. F. (2014). Higher diet quality is associated with decreased risk of all-cause, cardiovascular disease, and cancer mortality among older adults. *The Journal of Nutrition*, 144(6), 881–889. https://doi.org/10.3945/jn.113.189407.

Robertson, A. H. and Black, L. A. (1949). Standard methods for the examination of dairy products. *American Journal of Public Health and the Nation's Health*, 39(5 pt 2), 80–82. https://doi.org/10.2105/ajph.39.5_pt_2.80.

Rolls, B. J. (2009). The relationship between dietary energy density and energy intake. *Physiology & Behavior*, 97(5), 609–615. https://doi.org/10.1016/j.physbeh.2009.03.011.

Röös, E., Carlsson, G., Ferawati, F., Hefni, M., Stephan, A., Tidåker, P. and Witthöft, C. (2018). Less meat, more legumes: prospects and challenges in the transition toward sustainable diets in Sweden. *Renewable Agriculture and Food Systems*, s 1–14. https:// doi.org/10.1017/S1742170518000443.

Sabiu, S., Ajani, E. O., Abubakar, A. A., Sulyman, A. O., Nurain, I. O., Irondi, A. E., Abubakar, Y. A. and Quadri, D. F. (2015). Toxicological evaluations of *Stigma maydis* aqueous extract on hematological and lipid parameters of Wistar rats. *Toxicology Reports*, 2, 638–644.

Sabiu S, O'Neil FH, Ashafa AOT. (2016). Membrane stabilization and detoxification of acetam.

Sadaa, B. H. Amarteyb, Y. D. and Bako, S. (2013). An investigation into the use of groundnut shells fine aggregate replacement. *Nigerian Journal of Technology*, 30(1), 54–60.

Sandhu, S. and Lim, S. (2008). Structural characteristics and in vitro digestibility of mango kernel starches (Mangifera indica L.). *Food Chemistry* 107, 92–97.

Schonfeldt, H. and Hall, N. (2012). Dietary protein quality and malnutrition in Africa. *The British Journal of Nutrition*, 108, S69–S76. https://doi.org/10.1017/S0007114512002553.

Seid, M., Jafari, M. and Fathi, I. M. (2015). *Emerging Product Formation* (Chapter 13, pp. 293–317). Chania, Greece: Galanakis Laboratories, Academic Press.

Semih, O., Stella, D., Camelia, B. and Canan, K. (2015). *Food Waste Management, Valorization, and Sustainability in the Food Industry* (Chapter 1, pp. 337–360). Chania, Greece: Galanakis Laboratories, Academic Press.

Serna, L., Torres, C. and Ayala, A. (2015). Evaluation of food powders obtained from peels of mango (Mangifera indica) as sources of functional ingredients. *Information Technology*, 26, 41–50.

Serna-Cock, L., García-Gonzales, E. and Torres-León, C. (2016). Agro industrial potential of the mango peel based on its nutritional and functional properties. *Food Reviews International*, 32, 346–376.

Siaka, D. (2014). Potential of mango (Mangifera indica) seed kernel as feed ingredient for poultry – a review. *World's Poultry Science Journal*, 70, 279–288.

Slavin, J. and Carlson, J. (2014). Carbohydrates. *Advances in Nutrition (Bethesda, Md.)*, 5(6), 760–761. https://doi.org/10.3945/an.114.006163.

Sud, M., Fahy, E., Cotter, D., Dennis, E. A. and Subramaniam, S. (2012). LIPID MAPS-nature lipidomics gateway: an online resource for students and educators interested in lipids. *Journal of Chemical Education*, 89(2), 291–292. https://doi.org/10.1021/ed200088u.

Sulaiman, A., Othman, N., Baharuddin, A. S., Mokhtar, M. N. and Tabatabaei, M. (2014). Enhancing the halal food industry by utilizing food wastes to produce value added bio products. *Procedia – Social and Behavioral Science*, 121, 35.

Suresh, Y. and Das, U. N. (2006). Differential effect of saturated, monounsaturated, and polyunsaturated fatty acids on alloxan-induced diabetes mellitus. *Prostaglandins, Leukotrienes and Essential Fatty Acids*, 74(3), 199–213. https://doi.org/10.1016/j.plefa.2005.11.006.

Takenaka, A., Annaka, H., Kimura, Y., Aoki, H. and Igarashi, K. (2003). Reduction of paraquat-induced oxidative stress in rats by dietary soy peptide. *Bioscience, Biotechnology, and Biochemistry*, 67(2), 278–283. https://doi.org/10.1271/bbb.67.278.

Tanner, J. T. and Barnett, S. A. (1986). Methods of analysis for infant formula: Food and Drug Administration and Infant Formula Council collaborative study, phase III. *Journal – Association of Official Analytical Chemists*, 69(5), 777–785.

Tawakalitu, E. A., Ogugua, C. A., Akeem, O. R. and Aderonke, I. O. (2017). Protein enrichment of yam peels by fermentation with *Saccharomyces cerevisiae* (BY4743). *Annals of Agricultural Sciences*, 62(1), 33–37.

Thi, N. B. D., Kumar, G. and Lin, C. Y. (2015). An overview of food waste management in developing countries: current status and future perspective. *Journal of Environmental Management*, 157, 220–229.

Tomé, D. (2013). Digestibility issues of vegetable versus animal proteins: protein and amino acid requirements – functional aspects. *Food and Nutrition Bulletin*, 34(2), 272–274. https://doi.org/10.1177/156482651303400225.

Torres-León, C., Rojas, R., Serna, L., Belmares, R. and Aguilar, C. (2017). Extraction of antioxidants from mango seed kernel: optimization assisted by microwave. *Food and Bioproducts Processing*, 105, 188–196.

Trumbo, P., Schlicker, S., Yates, A. A. and Poos, M. (2002). Dietary reference intakes for energy, carbohydrate, fiber, fat, fatty acids, cholesterol, protein and amino acids. *Journal of the American Dietetic Association*, 102(11), 1621–1630. https://doi.org/10.1016/s0002-8223(02)90346-9.

Uribe, N. G., García-Galbis, M. R. and Espinosa, R. M. M. (2017). New advances about the effect of vitamins on human health: vitamins supplements and nutritional aspects. In: M. R. García-Galbis (Ed.), *Functional Food – Improve Health Through Adequate Food (Second Edition)* (p. Ch. 4). https://doi.org/10.5772/intechopen.69122.

van Itallie, T. B. (1957). Role of dietary fat in human nutrition. IV. Experimental & clinical evidence relating to the effect of dietary fat upon health in man. *American Journal of Public Health and the Nation's Health*, 47(12), 1530–1536. https://doi.org/10.2105/ajph.47.12.1530.

Varzakas, T., Zakynthinos, G. and Verpoort, F. (2016). Plant food residues as a source of nutraceuticals and functional foods. *Foods*, 5, 1–32.

Venkata M. S., Venkateswar Reddy, M., Amulya, K., Rohit, M. V. and Sarma, P. N. (2014). Valorization of fatty acid waste for bioplastics production using Bacillus tequilensis: integration with dark-fermentative hydrogen production process. *International Journal of Hydrogen Energy*, 39, 7616–7626.

Vera Zambrano, M., Dutta, B., Mercer, D. G., MacLean, H. L. and Touchie, M. F. (2019). Assessment of moisture content measurement methods of dried food products in small-scale operations in developing countries: a review. *Trends in Food Science and Technology*, 88, 484–496. https://doi.org/https://doi.org/10.1016/j.tifs.2019.04.006.

Verrijssen, T. A. J., Smeets, K. H. G., Christiaens, S., Palmers, S., Loey, A. M. and Van, H. M. E. (2015). Relation between in vitro lipid digestion and b-carotene bioaccessibility in b-carotene-enriched emulsions with different concentrations of l-a-phosphatidylcholine. *Food Research International*, 67, 60–66.

Wang, A., Han, J., Jiang, Y. and Zhang, D. (2014). Association of vitamin A and b-carotene with risk for age-related cataract: a meta-analysis. *Nutrition*, 30, 1113–1121.

Wang, Z., Deurenberg, P., Wang, W., Pietrobelli, A., Baumgartner, R. N. and Heymsfield, S. B. (1999). Hydration of fat-free body mass: review and critique of a classic body-composition constant. *The American Journal of Clinical Nutrition*, 69(5), 833–841. https://doi.org/10.1093/ajcn/69.5.833.

WHO. (2007). *Protein and Amino Acid Requirements in Human Nutrition: Report of a Joint FAO/WHO/UNU Expert Consultation*. https://apps.who.int/iris/handle/10665/43411.

Wilk, R. (2012). The limits of discipline: Towards interdisciplinary food studies. *Physiology and Behavior*, 107(4), 471–475. https://doi.org/https://doi.org/10.1016/j.physbeh.2012.04.023.

Wilson, K. and Walker, J. (2002). Principles and techniques of practical biochemistry. In: K. Wilsonand and J. Walker (Eds.), *Biochemistry and Molecular Biology Education (Fifth Edition)* (Vol. 30). https://doi.org/10.1002/bmb.2002.494030030062.

Wolf, W. and Hamly, J. (1984). Trace element analysis in food. In: J. Gilbert (Ed.), *Analysis of Food Contaminants* (pp. 157–206). London, UK: Elsevier Applied Science.

Wong, M. H., Mo, W. Y., Choi, W. M., Cheng, Z. and Man, Y. B. (2016). Recycle food wastes into high quality fish feeds for safe and quality fish production. *Environmental Pollution*, 219, 631–638.

Wu, A. C., Li, E. and Gilbert, R. G. (2014). Exploring extraction/dissolution procedures for analysis of starch chain-length distributions. *Carbohydrate Polymers*, 114, 36–42. https://doi.org/https://doi.org/10.1016/j.carbpol.2014.08.001.

Wu, G., Bazer, F. W., Dai, Z., Li, D., Wang, J. and Wu, Z. (2014). Amino acid nutrition in animals: protein synthesis and beyond. *Annual Review of Animal Biosciences*, 2, 387–417. https://doi.org/10.1146/annurev-animal-022513-114113.

Xiuzhu, Y., Frederick, R., van de, V. Z. and Li, T. Y. (2007). Proximate composition of the apple seed and characterization of its oil. *International Journal of Food Engineering*, 3(5), 1556–3758.

Xu, C., Wang, G., Xing, C., Matuana, L. M. and Zhou, H. (2015). Effect of grapheme oxide treatment on the properties of cellulose nanofibril films made of banana petiole fibers. *BioResources*, 10, 2809–2822.

Younis, E. M., Al-Quffail, A. S., Al-Asgah, N. A., Abdel-Warith, A.-W. A. and Al-Hafedh, Y. S. (2017). Effect of dietary fish meal replacement by red algae, Gracilaria arcuata, on growth performance and body composition of Nile tilapia Oreochromis niloticus. *Saudi Journal of Biological Sciences*, 25, 198–203.

Zhang, Y., Zhou, W.-E., Yan, J.-Q., Liu, M., Zhou, Y., Shen, X., et al. (2018). A review of the extraction and determination methods of thirteen essential vitamins to the human body: an update from 2010. *Molecules (Basel, Switzerland)*, 23(6), 1484. https://doi.org/10.3390/molecules23061484.

Zuluaga, R., Putaux, J. L., Restrepo, A., Mondragon, I. and Gañan, P. (2007). Cellulose microfibrils from banana farming residues: isolation and characterization. *Cellulose*, 14, 585–592.

Part IV: **Product development, sensory evaluation, and packaging**

Adeoluwa Iyiade Adetunji
12 Food trends and development of new food products

12.1 Introduction

This chapter focuses on consumers' food preferences driving approach to new food products development. It brings together important factors that influence the consumer food needs and demand, enabling the food science and technology experts to have a balanced view of their consumer market. Consumer food choices based on the preferences and expectations are driving new food innovations, in relation to the development of new food products that align with those needs. A few details on the emerging innovative food trends and important consumer concerns in the development of innovative food products are also discussed. This provides adequate information needed with regard to possible future trends, thereby enabling development of new food products that can effectively meet consumer food needs and expectations. In conclusion, this chapter essentially emphasizes on the impact of consumer food choice and its effect on the approach the food industry needs to adopt in developing food products that meet and address consumer preferences and expectations.

Continuous changes in the agricultural and food sectors have placed a huge demand on the entire food value chain; from the producers to traders as well as processors and other contributors. These changes focused more on improving the effectiveness of procedures involved in meeting consumer food demands. More specifically, the food industry is subjected to frequent trends, which have significant impact on the success or failure of this sector. A trend implies a line of direction of predominant disposition, preference, development, or movement over a measurable period. Food trends as influenced by consumer preferences and expectations are driving the approach to development of new food products and their sustainability. Maintaining a balance between consumer food choices and strategies of producing products that meet this need is challenging without a clear understanding of consumer needs and preferences (Schmidt, 2005). Over the years, food production and consumption have evolved and this trend continues to evolve. Continuous evolving of food production can be linked to food security challenges. These challenges relate to food production shortage, inadequate food production distribution, and lack of access to food due to poverty (Caracciolo and Santeramo, 2013; Otsuka, 2013). This has led to an increased effort in understanding

Adeoluwa Iyiade Adetunji, Labworld/Philafrica Foods (Pty) Ltd, Centurion, South Africa

https://doi.org/10.1515/9783110667462-012

trends, as it relates to food supply chains, changes in expectations of consumers, transformation of food industry, and potential impacts of food innovations (Parfitt et al., 2010; Gereffi and Lee, 2012).

Consumer food choices have been noted to have significant implications on consumer health, social, and financial status (Oyewole and Khan, 2018), resulting in continuous evolving nature of consumer food consumption pattern preferences. One of such food trend among consumers is the shift toward better tasting and ready-to-eat products such as breakfast cereals and convenient beverage products. A major driver in this regard is rapid urbanization, which has resulted in the huge growth of convenience foods from cereals, such as wheat and rice (Galati et al., 2014). This food trend is a direct effect of increasing incomes and changes in the lifestyle of consumers. Hence, trending food choices among consumers have greatly influenced new food products development strategies to keep up with the complexity of consumer food preferences. Evolving food trends have created a great deal of new food opportunities, as it relates to the development of new food products (Culinary Trend Tracking Series, 2014). Food opportunities combine prevailing trends in food choices with the approach of preparation and processing in meeting the need of consumers. The key factors influencing consumers' food choices were reviewed in relation to how they are driving development of new food opportunities in meeting consumer demands and needs.

12.2 Food trends

Food trends is a phenomenon that keeps changing and this can be linked to changes in the demand for agricultural and food products being fueled by growing populations, increasing incomes, and changing lifestyles. Over the years, several key indicators identified to influence consumers' awareness in relation to their food choices have been extensively reviewed (Mehmeti and Xhoxhi, 2014; McCluskey, 2015; Oyewole and Khan, 2018). Consumer food choices is a complex phenomenon due to increasing dynamics and heterogeneity of consumer demand coupled with increasing diversity of food product types that consumers need to choose from (Grunert, 2002). The factors influencing consumer preferences in view of their food choices are discussed under the following groups: changes in consumer demography, consumer expectations, and impact of global changes. Under each of these groups are variables that consumers take into cognizance, thereby shaping different food consumption patterns and preferences of consumers. These variables relate to consumer spending, health and wellness concerns, and convenience (Melia, 2011). Coupled with these, knowledge of nutritional information and gender was among other factors that have been reported to influence consumer choices (Oyewole and Khan, 2018).

12.2.1 Changes in consumer demography

Development in the global economy with regard to urbanization has significant influence on the urban population food needs. Its direct effect relates to the emergence of new consumer food choices and demands, which has resulted in significant changes in the lifestyle of consumers, as it relates to their consumption patterns, and these are reflected in consumer food choices. These changes range from elementary considerations, such as food safety concerns, shelf life, and reducing wastage, to increasing demand for foods with special features in terms of nutritional value, taste, and convenience (Winger and Wall, 2006). According to Meulenberg and Viaene (2005), consumers now place more value on self-fulfillment, better quality of life, and work to live. The key variables that tend to influence these consumer food patterns or changes are: age, income capacity, household size, social orientation, and educational background (Mehmeti and Xhoxhi, 2014; McCluskey, 2015).

12.2.1.1 Age

Consumer populations can be grouped into different age categories. These are characterized with different food consumption preferences and demand requirements, such as by the infants, younger adults, and older adults. For example, food preferences of older consumer population tend toward more healthier foods (McCluskey, 2015). On aging, the food requirements of older consumer population become very diverse and this is based on several factors related to aging. Sensory perception changes such that the development of food for the older consumer population category need specific attention to sensory attributes (Murphy and Vertrees, 2017). Similarly, the effect of aging also drives certain types of foods, particularly with the older consumer groups (Mehmeti and Xhoxhi, 2014). This relates to prevalence of health conditions associated with aging, such as cardiovascular diseases, osteoporosis, and diabetes (Bleiel, 2010), making it critical for the elderly consumers to be more concerned with their food choices (Herne, 1995).

12.2.1.2 Income capacity

Income capacity has been identified to play a significant role in influencing consumers' dietary behavior (Steenhuis et al., 2011). Consumer group can be categorized into two based on their income capacity: low-income and high-income groups. Given the improvement in consumers' income capacity, there has been a significant change in consumer food requirements, which tend toward higher product quality with specific attributes that correspond with their lifestyle (Mehmeti and Xhoxhi, 2014). Food trends associated with this factor, as reviewed by Mehmeti and Xhoxhi (2014), also

show a significant increase in the consumers' eating out pattern. This has been reported to have a huge impact in the growth of fast food service sector worldwide, while
forecasting a stronger growth in the coming years. Income capacity of consumers also
has a direct link with food production, with increased development of food products
possessing higher quality coupled with added benefits to attract consumers resulting
in higher prices. Steenhuis et al., (2011) noted that this might have a negative impact
on the low-income consumer group due to limited resources in terms of income to af-
ford such high-quality product. This necessitates the need to find a right balance in
meeting the increased demand of consumers for variety and high-quality food products
that fall within their buying capability, with increased consciousness on the part of
consumers, as it relates to prices of food products (Bleil, 2010). Consequently, this has
resulted in the development of new food products being produced and patterned along
with the trend of finding a right balance between demand of consumers for high quality and the price that they are willing to pay (McCarthy, 2010).

12.2.1.3 Social orientation

Another important key factor is consumer social orientation. Consumer dietary
choices tend to be influenced by those they are socially connected with (Pachucki
et al., 2011; Barclay et al., 2013; Haye et al., 2013). This can be linked to shared cultural expectations and environmental indications such as religious restrictions,
medication concerns, parental or personal preferences, ethical beliefs, environmental concerns, as well as pursuit of a healthy lifestyle, as they play a significant role
in influencing consumer food preferences and expectations (Fan et al., 2019). In
view of social connections, there is growing interest in understanding the link between social alignment, dietary patterns, and health challenges such as obesity
(Higgs and Thomas, 2016). These authors noted that the good aspect of social
norms can be harnessed in influencing consumer food choices toward healthier
preferences. For example, some of the social factors highlighted above have been
identified to drive a group of consumer food preferences towards vegetarianism
(Fan et al., 2019). This category of food trend has been reported to be increasingly
popular among consumers (Raisfeld and Patronite, 2017). A Global Data market research conducted in 2017 among American consumer populations showed that
the rate of consumers abstaining from food products of animal origin increased by
5% within 3 years, growing from 1% in 2014 to 6% in 2017. Fan et al. (2019) further
noted that this percentage may not include self-identified vegetarians. These self-
identified vegetarians are classified as having more flexible plant-based food
choices by occasionally consuming food products of animal origin (Ruby, 2012).
The impact of this type of social orientation regarding this category of consumer
food choice is shown to be linked to their prosocial and proenvironmental tendencies (Chan and Hawkins, 2010).

12.2.2 Consumer expectations

Exposure to education is increasing consumer tendency to try out new foods, and this has increasingly driven consumer food expectations. Some of the expectations relate to value for money, quality, standards, flexibility, and service experience, which combines some physical and perceived attributes relating to consistency and creativity (Jones, 2009). More importantly, healthy food options and safety of food products are important priorities for consumers in meeting their expectations, thereby driving new food trends in terms of consumer food choices. In addition, consumer expectation with regard to lower prices is increasingly driving their food choices (Pitta, 2010). The expectation of foods they consume as inclined toward how it directly contributes to their health is discussed further.

12.2.2.1 Health concern

Consumer lifestyles have been significantly influenced by urbanization and globalization with trends shifting toward more convenience, coupled with time-pressured consumers trying to save time and effort in their food preparation and consumption pattern (Grunert et al., 2019). These factors reflect consumer preferences toward consumption of fast foods (Mehmeti and Xhoxhi, 2014) and this consumption pattern is most prevalent in today's developed societies (Grunert, 2013). This resulted in a food trend that is characterized by high calories, saturated fat and sugar diets that are poor in essential nutrients (Thow and Hawkes, 2009). The resulting effect of these food trends is abandonment of traditional diets featuring vegetables and grains for the above-mentioned choices (Kearney, 2010). Consequently, these food choices have raised some huge health problems due to significant increase in diseases associated with the new lifestyles such as obesity, cardiovascular diseases, cancer, high blood cholesterol, and diabetes (Popkin, 1999).

In view of this, there are growing health concerns among consumers due to their food choices. They are now more aware of the health issues associated with consumed foods, such that their expectations in choosing food products that positively impact their health and well-being have increased and manufacturers are responding proactively by offering new products that meet these needs (Gray et al., 2003). Consumer well-being in the context of their food choice is now strongly linked with its impact on their physical health (Ares et al., 2014). Hence, health benefits of food have become an important motivation for consumer food choices more than seeking pleasure and sensory reward in food (Grunert et al., 2019). Food choices are now based on the inherent potentials of food products in enabling healthy lifestyle by preventing or reducing the risk of having certain health disorders (Asioli et al., 2017).

12.2.3 Global changes

The world at large is confronted with a lot of challenges such as environmental issues involving loss of biodiversity and pollutions (Reisch et al., 2013; Auestad and
Fulgoni, 2015), resulting from high rate of global industrialization and urbanization.
These challenges directly impact the sustainability of global food security, as it relates to food availability via production, and access to food; utilization, as it relates
to cultural and dietary requirements; and food provision stability (FAO, 2009). The
resulting food system due to these global changes is putting significant pressure on
the availability of food in meeting consumer food needs and demands (Bennett and
Jennings, 2013). These global changes impact consumer food needs and demands,
and thereby drive certain changes in the dietary patterns of consumers.

Climate change is a global phenomenon, which has a significant direct effect
on the agricultural food system. The impact of climate change relates to changes
in the weather conditions affecting agricultural production of foods (Kirby et al.,
2016). According to Maggio et al. (2015), the effect of climate change on agricultural activities in terms of production could be positive and negative. The negative
effects of climate change are changes in rain pattern and temperatures, resulting
in loss of water and land resources, floods, and droughts (Kurukulasuriya et al.,
2006; Mendelsohn, 2014). These negative impacts result in poor agricultural production stability and affect other aspects of food system such as storage, access,
and utilization (Wheeler and Von Braun, 2013). Maggio et al. (2015) also indicated
that climate change triggers population migration resulting in overpopulation of
some regions of the globe. This alters the supply of food in meeting consumer
needs and demands.

In view of the highlighted impact of climate change, there is a growing concern regarding challenges of food security, pertaining to food productivity and
supply (Arnell et al., 2004). These have manifested in high level of hunger, malnutrition, and poverty. While addressing these challenges, some of the approach
shifts toward incorporation of new food sources in meeting growing consumer
food needs and demands such as insects as an alternative high-quality protein
source (Van Huis, 2013; Aiking and de Boer, 2018), leading to emergence of new
food trends. For example, one of the direct effects of climate change on food production supply relates to drought. These impact negatively on the production of
major food crops such as maize and wheat (Ali et al., 2017). Addressing this challenge has led to increased research interest in heat- and drought-tolerant food
crop varieties. Increased trend of food products of incorporating some of the
drought-tolerant food crops has contributed and continue to influence research
development of new food products in mitigating the challenges of food security
globally, thereby resulting in new food choices for consumers.

12.3 Development of new food products

Development of new food product is a systematic process that is based on creating a marketable product designed for meeting an identified consumer need (Winger and Wall, 2006). This process involves the interactions of the following: consumer expectations and demands, market and technical experts, as well as application of new knowledge in food science and technology research (Winger and Wall, 2006). Major drivers of new food development in the food industry sector are changes in technological innovations and in consumer needs (Harmsen et al., 2000). Costa and Jongen (2006) noted that consumer food consumption behavior keeps changing at a rapid rate with the notion that their preferences are not on the same level as the development of technological innovation. Nevertheless, continuous improvement in the innovation technology is one key factor influencing growing consumer food needs and expectations (Halagarda, 2008). In taking advantage of meeting these needs and expectations, food industries continue to research into effective food product development management processes.

A food product can be defined as a specific food item developed to provide certain benefits that meet the need of the consumers. On the other hand, there is no specific definition that can capture the generic term "new food product." However, new food product can be categorized based on the specific need being targeted, which includes the following: new to the company, improvement of existing product, extension of product line, and new to the market (Horvat et al., 2019). For a new food product development to be successfully executed, a generic process involving several stages need to be followed as highlighted by Kelly et al. (2008). These include idea generation, formulation, process development, initial testing and viability assessment, product shelf-life testing, scale-up and consumer testing, as well as packaging and labeling. According to Winger and Wall (2006), these stages can be grouped into four keys aspects: (1) product strategy development, (2) product design and development, (3) product commercialization, and (4) product launch and post-launch. A schematic flow of these stages of new food product development process is shown in Figure 12.1.

Product Strategy Development

Initial screening
Preliminary market assessment
Detailed market research
Product concept development
Financial feasibility study

Outcomes

Decisions
go/no-go

Product design and process development

Prototype design
In-house testing
Consumer testing
Scaling-up

Outcomes

Decisions
go/no-go

Product commercialization

Trail production
Market test

Outcomes

Decisions
go/no-go

Product launch and postlaunch

Prelaunch business analysis
Production start-up
Market launch
Postlaunch operational and financial analysis

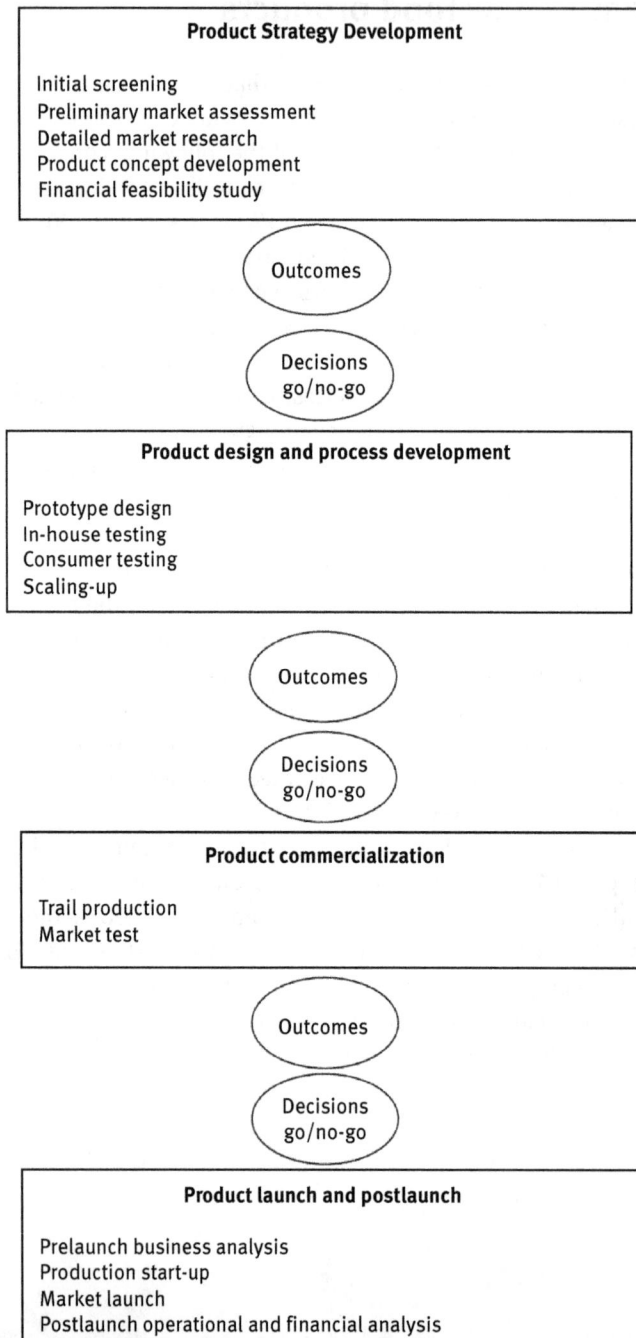

Figure 12.1: General flow diagram of new product development process (*adapted from Winger and Wall, 2006*).

12.3.1 Product strategy development

Adopting the right strategy is critical in the development of a new food product to be successful in the market. This requires conducting an appropriate consumer research that generates better understanding of consumer needs. According to Buck (2007), the adoption of this approach generates several important information in ensuring the success of the product as follows:

1. Generates information related to product capability in addressing genuine need of the consumer at an affordable price.
2. Generates information on the approach of communicating the benefits of the new product in addressing targeted consumers.
3. Information gathered assists in focusing the product development phase on delivering identified added benefits in meeting consumer expectations.
4. Finally, generates information that would assist in designing mechanisms of making the product accessible to the targeted consumers, which is critical in ensuring market success.

These highlighted key elements follow a series of steps as shown in Figure 12.2, which require critical evaluation for effective decision-making.

According to Horvat et al. (2019), the approach of consumer-oriented market research concept has increasingly been adopted in the development of a successful new product. This approach is based on optimum balancing of consumer needs with the new product being developed to ensure maximum consumer satisfaction. This involves consistent gathering and dissemination of market information in line with the current and future trends of consumer needs as emphasized by Costa and Jongen (2006). There are three major types of information that can be derived in adopting consumer-oriented market research depending on the approach: (1) consumer involvement, (2) food trend, (3) and environmental factor data (Horvat et al., 2019). Application of this concept in developing new products allows effective balancing of consumer needs with technical data (Costa and Jongen, 2006). This requires implementation of an effective information gathering and dissemination mechanism involving all the key actors, combined with appropriate technical capabilities. According to Stewart-Knox and Mitchell (2003), there is a need for considerable vertical incorporation of product innovation management in the implementation of an effective consumer-oriented product development strategy. This approach enables an effective transformation of subjective needs of consumers to an explicit objective product that satisfies those needs. Hence, this concept is gaining more interest as an approach that allows effective application of current and future needs of consumers, and it is a determining factor in developing innovative products that provide added value meeting the needs of the consumers. Similarly, this approach provides opportunity to understand the effective means

Consumer data   Technical data

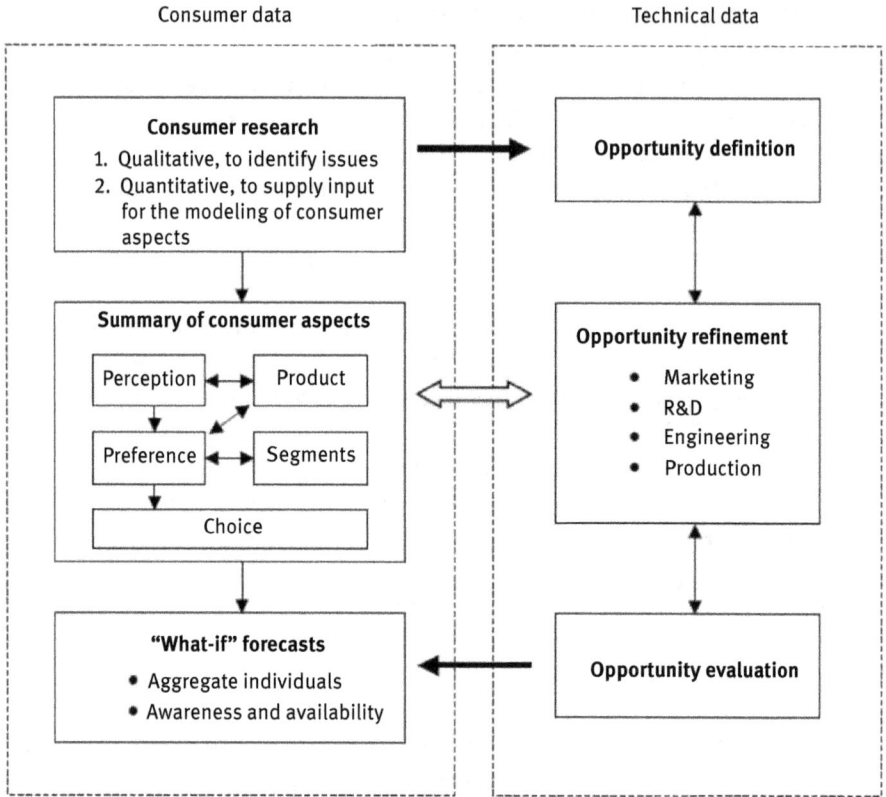

Figure 12.2: Procedure for generating consumer and technical data on new product development (*adapted from Costa et al., 2006*).

of communicating the true value a new product has to offer the consumer after its development (Costa and Jongen, 2006).

12.3.2 Product design

Based on the identified consumer needs, a product design is tailored accordingly. This process comprises product definition and features determination, which includes tangible and intangible features. The process of product design specifications starts with profiling the product features as defined by the consumer, the structure and composition, safety factors, convenience and aesthetics, as well as description of the production process and storage variables and their impact on the product qualities (Earle and Earle, 1999). The product design is based on definite and quantitative descriptions of the product in relation to the following: product attributes identification and screening, attributes measurement, complete product concept, and product concept evaluation. A successful product design process needs to be conducted at the consumer level rather than at the technical level only (Moskowitz et al., 2008). With the identification of specific consumer needs and expectations, based on the information gathered through market research, development of food product that meets those needs proceed with defining specific attributes that can be used in the identification of the product. The intangible features of the product relate to product attributes that cannot be seen nor touched physically, and serves as a key function that can be used in promoting its value. These are important inherent attributes that the product is specifically formulated to possessed. Coupled with this, it is pertinent to clearly consider the aspect of tangible attributes that make the product unique. These features help in giving a vivid description of the product and serve as key selling points to the targeted consumer group. Tangible attributes relate to physical features that can be seen and handled, such as shape, pack-size, color, and weight. These features can be used to describe the physical appearance of the new products being developed and careful attention needs to be given in addressing specific needs identified during market research and in meeting the expectations of the consumers. As a matter of fact, the tangible attributes of the product are fundamental in assuring the consumers about the main function, value, reliability, and consistency of the product.

12.4 Food choices influencing the development of new food products

Recent reviews of consumer food choices identified three main categories of food trends: foods that promote healthy lifestyles such as functional foods (Asioli et al., 2017); convenience foods that lessen consumer time and effort in preparation, consumption, and cleaning-up (Jackson and Viehoff, 2016); and food products that contribute to economic and environmental sustainability such as vegetarian (Grunert et al., 2014). According to Jnawali et al. (2016), development of products that address the consumer food choices in line with the identified trend must meet current demand and should be affordable. The derived health benefits stand out as the major determining factor that consumers emphasizes on. This has led to an increasing demand for food research and development in relation to sustainable production of food products of high quality, which can provide additional functionality in relation to prevention of lifestyle health problems such as cancer, obesity, diabetes, heart disease, and stroke (Wang and Bohn, 2012). Most of these research and development criteria are centered on functional food products, with a lot being focused on the application of ingredients with novel functionalities. Jnawali et al. (2016) emphasized that in designing these food products, the influence of consumer age group-specific requirements play a significant role.

12.4.1 Trending functional food types

Functional food is a nonspecific term used to describe food that has specific health benefits. According to Griffiths et al. (2009), functional foods are food types that offer more health benefits compared to those delivered by traditional nutrients in the food, thereby having inherent potential in preventing diseases or in promoting a better quality of life. In general, functional foods can be categorized as fortified, enriched, or enhanced foods with ingredients possessing biofunctional properties characterized to proffer certain health benefits along with provision of essential nutrients (e.g., vitamins and minerals), when they are consumed at efficient levels as part of a varied diet on a consistent basis (Hasler, 2002). Moreover, functional foods can be natural or processed foods that possess biologically active compounds, which are present in nontoxic amounts, capable of impacting clinically confirmed and documented health benefits by using specific biomarkers in preventing, managing, or treating chronic diseases or its symptoms (FFC, 2019). The acclaimed associated health benefits of these product types have led to increased trend in the consumption of functional food products. According to a study by Karelakis et al. (2019), consumer recognition of the inherent benefits of functional food products makes them to be willing to pay more for these products. This perception is based on functional foods contribution to a healthy and balanced diet in view of reduction in the risk of exposure to certain health problems. However, these authors also noted that consumers appeared to be anxious about beneficial claims communicated on the food label of such food products.

According to a market research, functional food products in the global market was estimated to grow from $ 168 billion market value in 2013 to above $ 300 billion in 2020 (Santeramo et al., 2018). This prediction in terms of market growth has led to significant investment in new functional food products development as alluded by Khan et al. (2014). However, sustainability of functional food products in the market is critical. Mellentin (2014) noted that the development of these products is a dicey activity due to the failure of a large number of new functional products introduced in the market. High rate of sustainability failure in the market was attributed to the development of these products mostly driven based on technical feasibility (Bleiel, 2010) and not by its potential consumer's acceptability (Van Kleef et al., 2005). This is due to lack of adequate understanding of consumer needs and preferences for these products. In other words, a perfect balance between understanding of consumer preferences and technical feasibility of producing these functional food products is essential for market sustainability.

In balancing consumers need with technical feasibility of functional foods, there is an increased interest among consumers regarding processes involved in the production of food products they consume. This awareness has led to further development of functional foods by natural fortification or enrichment process without the addition of synthetic chemical compounds (Caleja et al., 2016). Hence, functional food products development that is based on fortification or enrichment with functional ingredients possessing bioactive properties is critical for market success. These bioactive compounds are obtained from a variety of sources such as primary produce, marine sources, microorganisms, and inorganic raw materials (SAFIC, 2019). In addition to these sources, there is a growing trend in the development of functional food products, whereby bioactive ingredients applied in the development of these products are being derived from food process byproducts such as fruit and vegetable process waste as well as brewer spent grains (Lynch et al., 2016; Garcia-Amezquita et al., 2018). This approach provides additional economic benefits to food industries generating these waste products. In addition, this trend has led to increased research and development in converting food process wastes into premium raw materials, thereby contributing positively to environmental conservation. In view of emerging trend of functional foods, one major example of functional foods category that has gained more research interest in addressing specific consumer need in relation to health benefits is gluten-free products. Gluten-free products are categorized as functional food products based on the removal of harmful components as it relates to gluten-intolerance.

12.4.1.1 Gluten-free food products

Matos and Rosell (2015) noted an increased consumer interest in gluten-free products, which have both similar functional and nutritional attributes as their gluten-containing counterparts. The key factor driving consumer need in this direction relates to health concerns. Gluten-free diet is one of the trending food choices that consumers with gluten-related disorders have accepted as the only therapeutic approach in dealing with coeliac disease and for the nonceliac gluten-sensitive consumer group (Elli and Marinoni, 2019). This has become a popular food choice among these consumer population for its implied health benefits (Kim et al., 2016). Gluten-free foods can be categorized as new food products based on the replacement of some key ingredients, particularly gluten-containing ingredients such as wheat, rye, barley, and oats. This has led to increased research interest in optimizing gluten-free products, as it relates to functional and nutritional properties of nongluten raw materials in improving processing performance, sensory quality, and shelf life for their application in the formulation and development of gluten-free food products. The major focus is on addressing the challenge of finding suitable replacement for gluten (Gallagher et al., 2004).

One of such common gluten-free products that have gained significant research interest is gluten-free bread. In recent years, research and development of consumer-acceptable gluten-free bread products have grown significantly. Replacement of gluten in bread dough plays a major role in baking properties with regard to water absorption capacity, cohesivity, viscosity, and elasticity (Wieser, 2007). These baking properties directly impact on sensory attributes of gluten-free bread products, which serve as the major challenge to gluten-free bread bakers (Gallagher et al., 2004). Shelf-life stability of gluten-free bread product is another challenge, as it relates to high staling tendency compared to gluten-containing bread (Ahlborn et al., 2005). Examples of possible non-wheat flour ingredients for the development of gluten-free bread products include maize, sorghum, and cassava. As listed in Table 12.1, various research and development processes of gluten-free food products have considered methods of improving textural quality and nutritional value, coupled with improving consumer acceptability of replacing gluten with different ingredients, such as cassava starch (Onyango et al., 2011) and hydrocolloids (Moreira et al., 2013). However, a combination of factors is required for the development of safe and acceptable gluten-free functional food products in terms of selection of suitable gluten replacement flours (Nyembwe et al., 2018) coupled with application of appropriate processing techniques (Falade et al., 2014).

Table 12.1: Incorporation of gluten-free ingredients in the development of high-quality and consumer-acceptable gluten-free products.

Ingredient source	Product	Attributes targeted	References
Cassava starch and sorghum flour	Bread	Improved overall crumb property and nutritional quality	Onyango et al. (2011)
Fermented sorghum flour composited with wheat flour		Nutritional quality	Istianah et al. (2018)
Rice flour, potato starch, and fish surimi		Crumb and crust property, loaf volume, and nutritional quality	Gormley et al. (2003)
Defatted marama flour and cassava starch		Improved protein quality and fiber-rich bread	Nyembwe et al. (2018)
Rice, maize, sorghum, and pearl millet	Cookies	Nutritional quality and calorie value	Rai et al. (2014)

12.5 Conclusion

Important factors influencing consumer food preferences and demands, which also drive the development of new food products, are highlighted in this chapter. Increased consumer awareness about health benefits of food they consumed serve as a major driver of trending consumer food choices. Therefore, information regarding trending consumer food preferences and demand is critical in the development of new food products that addresses the needs of consumers. To ensure that consumer food needs and expectations are met, all the necessary steps required in the development of new food products must be followed and strictly adhered to. This would ensure that all the identified consumer- and product-related characteristics are well balanced and incorporated into the new food products being developed. Advancements in technology should also be adequately harnessed through research and development that is consumer-oriented. This approach to new food products development could effectively balance consumer preferences with technical feasibility and assures market sustainability of new food products after launch.

References

Ahlborn, G. J., Pike, O. A., Hendrix, S. B., Hess, W. M. and Huber, C. S. (2005). Sensory, mechanical, and microscopic evaluation of staling in low-protein and gluten-free breads. *Cereal Chemistry*, 82, 328–335.

Aiking, H. and de Boer, J. (2018). Protein and sustainability – the potential of insects. *Journal of Insects as Food and Feed*, 1–6.

Ali, S., Liu, Y., Ishaq, M., Shah, T., Abdullah, Ilyas, A. and Din, I. U. (2017). Climate change and its impact on the yield of major food crops: evidence from Pakistan. *Foods*, 6(39),1–19.

Ares, G., De Saldamando, L., Giménez, A. and Deliza, R. (2014). Food and well-being. Towards a consumer-based approach. *Appetite*, 74, 61–69.

Arnell, N. W., Livermore, M. J. L., Kovats, S., Levy, P. E., Nicholls, R., Parry, M. L. and Gaffin, S. R. (2004). Climate and socio-economic scenarios for global-scale climate change impacts assessments: characterising the SRES storylines. *Global Environmental Changes*, 14(1),3–20.

Asioli, D., Aschemann-Witzel, J., Caputo, V., Vecchio, R., Annunziata, A., Naes, T. and Varela, P. (2017). Making sense of the "clean label" trends: a review of consumer food choice behaviour and discussion of industry implications. *Food Research International*, 99, 58–71.

Auestad, N. and Fulgoni, V. L. (2015). What current literature tells us about sustainable diets: emerging research linking dietary patterns, environmental sustainability, and economics. *Advances in Nutrition*, 6(1),19–36.

Barclay, K. J., Edling, C. and Rydgren, J. (2013). Peer clustering of exercise and eating behaviours among young adults in Sweden: a cross-sectional study of egocentric network data. *BMC Public Health*, 13, 784.

Bennett, D. and Jennings, R. (2013). *Successful Agricultural Innovation in Emerging Economies: New Genetic Technologies for Global Food Production*. Cambridge, UK: Cambridge University Press.

Bleiel, J. (2010). Functional foods from the perspective of the consumer: how to make it a success? *International Dairy Journal*, 20(1),303–306.

Buck, D. (2007). Methods to understand consumer attitudes and motivations in food product development. In: *Consumer-Led Food Product Development* (Chapter 7, pp. 141–157). Woodhead Publishing Series in Food Science, Technology and Nutrition.

Caleja, C., Barros, L., Antonio, A. L., Carocho, M., Oliveira, M. B. P. and Ferreira, I. C. (2016). Fortification of yogurts with different antioxidant preservatives: a comparative study between natural and synthetic additives. *Food Chemistry*, 210, 262–268.

Caracciolo, F. and Santeramo, F. G. (2013). Price trends and income inequalities: will Sub-Saharan Africa reduce the gap? *African Development Review*, 25(1),42–54.

Chan, E. S. W. and Hawkins, R. (2010). Attitude towards EMSs in an international hotel: an exploratory case study. *International Journal of Hospitality Management*, 29(4),641–651.

Costa, A. I. A. and Jongen, W. M. F. (2006). New insights into consumer-led food product development. *Trends in Food Science & Technology*, 17, 457–465.

Culinary Trend Tracking Series. (2014). How to identify food trends. www.packagedfacts.com (accessed September 12, 2019).

Earle, M. D. and Earle, R. L. (1999). Product design and process development. In: *Creating New Foods. The Product Developer's Guide* (Chapter 5, p. 192). Oxford, UK: Chandos Publishing Ltd.

Elli, L. and Marinoni, B. (2019). Gluten rhapsody. *Nutrients*, 11, 1–4.

Falade, A. T., Emmambux, M. N., Buys, E. M. and Taylor, J. R. N. (2014). Improvement of maize bread quality through modification of dough rheological properties by lactic acid bacteria fermentation. *Journal of Cereal Science*, 60, 471–476.

Fan, A., Almanza, B., Mattila, A. S., Ge, L. and Her, E. (2019). Are vegetarian customers more 'green'? *Journal of Foodservice Business Research*, 22(5),467–482.

Food and Agricultural Organization (FAO). (2009). *Final Declaration of the World Summit on Food Security*. Rome. www.fao.org/fileadmin/templates/wsfs/Summit/Docs/Final_Declaration/WSFS09_Declaration.pdf.

Functional Food Centre (FFC). https://www.functionalfoodscenter.net (accessed September 10, 2019).

Galati, A., Oguntoyinbo, F. A., Moschetti, G., Crescimanno, M. and Settanni, L. (2014). The cereal market and the role of fermentation in cereal-based food production in Africa. *Food Reviews International*, 30, 317–337.

Gallagher, E., Gormley, T. R. and Arendt, E. K. (2004). Recent advances in the formulation of gluten-free cereal based products. *Trends in Food Science and Technology*, 15, 143–152.

Garcia-Amezquita, L. E., Tejada-Ortigoza, V., Serna-Saldivar, S. O. and Welti-Chanes, J. (2018). Dietary fibre concentrates from fruit and vegetable by-products: processing, modification, and application as functional ingredients. *Food and Bioprocess Technology*. https://doi.org/10.1007/s11947-018-2117-2.

Gereffi, G. and Lee, J. (2012). Why the world suddenly cares about global supply chains. *Journal of Supply Chain Management*, 48(3),24–32.

Gormley, T. R., Elbel, C., Gallagher, E., et al. (2003). Fish surimi as an ingredient in gluten-free breads. In: *Proceedings of the First Joint Trans-Atlantic Fisheries Technology Conference, Iceland* (pp. 246, 247).

Gray, J., Armstrong, G. and Farley, H. (2003). Opportunities and constraints in the functional food market. *Nutrition & Food Science*, 33(5),213–218.

Griffiths, J. C., Abernethy, D. R., Schuber, S. and Williams, R. L. (2009). Functional food ingredient quality: opportunities to improve public health by compendial standardization. *Journal of Functional Foods*, 1(1),128–130.

Grunert, K. G. (2002). Current issues in the understanding of consumer food choice. *Trends in Food Science &Technology*, 13, 275–285.

Grunert, K. G. (2013). Trends in food choice and nutrition. In: *Consumer Attitudes to Food Quality Products* (pp. 23–30). Wageningen: Wageningen Academic Publishers.

Grunert, K. G., do Canto, N. R., Liu, R. and Salnikova, E. (2019). *Well-Being as a Global Food Trend: Health, Sustainability and Authenticity*. https://danishfoodinnovation.dk/wp-content/up loads/2019/03/DFI-short-paper-finalwithrefs.pdf (accessed September 20, 2019).

Grunert, K. G., Hieke, S. and Wills, J. (2014). Sustainability labels on food products: consumer motivation, understanding and use. *Food Policy*, 44 (Suppl.C), 177–189.

Halagarda, M. (2008). New food products development. *Polish Journal of Commodity Science*, 4, 32–41.

Harmsen, H., Grunert, K. G. and Declerck, F. (2000). Why did we make that cheese? An empirically based framework for understanding what drives innovation activity. *R&D Management*, 30, 151–166.

Hasler, C. M. (2002). Functional foods: benefits, concerns and challenges – a position paper from the American council on science and health. *The Journal of Nutrition*, 132(12),3772–3781.

Haye, K., Robins, G., Mohr, P. and Wilson, C. (2013). Adolescents' intake of junk food: processes and mechanisms driving consumption similarities among friends. *Journal of Research on Adolescence*, 23(3),524–536.

Herne, S. (1995). Research on food choice and nutritional status in elderly people: a review. *British Food Journal*, 9(9),12–29.

Higgs, S. and Thomas, J. (2016). Social influences on eating. *Current Opinion in Behavioral Sciences*, 9, 1–6.

Horvat, A., Granato, G., Fogliano, V. and Luning, P. A. (2019). Understanding consumer data use in new product development and the product life cycle in European food firms – an empirical study. *Food Quality and Preference*, 76, 20–32.

Istianah, N., Ernawati, L., Anal, A. K. and Gunawan, S. (2018). Application of modified sorghum flour for improving bread properties and nutritional values. *International Food Research Journal*, 25(1),166–173.

Jackson, P. and Viehoff, V. (2016). Reframing convenience food. *Appetite*, 98, 1–11.

Jnawali, P., Kumar, V. and Tanwar, B. (2016). Celiac disease: overview and considerations for development of gluten-free foods. *Food Science and Human Wellness*, 5, 169–176.

Jones, M. T. (2009). A celebrity chef goes global. *Journal of Business Strategy*, 30(5),14–23.

Karelakis, C., Zevgitis, P., Galanopoulos, K. and Mattas, K. (2019). Consumer trends and attitudes to functional foods. *Journal of International Food & Agribusiness Marketing*. DOI: 10.1080/08974438.2019.1599760

Kearney K. (2010). Food consumption trends and drivers. *Philosophical Transactions of the Royal Society B: Biological Sciences*, 365(1554),2793–2807.

Kelly, A. L., Moore, M. M. and Arendt, E. K. (2008). New product development: the case of gluten-free food products. In: *Gluten-Free Cereal Products and Beverages, Food Science and Technology* (Chapter 18, pp. 413–431). Academic Press.

Khan, R. S., Grigor, J. V., Win, A. G. and Boland, M. (2014). Differentiating aspects of product innovation processes in the food industry. *British Food Journal*, 116(8),1346–1368.

Kim, H. S., Patel, K. G., Orosz, E., Kothari, N., Demyen, M. F., Pyrsopoulos, N. and Ahlawat, S. K. (2016). Time trends in the prevalence of celiac disease and gluten-free diet in the US population: results from the National Health and Nutrition Examination surveys 2009–2014. *JAMA Internal Medicine*, 176, 1716, 1717.

Kirby, J. M., Mainuddin, M., Mpelasoka, F., Ahmad, M. D., Palash, W., Quadir, M. E., Shah-Newaz, S. M. and Hossain, M. M. (2016). The impact of climate change on regional water balances in Bangladesh. *Climate Change*, 135, 481–491.

Kurukulasuriya, P., Mendelsohn, R., Hassan, R., Benin, J., Deressa, T., Diop, M., Eid, H. M. Fosu, K. Y., Gbetibuo, G., Jain, S., et al. (2006). Will African agriculture survive climate change? *The World Bank Economic Review*, 20(30),367–388.

Lynch, K. M., Steffen, E. J. and Arendt, E. K. (2016). Brewers' spent grain: a review with an emphasis on food and health. *Journal of Institute of Brewing*, 122, 553–568.

Maggio, A., Van Criekinge, T. and Malingreau, J. P. (2015). *Global Food Security 2030: Assessing Trends with a View to Guiding Future EU Policies. Joint Research Centre Science and Policy Report* (p. 41). Luxembourg: Publications Office of the European Union.

Matos, M. E. and Rosell, C. M. (2015). Understanding gluten-free dough for reaching breads with physical quality and nutritional balance. *Journal of the Science of Food and Agriculture*, 95, 653–661.

McCarthy. (2010). *Kerry Group. Deutsche Bank Global Consumer Conference, Paris, June 15*.

McCluskey, J. J. (2015). Changing food demand and consumer preferences. In: *Responding to Future Food Demands. Paper prepared for Agricultural Symposium, Federal Reserve Bank of Kansas City, July 14 –15*(p. 18).

Mehmeti, G. and Xhoxhi, O. (2014). Future food trends. *Annals of Food Science and Technology*, 15(2),392–400.

Melia, D. (2011). Trends in the food and beverage sector of the hospitality industry. In: *EuroCHRIE Conference, Dubrovnik, Croatia, October 2011*.

Mellentin, J. (2014). *Failures in Functional Foods and Beverages*. London, UK: New Nutrition Business.

Mendelsohn, R. (2014). The impact of climate change on agriculture in Asia. *Journal of Integrative Agriculture*, 13(4),660–665.

Meulenberg, M. T. G. and Viaene, J. (2005). Changing agri-food systems in western countries: a marketing approach. In: W. M. F. Jogen and M. T. G. Meulenberg (Eds.), *Innovation in Agri-Food Systems: Product Quality and Consumer Acceptance* (Chapter 2). Wageningen, The Netherlands: Wageningen Academic Publishers.

Moreira, R., Chenlo, F. and Torres, M. D. (2013). Rheology of gluten-free doughs from blends of chestnut and rice flours. *Food Bioprocess Technology*, 6, 476–1485.

Moskowitz, H. R., Porretta, S. and Silcher, M. (2008). *Concept Research in Food Product Design and Development* (p. 612). John Wiley & Sons.

Murphy, C. and Vertrees, R. (2017). Sensory functioning in older adults: relevance for food preference. *Current Opinion in Food Science*, 15, 56–60.

Nyembwe, P. M., De Kock, H. L. and Taylor, J. R. N. (2018). Potential of defatted marama flour-cassava starch composites to produce functional gluten-free bread-type dough. *LWT-Food Science and Technology*, 92, 429–434.

Onyango, C., Mutungi, C., Unbehend, G. and Lindhauer, M. G. (2011). Modification of gluten-free sorghum batter and bread using maize, potato, cassava or rice starch. *LWT-Food Science and Technology*, 44, 681–686.

Otsuka, K. (2013). Food insecurity, income inequality, and the changing comparative advantage in world agriculture. *Agricultural Economics*, 44(s1), 7–18.

Oyewole, P. and Khan, M. (2018). Factors influencing food-choice behaviour of African-American college students: implications for the foodservice industry. *Journal of Food Service Business Research*, 21(5),511–538.

Pachucki, M. A., Jacques, P. F. and Christakis, N. A. (2011). Social network concordance in food choice among spouses, friends, and siblings. *American Journal of Public Health*, 101, 2170–2177.

Parfitt, J., Barthel, M. and Macnaughton, S. (2010). Food waste within food supply chains: quantification and potential for change to 2050. *Philosophical Transactions of the Royal Society of London B: Biological Sciences*, 365(1554),3065–3081.

Pitta, D. A. (2010). Product strategy in harsh economic times: subway. *Journal of Product and Brand Management*, 19(2),131–134.

Popkin, B. M. (1999). Urbanization, lifestyle changes and the nutrition transition. *World Development*, 27(11),1905–1916.

Rai, S., Kau, A. and Singh, B. (2014). Quality characteristics of gluten-free cookies prepared from different flour combinations. *Journal of Food Science and Technology*, 51(4),785–789.

Raisfeld, R. and Patronite, R. (2017). A to V: the encyclopedia of vegan food. *New York Magazine*, November 2017, pp. 62–69.

Reisch, L., Eberle, U. and Lorek, S. (2013). Sustainable food consumption: an overview of contemporary issues and policies. *Sustainability: Science, Practice and Policy*, 9(2),7–25.

Ruby, M. B. (2012). Vegetarianism. A blossoming field of study. *Appetite*, 58(1),141–150.

Santeramo, F. G., Carlucci, D., De Devitiis, B., Seccia, A., Stasi, A., Viscecchia, R. and Nardone, G. (2018). Emerging trends in European food, diets and food industry. *Food Research International*, 104, 39–47.

Schmidt, J. B. (2005). What we still need to learn about developing successful new products: a commentary on Van Kleef, Van Trijp & Luning. *Food Quality and Preference*, 16, 213–216.

South Australian Food Innovation Centre (SAFIC). (2019). *Introduction to Functional Foods and Ingredients*. https://www.pir.sa.gov.au/__data/assets/pdf_file/0007/287683/Functional_ Foods_Guidance_-_What_are_Functional_Foods.pdf (accessed September 17, 2019).

Steenhuis, I. H. M., Waterlander, W. E. and de Mul, A. (2011). Consumer food choices: the role of price and pricing strategies. *Public Health Nutrition*, 14(12),2220–2226.

Stewart-Knox, B. and Mitchell, P. (2003). What separates the winners from the losers in new product development? *Trends in Food Science and Technology*, 14, 58–64.

Thow, A. M. and Hawkes, C. (2009). The implications of trade liberalization for diet and health: a case study from Central America. *Global Health*, 28, 5.

Van Huis, A. (2013). Potential of insects as food and feed in assuring food security. *Annual Review of Entomology*, 58, 563–583.

Van Kleef, E., van Trijp, H. C. M. and Luning, P. (2005). Functional foods: health claim-food product compatibility and the impact of health claim framing on consumer evaluation. *Appetite*, 44, 299–308.

Wang, L. and Bohn, T. (2012). *Health-Promoting Food Ingredients and Functional Food Processing, Nutrition, Well-Being and Health* (Dr. J. Bouayed, Ed.). InTech. ISBN: 978-953-51-0125-3.

Wheeler, T. and Von Braun, J. (2013). Climate change impacts on global food security. *Science*, 341, 508–513.

Wieser, H. (2007). Chemistry of gluten proteins. *Food Microbiology*, 24(2),115–119.

Winger, R. and Wall, G. (2006). Food product innovation: a background paper. In: *Food and Agriculture Organization of the United Nation Rome* (p. 35).

ARISE, Abimbola Kemisola, Akintayo Olaide

13 Understanding sensory evaluation of food

13.1 Introduction

One of the most important goals of the food industry is to determine how food affects the senses of consumers. This is also important for nutritionists and dietitians who develop healthier recipes. The benefits of a healthy diet can only be harnessed if our senses accept it. Therefore, the consumer's response perceived by the five senses is considered an essential measure of food development.

According to the Institute of Food Technologists, sensory evaluation is a scientific method used to evoke, measure, analyze, and interpret the responses to products as perceived through the senses of sight, hearing, touch, smell, and taste (Stone and Sidel, 2004, Sharif et al., 2017). Similarly, Watts et al. (1989) defined it as "a multidisciplinary science that uses human panelists and their senses of sight, smell, taste, touch, and hearing to measure the sensory characteristics and acceptability of food products, as well as many other materials." Although these definitions cover other materials apart from food, it is pertinent to note that they both make it abundantly clear that sensory evaluation involves all aspects of human senses rather than just sense of taste as being regarded in some climes (Stone and Sidel, 2004). Sensory assessment, since its advent in the 1940s, has been an exciting, dynamic scientific discipline, which continues to evolve and is now recognized as a renowned scientific field in its own right (Sharif et al., 2017).

Some of the areas where sensory evaluation finds application include product development, product improvement, quality control, storage studies, and process development. According to Watts et al. (1989), sensory evaluation can be consumer- or product-oriented. Consumer-oriented tests normally involve large number of untrained panelists and are used to reveal information about consumer likes and dislikes, preferences, and requirements for acceptability. On the other hand, product-oriented tests are used to obtain specific characteristics of a food, through the use of fewer but trained panelists whose sensory acuity is reliable.

Sensory evaluation can also be objective or subjective, that is, involving the use of machine or humans, respectively. Electronic devices are available for the simplification of sensory evaluation (Sharif et al., 2017); however, some schools of thought hold it that there is no instrument that can replicate or replace the human response. Sensory panel can, however, be treated as a scientific instrument to produce reliable

ARISE, Abimbola Kemisola, Akintayo Olaide, Department of Home Economics and Food Science, University of Ilorin, Kwara State, Nigeria

https://doi.org/10.1515/9783110667462-013

and valid results. This is possible when sensory test is conducted under controlled conditions, using appropriate experimental designs, test methods, and statistical analyses (Watts et al., 1989).

Hedonic scale of sensory evaluation rapidly gained popularity and application in the twentieth century in the wake of growth of food processing industries. With this technique, it was possible to precisely measure the human reactions to food-stuffs, which reveal how such foodstuffs are found appealing or otherwise (Sharif et al., 2017). The competition in the food industries for more space in the market continues to escalate yearly; hence, sensory analysis has become a vital part of food production. Sensory properties have been noted as the most vital attribute of food, as majority of consumers' complaints relate directly to sensory quality failures (Dzung et al., 2003).

13.1.1 Importance of sensory evaluation

Sharif et al. (2017) recently noted that sensory evaluation is an essential component of food product development, along with research and development as well as marketing departments. According to the authors, sensory testing can help in pinpoint-ing the imperative sensory characteristics driving acceptability during the early stages of product development. Target consumers, product competitors, and new ideas are prior assessed with sensory evaluation. "When a consumer buys a food product, they can buy nutrition, convenience, and image. Nevertheless, most impor-tantly, consumers are buying sensory properties, performance, and sensory consis-tency" (Dzung et al., 2003). Sensory evaluation of food and drinks is very important when a new product is expected to maintain the sensory attributes of an already ex-isting one. This is so because little difference sensed by consumers can greatly change their choices. Therefore, as far as marketing is concerned, it is more impor-tant to know what consumers think they perceive more than what they actually per-ceive (Singh-Ackbarali and Maharaj, 2013).

Sensory evaluation reduces the risk of product failure by ensuring that foods that are inferior from consumer point of view are not released in to the market (Sharif et al., 2017). It considers consumers' needs in line with those of producers in the devel-opment or optimization of a product (Singh-Ackbarali and Maharaj, 2013). Improved status and louder voices are some of the benefits that sensory evaluation has given marketing research and brand management professions (Stone and Sidel, 2004).

Understanding the relationships that exist between a few intrinsic properties of food and sensory attributes is a useful tool in the quantitative, qualitative, and stabil-ity analyses of the former. Some changes in food, which remain concealed even with chemical analyses, can be detected through sensory evaluation. For instance, Thomas et al. (1981) found that sensory evaluation was opted when chemical analyses failed to detect the deterioration of some developed medical foods during storage. Similarly,

Sharif et al. (2017) stated that chemical and physical properties of a product, which drive sensory attributes, could be ascertained by combining data obtained from sensory and instrumental testing.

13.1.2 How to conduct sensory evaluation

According to Van Oirschot and Tomlins (2002), the following are some useful guidelines in the conduct of sensory evaluations:

1. If consistent results with least possible bias are to be achieved, the design and use of sensory facilities including a controlled environment is required. In other words, there should be clean hygienic food preparation areas, similar testing booths, easy access, suitable temperature, a neutral color scheme, and good even lighting.
2. General principles for sampling are to be adopted to ensure that product tested is a representative of the entire population or batch. Most test methods compare two or more samples at once. Samples should be presented in a random, but consistent manner, to avoid any bias. The use of alphabetic symbols should be avoided. Codes that give the panelists no clue as to the differences in the food to be tested are used. Samples should be identical and water or crackers should be used for palate cleansing.
3. A panel comprising personnel who are willing to take part in the sensory evaluation exercise should be properly set up, and it is required that a personnel experienced in sensory testing is appointed as a leader.
4. The number of experts, trained, and untrained assessors are selected based on the type of adopted sensory method. Through training, the precision of would-be panelists is improved. The panel leader can then select the best assessors with the highest precision for the test method from the trained pool.
5. It is important that questionnaires are designed in such a way that they are easy to understand with minimal possibility for misinterpretations. If this cannot be guaranteed, direct questioning can be adopted.

13.1.3 Sensory properties of food

Apart from taste, sensory properties such as appearance, smell, sound, and texture influence what we select to eat. Food must taste delicious, certainly, but mouth-feel, texture, looks, and smell are also important for the overall eating experience. To better understand the principles of sensory evaluation, it is necessary to describe these aspects of food properties. These are appearance, flavor, aroma, texture, and sound.

13.1.3.1 Appearance

Appearance is the first characteristic perceived by the human senses, which plays an important role in the identification and final selection of food. This is the visual perception of food comprising color, shape, size, gloss, dullness, and transparency (Sharif et al., 2017). Appearance can be reduced to two principal factors: the physical and the psychological. The physical factors consist of the geometrical, the food dimensions of size, shape, and intrinsic characteristic variability in uniformity and mass; and the optical, surface gloss or dullness, the nature and degree of pigmentation, and the light-scattering power of the food structure (MacDougall, 2003). A product appearance impacts an individual's response to the product taste (Stone and Sidel, 2004). This is because we perceive the eating experience with our eyes before we even smell or taste. Lawless and Heymann (2010) stated that consumers often first assess the quality of food products, such as meats, fruits, and vegetables, by its appearance and color.

Sight gives information about the size, color, shape, and the texture of the products (Figure 13.1). The human visual system detects the light of the wavelengths from about 400 nm (violet) to about 700 nm (red). Visible wavelength ranges between 360 and 780 nm. The color mixing is created in the eye, and depends on stimulus of the incoming light (the receptors are stimulates).

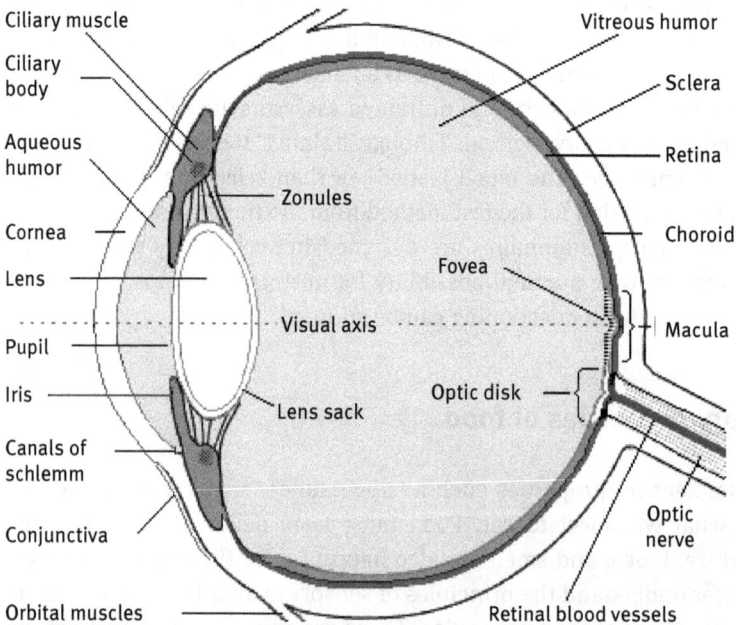

Figure 13.1: Parts of the human eye (Szabó, 2014).

Appearance of foods is very important as it influences how we perceive other sensory attributes such as flavor and taste. There is usually an acceptable specification or the color of every food product albeit this depends on certain factors such as those associated with the consumers, prevailing environmental conditions, as well as nature, pigmentation, and structure of the food itself. However, MacDougall (2003) noted that color specification alone is insufficient to define food appearance. Interaction between the structure of food and its variable light scatter and pigmentation could be very important in determining opacity, translucency, and color. Small changes in scatter can produce greater changes in the visual color appearance than are attributable to the normal range in pigment concentration in some products. Although it is important to distinguish between these, it is not always put into consideration during color measurement giving room to erroneous interpretation (MacDougall, 2003).

13.1.3.2 Flavor

The flavor of foods is usually defined as a combination of aroma and taste (Law, 1997). Sharif et al. (2017) further noted that it is a sensory phenomenon, which also denotes mouth-feel in addition to aroma and taste. The authors described in detail by stating that flavoring substances are aromatic compounds, which are conceived by the combination of taste and odor and perceived by the mouth and nose. In a more technical way, flavor is defined as a "complex combination of the olfactory, gustatory, and trigeminal sensations perceived during tasting" (Tournier et al., 2007). Although, it is understood that flavors may be associated with tactile, thermal, painful, and/or kinaesthetic effects (Tournier et al., 2007), the exact mechanisms involved in flavor perception are yet to be fully elucidated. These are attributable to the following reasons, namely, according to Taylor and Roberts 2004 (Tournier et al., 2007):
1. Flavor perception involves a wide range of stimuli.
2. The chemical compounds and food structures that activate the flavor sensors change as food is eaten.
3. The individual modalities interact in a complex way.

This sensory attribute depends on our taste buds. Taste buds are located on small bumps on the tongue called fungiform papillae (these are made up of about 50–150 taste receptor cells). On the surface of these cells, receptors are present that bind to small molecules related to flavor. Children have about 4,000–6,000 taste buds, adults have about 2,000–3,000, and the elderly have about 500–1,000 taste buds (Figure 13.2). The receptors in the tongue can distinguish four basic flavors, which are:
1. Sweet (sucrose)
2. Salt (NaCl): Salt receptor tastes the sodium ions

3. Sour (citric acid): Sour receptors detect the protons released by sour substances (acid)
4. Bitter (quinine sulfate)

The new flavors include metal (ferrous sulfate) and the umami (monosodium glutamate): a savory and subtle taste that is associated with a soupy or brothy note. The receptors on the tongue identify the glutamic acid residues in the food.

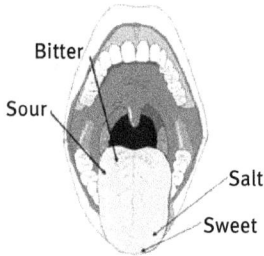

Figure 13.2: Taste buds of the human tongue (Szabó, 2014).

13.1.3.3 Aroma

Aroma compounds are volatile and they are perceived primarily with the nose. A specific aroma may be described for a particular food, for example, green, cheesy, and nutty. It is interesting to note that aroma and taste of food work together to produce flavor of the food. No wonder anyone who is suffering from cold may not be able to determine the flavors in food. Aromatic, floral, earthy, rotten, acrid, musty, fragrant, pungent, bland, rancid, tart, strong, weak, spicy, green, savory, oily, mild, scented, woody, citrus-like, leathery, creamy, buttery, and mossy are the usual odors or aroma perceptions of food.

13.1.3.4 Texture

One of the most important attributes of food is texture. This is because texture is observed by a combination of senses; for instance, it combines touch, mouth-feel, sight, and hearing. If a customer eats a sandy-textured ice cream, such customer will notice immediately and may not want to buy from such place again. Another example is frozen fish that can be tested if it is spoilt by its texture. Texture is a prerequisite to accepting many foodstuffs. For instance, when buying bread, the texture in the form of softness of the bread becomes a significant factor in choosing the bread to buy. Texture also includes the consistency, fragility, chewiness, thickness, size, and shape of particles in food. Texture analyzer is used to analyze the texture of food in the laboratory (Sharif et al., 2017).

Texture is clearly defined as a sensory attribute, and is thus only measurable directly by sensory means. It is only humans that can perceive, describe, and quantify texture. To analyze texture more completely, one should use sensory analysis methodology (Costell and Duran, 2002).

13.1.3.5 Sound

This is used to describe the sound made by food during mastication and ingesting. There is a certain sound that is expected of a particular product; if such is not heard, it is assumed that the product is no longer good for consumption, for instance, the cracking of crackers biscuit or fried chips. Therefore, we can conclude that in sensory analysis, the senses are used to measure, analyze, and interpret the organoleptic or sensory properties of food (Sharif et al., 2017).

13.1.4 Methods of sensory evaluation

Affective and analytical methods are the major sensory evaluation methods (IFT, 2007).

13.1.4.1 Affective testing

This method of testing is useful for preliminary investigations performed before consumer research and consumer-oriented testing. This method uses consumer panels to answer questions such as:
1. Which product do you prefer?
2. Which product do you like?
3. How well do you like this product?
4. How often will you buy or use these products?

Affective methods require the use of untrained assessors; at least 50–100 assessors are recommended. Separate sensory panels should be established for affective testing. Affective testing measures pleasure responses and the subjective feel about a product. Consumers render final judgment regarding whether they will purchase, prepare, or eat a food product. Unfortunately, proper consumer research conducted in compliance with recommended guidelines for best results can be quite costly. Whenever possible, most researchers choose to use laboratory tests as predictions of consumer responses. Data relationships between laboratory and consumer research help to reduce costs fairly, but some consumer response is always necessary to determine the success of any product treatments. Good consumer research requires attention to four areas: subjects, location, test design, and questionnaire design. A minimum of about 100 subjects who represent the consumer population, the population for whom the product is intended (users or potential users), is necessary. Consumer's age, gender, and so on should also be considered with regard to certain products, as it may affect responses and representation of the intended consumer population. The location(s) where the test is conducted should represent the current or potential market in terms of geographical area (Northeast, Midwest, Southeast, Southwest, West), rural, suburban, urban groups, or selected ethnic groups. The test design takes into account the number of products each consumer in each location can test and may require some statistical support to develop. The questionnaire asks the key questions, such as overall liking and liking questions for general appearance, flavor, and texture. A questionnaire may also include attribute diagnostics. All of these pieces of information are important and can be vital in developing data relationships to descriptive data to develop predictive models. An example of the type of questionnaire used in a sophisticated data relationship project by Murioz, Bonnans, and Chambers is shown in questionnaires 1 and 2.

Questionnaire 1

Name ————————————
Date ————————————
Sample ————————————

Hot Dogs

Please look and taste product and answer the following questions:

Overall opinion

Considering all the appearance, flavor, and texture characteristics of this hot dog, please circle one number below to indicate how much you liked or disliked the product overall

0   1   2   3   4   5   6   7   8   9   10

Dislike extremely   Neither like nor dislike   Like extremely

What, specifically, did you like and dislike about this hot dog? (Use words, not sentences)

Likes	Dislikes
—————————	—————————
—————————	—————————
—————————	—————————

Source: Munoz, Bonnans and Chambers (1994).

Questionnaire 2

OVERALL APPEARANCE

Considering all the **APPEARANCE** characteristics of this hot dog, please circle one number below to indicate how much you **LIKE** or **DISLIKE** the appearance of this product.

0	1	2	3	4	5	6	7	8	9	10

Dislike extremely　　　　　Neither like nor dislike　　　　　Like extremely

OVERALL FLAVOR

Considering all the FLAVOR characteristics of this hot dog, please circle one number below to indicate how much you LIKE or DISLIKE the flavor of this product.

0	1	2	3	4	5	6	7	8	9	10

Dislike extremely　　　　　Neither like nor dislike　　　　　Like extremely

OVERALL TEXTURE

Considering all the TEXTURE characteristics of this granola bar, please circle one number below to indicate how much you LIKE or DISLIKE the texture of this product.

0	1	2	3	4	5	6	7	8	9	10

Dislike extremely　　　　　Neither like nor dislike　　　　　Like extremely

13.1.4.2 Analytical methods

Most common analytical methods of sensory evaluation are discrimination (or difference) and descriptive methods. Discriminatory tests are used to determine whether a difference exists between samples. The panelist does not allow personal likes and dislikes to influence the response. Laboratory difference panels can be used to determine whether there is a difference among samples. Descriptive tests are used to determine whether there is a difference among samples. Descriptive tests are used to determine the nature and intensity of the differences.

13.1.4.2.1 Difference tests
Difference testing is a way to determine whether a sensory difference actually exists between samples. The degree or nature of the difference cannot be quantified. Descriptive tests are generally needed to truly define differences. The test to determine a difference between samples include the triangle test, the simple paired comparisons test, the duo-trio test, the multiple comparisons test, ranking, scoring, ratio scaling, and others. In this chapter, the major ones such as the triangle test, simple paired comparisons test, and duo-trio test are discussed.

13.1.4.2.1.1 Triangle test
In this type of test, the panelist receives three coded samples. He or she is told that two of the samples are the same and one is different and is asked to identify the odd sample. This method is very useful in quality control work to ensure that samples from different production lots are the same. It is also used to determine

whether ingredient substitution or some other change in manufacturing results in detectable difference in the product. The triangle test is frequently used for selecting panelists.

Analysis of the results of triangle tests is based on the probability that if there is no detectable difference, the odd sample is selected by chance one-third of the time. A sample of the questionnaire is shown below.

Questionnaire 3: Questionnaire for triangular tests

Name————————————————   Date————————-

Product ——————————————————

Two of these three samples are identical, the third is different.

1. Taste the samples in the order indicated and identify the odd sample

    Code Check odd sample

    314 ———————————

    628 ———————————

    542 ———————————-

1. Indicate the degree of difference between the duplicate samples and the odd sample

    Slight ————————————

    Moderate ———————————

    Much ——————————————

    Extreme ——————————————

3. Acceptability:

    Odd sample more acceptable ——————-

    Duplicates more acceptable ——————

4. Comments

    ——————————————————————————————-.

    ——————————————————————————————-.

    ——————————————————————————————

    Source: Ihekoronye and Ngoddy (1985).

13.1.4.2.1.2 Simple paired comparisons test

A pair of coded samples is presented for comparison based on some specific characteristics such as sweetness. This method has application similar to the triangle test. Fewer samples are required and there is less tasting but the statistical efficiency is not as great. The probability of a panelist selecting a sample by chance is 50%. Paired comparisons tests give no indication of the size of the difference between the two samples, but determine whether there is a detectable difference. A sample questionnaire for a simple paired comparisons test is shown below.

Questionnaire 4: Questionnaire for simple paired comparisons test

Name ————————————-·   Date ————————————-·

Product ————————————————————————————

Evaluate the sweetness of these two samples of canned mangoes. Taste the sample on the left first. Indicate which sample is sweeter.

581   716

——————————   ————————

Comments:

                ————————————————————————————-·

                ————————————————————————————-·

                ————————————————————————————-·

        Source: Ihekoronye and Ngoddy (1985).

13.1.4.2.1.3 Duo-trio test

In the duo-trio test, three samples are presented to the taster, one is labeled R (reference) and the other two are coded. One coded sample is identical to R and the other is different. The panelist is asked to identify the odd sample. The duo-trio test has the same applications as the triangle test, but is less efficient because the probability of selecting the correct sample by chance is 50%. This test is often used instead of the triangle test when tasting samples that have a strong flavor because less tasting is required.

It can be used in place of the paired comparisons test where the panelist is asked which sample has more of some specified characteristics, whereas in the duo-trio and the triangle tests, the panelist bases his/her judgment on any difference he/she can detect. An example of this questionnaire is shown.

Questionnaire 5: Questionnaire for duo-trio test

Name ————————————   Date ———————

Product ————————————

On your tray you have a marked control sample (R) and two coded samples. One sample is identical to R and the other is different. Which of the coded samples is different from R?

Samples Check odd sample

432   ————————-·

701   ————————

Comments

_____.

_____.

_____.

Source: Ihekoronye and Ngoddy (1985).

13.1.4.2.2 Descriptive sensory analysis

This analysis is a valuable tool in difference testing and in product development. The perceived levels (intensities) of each of the described attributes are measured (quantitative aspect). Methods of descriptive analysis can only be used by a highly trained (expert) panel, usually consisting of a minimum of six to eight assessors. The result is usually a sensory profile or fingerprint of each product. It provides a complete description of sample differences and guides the product developer in modifying product characteristics to meet consumer demands. The training of profile panels requires considerable time and members must possess a high degree of motivation and interest. Once trained, the panel can provide thorough and reliable descriptions of products in a short time. As the descriptive panel members work together as a group, potent members could have undue influence on the other panelists and hence, change the results. Although descriptive analysis methods have been in use for decades, this method still works better than some other methods. For instance, Oreskovich (1994), during her PhD, compared results from a conventionally trained descriptive panel with results from a free-choice profiling panel for flavor and texture analyses of pork. She concluded that the free choice, that is, untrained panel could discriminate and describe differences in pork treatment effects; however, the free-choice method did not describe attributes that characterize the unique properties of food (Oreskovich, 1994). Therefore, the free-choice profiling method provides less specific information about a product, as the results provided only relative differences among treatments.

13.1.4.2.2.1 Descriptive analysis with scaling

A method of sensory evaluation called quantitative descriptive analysis was developed at the Stanford Food Research Institute. During preliminary sessions, the sensory properties of the product are identified by trained panel. Samples are made up to show the different properties so that the panel agrees on the meaning of each term used. During these sessions, the judges work together as a group and discussion is encouraged. The panel members decide the sensory properties that should be evaluated and also develop the language to be used. During the evaluation sessions, the panelists work individually. Each judge records his/her evaluation by making a vertical line across the horizontal line at a point that best reflects his/her perception of the magnitude of that property. A sample questionnaire for descriptive analysis with scaling is as follows.

Questionnaire 6: Questionnaire for descriptive analysis with scaling

Name ———————————-·   Date ————————

Please evaluate the firmness and chewiness of these samples of dried meat.

1. Firmness: Make vertical lines on the horizontal line to indicate your rating of the firmness of each sample. Label each vertical line with the code number of the sample it represents.

   Please taste the samples in the following order:

   572   681   437   249

   Very soft   very firm

2. Chewiness: Make vertical lines on the horizontal line to indicate your rating of the chewiness of each sample. Label each vertical line with the code number of the sample it represents.

   Very mushy   Very rubbery

Comments

———————————————————————————-·

———————————————————————————·

———————————————————————————-·

Source: Ihekoronye and Ngoddy (1985).

13.1.4.2.2.2 Applications of descriptive profiling

Descriptive profiling can be used to define the sensory properties of a target product for new product development. It can be used to define the characteristics (specification) of a control or standard, as well as for quality assurance or quality control and research and development purposes. Furthermore, it can be used to monitor changes in the sensory properties of a product during shelf-life. In addition, description of product attributes before consumer testing can also be achieved by descriptive profiling.

13.1.5 Factors affecting sensory measurements

Apart from sensory gadgets, psychological or physiological factors can easily affect human decisions. To diminish or eradicate such biasness, panelists should pick right protocols and experimental design (Hough 2010). The possible sources of error and recommended approaches for tumbling associated effects are discussed further.

13.1.5.1 Psychological factors

1. **Expectation error:** The information about the samples to be assessed or objectives of the investigation can influence the decision of the panelists because the assessor is inclined to find what is desirable. Try to avoid including such persons in the panel who are familiar with the product. Furthermore, minimum possible information should be shared with the judges and do not reveal information concerning the samples unless it is indispensable due to religious integrity, especially while using innovative ingredients. During sample coding, use random three-digit numerics rather than colors or alphabets. Numbers like 15, 911, 1,122 have specific links, hence should be avoided. Similarly, codes such as "A," "1" or round numbers (e.g., 100, 250) can be related with a higher score.
2. **Suggestion:** Effective sensory evaluation should be conducted in designated facility to avoid the influence of commentaries or sounds on the judgment. There should be separate sensory booths for sample evaluation and judges should be discouraged for any discussion related to samples before or after assessment, except when instructed to do so.
3. **Stimulus and logical error:** Logical error happens when the stimulus is rationally allied with one or more of the attributes under appraisal. This occurs when evaluators use extra information while making a decision about the samples. For example, food products with intense color are supposed to be more intense in flavor. Similarly, a thin cream layer is assumed as poorer quality. Sometimes conducting sensory evaluation at an unusual time may prompt evaluators to consider about a manufacture problem. Similarly, the use of costly containers may lead appraisers to think that the foodstuff is of superior quality. Try to disguise irrelevant variations and ensure that sample characteristics are consistent using suitable colored illumination, coverings, and ear guards.
4. **Distractions:** Distractions such as listening to radio, chatting, and personal obsessions in the evaluation area can easily influence the panelists. Use of electronic as well as communication devices should be prohibited in the test area to create noise-free environment. Furthermore, create an atmosphere that inspires professionalism among the evaluators.
5. **Attribute dumping:** In designing sensory performa, efforts should be made to include all possible attributes for optimum evaluation. It has been noticed that if judges are not given the chance to rate all the traits they observe in the foods under assessment, they still exhibit their opinion using existing attributes. For example, if samples are varying in sweetness but no sweetness characteristic is included in the performa, panelists record this variation on a flavor intensity scale.
6. **Order effect:** This hedonic response of preceding sample can influence the attribute of the next sample, for example, a sample is considered less sweet if next one is with greater intensity. Furthermore, the sample position may affect the scoring, for example, sample placed at first position is generally assigned

higher scores in hedonic scale. Order effect can be reduced through randomization or balancing the order of sample presentation. Using a mock sample at position, one is suggested for effective evaluation.

7. **Acclimatization**: This happens where judges evaluate similar stuff repetitively. To diminish the influence of adaptation, present spiked samples from time to time or vary product. This is because continuous exposure of evaluator to a particular stimulus, particularly at high concentration for long time, leads to decrease in sensitivity (also called as fatigue). It is, therefore, desirable either to give sufficient time between the samples or use taste sanitizers, such as brine solutions, fruits, and mild acids. The taste sanitizers improve the taste sensitivity or bring it back to normal level.

8. **Halo effect and proximity error**: This usually happens while judging numerous traits in a single run, especially by the untrained assessors. In this case, scoring one attribute may influence the assessment of other characteristics, for example, sweet sample may be regarded as stickier. Wherever possible, assess one or at least a limited number of qualities at a time. Furthermore, try to assign trained panelist and randomize the order of characteristic when evaluating several traits deemed necessary at once.

9. **Contrast and convergence effects**: To reduce the contrast and convergence effects, randomize or balance the order of presentation of samples and consider eliminating outlying samples from the sample set.

10. **Motivation error**: An interested judge performs sensory evaluation more consistently. Usually assessors rate the samples based on their feeling about the food manufacturer or team leader. This can be a concern, especially if the evaluation is being carried out by the company employees. To minimize this error, try to conduct the sensory analysis in a professional manner by giving regular feedback and give due respect to the judges.

11. **Central tendency error**: This is more likely to happen with untrained evaluators or when they are not conversant with the product range. When using scales, judges mostly give scores to the middle of the scale. Encourage them to use broad scale to differentiate between the products and this is especially important when using unskilled assessors. There is a need for panelist training in the use of the scale and exposure of a wide range of products to cope with central tendency error.

13.1.5.2 Physiological factors

1. **Adaptation**: There is a strong association between reduction in sensitivity with the continuous exposure of a stimulus and reduction in response to other stimuli. Subsequently, sensory assessment is affected by the adaptation to a stimulus. These are known as "carry-over effects." To decrease the adaptation impacts, first confine the number of samples to be evaluated. Second, ensure sufficient

time interval before proceeding to the next sample to recover the sensory system. This duration ranges from a few seconds to hours, contingent with the stimulant, for example, cooling can take 10 minutes to retreat. The panelists should be provided appropriate palate cleansers to ensure cleaning of oral cavity, for example, milk may be desirable for some spicy compounds.

2. **Physical condition**: Sensory evaluation is strongly influenced by the age, health, nutritional disorders, hormonal state, stress level, and mood of the assessors. The evaluator should be physically and mentally in good health. The sensitivity for evaluator with regard to the sense of smell and taste should be normal. He/she should not be suffering from anosmia and ageusia. In terms of age, evaluators should preferably be in the age group of 18–50 years. Persons of younger age are unable to properly interpret and communicate the sensory results; whereas at older age, the memory decreases. Sharp memory of evaluator is considered highly useful in passing judgement. In addition, sensory sessions should be scheduled around a similar time each day preferably between 10 am and lunch. Furthermore, ask evaluators to restrain from eating for at least an hour before evaluation.

3. **Perceptual interactions between stimuli**: Certain stimuli can interact to cause suppression (existence of one ingredient diminishes the perceived concentration of another, e.g., sourness reduces apple flavor), potentiation (occurrence of one element surges the intensity of another, e.g., Chinese salt accelerates the meat flavor), and synergy (intensity of a mixture is greater than the intensity of the sum of the individual components, e.g., sweetness and sourness impact on strawberry flavor).

13.1.5.3 Cultural factors

Cultural factors are especially important when working with assessors from diverse regions or cultures. In some cultures, specific product codes may have significant associations; for example, eating in public is considered as a social offense; religious limitations may influence sample selection. In addition, the use of a scale can differ across the cultures, for example, some are inclined to score lower than "average" or much higher while using the hedonic scale. For effective sensory evaluation, be aware of cultural tendencies as these strongly influence many aspects of sensory testing such as products, protocols, scale use, and feedback.

13.1.6 Conclusion

Sensory evaluation encompasses a set of test methods and recognized techniques of product presentation, statistical methods, and strategies for interpretation of results. Accurate application of sensory technique involves correct corresponding method to the objectives of the tests followed by good communication between sensory experts and end-users of the test results. In food industries, the sensory evaluation department does not only interact with the product development department, but may also deliver information to quality control, packaging, marketing, and many other groups throughout the company. The main advantages of sensory information include development of food products in an economic way by lowering the risks in decisions about product development and strategies for meeting consumer needs. Note that sensory tests used with technical care, creativity, and a large dose of common sense can deliver more information for decision-making at a lower cost and in shorter time compared to other data collection methods.

References

Costell, E. and Duran, L. (2002). Food texture: sensory evaluation. *Food Engineering* (1), 1–18.

Dzung, N. H., Dzuan, L. and Tu, H. D. (2003). The role of sensory evaluation in food quality control, food research and development: a case of coffee study. *Proceeding of the 8th ASEAN Food Conference*, 862–866.

Hough Guillermo (2010). Sensory shelf-life estimation of food products. CRC Press, Taylor and Francis group, Boca Raton, London, New York.

IFT (Institute of food Technologists). (2007). *Sensory Evaluation Methods*. Chicago, IL: The Society for the Food Technologists.

Law, B. A. (1997). *Microbiology and Biochemistry of Cheese and Fermented Milk* (pp. 1–56). Springer Science and Business Media.

Lawless, H. T. and Heymann, H. (2010). *Sensory Evaluation of Food: Principles and Practices (Second Edition)* (pp. 563–566). New York, NY: Springer.

MacDougall, D. B. (2003). Sensory Evaluation: Appearance. In B. Caballero, L. Trugo, & P. M. Finglas (Eds.), Encyclopedia of Food Sciences and Nutrition (2 ed., pp. 5161–5166). USA: Academic Press.

Oreskovich, D. C. (1994). *A Comprehensive Approach to Determining the Nutritional and Sensory Properties of Today's Cooked Pork*. Ph.D. thesis. The University of Illinois at Urbana-Champaign.

Sharif, M. K., Masoos, S., Hafiz, R. and Muhammmad, N. (2017). *Sensory Evaluation and Consumer Acceptability*. Faisalabad: University of Agriculture.

Singh-Ackbarali, D. and Maharaj, R. (2013). *Sensory Evaluation as a Tool in Determining Acceptability of Processed Local Agricultural Products*.

Stone, H. and Sidel, J. L. (2004). *Introduction to Sensory Evaluation. Sensory Evaluation Practices* (pp. 1–19). Boston, MA: Elsevier Academic Press.

Szabó P. B. (2014). Sensory evaluation in food industry. In: TÁMOP-4.1.1.C-12/1/KONV-2012-0014 „Élelmiszerbiztonság és gasztronómia vonatkozású egyetemi együttműködés, DE-SZTE-EKF-NYME" projekt segítségével jött létre, 32.

Thomas, S., Caporaso, F., Miller, G. and Kiral, R. (1981). Importance of sensory evaluation in the stability testing of a medical food. *Journal of Food Science*, 46, 435–439.

Tournier, C., Sulmont-Rossé, C. and Guichard, E. (2007). Flavour perception: aroma, taste and texture interactions. *Food*, 1, 246–257.

Van Oirschot, Q. E. and Tomlins, K. I. (2002). Applying analytical sensory evaluation techniques, which translate qualitative perceptions to numerical data to research on development issues. In: *Conference on Combining Qualitative and Quantitative Methods in Development* Research, *Centre for Development* Studies, *University of Wales*, Swansea, *Citeseer* (pp. 1–2).

Watts, B. M., Ylimaki, G., Jeffery, L. and Elias, L. G. (1989). *Basic Sensory Methods for Food Evaluation*. Ottawa, Canada: IDRC.

Mlungisi Mtolo, Faith Ruzengwe, Oluwatosin Ademola Ijabadeniyi

14 Food packaging and packaging innovations

14.1 Introduction to packaging

Packaging is a coordinated system that allows the distribution of goods or products to the final consumer in an acceptable state as per the intended utilization (Kelsey, 1989). The Packaging Institute International describes packaging as the enclosure of products, items, or packages in a wrapped pouch, bag, box, cup, tray, can, tube, bottle, or other containers. These are developed so to accomplish several purposes such as containment, protection, preservation, communication, utility, and performance (Anonymous, 1988). Coles et al. (2003) have also described packaging as (1) a coordinated system of goods or products essential for the transportation, distribution, storage, retailing, and end-use; (2) a way to ensure safe delivery of product to the final consumer in a good condition at a cheap ideal cost; and (3) a techno-commercial function aimed at delivering maximized sales in a cost-effective manner which will then result in higher profit margins (Coles, et al., 2003).

There are three levels of packages which are primary, secondary, and tertiary. According to Robertson, 2006, these three levels are distinctly essential. However, in terms of food safety, the most important is the primary package because it is in direct contact with the product. The secondary package holds together the individual packets of the goods while the tertiary package protects the product, primary as well as the secondary packages.

Packaging forms an essential part of modern life through enabling the society to get their desired quantities of several products, consume fresh, safe foods, and beverages (IPSA, 2019). However, it can negatively affect the environment. About 3% of the waste sent to landfill, whether measured by weight or by volume, is from the different packages that are in use. The EIPRO 2008 (Environmental Impact of Products) reported that packaging accounts for 8% of the total energy requirement of packaged products on average. Hence, signifying that the resources used for the products which are inside the package are about 10 times greater compared to those used for the packaging (IPSA, 2019).

The packaging industry is continuously seeking ways to improve its environmental impact profile, according to Von Holdt (2011) these include:

Mlungisi Mtolo, Reckitt Benckiser South Africa (Pty) Ltd, Elandsfontein, South Africa
Faith Ruzengwe, Oluwatosin Ademola Ijabadeniyi, Department of Biotechnology and Food Technology, Durban University of Technology, South Africa

https://doi.org/10.1515/9783110667462-014

1. Innovations to utilize resources more efficiently also reduce the exhaustion of resources and pollution as well as reduction of costs;
2. Close-looped packaging industrial systems were introduced to assist with the reduction of material amount as well as the elimination of waste. These are in use for recycling and reuse systems where suitable. Some of these include the involvement of business-to-business packaging systems which the consumer does not notice.

14.2 Food packaging

Food packaging assists in enabling storage, handling, transportation, and preservation of food products as well as prevention of food waste. Food packaging is vital in the modern food industry, and for a food packaging technologist to successfully perform their professional duties, they must acquire knowledge from multiple disciplines including chemistry, microbiology, and physical science (Robertson, 2006).

Food packaging protects the food from environmental and element (such as odors, shocks, dust, temperature, physical damage, light, microorganisms, and humidity) contamination that can negatively influence the quality of food. Thereby, minimizing food losses and wastage, maintaining the quality and safety of food, while also ensuring the increase of the shelf life. Postharvest losses can also be reduced by packaging, which collectively with offering access to more markets, and permit producers to maximize their profits. This indicates that efficient packaging in developing countries has a significant impact on both the pattern of food consumption and the quantity of food consumed (Fellows, 2008).

Food packaging use started in the eighteenth century (Brody, et al., 2008). Several improvements in packaging technology appeared in the twentieth century; these include:
- Intelligent or smart packaging (IOSP)
- Time–temperature indicators (TTI)
- Gas indicators
- Microwave doneness indicators
- Radiofrequency identification (RFID)
- Active packaging (AP; like oxygen scavengers, moisture absorbers, and antimicrobials) (Brody, et al., 2008)

The introduction of these technologies enhanced the levels of food quality, food safety, and shelf life (Han, 2018). Developments within the food processing and food packaging industries also play a crucial role in keeping a safe food supply chain in the world. The benefits brought about by food processing are maintained by packaging, thereby enabling safe long-distance transportation of food products.

That is from their point of origin and production to the point of consumption without losing their wholesomeness.

Food packaging has numerous advantages; however, there is a need to balance the significant benefits associated with food protection with other issues. These issues include energy and material costs, high levels of social and environmental awareness, as well as strict regulations regarding pollution and removal of municipal solid waste (Marsh and Bugusu, 2007). Generally, the food industry spends approximately 15% of the total variable costs on packaging materials (Han, 2018). The environmental issues result from the high production volume, short usage time and challenges that are associated with waste management and littering. Hence, specific tools such as reduction, reuse, recycling, and redesign, which play a significant role in supporting the goals of the circular economy, have been in use. Such tools also have the probability of reducing the negative environmental impact of food packaging (Geueke, 2018).

14.3 Food product characteristics and packaging

Package design and development professionals need to have a strong knowledge of product characteristics. Such characteristics include deterioration mechanisms, distribution requirements, and possible interactions with the package (Gupta and Dudeja, 2017). Also, there is need to know the physical, chemical, biochemical, and microbiological nature of the food product. Materials that deliver the best protection of product quality and safety should be given preference. Likewise, the type of the packaging material to be utilized is depended on the distribution systems and conditions (Gupta and Dudeja, 2017).

In particular, the interaction of food and package plays a significant part in the appropriate selection of the actual packaging material for a certain application (IFT, 1988). The inherent properties of packaging materials differ with the material (e.g., rigidity and permeability to gases). These properties impact the selection of which material is best for a specific food, given the characteristics of that food (e.g., acidity and light sensitivity) (IFT 1988).

Food–package interaction comprises of the transportation of low-molecular-weight compounds such as gases or vapors and water from (1) the food through the package, (2) the environment through the package, (3) the food into the package, and (4) the package into the food. Also, it may include chemical changes in the food, package, or both. These interactions result in food contamination (a potential health issue), loss of package integrity (a potential safety issue), or reduced in quality (IFT 1988). According to Arvanitoyannis and Bosnea (2004), the interactions that are most common between food and package are the transfer of substances with low molecular weight, for example, stabilizers, plasticizers, antioxidants, monomers, and oligomers. These substances can be transferred from plastic packaging materials into food.

One of the essential criteria in the selection of food package is material quality and safety. It should be composed of materials that preserve the quality and safety of the food for a reasonably extended duration without degradation over time (Doyle, 1996). According to Marsh and Bugusu (2007), the food package must be attractive, convenient, easy to use, have all the appropriate information, sourced from renewable resources, not generate waste for disposal and be inexpensive. However, it is challenging for today's food packages to meet this high standard goal. Creation of a food package is an art–science because the food package needs to accomplish excellent overall outcome without reducing the standards in any single category to be below acceptable levels (Marsh and Bugusu, 2007).

Several materials are used in packaging, for example, glass, plastics, paper, and metal. Out of all primary packaging materials, the glass makes the best choice for most packaging applications from a unique product perspective; this is because of its inertness and absolute barrier properties (Foodservice Packaging Institute, 2006). However, the use of glass has some economic disadvantages and hence increases the usage of alternative materials such as plastic. Plastics are offering several advantages regarding properties and used in several food applications. Their permeability is not ideal – contrasting metal, which is entirely impervious to light, moisture, and air (Marsh and Bugusu, 2007). Efforts to having a package that has all the required properties can sometimes be resolved by mixing packaging materials – such as merging various plastics using co-extrusion or lamination – or by laminating plastics with foil or paper (Marsh and Bugusu, 2007). The consumer has the ultimate and significant influence in package design. Consumer requirements determine product sales, and the package is an essential sales instrument (US Environmental Protection Agency, 2015).

14.4 Functions of food packaging

Numerous functions performed by packaging include protection of the contents from contamination and spoilage, simplifying transportation and storage of goods and provision of a consistent amount of contents (Hine, 1995). Also, packaging made it possible to create and standardize brands, which allow the businesses to advertise meaningfully as well as distribute in large scale. Other unique features that came with packaging are, for instance, dispensing caps, sprays, and others assisted in improving the convenience of using packaged products (Doyle, 1996). Apart from that packaging help in product promotion as well as serving as a communication tool between the producer and the consumer. Promotion and convenience are closely related; hence, packages that are convenient drive sales (Doyle, 1996). In 1985, the Codex Alimentarius commission issued a definition of food package functions; this definition state that "Food is packaged to preserve its quality and freshness, add appeal to consumers and to facilitate storage and distribution."

Furthermore, these significant functions are interrelated and must be evaluated and considered concurrently in the packaging development process. The following functions such as containment, protection, convenience, and communication are discussed in detail further.

14.4.1 Containment

Containment is defined as the containing of products in order to enable their storage and movement (Shin and Selke, 2014). Even though the containment function of packaging is the most noticeable, it can be easily overlooked by the majority of packaging design stakeholders (Bureau and Multon, 1996). This function is essential because all products have to be contained before being moved from one place to another; an exception can apply to large and discrete products. The package must contain the product regardless of the size and nature of the product (Bureau and Multon, 1996). The lack of containment results in product loss as well as contamination of products. This function contributes mainly to the protection of the product from numerous environmental challenges that the product may encounter when being transported from one place to another on numerous instances each day in any modern society (Boyette, 1996). With containment comes portion control, a benefit that ensures that the consumer receives approximately the same amount of product as per expectations (Thyberg and Tonjes, 2016). Portion control also results in inconsistency of the packaged food product, which means there would be fewer consumer complaints. The benefits of having fewer consumer complaints are that the company reputation improves because it will have reasonable consumer expectation (Stancu, et al., 2016). Also, containment is a crucial function for other packaging functions.

14.4.2 Protection

Protection is also referred to as preservation. Food packaging must be able to offer physical, chemical, and biological protection. The protection function mainly affects the primary level of packaging. This includes the protection of its contents from the physical damage (such as shocks, vibrations, and compressive forces) and environmental damages occurring as a result of the exposure to water, moisture vapor, gases, odors, microorganisms, dust, as well as protection of the environment from the product (Robertson, 2006). Hence, a good packaging must be able to reduce and protect the food product from both physical and environmental damages.

Apart from the above, packaging also plays a significant role in energy utilization during the production and processing of the product (Lindh, et al., 2016) and is also essential for the preservation of foods. Packaging materials preserve the freshness and integrity of food or beverage products for specified shelf life (Alsaffar, 2016). The preservation of product contents through packaging results in minimization of food spoilage and wastage (Bureau and Multon, 1996).

14.4.3 Convenience

Convenience is sometimes known as a utility, and the features include ease of access, handling and disposal, product visibility, resealability, and microwave ability. The world has undergone significant changes which came as a result of modernization as well as industrialization. This has had a high level of impact on modern lifestyles, and the packaging industry has had to respond to those changes (IPSA, 2019). There have been lot of changes in the social structure, which include the change in the average family structure, the current reality is that a three-member family structure is the most popular and dominant family structure (Robertson, 2006). Hence, these changes and other factors like the trend towards eating snack meals regularly and on-the-go meals over regular meals generated a market for more convenience in household products (Brody, et al., 2008). There is a higher demand for food products that fit such lifestyles offering both simplification and convenience.

Packaging contributes to the convenient handling of product during the production, storage and distribution system, including easy opening, dispensing and resealing. Furthermore, it also simplifies the disposal process, recycling, and reuse (Hine, 1995). Packaging design has two other significant aspects of convenience that are essential in package design that is portion sizing and unitization (Recordati, 2015).

14.4.4 Communication

As a general idea, a package must protect what it sells and sell what it protects. One of the package purposes is to serve as a silent salesman (Judd, et al., 1989). Product development as well as packaging design professionals needs to develop with an understanding that, the current era is an information age, with consumers being educated, curious, and interested in knowing everything about the product (IPSA, 2019). In order for the modern methods of consumer marketing to work, the messages have to be communicated by the package (Robertson, 2006). Branding and labeling of products need to be distinctive enough to allow consumers instant recognition of products. If the packaging does not include communication about design, shopping experience for consumers at the supermarket would not be adequate, resulting in the lack of information regarding the products which will guide the consumer on deciding which product to purchase (Eldred, 1993).

Communication of relevant information on the package also improves the workflow in warehouses and distribution centers by the elimination of chaos, incomplete details, and unambiguity, especially in secondary and tertiary packages (Robertson, 2006). Such relevant information includes the use of the Universal Product Code (UPC) in both retail stores and warehouses. Also, attached to the secondary and tertiary packages is the RFID tags used by manufacturers to get better demand signals from their markets and customers (Shin and Selke, 2014).

There is also a marketing element in the packaging communication function, and this is because packaging attracts consumers to the product, which is part of the brand experience (Von Holdt, 2011). Packaging ensures that consumers can recognize the product, and through design, they can be able to differentiate a product on the shelf (Brody, et al., 2008). Part of the brand experience includes engaging a consumer in opening, accessing, and consuming a product and all these activities need to appeal to a number of a consumer's sense.

14.5 History of food packaging

In ancient times, the only materials used for packaging were natural. These materials comprised of leaves, animal skin, bark, coconut shells, dried vegetable skins, and others (Hirsch, 1991). As time went by, there were further developments on the type of materials used for packaging that is baskets of reeds, wooden boxes, wooden barrels, and woven bags. With those materials, there was a certain level of complexity compared to those used before them. Later, materials like pottery vases and water storage containers came into use, and these involved even more complexity on their manufacturing process compared to those previously used (Foodservice Packaging Institute, 2006). In the same way, in ancient times, packaging mainly focused on the covering and protection of the product.

The industrial revolution resulted in the development of new manufacturing processes and materials. Although most of the new manufacturing processes and materials were initially not planned for food products, however, they later became useful as food packaging materials (Risch, 2009). The industrial revolution led to the increase in demand for food packaging methods, primary packaging materials, bags, storage and transportation bins, and in-store packaging options.

Apart from that, the industrial revolution also brought about the use of metal cans as food packaging materials. Metal cans can maintain the moisture of the product through the provision of a barrier as well as protect the flavor of the food product (Risch, 2009). Before the use of metal can, glass bottles with corks, secured with a wire as the closure was used to contain food while heating; however, these glass bottles were fragile. Hence, the use of metal cans allowed the heat processing of food products for a more extended period, thereby increasing the shelf life and prevented product spoilage.

Folding cartons were first manufactured in the early 1800s from paperboard. Around the 1850s, the now most commonly used corrugated boxes as shipping containers were first developed (Risch, 2009). These were manufactured after Robert Gair discovered that by cutting and creasing the making of prefabricated paperboard boxes was made possible (Foodservice Packaging Institute, 2006).

Polyvinyl chloride (PVC) discovery has revolutionized food packaging. A seal could be formed without clinging to itself, food, or to the container. Other advantages included that PVC has a low permeability to oxygen, water vapors, and flavors. However, there have been concerns about its toxicity. Plastics such as cellulose nitrate, styrene, and vinyl chloride were discovered in the 1800s. In the twenty-first century, food packaging has evolved as a specialized industry (Foodservice Packaging Institute, 2006).

14.5.1 Traditional packaging materials

Traditional packaging materials used in the earliest times when domestic storage and local merchandising of the foods were the only available options (Hanlon, 1992). These materials were only used for the containment of foods as well as to keep them clean, and this is because the traditional packaging materials had poor barrier properties, except for glazed pottery.

Below is the discussion on the main types of traditional materials and their uses:

14.5.1.1 Leaves, vegetable fibers, and textiles

Leaf-based packaging came into use because it was inexpensive and easily accessible. Leaves, used as wrappers for cooked food products, were consumed in a short period. Examples of the leaves used include those from coconut and *palas* used for wrapping sticky rice, banana used as a wrapper for traditional cheese, rice, and fruit confectionery like guava cheese, casuarina for *tempe* wrapping, and bamboo for food and water holding (Fellows, 2008). Fibers from vegetables and plants such as *kenaf* and *sisal* were used for rope, cord, and string making. These can then be used for making net bags that can transport hard fruits. Textile containers examples are woven jute sacks, which are utilized in transportation of different types of bulk foods such as grain, flour, sugar, and salt (Fellows, 2008). However, although textile containers are used as packaging materials, they have poor gas and moisture barrier properties and appearance when compared to plastics.

14.5.1.2 Wood

Wood used as a packaging material for making wooden boxes, trays, and crates. These were traditionally used as shipping containers for a wide variety of solid foods, including fruits, vegetables, and bakery products (Fellows, 2008). Apart from that, wooden barrels were used for a wide range of liquid foods, including cooking oils, wine, beer, and juices. Nowadays, they are used for some wines and spirits. This is because flavor compounds from the wood improve the quality of the products; however, in other applications, they have been replaced by aluminum, coated steel, or plastic barrels (Fellows, 2008).

14.5.1.3 Leather

Camels, pig, and goatskins were used for making leather containers which are flexible, lightweight, and nonbreakable for storing water, milk, and wine. Leather cases and pouches have also been used for packing flour and solidified sugar. The use of leather has been stopped in commercial applications (Fellows, 2008).

14.5.1.4 Earthenware

Pottery is still being used worldwide, especially for domestic purposes. The uses include storage of liquid and solid foods such as yoghurt, beer, dried foods, and honey. Preserved fish, pickled foods, and vegetables are also stored in clay pots (Fellows, 2008). Nowadays, due to high weight and fragility, pottery has been replaced with plastic and glass containers.

14.6 Types of food packaging materials

Different food packaging materials were developed over the last 200–300 years (Fellows, 2008). Generally, the choice of the packaging material to be used is dependent on several factors such as food item, the process of production and quality of food, shelf and life desired transport considerations. Also, the other factors taken into consideration include the shape, size, color, stacking options, printing of labels, cost, environmental attributes (e.g., recyclability, carbon imprint), and handling properties. Various packaging materials are in use in the food industries, and these are classified into two groups: retail containers and shipping containers. Retail containers, also used for home storage, mainly protect the food product from damages and help advertise the product during the retail sale. On the other hand, shipping containers are for containing and protecting the food during distribution and transportation. Some examples are discussed below:

14.6.1 Glass

Glass is a popular packaging material made from sodium carbonate, silica, and calcium carbonate as well as other compounds that are added to add color, sparkle, and heat shock resistance. The advantages of using glass include the resistance to microorganisms, pests, moisture, oxygen, and odors, a certain level of being resistant to pressure possibility to sterilize, ability to mold into a variety of shapes and can be recycled.

Additionally, glass can be stacked without damage, and unlike metal cans, it is transparent and hence can display the contents. Also, glass is nontoxic, nonleaching, easy to clean, nonreactive to food/chemicals, nonporous, and relatively cheap (Fellows, 2008). However, several disadvantages are associated with glass as a packaging material, and these include that it is of higher weight than most other types of packaging; hence, higher costs are incurred during transportation and can be easily broken mainly when transported over rough roads and there are potentially severe hazards that come from the glass splinters or fragments. Glass containers are used for foods like juices, wines, beers, pickles/chutneys, and jams, especially in countries that have a glass-making factory (Fellows, 2008).

14.6.2 Aluminum

As the most abundant material available on earth, aluminum has extensive use as a food packaging material (Foodservice Packaging Institute, 2006). The advantages of using aluminum are that it is of lightweight, good strength ratio, and high-quality surface for decorating or printing, reasonably robust, long-lasting, and is recyclable. Apart from that, it provides a barrier to moisture, odors, light, microorganisms, and gases as well as high resistance to temperature fluctuations and hence can be able to extend the food product's shelf life. Aluminum is used for making cans, metallic trays, pouches, tubes, and foils. The two basic types of metal cans are: (a) those that are sealed using a "double seam," which are used for canned foods, and (b) those that have push-on lids or screw-caps used for packing dried foods (e.g., milk or coffee powder and dried yeast) or cooking oils, respectively. Aluminum foils can store food directly. Aluminum can also be used for paper and plastic lamination mainly for better strength, heat stability, and barrier against moisture, oils, air, and odors. Such materials are used for soups, herbs, and spices packaging. However, both the high metal and manufacturing costs make cans expensive compared to other containers.

Also, since metals are heavier than plastic containers, they, therefore, have higher transport costs (Foodservice Packaging Institute, 2006).

14.6.3 Plastic

Plastic is a generic term for a related wide range of synthetic materials that are commonly used for food packaging. Examples of packaging material made from plastics are bags, wraps, bottles, tubs, buckets, containers, resealable pouches, and bowls. Plastics are widely used mainly because of their low cost, high strength, long-lasting, lightweight, air-tight, and recyclable. Apart from that plastics can be quickly processed, formed, and are resistant to chemicals; hence, they can maintain both the shelf life and the freshness of the food product. However, sometimes this is dependable on the moisture content of the food. Generally, moisture-free foods can be stored in plastic bags for a long time. Although there are several advantages associated with plastics, their main disadvantage is that they are nonbiodegradable hence posing harm to the environment. In addition to that, plastics are also permeable to gas, vapor, and light; hence, they reduce the shelf life and affect the sensory properties of other food products. Another cause for concern is the leaching and diffusion of substances from plastics into food that may be carcinogenic (Foodservice Packaging Institute, 2006).

The two basic types of plastic packaging materials are flexible plastic and rigid/semi-rigid plastic. Flexible plastic films are of relatively low-cost, heat-sealable to prevent leakage of contents, add a minor amount of weight to the product and fit tightly to the shape of the food. Hence, they do not waste space during storage and distribution. In addition to this, they have good wet and dry strength and can be handled easily, thus offering convenience to the manufacturer, retailer, and consumer (Hirsch, 1991). The main disadvantages are that (except cellulose) these plastic films are produced from nonrenewable oil reserves and are not biodegradable.

There is a wide choice of plastic films made from different types of plastic polymers. Each has mechanical, optical, thermal, and moisture/gas barrier properties. It can be noted that they are produced by variations in film thickness, amount, and type of additives used in their production. The different types of flexible plastic films are discussed below:

Cellulose: This is a shiny transparent film, and its main advantages include that it is odorless, tasteless, and biodegradable. Although cellulose breaks easily, it is durable and puncture-resistant. The dead-folding properties of cellulose make it suitable for twist-wrapping (e.g., sugar confectionery). However, it is not heat-sealable, and the dimensions and permeability of the film vary with changes in humidity. Cellulose, because of its properties, has found application in foods that do not require a full moisture or gas barrier such as fresh bread and some types of sugar confectioneries (Fellows, 2008).

Low-density polyethene (LDPE): LDPE is the cheapest among all other films and hence has extensive use for flexible tubes, films, and some bottles. It has a sound barrier for moisture but is relatively permeable to oxygen and is a weak odor barrier. The properties of LDPE include that it is heat-sealable, inert, odor-free and has a low melting point; hence, it shrinks when heated. (Fellows, 2008).

High-density polyethene (HDPE): HDPE is widely used for bottles. The HDPE is stronger, thicker, less flexible, and more brittle than LDPE and has a better barrier to moisture and gases, although it is not sufficient for carbonated drinks. HDPE makes sacks that have high tear and puncture resistance and have good seal strength. HDPE also has a higher melting point; however, it is not ovenable. They are waterproof and chemically resistant, which is enhanced by fluorination and are increasingly used instead of paper sacks (Fellows, 2008).

Polypropylene (PP): PP is a transparent, glossy film, with a high strength though less brittle when compared to the other plastics and puncture resistance. Apart from that, PP has a moderate barrier to moisture, gases, and odors; hence, it is not affected by humidity variations. PP can stretch less than polyethene. Due to its high melting point, PP has found full application in food packaging products that are used in microwaves (Hirsch, 1991). PP is widely used in making packaging containers for yoghurts, maple syrup, cream, and sour cheese.

PVC and polyamides (or Nylons): PVC is a prevalent type of plastic that is very strong and is used in the formation of thin films. PVC is both biological and chemical-resistant, thereby providing high barrier to gas and water vapor. Containers that are made from PVC help maintain the integrity of food products. Additionally, PVC is heat-shrinkable and heat-sealable. However, the brown tint has limited its use in specific applications. Another group of films are those from polyamides (or Nylons), which are clear and strong over a wider temperature range. They also have a low permeability to gases and are greaseproof (Bureau and Multon, 1996). However, on the other side, these films are expensive to produce, and they require very high temperatures to seal. Hence, these are used with other polymers to make them sealable at low temperatures and improve their barrier properties and are used to pack meats and cheeses. Also, their permeability changes with changes in the storage humidity (Hanlon, 1992).

Other examples of films include coated films – films glazed with different polymers or in some cases, aluminum to enhance their barrier properties or to impart heat-sealability (Arvanitoyannis and Bosnea, 2004). Metallization (thinly coated aluminum) creates a significant obstacle to light, moisture, gases, odors, and oils. The metallized film is less expensive and more flexible than plastic/aluminum foil laminates (Lamberti and Escher, 2007). Laminated films – with lamination (bonding together) of two or more films – improve the appearance, barrier properties, or mechanical strength of a package (Hirsch, 1991). Coextruded films, the simultaneous extrusion of two or more layers of different polymers to make a film, are used for confectionery, snack foods, cereals, and dried foods (Hirsch, 1991). A wide range of plastic bottles, pots, jars, trays, and tubs made from single or coextruded plastics are used for processed foods. They are used as, cups or tubs for margarine, trays for meat products, bottles and jars for fruit juices, bottles for carbonated drinks, squeezable bottles and pots for mustard, trays for chocolates, and foam cartons or trays for eggs (Coles, et al., 2003).

14.6.4 Paper and paperboard

Paper and paperboard are age-old packaging items that are made from cellulose-based materials. Cellulose fibers are produced from the wood pulp through the use of sulfite and sulfate. Additives are also used to produce packaging materials with the required properties. The main advantage of paper and paper board is that they are biodegradable, burned (with energy recovery), or recycled, are permeable to air, water vapor, and gases (oxygen), and have excellent mechanical strength. The permeability to gases is as a result of the waxing that is done to the paper. Usually, the wax is laminated between layers of the paper and polyethene since the coatings can be easily damaged. Their applications include bread wrappers and inner liners for cereal cartons (Marsh and Bugusu, 2007). Other examples of packaging materials made from paper include a wide range of bags and boxes for different applications (Chiellini, 2008). Paperboards are thicker and have a higher weight per unit ratio than paper; hence, they are mainly used as shipping containers. There are different types of paperboard used in food packaging (Brody, et al., 2008):

- Whiteboard, which is made from several thin layers of the bleached chemical pulp, is suitable for contact with foods. It is often coated with wax or laminated with plastic to make it heat sealable. Whiteboard is used for making ice cream, chocolate, and frozen food cartons (Kerry, et al., 2006).
- Chipboards are made from recycled paper and are used usually for making outer cartons for tea or cereals. Since chipboard has impurities from the original paper, it does not come in contact with foods. Whiteboard can be used for lining the chipboard; this improves the appearance and strength. Examples of molded paperboard trays are those used for eggs, fruit, meat, or fish or egg cartons (Kerry, et al., 2006).
- Small paperboard tubs or cans are used for snack foods, confectionery, nuts, salt, cocoa powder, and spices (Kerry, et al., 2006).
- Solid board is made from a series of layers of bleached sulfate board. It possesses high strength and durability. Solid board is sometimes laminated with polyethene to make liquid cartons usually referred to as milk board. These can as well be used for fruit juices and soft drinks packaging (Raheem, 2013).
- Corrugated boards are widely known as cardboard; they are used in secondary packaging, and sometimes they come in direct contact with food, for example, pizza boxes. They are composed of three layers of paper that are the outside

liner, inside liner, and the fluting, which is the corrugated medium. Generally, the corrugated board resists impact, abrasion, and compression damage and is therefore used for shipping containers (Bureau and Multon, 1996). Corrugated boards are strong, as a result of the fluting used during their preparation. There are different types of corrugated boards; the smaller and more numerous corrugations give the rigidity, while larger corrugations or double- and triple-wall corrugated material provides cushioning and resists impact damage. The lining of the corrugated board with polyethene or a laminate of wax-coated greaseproof paper and polyethene results in packaging material for wet foods, chilled bulk meat, dairy products, and frozen foods (Boyette, 1996).

14.6.5 Other packaging materials or techniques

Aseptic packaging: This is a technique in which the food item and the packaging itself are sterilized separately, they are then combined, and the sealing will be done under the sterilized atmosphere (Sadiku, et al., 2019). An example is an ultrahigh temperature (UHT) milk. Aseptic packaging is mainly used in a food preservation method.

Tetra packs: These are tetrahedron-shaped plastic-coated paper carton, with aseptic packaging technology. They are used for the distribution and storage of dairy, beverages, cheese, ice creams, and prepared foods (Kelsey, 1989).

Pouches: There are various types of pouches prepared from high-quality material such as spout pouches, zipper pouches, reusable pouches, and printed stand-up pouches. These pouches are durable and environmental-friendly. Also, they have the food labeling like manufacturing date, expiry date, nutrient content, and logos (Eldred, 1993).

Retort packaging: The retort pouch was invented by the US Army Natick R&D Command, Reynolds Metals Company, and Continental Flexible Packaging (Foodservice Packaging Institute, 2006). A retort pouch is slowly replacing the use of cans in packaging due to their flexibility and effectiveness. Retort pouches are made up of plastic and metal foil laminate. A total of three or four wide seals usually created by aseptic processing usually are used. These allow for the sterile packaging of a wide variety of drinks ranging from water to fully cooked as well as thermo-stabilized meals. The produced packages are of lightweight and are less expensive to ship. In this technique, food is first prepared (raw or cooked) and after that sealed into the retort pouch. The pouch is heated to 240–250 °F (116–121 °C) for several minutes under high pressure, inside retort, or autoclave machines. The process reliably kills all commonly occurring microorganisms, preventing it from spoiling (Fellows, 2008). Table 14.1 presents a summary of the materials used for food packaging

Table 14.1: Materials used for food packaging.

Materials	Examples of use
Paper	Bags, boxes, cartons
Glass	Bottles, jars
Metal	Cans, aluminum foil
Plastics	Overwraps, pouches, cups, boxes, bottles
Laminates	Cartons for liquids, multi-layered plastics
Wood	Crates, pallets
Cloth	Sacks, special packages
Wax	Coating of fruits and vegetables, some cheeses
Edible containers	Particular uses (ice cream cones, cabbage leaves)

Reference (Jelen, 1985).

14.7 Packaging for various food products

Food products are categorized into several groups with the common groups being meat products, fresh fruits and vegetables, dairy products, cereals, snack foods, and confectionery and beverages. These different food groups have different packaging requirements, as discussed below.

14.7.1 Meat products

Freshly packed meat has been one of the significant meat products in the market since the early 1900s. The main reasons for meat packaging are for storage, preservation, and protection adversary environmental factors that would otherwise cause quality degradation (Barlow and Morgan, 2013). Hence, the shelf life of meat and meat products has been dramatically improved through the use of vacuum packaging (VP) and modified atmosphere packaging (MAP) along with refrigeration (Cachaldora, et al., 2013). Application of VP and MAP can be combined with overwrapped thermoforming films, and this has become a common practice in meat packaging (Arvanitoyannis and Stratakos, 2012). Overwrap packaging is being used for short-term chilled storage or display. However, these techniques do have some challenges, such as the potential negative impact on product sensory attributes, enormous costs, and difficulty in achieving standardized coating procedures for large-scale commercial operations. Hence, there is much research being done on improving the manufacturing and application processes of economically feasible edible coatings and films that are being intended for the meat industry (Fang, 2015). It can be noted that packaging intended for fresh meat provides a limited barrier to moisture; thus, surface desiccation is prevented (Faustman and Cassens, 1990).

14.7.2 Fresh fruits and vegetables

Fruits and vegetables can be the package for easy handling, transporting, and marketing as well as for maintaining the minimum shelf life. As a result, bags, crates, baskets, cartons, bulk bins, and palletized containers are convenient containers (Boyette, 1996). Corrugated fiberboard, manufactured in many different styles and weights, is the most common type of packaging used for fruit and vegetables (Williams, et al., 2012). Aged due to abrasion is reduced through the addition of internal packaging such as paper wraps, trays, cups, and pads. Another fresh fruits and vegetable package is paper and mesh bags used for packing potatoes, onions, cabbage, turnips, citrus, and some specialty items. Mesh bags are mainly used because of their low cost and the uninhibited airflow (Yam, et al., 2005). Other packages for products include plastic bags (polyethene film), shrink wrap, and rigid plastic packages (Boyette, 1996). However, polyethene and PP have a low water vapor permeability; hence, both films and bags are perforated to allow for the breathing of the product.

14.7.3 Dairy products

The primary role for packaging milk and the milk products apart from maintaining the product quality is to provide a physical barrier which prevents various damages such as mechanical, physical, and microbial contamination (Ščetar, 2018). Packaging materials for milk and dairy products include paper and paper-based products, glass, tinplate, aluminum foil, polymers, metal, and laminates. Materials such as glass used in the form of bottles, tumblers, jars, and jugs are used for the packaging of fresh milk, yoghurt, and cream (Arvanitoyannis and Stratakos, 2012). Products such as powdered, condensed and evaporated milk are usually packaged in metals (Fellows, 2008), while cheese is packaged in films such as polyvinylidene chloride (PVDC), and cellophane is nitrocellulose coated (Dermiki, et al., 2008). LDPE plastic is used in liquid milk (sachet packs) and condensed milk (squeeze bottles) packaging. HDPE (plastic) is employed in the packaging of milk, yoghurt, sour cream, and ice cream (Duncan, et al., 2009). Laminated papers combined through a combination of two or more webs of different surfaces are used to pack mostly butter products. Coextrude-laminated (aseptic) paper, excellent barrier against moisture, gas, odor, light, and UV light, is mostly used in milk, milk powder, as well as yoghurt packaging (Alvarez and Pascall, 2011). Also, waxed paper is suitable for ice cream or cream and butter (Alvarez and Pascall, 2011).

14.7.4 Cereals, snack foods, and confectionery

14.7.4.1 Packaging of cereals

Cereals such as wheat, rye, sorghum, maize, barley, millet, oats, and rice are also known as grains. Wheat is used to produce breakfast cereal, meal, and flour for bakery products. The most crucial property of cereals that can be lost due to moisture gain is crispiness (Chinnadurai and Sequeira, 2016). Hence, in order to maintain this feature cereals are usually packaged in a combination of plastic films. Layers like PP, laminated PP, and aluminum-metalized polyester films form good moisture barriers and offer seal performance (Shewry, 2009). Apart from that, both primary and secondary packages are used in packaging cereal products.

14.7.4.2 Packaging of snack foods

Snack food packages are supposed to have the following requirements: prevent rancidity, avoid loss of crispness; prevent them from physical, chemical, and biological contaminants; protection from tempering, and enhance appearances to attract the consumers (Chinnadurai and Sequeira, 2016). Snack foods are usually packaged in flexible packages such as sachets, pouches, and composite cans. Paperboard cans which will be lined with aluminum foil–LDPE and flushed with nitrogen are used for packaging fried and extruded snacks while recycled cellulose fiber (RCF) coated with LDPE or PVDC copolymer with a layer of glassine is used for biscuits (Robertson, 2006).

14.7.5 Packaging of confectionery

Confectionaries are packaged depending on their water activity. That is the appropriate packaging for use is depended on how the product will gain or loss of moisture from the relative humidity of the surrounding atmosphere (Chinnadurai and Sequeira, 2016). For instance, chocolate, toffees, and candies are wrapped individually to protect them from atmospheric moisture, to prevent them from sticking, and to avoid intermixing of flavors. Cast PP is a widely used package confectionery since it twists and holds it easily. Apart from that, waxed paper, waxed glassine, and waterproof plasticized RCF are also used. However, these packaging materials offer little water vapor resistance. The secondary outer packages are to act as a moisture barrier. These include metal containers, glass jars, foil, metallized laminates, paperboard cartons, plain or waxed glassine liner, or heat-sealed bags made from coated RCF (Robertson, 2006).

Nowadays, there has been a shift in confectionery packaging towards the use of flexible packaging. The used pouches are provided with zippers, thus making them easy to be resealed, as a result, they provide firm and clean barrier properties. The other advantage is that they are strong to prevent punctures and tears. The flexible pouches are lighter, easier to dispose, are associated with low transportation costs and reduced fuel emissions since more packages can be transported at a time. Another package that is gaining attention is the stand-up pouches. These pouches are aesthetically appealing to the consumer as visibility of brand logo is increased (Chinnadurai and Sequeira, 2016).

14.7.6 Beverages

Beverage packaging requirements include a leak-proof that is able to prevent contamination, chemical deterioration, pick up of external flavors, and retain the carbonation in the case of carbonated beverages, economical, easy to use, and dispose of good aesthetic appearance (ICPE, 2012).

14.7.6.1 Packaging materials for non-alcoholic beverages

Fruit Beverages: Glass is an ideal packaging material for high-quality fruit beverages. The most common cans in fruit beverages packaging are tinplate cans, especially those made of low carbon mild steel of 99.75% purity, coated with tin with easy-open ends are used. Hot filling applications require the use of PET bottles (Kern, 2002).

Coffee: Tinplate containers, composite containers, glass jars, and flexible plastic pouches are usually used for coffee packaging. The flexible laminates most widely used are 12 μ PET/2 μ Al Foil/70 μ LDPE and MET PET/LDPE. Aluminum foil–lined plastic pouches are also popular (Rashmi and Smita, 2003).

Tea: Paperboard carton with a liner or an overwrap of PP or regenerated cellulosic film are common in tea packaging. Other types are plastic jars, bottles, pouches, strips, and envelopes (Kumar, 2002).

Carbonated drinks: Among the plastic containers, PET bottles are the most preferred packaging material for packaging of soft drinks (Kern, 2002).

14.7.6.2 Packaging materials for alcoholic beverages

Beer: Glass bottle sealed with a crown closure is the traditional package for beer. However, recent development has resulted in the use of different types of PET bottles for beer packaging (ICPE, 2012). These include the non-tunnel pasteurized, one-way tunnel pasteurized, and returnable/refillable bottles. Beer needs high performance in both CO_2 and O_2 barrier compared to PET used in carbonated soft drinks (CSD) applications. The level required depends on the type of beer, container size, distribution channels, and environmental conditions (storage time, temperature, and humidity levels) (ICPE, 2012).

Wine: Glass bottles sealed with natural cork are used in wine packaging. The preferred glasses are the colored ones. This is because wines are affected by sunlight glass (ICPE, 2012). A barrier to the ingress of oxygen is provided by keeping the bottled wine in a horizontal position so as to keep the cork moist. Wines are packaged in PET bottles and stand-up pouches of metallized polyester laminates. The extensive change in the packaging of wine resulted from the development of the bag-in-box package (ICPE, 2012).

Brandy and Whisky: These are mostly packed in glass bottles since they have high alcohol percentages and in order to increase their storage time even after opening. However, to prevent alcohol from evaporating and to protect the contents of the bottles from dirt and dust, such bottles are sealed (Industry Indian Food, 2003).

14.8 Future direction of food packaging

There are several innovative packaging techniques with the aim still being to maintain and improve the safety and quality of food, extend shelf life, and reduce the environmental burden of food packaging being driven by several reasons such as:
- change in consumer demands,
- industrial production trends,
- retailing practices,
- lifestyles of customers (Gupta and Dudeja, 2017).

Hence within the past 20 years, a series of packaging techniques have been developed. These include IOSP, AP, and sustainable or green packaging (SOGP) (Brody, et al., 2008).

14.8.1 Intelligent or smart packaging

This is defined as packaging that has an external or internal indicator. These indicators provide required information on package history and the packaged food quality (Robertson, 2006). Hence, decisions regarding shelf life, safety, and quality, as well as alert people to possible problems with food can be conveyed (Yam et al., 2005). It can then be noted that intelligent packaging can be essential in information communication; hence, it is an extension of the communication function. The sensors or indicators include electronic, chemical, and mechanical triggers. In an intelligent packaging system, the indicators used are put within the primary packaging; hence, they have direct contact with the food product or headspace (Han, 2018).

There are two groups of smart devices in IOSP that are classified as indirect or direct indicators of food quality. The indirect indicators cannot provide direct information to help consumers judge the quality and edibility of food. However, they can evaluate the effects of the environment surrounding the food on the shelf life and quality of the food which might lead to a hidden danger to the consumers (Wang, et al., 2017). On the other hand, direct indicators can directly present some information about the freshness, edibility, quality, and safety of food to consumers. Intelligent packaging can be used to check AP effectiveness and integrity (Hutton, 2003).

14.8.2 Time–temperature indicators

TTIs were developed to monitor time- and temperature-dependent changes, thereby maintaining the food product quality and safety. Slight temperature changes in a food product result in product safety and quality changes. TTIs are indirect indicators, commonly used in the food industry due to their small size, cost-effectiveness, and being user-friendly compared to other temperature-monitoring devices (Han, 2018). TTI communicate temperature changes that is tampering and abuse to the consumer. TTIs can identify irreversible responses such as enzymatic, electronic, chemical, nanoparticle, or biological changes after a product has been exposed to a higher temperature (Han, 2018). For instance, TTI use in shipping containers or individual packages, irreversible changes such as color changes in the product occurs when TTI has been tampered with. TTIs can be classified into three types based on their capabilities: (1) critical temperature indicators show if a food product has been exposed to a different temperature to the reference that is either a higher or sometimes lower; (2) critical TTIs will indicate the additive effect of the time–temperature changes on product quality or safety after exposure to temperatures above a reference temperature; and (3) full history indicators are used for the continuous monitoring of the food product in cases where the temperature varies with time throughout a product's history (Han, 2018).

Table 14.2: Examples of external and internal indicators and their working principles in intelligent packaging (adapted from Han et al., 2005).

Technique	Principles/reagents	Information given	Application
Time-temperatures indicators (external)	Mechanical, chemical, enzymatic	Storage conditions	Foods stored under chilled and frozen conditions
Oxygen indicators (internal)	Redox dyes, pH dyes, enzymes	Storage conditions Package leaks	Foods stored in packagers with reduced oxygen concentration
Carbone dioxide (internal)	Chemical	Storage conditions Package leak	Modified or controlled atmosphere food packaging
Microbial growth indicators (internal/external) and freshness indicators	pH dyes, all dyes reacting with certain metabolites	Microbial quality of food (i.e. spoilage)	Perishable foods such as meat, fish, and poultry
Pathogen indicator (internal)	Various chemical and immunochemical methods reacting with toxins	Specific pathogenic bacteria such as *Escherichia coli* O:157	Perishable foods such as meat, fish, and poultry

14.8.3 Radiofrequency identification

RFID tags are based on wireless communication (magnetic field or electromagnetic wave) with the capability of providing real-time data to be able to trace and identify the product. Such information can be about the temperature, relative humidity, and nutritional and supplier information as the product moves through the supply chain, (Bibi, et al., 2017). Generally, the reader will emit a radio signal so that data are captured from an RFID tag, which is then passed to a computer for analysis. RFID is essential for product identification than traditional labels and barcodes since it has a relatively large data storage capacity, a more extended reading range and does not require visual contact. Usually, the RFID tags are embedded in an item, placed inside food packing. Therefore, RFID tags are considered to be a replacement for barcodes. However, RFID usage is limited because of their relatively high cost. To overcome such obstacles, future studies are expected to be aimed at reducing the cost of RFID tags as much as possible (Ruiz-Garcia and Lunadei, 2011).

14.8.4 Gas indicators

Gas indicators were developed to monitor changes in gas composition within the package. These changes in the composition of gas within a packaged food product can be due to food interaction with its environment. Hence, gas indicators monitor the quality and safety of food products. There are different types of gas indicators used in food packaging in the form of package labels or package printing films with the ability to detect oxygen, ethanol, hydrogen sulfide ($H_2 S$), water vapor, carbon dioxide, or other gas components (Fang, et al., 2017). For instance, oxygen indicators show color changes when it is present, thereby indicating an improper sealing, a leak, or tampering of the package.

14.8.5 Active packaging

AP has been defined as the packaging made together with subsidiary constituents to improve the performance of the packaging system (Robertson, 2006). The primary purpose of AP is to extend the shelf life, maintain, or improve properties of the food product. This should be based on the interactions between active compounds and food and packaging headspace (Elsamahy, et al., 2017). Hence, AP is associated with the protection function of a package. It is used to protect against oxygen and moisture.

There are different types of APs; they are as follows:

1. Oxygen scavenging agents: these react with oxygen, thereby reducing its concentration. This minimizes chances of oxidative degradation of food (e.g. fruits), hence improving their shelf life. Inactive packaging systems, the oxygen can even be scavenged by an outside source such as UV light.
2. Carbon dioxide absorbers and emitters: these are added for suppression of microbial growth, for instance, in fresh meat, poultry, cheese, and baked goods.
3. Hygroscopic agents: these help to control moisture and water activity, hence minimizing the growth of microbes in the food products, for example, in sweets and candy.
4. Antimicrobials help eliminate or reduce surface contamination of processed food, thus enhancing quality and safety. The packaging material is made in such a way that it controls the release of antimicrobials. For instance, in an AP system called BioSwitch, the antimicrobials will be released when there is any bacterial growth. These antimicrobial compounds can be encapsulated in polysaccharide particles; hence, they are released upon command, and the system is only active at certain specific conditions (Lopez-Cervantes, et al., 2003).

14.8.6 Sustainable or green packaging

SOGP employs the use of manufacturing methods and materials for product packaging that do not pose or have a low impact on the consumption of energy and the environment. This packaging technique has become famous among researchers since it seeks to improve environmental sustainability and minimize the environmental impact of the entire product-packaging chain (Peelman, et al., 2013). SOGP can be achieved at three levels as follows:

i. Raw materials level: This is achieved through the use of recycled materials and renewable resources. This is because these reduce CO_2 emissions and higher dependency on fossil resources.

ii. Production process level: SOGP utilizes lighter and thinner packaging produced through the use of relatively energy-efficient processes.

iii. Waste management level: It is essential to reuse or recycling of food packaging material as this eliminated the problem of municipal solid waste (Licciardello, 2017).

14.8.7 Nanopackaging

Nanotechnology is defined as the use of engineered nanomaterial that has been intentionally synthesized or incidentally produced to exploit functional properties exhibited on the nanoscale (FSA1, 2008). The International Standards Organization has also defined nanotechnology as "understanding and control of matter and processes at the nanoscale, typically, but not exclusively, below 100 nanometres in one or more dimensions where the onset of size-dependent phenomena usually enables novel applications, where one nanometer is one thousand millionth of a metre" (ISO/TC 229, 2008). Nanotechnology involves characterization, fabrication, and manipulation of structures, devices, or materials that are in 1–100 nm range of length (Foodservice Packaging Institute, 2006). As a result of this, the polymer barrier properties, material strength, flame resistance, thermal properties and surface wettability, and hydrophobicity are enhanced. Thus, nanotechnology innovation has produced new packaging concepts ideal for barrier and mechanical properties, pathogen detection, and active and intelligent packaging.

The combination of intelligent packaging and AP is vital in food packaging innovation. Future developments in food packaging have to be directed towards "ultimate" packaging. Ultimate packaging combines IOSP, AP, and SOGP; hence, all the benefits are also combined to improve environmental sustainability. Technological innovations in food packaging have the potential to enhance the safety and quality of food products, safety concerns, and limitations must be considered (Han, 2018).

14.8.8 Packaging and food safety

It is vital to ensure that packaging materials in contact with food are free of small amounts of synthetic chemicals which can leach into food or drink bottles, resulting in health issues. Can coatings, the laminate on beverage cartons, caps and closures on glass jars contain chemicals that can migrate into food (Packaging-Gateway, 2014). Furthermore, some chemicals such as Bisphenol A, tributyltin, triclosan, and phthalates used in food and drink packaging have been reported to interrupt hormone production (Packaging-Gateway, 2014). Because of these reasons, all packaging used in food contact must have specific approval from regulatory agencies for the intended use. It is also vital that food manufacturer get written assurance from the packaging manufacturer that their container meets all requirements for use in food contact (Potter and Hotchkiss, 1995).

14.9 Conclusion

Packaging is essential to the viability of the food industry. The incredible achievement of the industry has been made possible because of packaging development, design, and innovation. Packaging protects the food products from the environment and numerous factors that can compromise the quality, safety, and wholesomeness. This helps to maintain the product's shelf life along the value chain. Concerns regarding cost and environmental pollution have to be always considered while selecting a particular food package. Investments in packaging material to reduce the food waste of products with high environmental impact are justifiable. Massive technological developments are on the pipeline, and these will make the food packaging almost as intelligent as the consumer.

Acknowledgements: This work is based on research supported in part by the National Research Foundation of South Africa, SA (NRF)/Russia (RFBR) Joint Science and Technology Research Collaboration (Grant Number: 118910).

References

Alsaffar, A. A. (2016). Sustainable diets: the interaction between food industry, nutrition, health and the environment. *Food Science and Technology International*, 22(2),102–111.
Alvarez, V. and Pascall, M. (2011). *Packaging in Encyclopedia of Dairy Sciences (Second Edition)*. San Diego, CA: Academic Press.
Anon. (1988). *Glossary of Packaging Terms*. Stamford, CT: The Packaging Institute International.
Anon. (n.d.). Food distribution and. In: *Mayor's Office of Sustainability* (Chapter 3). s.l.: s.n.

Anonymous. (1988). *Glossary of Packaging Terms*. Stamford, CT: The Packaging Institute International.

Arvanitoyannis, I. and Bosnea, L. (2004). Migration of substances from food packaging materials to foods. *Critical Reviews in Food Science and Nutrition*, 44, 63–76.

Arvanitoyannis, I. and Stratakos, A. (2012). Application of modified atmosphere packaging and active/smart technologies to red meat and poultry: a review. *Food and Bioprocess Technology*, 5, 1423–1446.

Barlow, C. and Morgan, D. (2013). Polymer film packaging for food: an environmental assessment. *Resources, Conservation and Recycling*, 78, 74–80.

Bibi, F., Guillaume, C., Gontard, N. and Sorli, B. (2017). A review: RFID technology having sensing aptitudes for food industry and their contribution to tracking and monitoring of food products. *Trends in Food Science & Technology*, 62, 91–103.

Boyette, M. (1996). *Packaging Requirements For Fresh Fruits and Vegetables*. https://content.ces. ncsu.edu/packaging-requirements-for-fresh-fruits-and-vegetables (accessed December 17, 2019).

Brody, A. L., et al. (2008). Innovative food packaging solutions. *Journal of Food Science*, 73, R107–16.

Bureau, G. and Multon, J. (1996). *Food Packaging Technology (Second Edition)*. New York, NY: VCH Publishers.

Cachaldora, A., Garcia, G., Lorenzo, J. and Garcia-Fontan, M. (2013). Effect of modified atmosphere and vacuum packaging on some quality characteristics and the shelf-life of "morcilla", a typical cooked blood sausage. *Meat Science*, 220–225, 93.

Chiellini, E. (2008). *Environmentally-Compatible Food Packaging*. Cambridge, UK: Woodhead Publishing.

Chinnadurai, K. and Sequeira, V. (2016). *Packaging of Cereals, Snacks, and Confectionery*. Menomonie, WI: University of Wisconsin.

Coles, R., McDowell, D. and Kirwan, J. M. (2003). *Food Packaging Technology*. Boca Raton, FL: CRC Press.

Dermiki, M., et al. (2008). Shelf-life extension and quality attributes of the whey cheese "Myzithra Kalathaki" using modified atmosphere packaging. *LWT Food Science and Technology*, 41, 284–294.

Doyle, M. (1996). *Packaging Strategy: Winning the Consumer*. Lancaster, PA: Technomic Publishing.

Duncan, S., Webster, J. and Steve, L. (2009). Sensory impacts of food packaging interactions. *Advances in Food and Nutrition Research*, 56, 17–64.

Economywatch. (2010). *Food Industry, Food Sector, Food Trade*. http://www.economywatch.com/ worldindustries/ (accessed June, 2010).

Eldred, N. (1993). *Package Printing*. Plainview, NY: Jelmar Publishing.

Elsamahy, M. A., Saa, M., Abdel Rehim, M. H. and Mohram, M. E. (2017). Synthesis of hybrid paper sheets with enhanced air barrier and antimicrobial properties for food packaging. *Carbohydrate Polymers*, 168, 212–219.

Fang, Z. (2015). *Current Practice and Innovations in Meat Packaging*. North Sydney: Australian Meat Processor Corporation.

Fang, Z., Zhao, Y., Warner, R. D. and Johnson, S. K. (2017). Active and intelligent packaging in meat industry. *Trends in Food Science & Technology*, 61, 60–71.

Farm Radio International. (2006). *Voices Newsletter*, s.l.: s.n.

Faustman, C. and Cassen, R. G. (1990). Influence of aerobic metmyoglobin reducing capacity on colour stability of beef. *Journal of Food Science* 55, 1279–1283.

Fellows, P. (2008). *Practical Action*. https://answers.practicalaction.org/our-resources/collection/packaging-and-bottling-1 (accessed November 2, 2019).

Foodservice Packaging Institute. (2006). *A Brief History of Foodservice Packaging*. s.l.: s.n.

FSAI. (2008). *The Relevance for Food Safety of Applications of Nanotechnology in the Food and Feed Industries* (82pp.). Dublin: Food Safety Authority of Ireland Abbey Court.

Garber C. (2007). *Nanotechnology Food Coming to a Fridge Near You*. http://www.nanowerk.com/spotlight/spotid=1360.php (accessed June 11, 2011).

Gennadios, A., Hanna, M. A. and Kurth, L. B. (1997). Application of edible coatings on meats, poultry and seafoods: a review. *Lebensm-Wiss u-Technol*, (30), 337–350.

Geueke, B. (2018). Food packaging in the circular economy: overview of chemical safety aspects for commonly used materials. *Journal of Cleaner Production*, 491–505.

Gupta, R. K. and Dudeja, P. (2017). Food packaging. In: *Food Safety in the twenty-first Century* (Chapter 46, pp. 547–553). s.l.: s.n.

Han, J.-W. (2018). Food packaging: a comprehensive review and future trends. *Comprehensive Reviews in Food Science and Food Safety*, 17.

Han, J. H., Ho, C. H. L and Rodrigues, E. T. 2005. Intelligent packaging. In J. H. Han (Ed)., Innovations in food packaging. Elsevier Academic Press.

Hanlon, J. (1992). *Handbook of Package Engineering (Second Edition)*. Lancaster, PA: Technomic Publishing.

Hine, T. (1995). *The Total Package: The Evolution and Secret Meanings of Boxes, Bottles, Cans and Tubes*. New York, NY: Little Brown.

Hirsch, A. (1991). *Flexible Food Packaging*. New York, NY: Van Nostrand Reinhold.

Hrnjak-Murgic, Z., 2015. Nanoparticles in Active Polymer Food Packaging. Smithers Pira.

Hutton, T. (2003). Food packaging: An introduction. Key topics in food science and technology. Chipping Campden: Gloucestershire, UK.

ICPE. (2012). *Packaging of Beverages*. s.l.: Indian Centre for Plastics in the Environment.

Industry Indian Food. (2003). Category liquor. *Industry Indian Food*, 22(3).

IPSA. (2019). *A Handbook of Packaging Technology (Eleventh Edition)*. Sandton, South Africa: The Institute of Packaging (South Africa).

ISO TC 229. (2008). *Draft Standard on Nanotechnologies – Terminology and Definitions for Nanoparticles*. http://www.iso.org/iso/iso_catalogue/catalogue_tc/catalogue_detail.htm?csnumber=44278 (accessed June 17, 2011).

Jelen, P. (1985). *Introduction to Food Processing (First Edition)*. USA: Prentice-Hall.

Judd, D., Aalders, B. and Melis, T. (1989). *The Silent Salesman, Octogram Design*. Singapore: Octogram Design.

Kelsey, R. J. (1989). *Packaging in Today's Society (Third Edition)*. Lancaster, PA: Technomic Publishing.

Kern, C. (2002). *The Right PET for the Product*. s.l.: Beverage and Food World.

Kerry, J., O'Grady, N. and Hogan, S. (2006). Past, current and potential utilization of active and intelligent packaging systems for meat and muscle-based products: a review. *Meat Science* 74, 113–130.

Kumar, K. (2002). *Packaging Aspects of Fruits Beverages*. s.l.: Beverage and Food World.

Lamberti, M. and Escher, F. (2007). Aluminium foil as a food packaging material in comparison with other materials. *Food Reviews International*, 23(4),407–433.

Licciardello, F. (2017). Packaging, blessing in disguise. review on its diverse contribution to food sustainability. *Trends in Food Science & Technology*, 65, 32–39.

Lindh, H., Williams, H., Olsson, A. and Wikström, F. (2016). Elucidating the indirect contributions of packaging to sustainable development: a terminology of packaging functions and features. *Packaging Technology and Science*, 29, 225–246.

Lopez-Cervantes, J., et al. (2003). Evaluating the migration of ingredients from active packaging and development of dedicated methods: a study of two iron-based oxygen absorbers. *Food Additives and Contaminants*, 20, 291–299.

Marsh, K. and Bugusu, B. (2007). Food packaging – roles, materials, and environmental issues. *Journal of Food Science*, 72(3),29–55.

Packaging-Gateway. (2014). *Use of Synthetic Chemicals in the Processing, Packaging and Storing of Food Products may Lead to Long-Term Health Damage for Consumers*. http://www.packaging-gateway.com/news/newsenvironmental-scientists-warn-over-usage-of-chemicals-in-food-packaging-4181118 (accessed February 24, 2014).

Peelman, N., et al. (2013). Application of bioplastics for food packaging. *Trends in Food Science & Technology*, 32, 128–141.

Potter, N. N. and Hotchkiss, J. H. (1995). *Food Science (Fifth Edition)*. UK: Springer.

Raheem, D. (2013). Application of plastics and paper as food packaging materials - An overview. *Emirates Journal of Food Agriculture* 25 (3): 177–188.

Rashmi, M. and Smita, L. (2003). *Plastic Films for Processed Foods – Special Requirements*. s.l.: Packaging India.

Recordati, G. B. P. (2015). *The Food Industry: History, Evolution and Current Trends* (p. 106).

Ribeiro-Santos, R., Andrade, M., Melo, N. and Sanches-Silva, A. (2017). Use of essential oils in active food packaging: recent advances and future trends. *Trends in Food Science & Technology*, 61, 132–140.

Risch, S. J. (2009). Food packaging history and innovations. *Journal of Agricultural and Food Chemistry*, 57, 8089–8092.

Robertson, G. L. (2006). *Food Packaging: Principles and Practice (Second Edition)*. Boca Raton, FL: CRC Press.

Ruiz-Garcia, L. and Lunadei, L. (2011). The role of RFID in agriculture: applications, limitations and challenges. *Computers and Electronics in Agriculture*, 79, 42–50.

Sadiku, M., Musa, S. and Ashaolu, T. (2019). Food industry: an introduction. *International Journal of Trend in Scientific Research and Development (IJTSRD)*, 3(4).

Ščetar, M. (2018). Packaging perspective of milk and dairy products. *Mljekarstvo*, 1(69),3–20.

Shewry, P. (2009). Wheat. *Journal of Experimental Botany*, 60(6),1537–1553.

Stancu, V., Haugaard, P. and Lahteenmaki, L. (2016). Determinants of consumer food waste behaviour: two routes to food waste. *Appetite*, 96, 7–17.

Thyberg, K. and Tonjes, D. (2016). Drivers of food waste and their implications for sustainable policy development. *Resources, Conservation & Recycling*, 106, 110–123.

US Environmental Protection Agency. (2015). *Reducing Wasted Food and Packaging: A Guide for Food Services and Restaurants*. s.l.: US Environmental Protection Agency.

Von Holdt, V. (2011). *Packaging Council of South Africa (PACSA)*. s.l.: s.n.

Wang, Y. C., Lu, L. and Gunasekaran, S. (2017). Biopolymer/gold nanoparticles composite plasmonic thermal history indicator to monitor quality and safety of perishable bioproducts. *Biosensors and Bioelectronics*, 92, 109–116.

Williams, H. et al. (2012). Reasons for household food waste with special attention to packaging. *Journal of Cleaner Production*, 24(3),141–148.

Yam, K., Takhistov, P. and Miltz, J. (2005). Intelligent packaging: concepts and applications. *Journal of Food Science*, 70, 1–10.

Part V: **Food innovations and nonthermal processing**

Dele Raheem

15 Blockchain and food traceability

15.1 Introduction

Industrial revolution led to the production of food on a massive scale largely increasing the use of fossil fuel, which contributed to global warming (Mgbemene et al., 2016). The current global food system needs to change by taking advantage of the current fourth industrial revolution in ensuring that the impacts of climate change are mitigated. Food security, global warming, and its twin evil ocean acidification are serious threats that humanity face (Ahmed et al., 2019). These threats have driven a large interest in digitalization to produce food in a sustainable way and ensure the efficient utilization of natural resources. Digitalization permeates many sectors of our lives in creating smart cities, personal medicine, precision agriculture, and the Internet of things (IoT) with increased connectivity of gadgets and processes. Innovations in the food sector will benefit from the application of digital solutions such as blockchain technology.

A blockchain is the record of distributed database that are encrypted as "blocks," shared among the participating parties with the possibility for verification in the future. Blockchain is not owned by any single entity since its time-stamped series of immutable data are managed by a cluster of computers. It was reported to have emerged in the wake of the 2008 global economic crisis, when Satoshi Nakamoto released a new protocol for "A Peer to-Peer Electronic Cash System" by using cryptocurrency called bitcoin (Nakamoto, 2008). For the agrifood sector, the blockchain market is already predicted to increase from US $60.3 million in 2018 to US $429.7 million by 2023 (Addison, 2019). Blockchain as an emerging technology will offer companies the opportunity to transact and move assets globally in a more secure manner. The chain that connects the blocks breaks if the data in a block is tampered with, which makes blockchain to be unique in providing security.

Blockchain provides transparency to inefficient and corrupt business practices by enabling equitable participation for farmers and other stakeholders on the global food value chain, leading to greater prosperity for developing world agricultural workers (Townsend et al., 2018). The food industry has been plagued with data from disparate and disconnected sources. These data can be connected by encrypted blockchain using heavy-duty encryption that involves keys that are both public and private; the two-key system helps to maintain virtual security (Tapscott and Tapscott, 2017).

The global food safety of our twenty-first century will need to expand beyond improving nutritional profile. It must also ensure the transparency of food ingredients,

Dele Raheem, Arctic Centre (NIEM), University of Lapland, Rovaniemi, Finland

https://doi.org/10.1515/9783110667462-015

regular monitoring, and surveillance of food products to guarantee the prevention of food-borne illnesses and public well-being (Silver and Bassnet, 2008). Henceforth, global supply chain participants will be able to gain permissioned access to trusted information regarding food provenance by using blockchain. The participants could then access data on the blockchain network to trace contaminated products expeditiously, stemming public health outbreaks, and potentially saving lives (del Castillo, 2017). Blockchain as a decentralized database will rely on consensus among its participants to be successful. In this chapter, as part of food innovation and nonthermal processing, the literature on the innovative role blockchain technology can play in food traceability is reviewed.

The next section highlights global concerns on food safety; Section 15.3 addresses the disruptive effect of blockchain in the food sector; Section 15.4 addresses traceability as a major issue in the food system, and how blockchain technology can be a solution; Section 15.5 identifies the pitfalls of blockchain as an emerging technology and its outlook in the future. Finally, Section 15.6 concludes the chapter.

15.2 Global food safety concerns

Due to the increasing interconnectedness of our existence, foods cross borders frequently, which can be a major cause for safety concerns. The safety of food is essential to human health as well as being a part of human right. Therefore, we need to identify, assess, and manage the risks that are associated with food safety, with science-based information and harmonized global procedures (FAO, 1997). The contamination of food products with chemical and microbial pollutants as major food safety concerns in the twenty-first century was reviewed by Fung et al. (2018). The authors highlighted the economic burden of food safety and identified four main related challenges as chemical safety, microbiological safety, personal hygiene, and environmental hygiene.

Another worrisome issue for global food safety is food fraud, which accounts for US $40 billion financial loss annually (Tripoli, 2019). Food fraud is often characterized with the intention to make economic gains. There are many aspects of food fraud; it encompasses deliberate and intentional substitution, addition, tampering, or the misrepresentation of food, food ingredients, food packaging, or false or misleading statements about a food product with the intention of making economic gain (Spink and Moyer, 2011). A subcategory of food fraud referred to as "economically motivated adulteration (EMA)," was identified as a cause of public health food risks (FDA 2009; Spink 2009a). Globally, an accurate number of food recalls is difficult to ascertain due to lack of global harmonization especially in many developing countries. For instance, in the 2018 fiscal year, the USA reported 1,928 recalls of food and cosmetic, over half of these recalls were due to supply chain hazards resulting from food fraud, microbes, allergens, and chemicals (FDA, 2018).

The food system in developing countries faces a significant share of the global burden of food-borne illnesses and a well-managed traceability system will be beneficial. Since our modern food system has become more complex, the risk of food fraud has increased to the global populations (Spink et al., 2010).

The global food supply chain is highly complicated with lack of transparency and trust. The increasing emergence of consumer interest in the origin of food and its authenticity, partly in response to major food crises, such as the horse gate scandal in Europe in 2013, needles in strawberries in Western Australia in 2018 has promoted transparency as being essential criterion for any brand, whether it is a large global or a local food provider (Elliott, 2014; Mortimer and Grimmer, 2018).

The most notorious food epidemic was the Minamata disease (methylmercury poisoning) first discovered in 1956 around Minamata Bay in Kumamoto Prefecture, Japan (Harada, 1995). A second epidemic occurred in 1965 along the Agano River, in Niigata Prefecture, Japan. Symptoms of this disease included cerebellar ataxia, sensory disturbance, narrowing of the visual field, and hearing and speech disturbances. The discharged methyl mercury accumulated in fishes and shellfishes and caused poisoning on consumption (Eto, 2000; Shimohata et al., 2015).

Contaminated foods may be due to negligence in the food supply chain or as a deliberate fraudulent attempt. In order to put into perspective the importance of food traceability, a catalogue of food categories with reported cases of food fraud is summarized further (Johnson, 2014):

1. Olive oil is substituted with a cheaper alternative, instead of a higher priced extra virgin olive oil (Aparicio et al., 2013). Such fraud was related to fraudsters making financial gains from the European Union's (EU) farm support program that give subsidies to farmers that grow olive oil, as part of the Common Agricultural Policy (CAP).
2. Fish of higher value was substituted with cheaper and readily available fish. For example, it was reported that fish samples bought from grocery stores, restaurants, and sushi bars were often mislabeled (Jacquet and Pauly, 2008).
3. Bovine milk from cows was frequently adulterated with milk from other sheep, buffalo, goats, and antelopes. They are sometimes watered down or supplemented with melamine to raise their protein content. The Chinese incidents of melamine in 2007 and 2008 showed how adulteration affects food safety from the hazards of EMA. In these incidents, criminals focused on fooling total nitrogen measurements in protein for economic gains and were not interested in assessing the safety of the fraudulent ingredients (Moore et al., 2010).
4. The quality of honey is compromised with added sugar syrup, corn syrup, glucose, fructose, or high-fructose corn syrup, and beet sugar, without being disclosed on the label (Perkins and van den Berg, 2009).

5. Fruit juices may be diluted with water, or a more expensive juice (e.g., pomegranate juice) might be cut with a cheaper juice (such as apple juice). In some extreme cases, juice may contain water, dye, and sugary flavorings without any fruit, despite fruit being listed as an ingredient on the label (Johnson, 2014).
6. Saffron, the world's most expensive spice, has been adulterated with glycerin, sandalwood dust, tartrazine, barium sulfate, and borax (Johnson, 2014). Similarly, there were cases of ground black pepper having added starch, papaya seeds, buckwheat, flour, twigs, and millet (Johnson, 2014). Other spices such as vanilla extract, turmeric, star anise, paprika, and chili powder are also prone to fraud (Parvathy et al., 2014). Sudan red dyes were used to color paprika, chili powders, and curries, despite the fact that they are carcinogenic and banned for use in foods (Johnson, 2014; Oliviera, 2019).
7. Ground coffee may contain leaves and twigs, roasted corn, ground roasted barley, and roasted ground parchment. In some cases, to instant coffee was added chicory, cereals, caramel, more parchment, starch, malt, and figs. Leaves from other plants, color additives, and colored saw dust may also be added to dried and packed tea (Johnson, 2014).

There were also cases where fraudulent certificates were obtained to market, label, or sell nonorganic agricultural products as certified organic products. They were fraudulently labeled as "organic" premium products (Baksi et al., 2017). Sometimes, fraudsters are able to use clouding agents or food processing aids in order to enhance the appeal of a food or food component.

The safety of our food supply is a continuous challenge that needs attention from all stakeholders. Ensuring food safety by eliminating food fraud and adulteration with digital tools can be effective in tracing food from farm to plate.

In order to enable an ecosystem with a single trusted source of information will require hyperconnected value chains wherein information is dynamically fed to incrementally improve systems through enhanced human-digital interaction, efficiency, reduction of waste, and greater automation.

The Food Safety Modernization Act (FSMA) of 2019 having recognized the vulnerability of food suppliers emphasized that minimum standards are followed at food facilities that are regulated by the US Food and Drug Administration (FSMA, 2019). The standards are to consider the hazards that occur naturally and those that are introduced either intentionally or unintentionally.

The food industry sector is one of the largest and most important manufacturing sectors in the EU. The industry relies on over 70% of agricultural goods to be converted into food industry products. Food safety issues have become prominent across Europe in the last 20–30 years, which led to many countries establishing their "National Food Safety Agencies," followed by the European Food Safety Authority (EFSA) in 2002. The EFSA is primarily responsible for EU risk assessments regarding food and feed safety. This has helped to enhance the science underlying food safety resulting in a safer food supply.

At the international level, the expert bodies of the FAO/WHO expert in 2002 recommended the promotion of food safety through:

- a better awareness on nutritional requirements of healthy diets and the adoption of nutrition patterns that can minimize the risk of diet-related diseases;
- a comprehensive approach to the health of plants and animal and food safety, by fostering increased synergy within an international regulatory framework;
- periodic regional and global fora for food safety regulators on risk management and encourage partnership alliances among countries to resolve any issue related to food safety and trade;
- ensure good practices in the food chains by promoting research and the use of appropriate technologies at farm levels during food handling and processing to meet the safety and quality requirements of the consumers in cost-effective ways;
- communication systems that are interactive by making knowledge available on food-borne diseases and on nutritious diets, through the Internet and other channels;
- improve the effectiveness of worldwide information exchange through an international rapid alert system on food safety hazards;
- international financial and technical assistance to build capacity that can enable developing countries to strengthen their ability to control the safety of their foods, for export as well as for domestic consumption, and to participate more actively in international regulatory systems (FAO, 2002).

There are efforts by regulators to improve global food safety measures through joint collaboration. For instance, the FDA and the USDA collaborated efforts to implement the FSMA in the USA, which will increase the controls and preventative measures on food imports. Furthermore, there are efforts to ensure the safety of food exports through the Foreign Suppliers Verification Programs (FSVP).

Food production is geographic dependent and it is distributed across border; the level of visibility from supply source to retail will need to be enhanced by hyperconnectivity. Digital solutions that employ devices and sensors that are connected and can interact over distances with real-time feedback will help ensure food safety globally.

The EU through joint collaboration implements the Trade Control and Expert System (TRACES) for food traceability and risk-management functions, which can trace traded goods throughout the production chain for animals, plants, or animal-based products (EU-TRACES, 2019). The EU Rapid Alert System for Food and Feed (RASFF), as a rapid early warning system that reacts quickly to stem outbreaks and food-borne illness in the EU (Papapanagiotou, 2017), can benefit immensely by integrating blockchain technology to enhance these systems with high potential and probability.

15.3 Blockchain technology and its disruptive effect

Food system digitalization refers to the application of innovative technology that enhances harvesting, processing, distribution, and storage operations along the agrifood value chain (Raheem et al., 2019). These operations can be enhanced by digital solutions such as blockchain to verify the authenticity of food ingredients and products.

A blockchain is a useful tool in supply chain integrity due to its use of encryption and distributed computing technology to build a ledger system that is almost tamper-proof. For example, in the fishing industry, workers along the supply chain – from the fishermen to the packagers, transporters, distributors, and retailers – will be able to record details of the size, quality, and condition of the product as it passes through their hands. However, the information cannot be altered by any single link in the supply chain.

Blockchain technology has the potential to disrupt the supply chain and many other domains of an organization. A blockchain allows secure exchange of data in a distributed manner, it impacts the way organizations are governed, and the relationships in a supply chain are structured while transactions are conducted. By integrating blockchain technology with other technologies, like the IoT, the blockchain can create a permanent, shareable, and actionable record in every moment of a product's trip throughout the entire supply chain thus capable of creating efficiencies throughout the global economy (Wang et al., 2019).

Disruptive providers who utilize blockchain are emerging in the digital currency sector, where they ensure that all vendors and buyers know how much is paid and what is received. The alternative does not guarantee on transactions, contracts, or diligence in a similar way. Blockchain provides a guarantee by linking verifiable sources of capital to business transactions. The blocks of information are chained so that trust within a transaction can be ascertained. Bitcoin as a currency have no central banking mechanism of underwriting risk and blockchains have provided this endorsement of trust.

There has been little consideration of a potential link between blockchains and fast-moving consumer goods (Dujak and Sajter, 2019). However, the authors noted that at the end of 2018, the investments in food-related blockchain technologies and platforms have overshadowed the investment in developing any other technology in the food sector. This unprecedented situation affirms the potential applications of blockchain as a paradigm shift that is enabled by the other facets of the fourth industrial revolution.

In technical terms, the blockchain architecture has a 160-bit address space, as opposed to IPv6 address space which has 128-bit address space (Antonopoulos, 2014). The address of a blockchain is 20 bytes or a 160-bit hash of the public key generated by Elliptic Curve Digital Signature Algorithm (ECDSA). The 160-bit address is capable of generating and allocating addresses offline for around 1.46×10^{48} IoT devices (Khan and Salah, 2018). The probability of address collision is approximately 10^{48}, which can sufficiently secure and provide a Global Unique Identifier (GUID). Hence, there is no need for registration or uniqueness verification when assigning and allocating an address to an IoT device. A centralized governance and authority such as the Internet Assigned Numbers Authority (IANA) is eliminated with blockchain. Blockchain can provide 4.3 billion addresses more than IPv6, which makes blockchain more scalable as a solution for IoT than IPv6 (Khan and Salah, 2018).

The role of blockchain in providing seamless networks has been described by many authors (Bonino and Vergori, 2017; Xu et al., 2017; Wang et al., 2018), its ability to provide an entire visibility (Li et al., 2017) and provides symmetric information to all (Nakasumi, 2017). Such seamless reliability and connectivity are useful for business ethics and social responsibility. Due to the increasing concerns and reports on food safety, for example, horsemeat scandal in Europe and toxic milk powder scandal in China (Tian, 2016; 2017), there are also concerns on sustainability related issues (e.g., child labor, fairtrade, and organic products) (Abeyratne and Monfared, 2016). This has prompted many consumers to pay more attention to the legitimacy and authenticity of the products they want to purchase.

Legitimacy can be guaranteed when a computerized transaction protocol automatically executes the terms of a contract upon a blockchain as a "smart contract." The general objectives are to satisfy common contractual conditions, while the costs and delays associated with traditional contracts are reduced (Gupta, 2017). Immutability is achieved when each block holds an ordered sequence of transactions, which corresponds to a block chain. Ultimately, the risk of data tampering and single point of failure (SPOF) from hackers and powerful admins are eliminated since any user can issue an asset with the asset-issuance permission or transfer an asset with the asset-transfer permission.

Smart contracts when incorporated with blockchains could automate the transfer of the various types of ownership of assets, property, and value. Because of this, it can facilitate process design for business operations that will lead to a more visible and less-intermediated working scheme (Chang et al., 2019). A smart contract runs on blockchain that is executed by all consensus nodes. It consists of program code and a storage file. It is possible for any user to create a contract by posting a transaction to the blockchain. When the contract is created, the program's code of a contract is fixed and cannot be altered (Delmolino et al., 2016). The auto-execution of contracts will help to monitor intellectual property right and reduce the frictions that may occur at the border for international trade (CTA, 2019). The notion of a smart contract may be the most transformative blockchain application for supply chains. While it is promising in raising transparency in the food industry, there is much yet to be tested and validated on its real-world application within the food chain.

15.4 Traceability in the food system

There are several suggested definitions for traceability in a food system. A commonly used definition of traceability by the International Standardization Organization (ISO) 8402 defines traceability as "The ability to trace the history, application or location of an entity by means of recorded identifications" (ISO, 1994).

Traceability is the ability to track a product batch and its history throughout the whole, or part, of a production chain from harvest through storage, processing, transport, distribution, and sales (Moe, 1998). On the other hand, the EU General Food Law defines traceability as "the ability to trace and follow a food, feed, food-producing animal or substance intended to be, or expected to be incorporated into a food or feed, through all stages of production, processing and distribution" (EU, 2002). Many scientific articles refer to this definition, and it is quite detailed with respect to what should be traced and followed, and where. However, it is less detailed in describing what types of properties are relevant or how the traceability might be implemented? Traceability has also been defined as "the ability to access any or all information relating to that which is under consideration, throughout its entire life cycle, by means of certain recorded identifications" (Olsen and Borit, 2013). According to the international Codex Alimentarius Commission Procedural Manual, traceability is defines as "the ability to follow the movement of a food through specified stage(s) of production, processing and distribution" (FAO/WHO, 1997).

The food industry needs to ensure that traceability is achieved at all stages of the food chain. This can be classified as "product-process" links for enhanced traceability (CTA, 2019). The use of radio frequency identification (RFID) as tags in packaged foods, quick response (QR) codes in packaged foods, crypto-anchors, facial recognition in livestock, DNA markers, and artificial intelligence in processing techniques can help to ensure transparency and traceability (2019).

Blockchain does not store all information in a single place compared to a centralized data management system, where information is distributed equally among the collaborating mobile devices. In blockchain, there is no SPOF, even if several mobile devices encounter errors or a limited performance, the possibility of encountering malicious threats by a messaging records management (MRM) infrastructure that use blockchain is very slim (Mohsin et al., 2019).

Blockchain technology is able to quickly identify and detect specific products thereby improving transparency, efficiency, security, and safety in the food sector (Pizzuti and Mirabelli, 2015). The attributes of the food can be captured as a "bundle." All the information that are supplied by stakeholders over the life time of a food item will be contained in the bundle.

Table 15.1 refers to a summary of some traceability problems and the solutions proffered by blockchain technology.

Table 15.1: Blockchain solutions to traceability problems.

Problem	Solution	Reference
Coordination of individual activities over the Internet without secured centralized storage	Blockchain links block to each other by using chronologically distributed databases in a proper linear manner and they cannot be deleted.	Anderson (1996) Benkler (2007) Nakamoto (2008)
Validation of entries without a central authority to verify that a transaction is not fraudulent or invalid	Blockchain uses a probabilistic approach. It makes information more transparent and verifiable by using mathematical problems that require substantial computational power to solve.	Schneier and Kelsey (1998) Nakamota (2009) Bonneau et al. (2015) Wright and Filippi (2017)
Ensure that only legitimate transactions are recorded into a blockchain	A new block of data is added to the end of the blockchain only after a consensus is reached by the computers on the network as to the validity of the transaction. Consensus can be achieved through different voting mechanisms.	Franco (2014) Bonneau et al. (2015) Wright and Filippi (2017)
Preservation of historic records	When a block has been added to a blockchain, it can no longer be deleted. The transactions it contains can be accessed and verified by everyone on the network. It becomes a permanent record that all computers on the network can use to coordinate an action or verify an event.	Bonneau et al. (2015) Wright and Filippi (2017)

For recent practical applications of blockchain technology in the food industry, Nestlé S.A., Switzerland, announced a collaboration with OpenSC, a blockchain platform that will allow consumers to track their food right back to the farm (Food Ingredients, 2019). Nestlé is the first major food and beverage company to pilot open blockchain technology; the pilot programs will trace milk from farms and producers in Nestlé factories and warehouses that are located in New Zealand and Middle East (Food Ingredients, 2019). Open blockchain technology will allow Nestle to share reliable information with consumers in an accessible way. IBM's Food Trust ecosystem connects supply chains such as Walmart's and other major retailers including global companies like Carrefour, Dole, Golden State Foods, Driscoll's, and Nestlé without sharing any information they have not intended to share. Such an ecosystem of transparency and accountability fostered by IBM's blockchain network can transform stakeholders' work in their supply chains as well as trace the sources of food-borne illnesses (IBM, 2018). The IBM Food Trust network connects participants across the food supply through a permissioned, permanent, and shared record of food system data (IBM, 2018). The IBM Trust owns certifications for organizations to digitally manage and share food certifications at the IBM website and allow French retailer, Carrefour SA, to have a boost in their sales by using blockchain ledger technology to track their foods (de Bruin, 2019). Similarly, IBM is collaborating with Maersk during import and export of goods to help track shipments of containers.

An illustration to depict a traceability system that engages authority organizations, producers, processors, storage, distributors, retailers, and consumers in a food sector is shown in Figure 15.1.

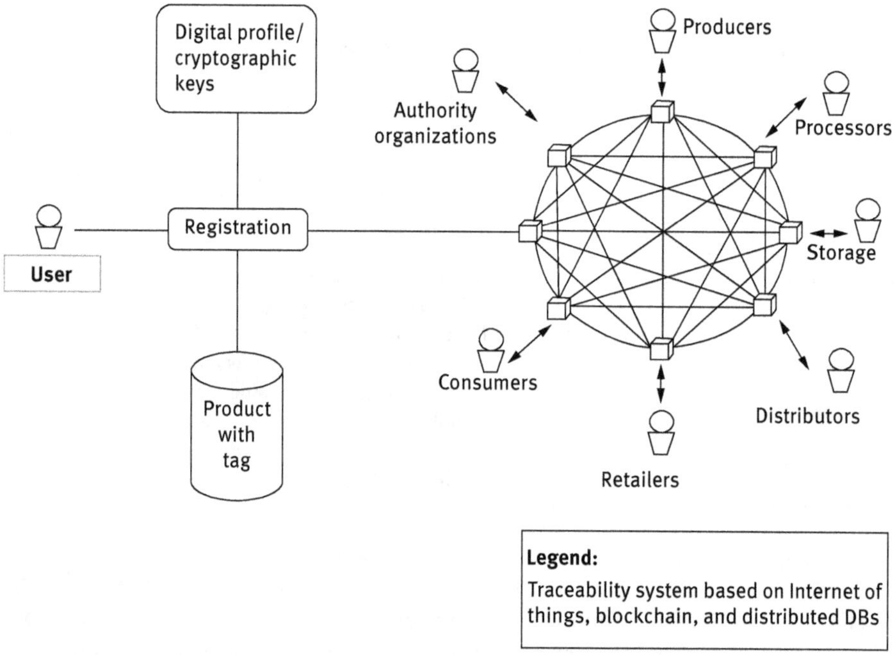

Figure 15.1: Conceptual framework for a proposed traceability system (adapted from Feng, 2017).

Figure 15.1 represents the proposed system of a typical decentralized distributed system, which uses the IoT (RFID, WSN, GPS) to store and manage relevant data of products in food supply chains by relying on BigchainDB and allows developers and enterprises to deploy blockchain proof-of-concepts, platforms, and applications with a blockchain database (de la Rocha, 2018).

The decentralized control can be achieved through the nodes in the system with voting processing known as a super-peer P2P network (Özsu et al., 2011). There are many members in the supply chain that includes suppliers, producers, manufacturers, distributors, retailers, consumers, and certifiers. Each product is attached to a tag (RFID) that represents a unique digital cryptographic identifier that connects the physical items to their virtual identity in the system. Once a member is registered as a user in the system, they can add, update, and check the information about the product on the BigchainDB. The virtual identity can be seen as the product information profile. Users in the system also have their digital profile, gather the information about their interaction, certifications, and association with products.

The private sector food businesses will benefit from technological advances that can make their supply chains more cost-effective and increase competitiveness. These supply chains are being digitized with technologies such as cloud computing, artificial intelligence, and the IoT, the greatest potential to increase efficiency and transparency in agricultural supply chains blockchain (Tripoli and Schmidhuber, 2018).

15.5 Drawbacks and concerns

When datasets are publicly available, the privacy of datasets becomes a major concern. To preserve privacy and avoid the violation of any data privacy regulations, there is a need for an automated tool to anonymize datasets prior to their release (Banerjee et al., 2018).

Another challenge is the lifetime of datasets, as the owners of datasets may not want to share them permanently. While this is a strong security property, it may not be conducive to sharing if any record needs to be removed. Most existing firmware protection techniques are based on integrity checking. Starting from a bootloader, the integrity of the next level firmware (operating system and application) is checked before it is executed. The bootloader is stored as a "root of trust" in secure read-only storage, so that it cannot under any circumstances be modified.

In ensuring a reliability of the execution or activity, the integrity of the reference integrity metric (RIM) itself is very important. When datasets are shared among the research and practitioner communities, their integrity needs to be maintained. In order to ensure integrity of the datasets, a RIM for the dataset is maintained using blockchain (Banerjee et al., 2018). If firmware cannot be updated, then the RIM should be stored in read-only memory. However, for reasons such as security patches and upgrades of services, updates are usually allowed. When the firmware has been updated, its corresponding RIM should also be updated.

The Global Food Traceability Centre raised concerns that the preferences of consumers shift rapidly and there are conflicting demands from national regulators around the world. Hence, traceability can vary among food products and industry, the current internal systems does not provide a means for reliable and rapid response to trace back data across the food chain (Tian, 2017).

As with any emerging technology, the drawbacks and concerns need to be addressed to ensure the smooth transition of blockchain technology.

Another major problem with blockchain technology is the difficulty of getting data into decentralized blockchains in a noncentralized way. Without a secure and reliable way to get data into smart contracts running on blockchains, the security and reliability of the blockchain and thus the advantage of the entire system can be lost.

Furthermore, Zhao et al. (2019) identified six main challenges for the application of blockchain technology in food traceability. They are storage capacity and scalability issue, privacy issue, regulation problem, high-cost problem, throughput and latency issue (i.e., time required to add a block of data to the chain), and lack of skills.

For blockchain to be effective, there must be participation from all parties and the points of contact that are involved. Additionally, data integrity lies in the hands of the data collectors and needs a system of validation to avoid tampering. There are also some concerns on how far the tracking goes? Can all evidence of employed migrant labor be verified with blockchain? Can price manipulation be detected? How to tackle issues of cybersecurity were all discussed at a recent EU Brussel briefings (Mooney, 2019). Without standardized data elements and collection practices, digitization can confine data to silos. Ultimately, the systems need to meaningfully transmit, receive, and interpret data.

15.6 Future outlook

Blockchain as an innovative technology creates an enabling environment that can protect smart devices to be connected in the "IoT." Blockchain can make food as safe as possible and empower people to make informed decisions about the food they eat. The European Commission launched the EU Blockchain Observatory and Forum in 2018 to further promote the awareness and engagement of blockchain technologies throughout European industries in the future (DSM, 2018).

Chang and coresearchers recently highlighted the need for further efficiency improvement that explore alternative blockchain configurations such as (a) blockchain platforms (e.g., Bitcoin, Ethereum, and Nxt); (b) consensus protocols (e.g., Proof-of-Work, Proof-of-Stake, and Byzantine Fault Tolerance); (c) on-chain/off-chain data storage and computation; (d) block sizes; and (e) degrees of centralizations. Further research into such configurations has the potential to achieve higher efficiency, especially for creating desirable blockchain-based systems (Chang et al., 2019). Blockchain has been designed as a decentralized system when used to pay for food and other goods. Nevertheless, there is a trend that miners are centralized in the mining pool. For example, it was gathered that the top five mining pools together owns larger than 51% of the total hash power in the bitcoin network and they could get a larger share of the revenue (Eyal and Sirer, 2014; Zheng et al., 2018). As blockchain is not intended to serve a few organizations, some methods should be proposed to solve this problem in the future (Zheng et al., 2018).

Furthermore, transactions on blockchain could be used for big data analytics, platforms can be developed for smart contract that could achieve more functionalities (Christidis and Devetsikiotis, 2016). Recent developments in blockchain technology are also creating new opportunities for the applications of artificial intelligence (Omohundro, 2014).

As the agrifood sector embraces the benefits of blockchain technology, the need to stay competitive is what will ultimately drive the migration to blockchain.

15.7 Conclusions

As an emerging technology, there are many opportunities for blockchain in food innovation, food safety, and food traceability. Data capturing is the essence of traceability; therefore, blockchain technology can also help to support food manufacturers in developing countries in collating data to be engaged in the global supply chain. However, concerns that bothers on security, cost, and the issues of digital divide for large and small food entrepreneurs will need to be addressed.

References

Abeyratne, S. A. and Monfared, R. P. (2016). Blockchain ready manufacturing supply chain using distributed ledger. 5(9),1–10.

Addison, C. (2019). Blockchain applications for ACP sustainable agriculture. Brussels Briefing n. 55 "Opportunities of blockchain for agriculture", April 26, 2019. Brussels, Belgium.

Ahmed, N., Thompson, S. and Glaser, M. (2019). Global aquaculture productivity, environmental sustainability, and climate change adaptability. *Environmental Management*, 63(2),159–172.

Anderson, R. J. (2017). The eternity service. In: *Proceedings of Pragocrypt.Arc-net, 1996.* http://arc-net.io/.

Antonopoulos, A. M. (2014). *Mastering Bitcoin: Unlocking Digital Crypto-Currencies (First Edition).* O'Reilly Media, Inc.

Aparicio, R., Morales, M. T., Aparicio-Ruiz, R., Tena, N. and García-González, D. L. (2013). Authenticity of olive oil: mapping and comparing official methods and promising alternatives. *Food Research International*, 54(2),2025–2038.

Baksi, S., Bose, P. and Xiang, D. (2017). Credence goods, misleading labels, and quality differentiation. *Environmental and Resource Economics*, 68(2),377–396.

Banerjee, M., Lee, J. and Choo, K. K. R. (2018). A blockchain future for internet of things security: a position paper. *Digital Communications and Networks*, 4(3),149–160.

Benkler, Y. (2007). *The Wealth of Networks: How Social Production Transforms Markets and Freedom*. University of Yale Press. https://www.fruugo.es/the-wealth-of-networks-by-yochai-benkler/p-8035637-17343401?gclid=EAIaIQobChMIkenGk_T72AIVkQrTCh1OfgDvEAYYASABEglpn_D_BwE.

Bonino, D. and Vergori, P. (2017). Agent marketplaces and deep learning in enterprises: the COMPOSITION project. In: *2017 IEEE 41st Annual Computer Software and Applications Conference (COMPSAC)* (pp. 749–754).

Bonneau, J., Miller, A., Clark, J., Narayanan, A., Kroll, J. A. and Felten, E. W. (2015). *Research Perspectives and Challenges for Bitcoin and Cryptocurrencies*. IEEE Security and Privacy. http://www.jbonneau.com/doc/BMCNKF15-IEEESP-bitcoin.pdf.

Chang, S. E., Chen, Y. C. and Lu, M. F. (2019). Supply chain re-engineering using blockchain technology: a case of smart contract based tracking process. *Technological Forecasting and Social Change*, 144, 1–11.

Christidis, K. and Devetsikiotis, M. (2016). Blockchains and smart contracts for the internet of things. *IEEE Access*, 4, 2292–2303.

CTA. (2019). Brussels Briefing n. 55 "Opportunities of blockchain for agriculture", April 26, 2019. Brussels, Belgium.

de Bruin, L. (2019). Blockchain supporting food systems: private sector perspective. Brussels Briefing n. 55 "Opportunities of blockchain for agriculture", April 26, 2019. Brussels, Belgium.

de la Rocha. (2018). *BigchainDB, a Database on Blockchain Steroids*. https://medium.com/coin monks/bigchaindb-a-database-on-blockchain-steroids-fb52f3cd0d56 (accessed July 22, 2019).

del Castillo, M. (2017). Walmart, Kroger & Nestle Team with IBM blockchain to fight food poisoning. *Coindesk*, August 22. www.coindesk.com/walmartkroger-nestle-team-with-ibm-blockchain-to-fight-food-poisoning.

Delmolino, K., Arnett, M., Kosba, A., Miller, A. and Shi, E. (2016). Step by step towards creating a safe smart contract: lesson and insights from a cryptocurrency lab. *International Conference on Financial Cryptography and Data Security*, 79–94.

DSM. (2018). *Digital Single Market*. EU Blockchain Observatory and Forum. https://ec.europa.eu/digital-single-market/en/eu-blockchain-observatory-and-forum

Dujak, D. and Sajter, D. (2019). Blockchain applications in supply chain. In: *SMART Supply Network* (pp. 21–46). Cham: Springer.

Elliott, C. (2014). *Review into the Integrity and Assurance of Food Supply Networks: Final Report*. https://www.gov.uk/government/publications/elliott-review-into-the-integrity-and-assur ance-of-food-supply-networks-final-report.

Eto, K. (2000). Minamata disease. *Neuropathology*, 20, 14–19.

EU, 2002. European Commission, 2002. Regulation (EC) No 178/2002 of the European Parliament and of the Council of 28 January 2002 laying down the general principles and requirements of food law, establishing the European Food Safety Authority and laying down procedures in matters of food safety. *Official Journal of the European Communities*, 31(01/02/2002), pp. 1–24.

EU-TRACES. (2019). *How Does Traces Work?* https://ec.europa.eu/food/animals/traces/how-does-traces-work_en (accessed August 18, 2019).

Eyal, I. and Sirer, E. G. (2014). Majority is not enough: Bitcoin mining is vulnerable. In: *Proceedings of International Conference on Financial Cryptography and Data Security, Berlin, Heidelberg* (pp. 436–454).

FAO. (1997). Risk management and food safety. Report of a Joint FAO/WHO Consultation 27 to 31 January 1997. Rome, Italy. *Technology*, 9(5),211–214. http://doi.org/10.1016/S0924-2244(98) 00037-5.

FAO. (2002). Safe food and nutritious diet for the consumer. In: *World Food Summit, Rome, June 10–13*. http://www.fao.org/worldfoodsummit/sideevents/papers/y6656e.htm#TopOfPage.

FAO/WHO. (1997). *Codex Alimentarius. Joint FAO/WHO Food Standards Programme*. Rome: Codex Alimentarius Commission, Rome.

FDA. (2018). *Food and Drug Administration Database*. https://datadashboard.fda.gov/ora/cd/re calls.htm (accessed September 2, 2019).

FDA, Food and Drug Administration. (2009). Economically Motivated Adulteration; Public Meeting; Request for Comment [Docket No. FDA-2009-N-0166] [Electronic Version]. Federal Register, 74, 15497. http://edocket.access.gpo.gov/2009/pdf/E9--7843.pdf (accessed July 2, 2019).

Feng, T. (2017). *A Supply Chain Traceability System for Food Safety Based on HACCP, Blockchain & Internet of Things* (pp. 1–6). Vienna University of Economics and Business.

Food Ingredients. (2019). *Open Blockchain Pilot: Nestlé Tests Technology to Trace Origin of Its Products*. https://www.foodingredientsfirst.com/news/open-blockchain-pilot-nestl%C3%A9-tests-technology-to-trace-origin-of-its-products.html (accessed August 28, 2019).

Franco, P. (2014). *Understanding Bitcoin: Cryptography, Engineering and Economics*. UK: Wiley & Sons. https://www.wiley.com/en-es/Understanding+Bitcoin:+Cryptography,+Engineering +and+Economics-p-9781119019169.

FSMA. (2019). *Food Safety Modernization Act*. https://www.fda.gov/food/food-safety-moderniza tion-act-fsma/fsma-rules-guidance-industry#Guidance (accessed July 18).

Fung, F., Wang, H. S. and Menon, S. (2018). Food safety in the twenty-first century. *Biomedical Journal*, 41(2),88–95.

Gupta, M. (2017). *E-Book: Blockchain for Dummies*. IBM. https://www-01.ibm.com/common/ssi/ cgi-bin/ssialias?htmlfid=XIM12354USEN (accessed June 30, 2019).

Harada, M. (1995). Minamata disease: methylmercury poisoning in Japan caused by environmental pollution. *Critical Reviews in Toxicology*, 25(1),1–24.

IBM. (2018). *Focus on Supply Chain Efficiencies*. https://www.ibm.com/downloads/cas/LR8VR8YV (accessed July 16, 2019).

ISO/TC 176/SC 1 8402:1994, Quality Management and Quality Assurance – Vocabulary.

ISO/TC 176/SC 1 9000:2000, Quality Management Systems – Fundamentals and Vocabulary.

Jacquet, J. L. and Pauly, D. (2008). Trade secrets: renaming and mislabeling of seafood. *Marine Policy*, 32(3),309–318.

Johnson, R. (2014). Food fraud and economically motivated adulteration of food and food ingredients. Congressional Research Service (CRS) Report. Source: CRS compilation from information reported by USP, Michigan State University, NCFPD and researchers at the University of Minnesota, Oceana, Consumers Union, Food Chemical News, and the Rodale Institute.

Khan, M. A. and Salah, K. (2018). IoT security: review, blockchain solutions, and open challenges. *Future Generation Computer Systems*, 82, 395–411.

Li, Z., Wu, H., King, B., Miled, Z. B., Wassick, J. and Tazelaar, J. (2017), On the integration of event based and transaction based architectures for supply chains. In: *IEEE 37th International Conference on Distributed Computing Systems Workshops* (pp. 376–382).

Mgbemene, C. A., Nnaji, C. C. and Nwozor, C. (2016). Industrialization and its backlash: focus on climate change and its consequences. *Journal of Environmental Science and Technology*, 9(4),301–316.

Moe, T. (1998). Perspectives on traceability in food manufacture. *Trends in Food* Science & *Markets Watch, 2018. Economic Forecast for Blockchain*. www.ResearchandMarkets.com.

Mohsin, A. H., Zaidan, A. A., Zaidan, B. B., Albahri, O. S., Albahri, A. S., Alsalem, M. A. and Mohammed, K. I. (2019). Blockchain authentication of network applications: taxonomy, classification, capabilities, open challenges, motivations, recommendations and future directions. *Computer Standards & Interfaces*, 64, 41–60.

Mooney, P. R. (2019). Critical views on blockchain development: control and sovereignty. Brussels Briefing n. 55 "Opportunities of blockchain for agriculture", April 26, 2019. Brussels, Belgium.

Moore JC, DeVries JW, Lipp M, Griffiths JC, Abernethy DR. 2010. Total protein methods and their potential utility to reduce the risk of food. protein adulteration. *Compr Rev Food Sci Food Safety* 9: 330–57.

Mortimer, G. and Grimmer, L. (2018). Growers are in a jam now, but strawberry sabotage may well end up helping the industry. *The Conversation*, (27).

Nakamoto, S. (2008). *Bitcoin: A Peer-to-Peer Electronic Cash System.* http://www.academia.edu/ download/32413652/BitCoin_P2P_electronic_cash_system.pdf (accessed June 21, 2019).

Nakasumi, M. (2017). Information sharing for supply chain management based on block chain technology. In: *2017 IEEE 19th Conference on Business Informatics* (pp. 140–149).

Nakomoto, S. (2008). *Blockchain, Bitcoin Foundation Wiki.* https://en.bitcoin.it/wiki/Block_chain.

Nakomoto, S. (2009). *Bitcoin: A Peer-to-Peer Electronic Cash System* (Vol. 3). Bitcoin Org. https://bitcoin.org/bitcoin.pdf.

Oliveira, M. M., Cruz-Tirado, J. P. and Barbin, D. F. (2019). Nontargeted analytical methods as a powerful tool for the authentication of spices and herbs: a review. *Comprehensive Reviews in Food Science and Food Safety*, 18(3),670–689.

Olsen, P. and Borit, M. (2013). How to define traceability. *Trends in Food Science & Technology.* http://www.sciencedirect.com/science/article/pii/S0924224412002117.

Omohundro, S. (2014) Cryptocurrencies, smart contracts, and artificial intelligence. *AI Matters*, 1(2),19–21.

Özsu, M. T. and Valduriez, P. (2011). *Principles of Distributed Database Systems (Third Edition).* Berlin, Germany: Springer Science & Business Media.

Papapanagiotou, E. (2017). Foodborne norovirus state of affairs in the EU rapid alert system for food and feed. *Veterinary Sciences*, 4(4), 61.

Parvathy, V. A., Swetha, V. P., Sheeja, T. E., Leela, N. K., Chempakam, B. and Sasikumar, B. (2014). DNA barcoding to detect chilli adulteration in traded black pepper powder. *Food Biotechnology*, 28(1),25–40.

Perkins, T. D. and van den Berg, A. K. (2009). Maple syrup – production, composition, chemistry, and sensory characteristics. *Advances in Food and Nutrition Research*, 56, 101–143.

Pizzuti, T. and Mirabelli, G., 2015. The Global Track&Trace System for food: General framework and functioning principles. *Journal of Food Engineering*, 159, pp. 16–35.

Raheem, D., Shishaev, M. and Dicovitsky, V. (2019). Food system digitalisation as a means to promote food and nutrition security in the Barents region. *Agriculture*, 9, 168. https://doi.org/ 10.3390/agriculture9080168. Switzerland: Multidisciplinary Digital Publishing Institute (MDPI).

RFF. (2019). *Refrigerated and Frozen Foods Magazine*, July, 2019. https://www.refrigeratedfrozen food.com/articles/97634-nestlé-announces-blockchain-pilot?id=97634-nestl%C3%A9-announ ces-blockchain-pilot&oly_enc_id=3136G3707801F0X (accessed July 16, 2019).

Schneier, B. and Kelsey, J. (1998). Cryptographic support for secure logs on untrusted machines. In: *USENIX Security Symposium, January* (Vol. 98, pp. 53–62).

Shimohata, T., Hirota, K., Takahashi, H. and Nishizawa, M. (2015). Clinical aspects of the Nigata Minamata disease. *Brain and Nerve= Shinkei kenkyu no shinpo*, 67(1),31–38.

Silver, L. and Bassett, M. T. (2008). Food safety for the twenty-first century. *Jama*, 300(8),957–959.

Spink, J. (2009). Defining food fraud and the chemistry of the crime. In: *Proceedings of the FDA Open Meeting, Economically Motivated Adulteration, College Park, MD, USA* (Vol. 1).

Spink, J., Helferich, O. K. and Griggs, J. E. (2010). Combating the impact of product counterfeiting. *Distribution Business Management Journal*, 10(6).

Spink, J. and Moyer, D. C. (2011). Defining the public health threat of food fraud. *Journal of Food Science*, 76(9),R157–R163.

Tapscott, D. and Tapscott, A. (2017). Realizing the potential of blockchain: a multistakeholder approach to the stewardship of blockchain and cryptocurrencies. *World Economic Forum, White Paper*, June. http://www3.weforum.org/docs/WEF_Realizing_Potential_Blockchain.pdf.

Tian, F., 2016, June. An agri-food supply chain traceability system for China based on RFID & blockchain technology. In *2016 13th international conference on service systems and service management (ICSSSM)* (pp. 1–6). IEEE.

Tian, F. (2017). A supply chain traceability system for food safety based on HACCP, blockchain & internet of things. In: *2017 International Conference on Service Systems and Service Management, Dalian* (pp. 1–6). DOI: 10.1109/ICSSSM.2017.7996119.

Townsend, R., Ronchi, L., Brett, C. and Moses, G. (2018). *Future of Food: Maximizing Finance for Development in Agricultural Value Chains*.

Tripoli, M. (2019). Opportunities and challenges for blockchain in the agri-food industry. Brussels Briefing n. 55 "Opportunities of blockchain for agriculture", April 26, 2019. Brussels, Belgium.

Tripoli, M. and Schmidhuber, J. (2018). *Emerging Opportunities for the Application of Blockchain in the Agri-food Industry*. Rome and Geneva: FAO and ICTSD. Licence: CC BY-NC-SA, 3.

Wang et al., 2018. Reference details: Wang, S., Zhang, Y. and Zhang, Y., 2018. A blockchain-based framework for data sharing with fine-grained access control in decentralized storage systems. *IEEE Access*, 6, pp.38437–38450.

Wang, S., Zhang, Y. and Zhang, Y., 2018. A blockchain-based framework for data sharing with fine-grained access control in decentralized storage systems. *IEEE Access*, 6, pp.38437–38450.

Wang, Y., Han, J. H. and Beynon-Davies, P. (2019). Understanding blockchain technology for future supply chains: a systematic literature review and research agenda. *Supply Chain Management: An International Journal*, 24(1),62–84. https://doi.org/10.1108/SCM-03-2018-0148

Wright, A., De Filippi, P. (2017). Decentralized blockchain technology and the rise of Lex Cryptographia. *SSRN*, p. 58. https://papers.ssrn.com/sol3/papers.cfm?abstract_id=2580664.

Xu, L., Chen, L., Gao, Z., Lu, Y. and Shi, W. (2017). *CoC: Secure Supply Chain Management System Based on Public Ledger, in 26th International Conference on Computer Communication and Networks (ICCCN)* (pp. 1–6).

Zhao, G., Liu, S., Lopez, C., Lu, H., Elgueta, S., Chen, H. and Boshkoska, B.M., 2019. Blockchain technology in agri-food value chain management: A synthesis of applications, challenges and future research directions. *Computers in Industry, 109*, pp. 83–99.

Zheng, Z., Xie, S., Dai, H.-N., Chen, X. and Wang, H. (2018). Blockchain challenges and opportunities: a survey. *International Journal of Web and Grid Services*, 14(4), 352–375.

Ajibola Bamikole Oyedeji, Olayemi Eyituoyo Dudu,
Oluwatosin Ademola Ijabadeniyi, Oluwafemi Ayodeji Adebo

16 Big data and the food industry

16.1 Introduction

Data generation and recording has evolved from paper documentations and form-filing into the use of computer codes and languages as well as physical and virtual data storage systems. Although, this shift has drastically reduced the stress and fatigue involved in data recording and the use of papers and physical spaces for data storage. It has brought about the generation of very huge and diverse amount of data that require specialized computer knowledge and infrastructure for processing and a different form of complex systems for data storage and retrieval. This vast and diverse amount of data are called "big data" and they are produced per second in every sector of the society, including government, banking, healthcare, engineering natural sciences, and, more recently, in the food industry (Marvin, Janssen, Bouzembrak, Hendriksen, and Staats, 2017).

Big data have been defined as such that possesses high volume (huge), high velocity (fast), and high variety (versatile) information requiring advanced computational tools to bring about discovery, data manipulation, process optimization, and decision-making (Laney, 2012). Data are rapidly generated, sometimes in split seconds and could require terabyte, petabyte, and even higher capacities for storage (Marvin et al., 2017; Ward and Barker, 2013). They thus require specialized software and programming languages for handling and interpretation to reveal patterns, trends, and other desired outcomes. From local food processors using stalls and shops to global brands with integrated supply chains across different continents of the world, the food industry is indeed very vast and diverse. More recently, farming systems as well as the logistics and supply chain of farm input and produce involved in the food production value chain have been recognized as a part of the food industry through the Food Safety Management System of the International Standards Organization. The vastness of the food industry also implies the production of big data. Consumers have become highly aware of their food and are interested in every process their food has undergone. This means that data generated during every single process involved in the value chain of food production, including farming operations, food processing unit operations, conditions of raw and processed foods transportation, storage and

Ajibola Bamikole Oyedeji, Oluwafemi Ayodeji Adebo, Department of Biotechnology and Food Technology, University of Johannesburg, Doornfontein Campus, Gauteng, South Africa
Olayemi Eyituoyo Dudu, Department of Food Science and Engineering, Harbin Institute of Technology, Harbin, China
Oluwatosin Ademola Ijabadeniyi, Department of Biotechnology and Food Technology, Durban University of Technology, Durban, South Africa

https://doi.org/10.1515/9783110667462-016

retail activities must be kept untampered and be easily retrievable. Also, considering the economic importance of food safety and the global adverse effects of food-borne illnesses, which may include loss of lives and closure of multi-million-dollar processing plants, it has become extremely important to employ the versatility and velocity, as well as the enormity of big data generated at every stage of food production. This is to devise means to quickly control the spread and combat the effects of food-borne pathogenic microorganisms.

With the accumulation of big data in various repositories, stored data must be retrievable and accessible to concerned stakeholders within each sector of the food value chain. This should be achievable in real time. Various data storage repositories therefore need to be connected, and this connectivity is currently best achieved through the Internet. The concept of Internet of things (IoT) involves the creation of network of connectivity of data and objects anytime and anywhere with anything and anyone who can connect to the network (Maksimović, Vujović, and Omanović-Mikličanin, 2015). This implies that data uploaded and updated in various repositories within a network connected by the Internet is accessible to everyone (with permission) on this network. Therefore, as big data/information about the various activities within the value chain of food production and consumption are generated and stored in various repositories, they could be tracked, traced, and verified by concerned stakeholders of the value chain in real time.

Such big data within the value chain in the food industry could exist in different repositories in structured (with specific and discrete values such as numbers, codes, and signs) or non-structured (opinions, comments, and reports) forms. These repositories include online databases, gene banks, and genomic databases and the Internet including social media, from where stored data can be pooled, analyzed, and interpreted for informed decisions. There are potentials for application of big data and its combination with IoT in the food value chain. This includes food safety assurance and traceability, food processing, storage, fresh produce management, intelligence packaging, and food logistics and supply. In this chapter, the various big data repositories applicable in the food production, processing, storage, and analysis as well as the potentials of application of big data in the food value chain are discussed.

16.2 Different big data repositories of interest in the food industry

16.2.1 The Internet

A huge amount of data that are of interest to the food industry are constantly being generated and deposited in various Internet repositories. These data could be either structured or non-structured. There are various Internet-based repositories of big data hosted by private and government corporations and are freely accessible by consumers. Internet data such as weather forecasts, traffic information, satellite imageries, online map information about distances and changes in route, opinion polls from different online questionnaires and many more are ever available on the Internet and could be analyzed and utilized in the food industry throughout the value chain. The direct opinions and sentiments of consumers for brands and raw materials within and outside specific regions written by them on various social media platforms could be obtained through various web crawling and information-scouting softwares.

Marvin et al. (2017) reported of MedISys web crawling tool developed by the European Commission's Joint Research Centre, which collects information about various food safety publications available on different databases and can be used to analyze and predict possible outbreaks of food-borne illnesses, in order to engage early warning systems (Rortais, Belyaeva, Gemo, van der Goot, and Linge, 2010). There are many internet-based repositories of data that are free to access and can be useful for decision-making in the food industry. Also, many food production, safety, and logistics-related mobile applications have been developed by different individuals or corporations and are available on various mobile application stores and are downloadable on mobile devices enabled by different operating systems. These mobile applications are usually connected to the Internet and provide vast amount of real-time data that can be used to make informed decisions. Mobile phone cameras and other connectivity-enabling sensors such as Bluetooth, infrared, and wireless fidelity (WiFi) can obtain information from the bar codes, quick response (QR) scan codes, and the storage environment of foods. They can then generate information with the aid of the food-related mobile apps connected to them to make real-time data available about the status and condition of foods as well as their environments.

16.2.2 Omics

The concept of omics involves the technologies adopted for the identification, differentiation, characterization of relationships, and roles of various molecules and biological components that make up a system. High-throughput technologies and associated bioinformatics tools adopted for the omics have revolutionized data collection and interpretation. Such omics techniques include genomics, metabolomics, transcriptomics, and proteomics, as well as research with an extension for application in other spheres. The omics technologies of metabolomics and metagenomics are techniques incorporated into food science and technology research. The outcomes of these applications are often of huge significant importance to the food industry.

Genomics is primarily based on the premise that every living organism (plant and animals), including microorganisms, are distinctly coded genetically and this serves as the basic unit for their uniqueness and, hence, their identification and differentiation. It thus involves the identification of genes of biological materials as coded by their DNA information, while proteomics is the functional analysis of genetic products, including identification and localization of genetic materials of protein moieties (Pandey and Mann 2000; Nair and Zhai, 2020). Transcriptomics on the other hand involves the study of RNA through transcription of DNA information of cells, tissues, or organisms at a particular time (Davies, 2010; Lamas et al., 2019), while metabolomics involves the analysis of the changes in metabolites produced as a result of the activity of biological components or other intrinsic or extrinsic factors in a biological system (Davies, 2010; Adebo et al., 2019). Other emerging fields of application of omics technology include pharmacogenomics, physiomics, phylogenomics, and interactogenomics (Schneider and Orchard, 2011; Ovejero-Benito et al., 2018; Hwang et al., 2019).

For generation of the big data associated with omics techniques, high or automated throughput technologies and infrastructure are usually needed. This involves the automation of the vast amount of materials passing through analytical equipment with subsequent enormous amount of information or data generated. High-throughput DNA sequencing methods, also called next-generation sequencing (NGS), have been applied in genomics, making it possible for the generation of sequences from amplified DNA fragments, thereby enhancing their identification and application (Schneider and Orchard, 2011). High-throughput messenger RNA sequencing (mRNA-Seq) methods have been applied in transcriptomics to understand the various stages and physiological conditions of a cell through the identification and quantification of transcripts or transcriptomes of cells (Wang, Gerstein, and Snyder, 2009). This approach generates huge amount of data and is superior to previous transcriptomic analytical approaches, including hybridization methods. For proteomics, high-throughput methods for identification and quantification of protein moieties include assisted mass spectrometry methods. This involves the extraction and digestion of soluble proteins extracted from samples, digested on silica gels, and subsequent ionization or desorption in highly sensitive mass spectrometers to obtain the sequences of peptides that can be run through databases to obtain their identities (Oyedeji, Mellem, and Ijabadeniyi, 2018; Schneider and Orchard, 2011).

Genetic information obtained from these various omics technologies are stored in different repositories, which could be specific for a product or material, for example, the "*Glycine max*" database for soybean proteomes or general for different biological materials, "UniProt" for proteomes of different plant and animal materials. Databases for genomics, such as GenBank, PubMed, and PMC, are hosted in the National Centre of Biotechnology Information (NCBI) repositories, which is hosted by the US Department of Health and Human Services. Likewise are human metabolome database, BinBase, biochemical, genetic, and genomic databases, Kyoto encyclopedia of genes and genomes and Metlin databases for metabolomics as well as coral transcriptome databases, database for gene expression and evolution (Bgee), European Union microbe databases, and mammalian transcriptomic databases for transcriptomics studies.

16.2.3 Others

Apart from the Internet and omics data repositories, there are other stand-alone big data storage systems which can be accessed and used in the food industry. Many farms, food processing and production industries, logistic industries related to food transportation and delivery and chain of retail outlets have recently become smart and have been involved with the collection and storage of data within the food industry value chain in different repositories. Information about food products are monitored and stored real time in these repositories, hosted by food industries or a third-party data management and storage center. These information are stored in such a way that they could be retrieved easily and/or shared with all the firms within the value chain of food production, transportation, sale, and delivery. With the aid of bar codes, scan codes, QR codes, and more recently, the radio frequency identification chips (RFID) imprinted or inserted on food packaging materials, various conditions of the food materials could be monitored from one end of the value chain to the other. This help to ensure product integrity and minimize wastage and spoilage (Doinea, Boja, Batagan, Toma, and Popa, 2015). These data can be monitored and shared within the various actors of a value chain system and communications could be established.

For the complexities around data security, blockchain technology provides huge tamper-proof assurance for data that are stored within the blockchain systems. Information within the chain is retrievable by the all users, that is, everyone within the blockchain in question. This is equally applicable to the value chain of production and delivery of the food commodity. Deep learning is also another big data-driven technique, gradually gaining prominence for use in the food industry. One of such is through computer vision, which refers to an automated technique involving the extraction, analysis, and consequent understanding/interpretation of useful information from an image or images.

16.3 Some application of big data in the food industry

16.3.1 Agriculture and raw produce farming

The application of big data to farming systems has a lot of potentials to help and influence correct decision-making. There are different sources of big data available for farming systems, including ground sensors such as biosensors, chemical sensors, weather stations, historical data by government or research institutes, online repositories and web services, airborne sensors, real-time data, mobile phones, and social media (Kamilaris, Kartakoullis, and Prenafeta-Boldú, 2017). Table 16.1 shows the

Table 16.1: Different sources of big data in farming and their classification.

No	Agri-area	No. of papers	Volume	Velocity	Variety	Ref.
1.	Weather and climate change	4	M	M	H	Tripathi et al. (2006), Fuchs and Wolff (2011), Schnase et al. (2014), and Tesfaye et al. (2016)
2.	Land	5	H	L	M	Barrett et al. (2014), Schuster et al. (2011), Galford et al. (2008), Wardlow et al. (2007) and Thenkabail et al. (2007)
3.	Animals' research	4	M	H	L	McQueen et al. (1995), Kempenaar et al. (2016), Chedad et al. (2001), and Pierna et al. (2014)
4.	Crops	3	M	M	L	Waldhoff et al. (2012), Sakamoto et al. (2005), and Urtubia et al. (2007)
5.	Soil	2	M	L	L	Armstrong et al. (2007), and Meyer et al. (2004)
6.	Weeds	1	L	H	L	Gutiérrez et al. (2008)
7.	Food availability and security	4	M	L	M	Frelat et al. (2016), Jóźwiaka et al. (2016), Lucas and Chhajed (2004)
8.	Biodiversity	1	M	L	H	Marcot et al. (2001)
9.	Farmers' decision-making	2	H	M	H	Sawant et al. (2016), and Field to Market (2015)
10.	Farmers' insurance and Finance	5	H	M	M	Global Envision (2006), Syngenta (2010) and Akinboro (2016)
11.	Remote-sensing	3	H	M	M	Becker-Reshef et al. (2010), Nativi et al. (2015)

Source: Kamilaris et al. (2017). H: high, M: medium, L: low.

different sources of big data available to farms and their classification as regards the different elements of big data. The analysis of big data plays important roles in the improvement of supply chain efficiency, reduction of production downtimes, and decrease of food security concerns, as they make future predictions practicable to farmers (Wolfert, Ge, Verdouw, and Bogaardt, 2017). Management of farms and farming systems cannot be reliably based on general knowledge, because of the specific conditions that are applicable from farm to farm. Therefore, big data analysis and subsequent application of such gives an opportunity for utilization of farm-specific information such as climatic conditions, rainfall patterns, and other environmental factors unique to the farms in decision-making (Lesser, 2014; Wolfert et al., 2017).

Precision agriculture and smart farming are currently the different forms of application of big data in farming systems. Precision agriculture involves the exact monitoring of farms using information from various big data repositories in determining the variations that may occur in-field and to decide the required inputs needed to achieve precision of desired result. An example of this is the use of satellite imagery in the understanding of the precise expanse of land covered by plants on a farm. These data will inform the precise application of fertilizers and pesticides, in order to achieve the efficiency of application, reduce cost of input, reduce environmental pollution, and also reduce food poisoning (Sundmaeker, Verdouw, Wolfert, and Pérez Freire, 2016). This farming system is however expensive and is only available to farmers practicing intelligent agriculture farmers. Smart farming, on the other hand, relies on the information about context and situation made available by huge amount of real-time generated data and analyses of such data help in management of tasks and decision-making (Sundmaeker et al., 2016). Real-time data used in smart farming help dynamism of decision-making, especially with the unpredictable possibilities of changes in conditions, such as weather, conditions, market conditions, currency fluctuations, and so on.

In developing countries, there are examples of the utilization of big data, generated and hosted by various agencies. This helps in ameliorating and, sometimes, completely eradicating the challenges faced by small-holder farmers in food and cash-crop production. In Kenya, partnerships exist between mobile network providers, data analysis firms, and financial institutions in providing credit facilities for small-holder farmers in rural areas, who have no access to banking facilities. Big data were generated from call and text message details of farmers and merged with their registration details to assess their credit worthiness through data analyses and then transferred to financial institutions for possibility of provision of credit facilities to farmers (Protopop and Shanoyan 2016).

16.3.2 Food safety

Throughout the value chain of food production and delivery to consumers, it is important to track and trace foods from farmers to the final consumer. This makes it easy to identify areas of concern and ensure quick recalls in case of pathogenic contamination, before resulting into disease outbreak. Big data can be used to predict the growth of pathogenic organisms on foods. This can be achieved by combining environmental data within the food production facility, whether farms or processing plants with the favorable conditions of growth of food-borne pathogens, and hence preventing the entry of such susceptible food into the value chain (Marvin et al., 2017). Structured and unstructured data about the conditions of food handlers obtained through employment data, task information, and video-captured data of food production facilities could be mined for relevant food safety information (Wiedmann, 2015).

In the prevention of food-borne disease outbreak, genomic repositories are very important in rapid identification and source tracking of the specific strain of pathogenic microorganisms that may be responsible for such potential outbreak. Due to the collection of huge amount of food samples during an outbreak, big volume analytical data are generated and are compared with existing repositories to adequately identify the causal pathogen. With the development of high-throughput genomic-sequencing methods, such as NGS and high-precision bioinformatics tools, these huge volume of data can be put into perspective to adequately narrow down to the desired result. Without the generation of large amount of data and prior existence of continually updated genomic repositories, the identification of pathogenic microorganisms and source tracking will be practically difficult and time-consuming. This would lead to fatal consequences, including adverse economic implications and in extreme cases, loss of lives. For example, the particular strain of *Listeria monocytogenes* responsible for the recent listeriosis outbreak in South Africa was identified as *Listeria monocytogenes* subtype 6 (ST6) through whole genome-sequencing approach to multilocus sequence typing (MLST) of about 621 clinical isolates (Allam et al., 2018). This identification was possible due to the existence of genomic repositories and it is important to rapidly identify and store the identity of this pathogenic strain in genomic repositories to enhance future identification, source tracking, and to prevent future fatal occurrences.

16.3.3 Food traceability, intelligent packaging, and transportation

Apart from the identification of food-borne pathogenic microorganisms, it is important to be able to track them to the source of primary contamination and monitor their effects throughout the food supply value chain. Before the application of big data in the food industry, the traceability of pathogenic microorganisms was very laborious and time-consuming and this can possibly sustain the various effects of disease outbreaks. Also, vital and critical control points may be left unattended. The use of bar codes, scan QR codes, and RFID individually or together has since replaced the traditional traceability methods, giving access of the value chain history of food products to stake holders in the food chain. These technologies have also been used together in the wheat flour milling traceability system (Figure 16.1). Barcodes are capable of storing information about the different stages processing and dates, usually depicted in digits. On the other hand, information about different stages of processing, dates, product name, enterprise name, product level, testing, or quality parameters as well as the web-links or uniform resource locator of a trace website for the product can all be stored in a QR code (Qian et al., 2012). As handling, processing,

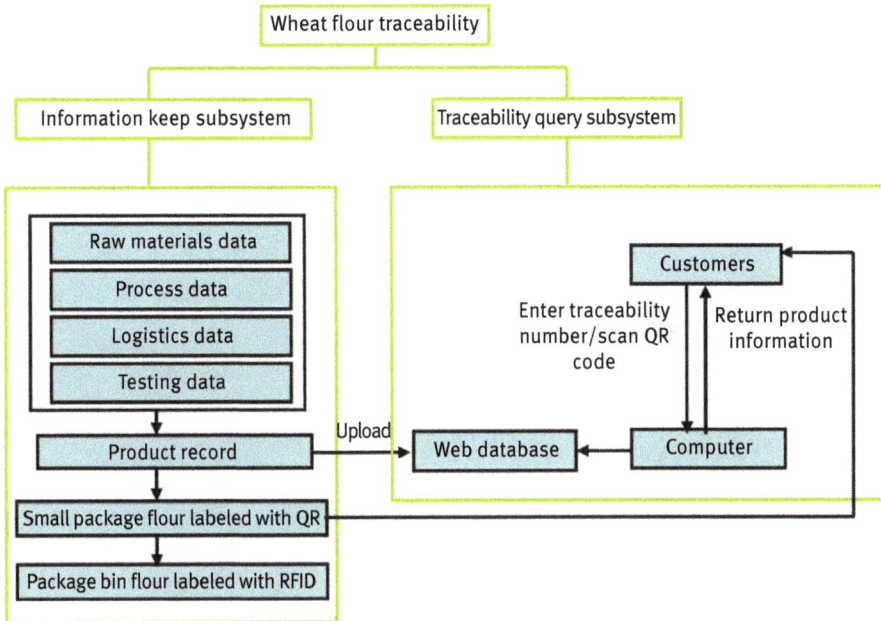

Figure 16.1: Wheat flour milling traceability system (adapted from Qian et al., 2012).

and transformation of raw materials into finished products continue, data are generated describing the details of processes, method, space, and time. Each process is assigned unique codes, linked to the products and stored in designated databases accessible to each stakeholder in the value chain through designated Internet protocols (IPs) (Šenk, Ostojić, Tarjan, Stankovski, and Lazarević, 2013).

RFIDs are more advanced and complex than bar and QR codes. Apart from the possibility of storing a greater amount of information on an RFID chip, they are more durable and tamper-proof (Qian et al., 2012; Šenk et al., 2013). A simple schematic depiction of the use of RFID in tracking and tracing of food products is shown in Figure 16.2. Maksimović et al. (2015) proposed RFID as a technology that could be connected with low-cost Raspberry Pi computer boards, which is capable of Internet connectivity through different Internet platforms including WiFi and sensors that could detect different environmental conditions within packaging materials, including temperature and humidity. This set up is capable of monitoring the environmental conditions of packaged foods during transportation and storage in real-time and sending such information, through the Internet to dedicated databases. This will ensure adequate monitoring of the integrity of different conditions of storage and transportation, especially for highly sensitive foods such as milk, fresh produce, and other high moisture foods that are highly prone to infestation of pathogenic microorganisms.

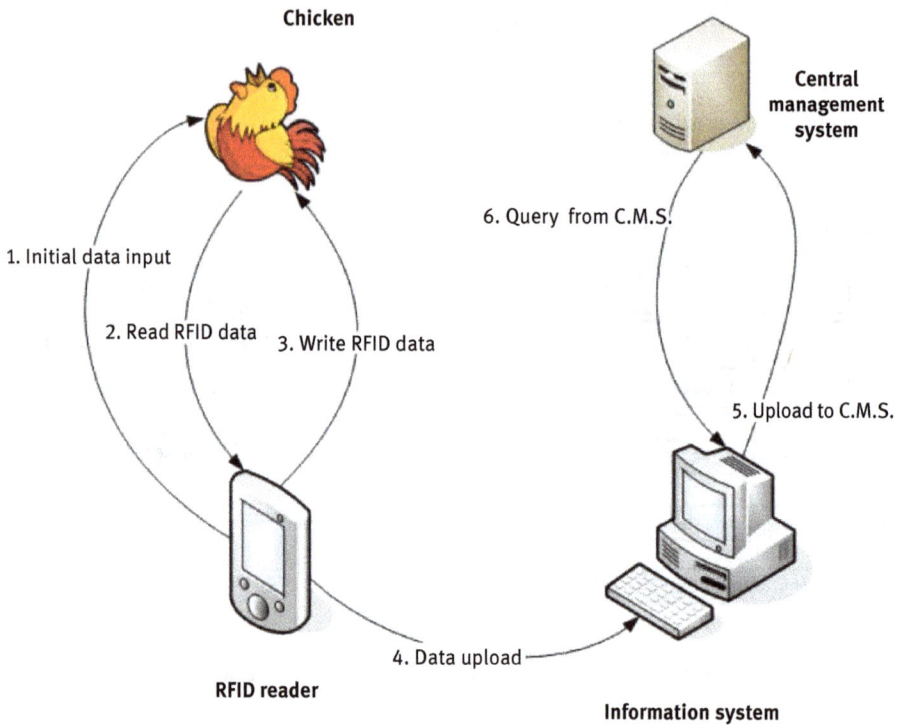

Figure 16.2: A simple schematic representation of the use of RFID technology for food tracking and tracing. Source: Chen et al. (2008).

These databases are repositories that are capable of hosting huge amount of data and are readily retrievable real time by every section of the value chain, including the end users.

Also crucial to food traceability is the use of blockchain technology. Blockchain technology has a wide application in food transportation, food delivery, food online transaction data, quality, and safety assurance (Kumar and Mallick, 2018). The use of RFID and near-field communication (NFC) systems in combination with blockchain technology and IoT in food traceability has been proposed to enhance food quality and safety in food logistics and transportation. RFID and NFC are used to acquire and share data into repositories shared with the public or verified users within the food value chain and blockchain is used to ensure that the data shared are tamper-proof, guaranteeing their authenticity and reliability (Zhao et al., 2019). Recent efforts and campaigns toward food traceability by the US Department of Agriculture and the World Economic Forum have emphasized the use of blockchain technology, aided by sensors and IoT in assuring the transportation of safe foods highly efficient traceability possibilities (Roy, 2019; USFDA, 2019). This has earlier been emphasized by the

Institute of Food Technologists, offering traceability guidance, expertise and solutions through their Global Food Traceability Centre (Nunes, 2013).

16.3.4 Food marketing and delivery

Big data from different repositories, including satellites, weather forecasts, maps, travel histories, traffic information, and construction information can be analyzed and used to advantage in food logistics, transportation, and delivery. Delivery of foods, especially those that are highly sensitive and prone to contamination with changes in their transportation conditions, can be planned and adequately executed by relying on information from these various big data repositories. With their reliability on the Internet to receive orders and feedbacks from customers, food delivery chains, grocery stores, restaurants, and cafeteria generate data including orders, location of delivery, tweets, updates, and images within their delivery platforms and other different media outlets to which they are attached to. These data could be processed to understand consumers' experiences with average wait time, delivery experience, feedbacks on customer satisfaction with foods, and many more (De Zyre, 2016). These data are generated in a very high volume, per second and in real time and represents the true dispositions of consumers to food products. Consumer sentiments and perceptions of a product can be obtained using social media platforms and blogs. The use of hashtags brings the information by consumers' sentiment into easy accessibility. These information/sentiments could be analyzed using different software-based analytical tools such as Textblob and Natural Language Toolkit to have deeper insights into these sentiments, collect the relevant information from vast perceptions and sentiments and make informed decision on sales, marketing, and delivery (Skyes, 2018).

16.3.5 Food analysis

The bedrock of food quality and safety assurance is in the analytical determination of the characteristics of foods. Samples must be taken from bulk food lots and subjected to different laboratory analytical methods to confirm their integrity. Application of big data in food analyses includes the use of different big data repositories in the qualitative and quantitative identification of food analytes. The different omics-based information hosted by government, university, and private repositories, described in section 2.2, are relied upon in the characterization of genomes, proteins and proteomes, transcriptomes, and other forms of genetically coded information relevant to the identification and quantification of these analytes. Other laboratory analytical methods, including different forms of digestion, extraction, polymerase chain reactions, electrophoresis, spectrometry and so on, are used to obtain purified forms of analytes, which are then sequenced to know their genetic and metabolic "languages" and compared with existing repositories for accurate identification. Since huge data are generated during sequencing, the knowledge and application of bioinformatics are crucial in separating irrelevant data and putting the relevant ones in perspective. For example, different softwares such as Bionumerics (Applied Maths, Belgium), CLC Genomic Workbench (QIAGEN Bioinformatics), Fast_q (Illumina), and Quantitative Insight into Microbial Ecology (QIIME) are designed and used to process huge data generated during polymerase chain reaction and sequencing, including high through of genomes, to identify genomic information.

16.4 Conclusion

The continuous upsurge in the sensitivity and awareness of consumers toward the quality and safety of foods makes the application of big data in the food industry inevitable. This is further compounded by the upsurge of the world population, estimated to reach 9.8 billion by 2050. Food industry is amongst the biggest in the world, and so is the amount of data generated from different food outlets, from cottage scale to big international brands. Big data repositories, primarily linked with the food industry, hosted by private and government agencies and those that are of secondary application are continuously updated. Big data generation by different sensors, with the application of IoT, has continued to expand the scope of different databases and enhance real-time data storage and retrieval. Information from these different repositories, whether structured or unstructured, must be stored with high reliability and integrity, and be accessible by the public or certified users linked to specific food value chain. These data must also be retrievable at any desired point and location for reference. Food industries must be ready to shift the paradigm into investment in the generation and analyses of big data, as well as adoption and utilization of already existing data in their business decisions. As the fourth industrial revolution unfolds, big data are very vital for food industries to keep up to pace and better serve the constantly escalating population. While the developed countries are obviously way ahead in the application of these big data in decision-making, much more concerted efforts should also be directed toward developing and underdeveloped countries. This should indeed be the case considering the huge market potential of these nations with substantial chunk of the world population coming from them.

Acknowledgments: Funding provided by Faculty of Science and the GES 4.0 Catalytic Funding Initiative of the University of Johannesburg is thankfully acknowledged.

References

Adebo, O. A., Kayitesi, E., Tugizimana, F. and Njobeh, P. B. (2019). Differential metabolic signatures in naturally and lactic acid bacteria (LAB) fermented *ting* (a Southern African food) with different tannin content, as revealed by gas chromatography mass spectrometry (GC–MS)-based metabolomics. *Food Research International*, 121, 326–335.

Allam, M., Tau, N., Smouse, S. L., Mtshali, P. S., Mnyameni, F., Khumalo, Z. T. H., et al. (2018). Whole-genome sequences of Listeria monocytogenes sequence type 6 isolates associated with a large foodborne outbreak in South Africa, 2017 to 2018. *Genome Announcements*. https://doi.org/10.1128/genomeA.00538-18.

Akinboro, B., 2016. Bringing Mobile Wallets to Nigerian Farmers. [Online] Available at: http://www.cgap.org/blog/bringing-mobile-wallets-nigerian-farmers (accessed 2017).

Armstrong, L., Diepeveen, D., & Maddern, R. (2007). The application of data mining techniques to characterize agricultural soil profiles. Sixth Australasian Data Mining Conference (AusDM 2007) (pp 85–100).

Barrett, B., Nitze, I., Green, S., & Cawkwell, F. (2014). Assessment of multi-temporal, multi-sensor radar and ancillary spatial data for grasslands monitoring in Ireland using machine learning approaches. Remote Sensing of Environment, 152, 109–124.

Becker-Reshef, I., Justice, C., Sullivan, M., Vermote, E., Tucker, C., Anyamba, A., ... & Hansen, M. (2010). Monitoring global croplands with coarse resolution earth observations: The Global Agriculture Monitoring (GLAM) project. Remote Sensing, 2(6), 1589–1609.

Chedad, A., Moshou, D., Aerts, J. M., Van Hirtum, A., Ramon, H., & Berckmans, D. (2001). AP—animal production technology: recognition system for pig cough based on probabilistic neural networks. *Journal of Agricultural Engineering Research*, 79(4), 449–457.

Chen, R. S., Chen, C. C., Yeh, K. C., Chen, Y. C. and Kuo, C. W. (2008). Using RFID technology in food produce traceability. *WSEAS Transactions on Information Science and Applications*, 5(11), 1551–1560.

Davies, H. (2010). A role for "omics" technologies in food safety assessment. *Food Control*. https://doi.org/10.1016/j.foodcont.2009.03.002.

De Zyre. (2016). https://www.dezyre.com/article/how-food-delivery-apps-are-leveraging-big-data-analytics/197 (accessed September 18, 2019).

Doinea, M., Boja, C., Batagan, L., Toma, C. and Popa, M. (2015). Internet of things based systems for food safety management. *Informatica Economica*. https://doi.org/10.12948/issn14531305/19.1.2015.08.

Field to Market, 2015. Fieldprint Calculator. [Online] Available at: https://www.fieldtomarket.org/fieldprint-calculator/ (accessed 2019).

Food and Drug Administration, United States. (2019). https://www.fda.gov/food/food-industry/new-era-smarter-food-safety (accessed September 18, 2019).

Frelat, R., Lopez-Ridaura, S., Giller, K. E., Herrero, M., Douxchamps, S., Djurfeldt, A. A., ... & Rigolot, C. (2016). Drivers of household food availability in sub-Saharan Africa based on big data from small farms. Proceedings of the National Academy of Sciences, 113(2), 458–463.

Fuchs, A., & Wolff, H. (2011, May). Drought and retribution: Evidence from a large scale rainfall index insurance in Mexico. In Economics of Climate Change Conference (pp. 13–14).

Galford, G. L., Mustard, J. F., Melillo, J., Gendrin, A., Cerri, C. C., & Cerri, C. E. (2008). Wavelet analysis of MODIS time series to detect expansion and intensification of row-crop agriculture in Brazil. Remote sensing of environment, 112(2), 576–587.

Global Envision, 2006.Unleashing Ugandan farmers'potential through mobile phones.[Online] Available at: https://www.mercycorps.org/research-resources/unleashing-ugandan-farmers-potential-through-mobile-phones (accessed 2019).

Gutiérrez, P. A., López-Granados, F., Peña-Barragán, J. M., Jurado-Expósito, M., & Hervás-Martínez, C. (2008). Logistic regression product-unit neural networks for mapping Ridolfia segetum infestations in sunflower crop using multitemporal remote sensed data. Computers and Electronics in Agriculture, 64(2), 293–306.

Hwang, M., Leem, C. H. and Shim, E. B. (2019). Toward a grey box approach for cardiovascular physiome. *Korean Journal of Physiology and Pharmacology*, 23, 305–310.

Jozwiak, A., Milkovics, M., & Lakner, Z. (2016). A network-science support system for food chain safety: a case from hungarian cattle production. International Food and Agribusiness Management Review, 19(1030-2016-83145), 1–26.

Kamilaris, A., Kartakoullis, A. and Prenafeta-Boldú, F. X. (2017). A review on the practice of big data analysis in agriculture. *Computers and Electronics in Agriculture*. https://doi.org/10.1016/j.compag.2017.09.037.

Kempenaar, C., Lokhorst, C., Bleumer, E. J. B., Veerkamp, R. F., Been, T., van Evert, F. K., ... & van Bekkum, M. (2016). *Big Data analysis for smart farming: results of TO2 project in theme food security* (Vol. 655). Wageningen University & Research.

Kumar, N. M. and Mallick, P. K. (2018). Blockchain technology for security issues and challenges in IoT. *Procedia Computer Science*. https://doi.org/10.1016/j.procs.2018.05.140.

Lamas, A., Regal, P., Vázquez, B., Miranda, J. M., Franco, C. M. and Cepeda, A. (2019). Transcriptomics: a powerful tool to evaluate the behavior of foodborne pathogens in the food production chain. *Food Research International*, 125, 108543.

Laney, D. (2012). The importance of 'big data': a definition. *Gartner. Retrieved*, 21, 2014–2018.

Lesser, A. (2014). *Big Data and Big Agriculture*.

Lucas, M. T., & Chhajed, D. (2004). Applications of location analysis in agriculture: a survey. Journal of the Operational Research Society, 55(6), 561–578.

Maksimović, M., Vujović, V. and Omanović-Mikličanin, E. (2015). Application of internet of things in food packaging and transportation. *International Journal of Sustainable Agricultural Management and Informatics*. https://doi.org/10.1504/IJSAMI.2015.075053.

Marcot, B. G., Holthausen, R. S., Raphael, M. G., Rowland, M. M., & Wisdom, M. J. (2001). Using Bayesian belief networks to evaluate fish and wildlife population viability under land management alternatives from an environmental impact statement. Forest Ecology and Management, 153(1-3), 29–42.

Marvin, H. J. P., Janssen, E. M., Bouzembrak, Y., Hendriksen, P. J. M. and Staats, M. (2017). Big data in food safety: an overview. *Critical Reviews in Food Science and Nutrition*. https://doi.org/10.1080/10408398.2016.1257481.

McQueen, R. J., Garner, S. R., Nevill-Manning, C. G., & Witten, I. H. (1995). Applying machine learning to agricultural data. *Computers and Electronics in Agriculture*, 12(4), 275–293.

Meyer, G. E., Neto, J. C., Jones, D. D., & Hindman, T. W. (2004). Intensified fuzzy clusters for classifying plant, soil, and residue regions of interest from color images. Computers and Electronics in Agriculture, 42(3), 161–180.

Nair, M. N. and Zhai, C. (2020). Application of proteomic tools in meat quality evaluation. In: A. K. Biswas and P. K. Mandal (Eds.). *Meat Quality Analysis* (pp. 353–368). The Netherlands: Elsevier.

Nativi, S., Mazzetti, P., Santoro, M., Papeschi, F., Craglia, M., & Ochiai, O. (2015). Big data challenges in building the global earth observation system of systems. Environmental Modelling & Software, 68, 1–26.

Nunes, K. (2013). *I.F.T. Launches Food Traceability Center*. https://www.foodbusinessnews.net/articles/2621-i-f-t-launches-food-traceability-center (accessed September 25, 2019).

Ovejero-Benito, M. C., Muñoz-Aceituno, E., Reolid, A., Saiz-Rodríguez, M., Abad-Santos, F. and Daudén, E. (2018). Pharmacogenetics and pharmacogenomics in moderate-to-severe psoriasis. *American Journal of Clinical Dermatology*, 19, 209–222.

Oyedeji, A. B., Mellem, J. J. and Ijabadeniyi, O. A. (2018). Potential for enhanced soy storage protein breakdown and allergen reduction in soy-based foods produced with optimized sprouted soybeans. *LWT.* https://doi.org/10.1016/j.lwt.2018.09.019.

Pandey, A., & Mann, M. (2000). Proteomics to study genes and genomes. Nature, 405(6788), 837–846.

Pierna, J. F., Baeten, V., Renier, A. M., Cogdill, R. P., & Dardenne, P. (2004). Combination of support vector machines (SVM) and near-infrared (NIR) imaging spectroscopy for the detection of meat and bone meal (MBM) in compound feeds. Journal of Chemometrics: *A Journal of the Chemometrics Society*, 18(7-8), 341–349.

Protopop, I., & Shanoyan, A. (2016). Big data and smallholder farmers: big data applications in the agri-food supply chain in developing countries. International Food and Agribusiness Management Review, 19(1030-2016-83148), 173–190.

Qian, J. P., Yang, X. T., Wu, X. M., Zhao, L., Fan, B. L. and Xing, B. (2012). A traceability system incorporating 2D barcode and RFID technology for wheat flour mills. *Computers and Electronics in Agriculture.* https://doi.org/10.1016/j.compag.2012.08.004.

Rortais, A., Belyaeva, J., Gemo, M., van der Goot, E. and Linge, J. P. (2010). MedISys: an early-warning system for the detection of (re-)emerging food- and feed-borne hazards. *Food Research International.* https://doi.org/10.1016/j.foodres.2010.04.009.

Roy, L. (2019). https://www.weforum.org/agenda/2019/09/5-ways-traceability-technology-can-lead-to-a-safer-more-sustainable-world/ (accessed September 18, 2019).

Sakamoto, T., Yokozawa, M., Toritani, H., Shibayama, M., Ishitsuka, N., & Ohno, H. (2005). A crop phenology detection method using time-series MODIS data. Remote sensing of environment, 96(3-4), 366–374.

Sawant, M., Urkude, R., & Jawale, S. (2016). Organized data and information for efficacious agriculture using PRIDE™ model. International Food and Agribusiness Management Review, 19(1030-2016-83147), 115–130.

Schnase, J. L., Duffy, D. Q., Tamkin, G. S., Nadeau, D., Thompson, J. H., Grieg, C. M., ... & Webster, W. P. (2017). MERRA analytic services: Meeting the big data challenges of climate science through cloud-enabled climate analytics-as-a-service. Computers, Environment and Urban Systems, 61, 198–211.

Schneider, M. V. and Orchard, S. (2011). Omics technologies, data and bioinformatics principles. In: *Bioinformatics for Omics Data* (pp. 3–30). Springer.

Schuster, E. W., Kumar, S., Sarma, S. E., Willers, J. L., & Milliken, G. A. (2011, November). Infrastructure for data-driven agriculture: identifying management zones for cotton using statistical modeling and machine learning techniques. In 2011 8th International Conference & Expo on Emerging Technologies for a Smarter World (pp. 1–6). IEEE.

Šenk, I., Ostojić, G., Tarjan, L., Stankovski, S. and Lazarević, M. (2013). *Food Product Traceability by Using Automated Identification Technologies BT – Technological Innovation for the Internet of Things* (L. M. Camarinha-Matos, S. Tomic and P. Graça, Eds.). Berlin, Heidelberg: Springer.

Skyes N. (2018). https://blog.kolabtree.com/5-uses-of-big-data-in-the-food-industry/ (accessed September 18, 2019).

Sundmaeker, H., Verdouw, C., Wolfert, S. and Pérez Freire, L. (2016). Internet of food and farm 2020. In: O. Vermesan and P. Friess (Eds.), *Digitising the Industry-Internet of Things Connecting Physical, Digital and Virtual Worlds* (pp. 129–151).

Syngenta Foundation for Sustainable Agriculture, 2016. FarmForce. [Online] Available at: http://www.farmforce.com/ [accessed 2017].

Tesfaye, K., Sonder, K., Caims, J., Magorokosho, C., Tarekegn, A., Kassie, G. T., ... & Erenstein, O. (2016). Targeting drought-tolerant maize varieties in southern Africa: a geospatial crop modeling approach using big data. International Agribusiness Management Review, 19(A), 1–18.

Thenkabail, P., GangadharaRao, P., Biggs, T., Krishna, M., & Turral, H. (2007). Spectral matching techniques to determine historical land-use/land-cover (LULC) and irrigated areas using time-series 0.1-degree AVHRR pathfinder datasets. Photogrammetric Engineering & Remote Sensing, 73(10), 1029–1040.

Tripathi, S., Srinivas, V. V., & Nanjundiah, R. S. (2006). Downscaling of precipitation for climate change scenarios: a support vector machine approach. Journal of Hydrology, 330(3-4), 621–640.

Urtubia, A., Pérez-Correa, J. R., Soto, A., & Pszczolkowski, P. (2007). Using data mining techniques to predict industrial wine problem fermentations. Food Control, 18(12), 1512–1517.

Waldhoff, G., Curdt, C., Hoffmeister, D., & Bareth, G. (2012). Analysis of multitemporal and multisensor remote sensing data for crop rotation mapping. ISPRS Annals of the Photogrammetry, Remote Sensing and Spatial Information Sciences, 1, 177–182.

Wang, Z., Gerstein, M. and Snyder, M. (2009). RNA-Seq: a revolutionary tool for transcriptomics. Nature Reviews Genetics. https://doi.org/10.1038/nrg2484.

Wardlow, B. D., Egbert, S. L., & Kastens, J. H. (2007). Analysis of time-series MODIS 250 m vegetation index data for crop classification in the US Central Great Plains. Remote Sensing of environment, 108(3), 290–310.

Ward, J. S. and Barker, A. (2013). Undefined by data: a survey of big data definitions. ArXiv Preprint ArXiv:1309.5821.

Wiedmann, M. (2015). Can big data revolutionize food safety? Food Quality & Safety.

Wolfert, S., Ge, L., Verdouw, C., and Bogaardt, M. J. (2017). Big data in smart farming – a review. Agricultural Systems. https://doi.org/10.1016/j.agsy.2017.01.023.

Zhao, G., Liu, S., Lopez, C., Lu, H., Elgueta, S., Chen, H. and Boshkoska, B. M. (2019). Blockchain technology in agri-food value chain management: a synthesis of applications, challenges and future research directions. Computers in Industry, https://doi.org/10.1016/j.compind.2019.04.002.

Samson A. Oyeyinka, Ocen M. Olanya, Beatrice I.O. Ade-Omowaye,
Brendan A. Niemira

17 Nonthermal processing techniques for innovative food processing

17.1 Introduction

Nonthermal food processing (NFP) is an innovative processing technology that exerts a minimal impact on the nutritional and sensory properties of foods and extends shelf life by inhibiting, inactivating, or killing microorganisms. The NFP technologies, including pulsed electric field (PEF), pulsed light (PL) treatment or light technology, ultrasound and high-pressure processing (HPP) are processing techniques that do not involve the use of thermal energy during food processing for the destruction or inactivation of microorganisms in food substrates, foods, or processing environments. However, more technically, it is often used to describe technologies that are effective at ambient or sub-lethal temperatures (Pereira and Vicente, 2010). Perhaps, the use of the term nonthermal stems from the fact that the heat generated during processing is not high enough to cause significant heating of the product, but the processing is sufficient to inactivate or destroy possible organisms of interest that could pose food safety threats.

The NFP technologies or methods have been suggested as possible alternatives to some thermal processing methods in foods for various reasons. For example, nonthermal processing preserves the nutritional value of foods to a greater extent compared to thermal methods. Currently NFP methods are receiving attention from consumers, producers and researchers possibly because they have the ability to inactivate microorganisms at near-ambient temperatures, avoiding thermal degradation of the food components, and consequently preserving the sensory and nutritional quality of minimally processed produce and fresh-like food products (Pereira and Vicente, 2010). The consideration of NFP in food processing is attributed to optimal microorganisms inactivation, enhanced ability to achieve faster process, enhanced lower temperatures, relatively lower energy consumption and requirement with subsequent lower overall costs, its potential to combat climate change due to their low

Samson A. Oyeyinka, School of Agriculture and Food Technology, University of the South Pacific, Alafua Campus, Fiji
Ocen M. Olanya, Brendan A. Niemira, USDA-ARS, Eastern Regional Research Center, Food Safety & Intervention Technologies Research Unit, Wyndmoor, PA, USA
Beatrice I.O. Ade-Omowaye, Department of Food Science, Ladoke Akintola University of Technology, Nigeria

https://doi.org/10.1515/9783110667462-017

carbon footprint, guaranteed high quality product and adequate sensory properties retention (Jambrak et al., 2018). Similarly, scalability of NFP for low, medium and large-scale industrial food processing facilities is advantageous as well. Other benefits derived from NFP are the potential for additional market opportunities since it is amendable to value-added food products. The advantages of these NFP suggest their potentials in future applications in the food industry. This chapter described the working principles of the NFP including PEF, HPP, ultrasound technology, irradiation, and bio-control. Furthermore, the chapter highlighted the importance of these processing techniques in food processing. Emphasis was based on plants and animal food products, with regards to improved processing and food-borne pathogen inactivation or its destruction.

17.2 Pulsed electric fields (PEF)

PEF is a nonthermal method of food preservation that uses short pulses of electricity for the inactivation of microbes, but causes minimal detrimental effect on food quality attributes (Mohamed and Eissa, 2012). It involves the application of pulses of high voltage, typically 20 and 80 kV/cm for relatively short period of time (<1 s) to fluid foods placed between two electrodes (Señorans et al., 2003). The concepts of this PEF were introduced to the food industry about 50 years ago; however, PEF can be still considered an emerging technology due to the recent developments in its industrial applications (Gómez et al., 2019). PEF technology involves the generation of high electric field intensities, the design of chambers that impact uniform treatment to foods with minimum increase in temperature and the design of electrodes that minimize the effect of electrolysis (Olajide et al., 2006). According to these authors, the large field intensities are achieved by storing a large amount of energy in a series of capacitors from a direct current power supply, which is then discharged in the form of high-voltage pulses to the food. The high-voltage pulses are transmitted within the food due to the presence of several ions, giving the food a certain degree of electrical conductivity (Zhang et al., 1995). Application of PEF for the inactivation of microbes and subsequent preservation of different food commodities is well-documented in the literature (Barba et al., 2017; Gómez et al., 2019; Qin et al., 1995; Simonis et al., 2019). Furthermore, PEF, an innovative technology, can be used in other various areas of food processing including extraction of valuable substances from plant materials, for improving peeling behavior and as a preprocessing step in potato-processing snack industry and in dehydration for the production of dried products with superior esthetic quality to those dried with conventional methods (Siemer et al., 2018; Ade-Omowaye, 2019). The thrust of PEF application in food processing has focused on inactivation of microorganisms (Jeyamkondan et al., 1999). PEF has reportedly resulted in the inactivation of microorganisms due to damage on the cell membrane (electroporation), which affects its functioning and may lead to cell death (Pereira and Vicente, 2010; Sale and Hamilton, 1968). However, it appears that PEF application now includes other aspects of food processing such as PEF-induced inactivation of enzymes and permeabilization of plant and animal membranes (Knorr and Angersbach, 1998).

17.2.1 Improvement in yield of juice and bioactive compounds

The emergence and application of PEF in the extraction of plant materials such as juice and other bioactive compounds present a suitable alternative to conventional thermal or enzyme processing. Several researchers have documented the successful application of PEF in increasing the yield of juice from fruits and vegetables as well as higher nutrient retention due to the low or sub-lethal temperature involved during processing. PEF has been generally used to increase juice yield and extraction of bioactive compounds from apple (Bazhal et al., 2001; Grimi et al., 2011), white grapes (Praporscic et al., 2007), alfalfa (Gachovska et al., 2006), sugar beet (Bouzrara and Vorobiev, 2000), orange (El Kantar et al., 2018; Rivas et al., 2006), paprika (Ade-Omowaye et al., 2001), carrot (Rivas et al., 2006), and blackcurrant (Gagneten et al., 2019). El Kantar et al. (2018), reported 25%, 37%, and 59% increase in the yield of juice from orange, pomelo, and lemon, respectively, after PEF treatment. Furthermore, the concentrations of polyphenols in the extracted juice were found to be significantly enhanced by the treatment of citrus peels with PEF at high electric field strength (El Kantar et al., 2018). Gagneten et al. (2019) optimized the extraction of bioactive compounds from blackcurrant using PEF and reported increase in total polyphenol contents (19%), antioxidant activity (45%), and total monomeric anthocyanins (6%) in the optimized juice at a field strength of 1.318 kV/cm and 315 pulses. PEF has also been reported to be used in the processing of foods with minimal loss of flavor, color, and bioactive compounds such as anthocyanins (Cserhalmi et al., 2006; Hodgins et al., 2002; Yeom et al., 2002). PEF reportedly increase the anthocyanin concentration in grape juice when used as a pre-treatment prior to extraction (Knorr, 2003). The increased juice yield and better extraction of bioactive compounds has been associated with better pore formation after PEF treatment which enhanced the separation of the juice from the cells (Ade-Omowaye et al., 2001).

The application of sufficiently high electric fields reportedly results in pore formation and breakage of cell membranes and consequently local structural changes in the cell membranes thus favoring mass transfer rates in cellular materials (Ho and Mittal, 1996).

17.2.2 Preservation

With regards to preservation, PEF has been used to inactivate microorganisms by a mechanism called electromechanical instability of the cell membrane (Castro et al., 1993; Jeyamkondan et al., 1999). PEF has been effectively used against various pathogenic and spoilage microorganisms and enzymes in food systems. For instance, PEF application was found to result in two logarithmic cycles of microbial count reduction (2D reduction) of *Escherichia coli* inoculated in skim milk and exposed to 60 pulses of 2 µs width at 45 kV/cm and 35 °C (Zhang et al., 1994). Several factors including process parameters, nature of product and properties of microbial cells influences the effectiveness of PEF in microbial inactivation (Syed et al., 2017; Wouters et al., 2001). These factors, which play a dynamic role to attain the optimum performance of PEF treatment, were reviewed by Syed et al. (2017). Process parameters that affect the ability of PEF to reduce microbial population in food include electric field strength, pulse length and shape, number of pulses, and temperature (Raso et al., 2000; Syed et al., 2017). The mechanism of microbial destruction during PEF treatments as explained by Tsong (1990) involves the permeability of cell membranes to small molecules leading to swelling and the eventual rupture of the cell membrane (Figure 17.1). Although PEF technology enables inactivation of bacterial and yeast vegetative cells in various foods, bacterial spores are resistant to PEF (Syed et al., 2017). Few successes have however been reported on the potential of PEF in combination with a preheating step to approximately 80 °C for bacterial spore inactivation. Inactivation of *A. acidoterrestris* spores in orange juice and carrot juice using PEF in combination with thermal energy has been documented (Siemer et al., 2018). Thus, future applications may require synergistic combinations of PEF with other processing methods for the destruction of spores in food and perhaps in producing juice with superior sensory quality. For example, treatment of apple juice with PEF in combination with ultraviolet light and high intensity light pulses resulted in juice with no adverse effect on color, odor, and flavor when compared with conventional methods of heating (Caminiti et al., 2011).

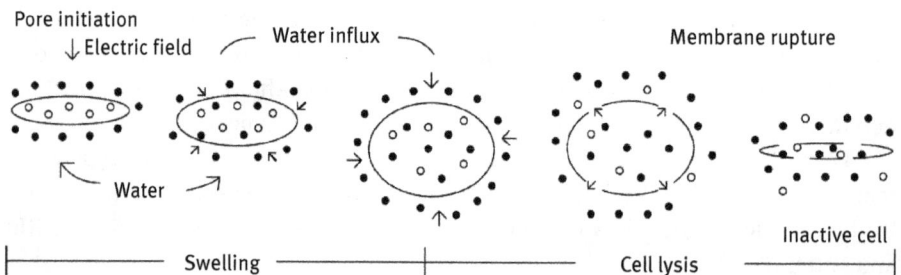

Figure 17.1: Mechanism of cell inactivation during PEF treatment.
Source: Tsong, 1990.

17.3 High pressure processing (HPP)

The application of high hydrostatic pressure (HHP) processing or HPP as a nonthermal food preservation technique for microbial and enzyme inactivation is on the increase, due to their reduced effects on nutritional and quality parameters when compared to conventional thermal treatments (Tiwari et al., 2009). HHP offers process advantages such as independence of size and geometry of the samples during processing, possibilities for low temperature treatment, and the availability of a waste-free, environmental-friendly technology (Fellows, 2000). Other advantages of HPP include no evidence of toxicity, preservation of nutrients, colors, and flavors, reduced processing times, uniformity of treatment throughout the food, potential for reduction or elimination of preservatives, and positive consumer appeal (Muntean et al., 2016). Although HPP has several benefits as highlighted above, it also has its own limitations which include amongst others, the use of expensive equipment, limited packaging options, foods should have 40% free water for antimicrobial effect, operation can only be done in batches and some level of microbial survival has been reported (Muntean et al., 2016). According to Huang et al. (2017), HPP is the most successfully commercialized nonthermal processing technology and eliminates food pathogens at room temperature. In addition, HPP extends the shelf life of foods circulated through the cold chain. HPP involves the application of high pressure (100–600 MPa) at room temperature to a food that is hermetically sealed in a flexible container (Balasubramaniam et al., 2015). The high pressure applied subjects the interior and surface of the food to pressure sufficient to achieve pasteurization (Balasubramaniam et al., 2015). This leads to modifications of cellular membranes and interrupts cellular functions which are responsible for reproduction (Torres and Velazquez, 2008). HPP ensures the microbial safety of food without the use of preservatives and allows the processed food to maintain the natural flavors and nutritional value of the original food material (Huang et al., 2017). Hence, HPP technology is recognized as a minimal processing technology that ensures food safety and guarantees better flavor retention (Huang et al., 2017). The efficiency of microbial inactivation by HPP is influenced by several factors including the type and number of microorganisms, the magnitude of pressure, treatment time, time to achieve pressure (come-up time), time of decompression, process temperature, pH, water activity, package integrity, product temperature, vessel temperature, and composition of the foods (Barbosa-Cánovas et al., 1998; Hoover, 1993; Morris et al., 2007). In comparison with conventional thermal processing methods, HPP reduces energy consumption as processing is done at room temperature and prevents postprocessing contamination since foods are processed in packaged form and does not have direct contact with processors or the processing device (Huang et al., 2017). Different designs of HPP system exist the world over; however, the system has the same working principle. Huang et al. (2017) in a review reported that there are more than 300 sets of HPP system that have been operating for mass production worldwide, with larger operations recorded in North America (54%), followed by Europe (25%) and Asia (12%). A simple horizontal-type HPP system from Multivac Inc., Germany, is presented in Figure 17.2 (Huang et al., 2017).

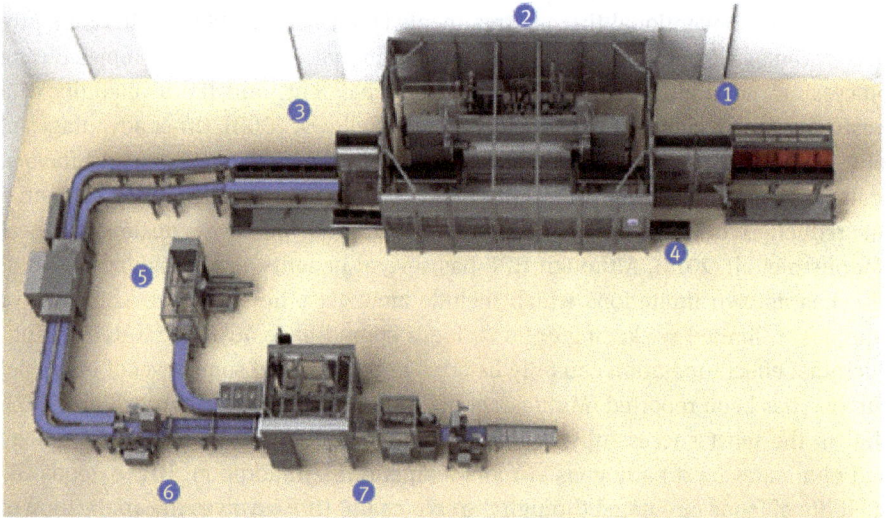

(1) Loading of baskets   (2) HPP   (3) Unloading of baskets

(4) Return conveyor for baskets   (5) Drying of packs   (6) Labelling of packs

(7) End-of-line processing of packs

Figure 17.2: A horizontal HPP system.
Source: Huang et al., 2017.

17.3.1 Preservation

As a nonthermal processing technology, HPP has been widely used to extend the shelf life of foods through microbial inactivation. The level of inactivation has been found to vary with the type of food as well as the microbes of interest. In general, yeasts and molds are more susceptible to HPP than bacteria, while bacterial spores are reportedly extremely resistant to HPP (Nakayama et al., 1996). The application of HPP above 200 MPa has been reported to inactivate vegetative bacteria, yeast, and molds (Daher et al., 2017). According to Wilson et al. (2008), the resistances of spores are different even amongst the same species. Hence, spore inactivation is achieved through a combination of HPP with subsequent heat treatment. For instance, HPP alone at 400 MPa and 20 °C for 15 min resulted in 1-log decimal reduction of *E. coli* 0157:H7 in poultry meat, but a higher reduction (6-log cycle) was achieved when the HPP condition was combined with a temperature of 50 °C (Patterson and Kilpatrick,

1998). There are evidences that differences in the cell wall composition may also influence the susceptibility of the bacterium to HPP. Fonberg-Broczek et al. (2005) reported that Gram-positive bacteria are more resistant than Gram-negative bacteria, while cocci-shaped bacteria are more resistant than the rod. Most yeast and mold spores are destroyed by a pressure of about 400 MPa. Heinz and Knorr (2002) reported that spores between strains of the same species respond differently to HPP. For example, spores of *Clostridium sporogenes* in fresh chicken breast were reported to be inactivated (5-log) using a pressure of 680 MPa for 1 h (Crawford et al., 1996), while a higher pressure (1,500 MPa) and lower log reduction (1.5-log) were reported for *Clostridium sporogenes* in liquid media (Maggi et al., 1996). The application of HPP to different meat and meat products (Table 17.1) showed varying levels of microbial log reduction, which could be associated with the different processing conditions as well as the nature and resistance of the microorganism involved. High pressure application is of special interest in food processing technology since pressure represents an alternative to thermal processing, especially for foods such as meat and poultry whose nutritional and sensorial characteristics are thermosensitive. Hence, the application of HPP in food preservation will require an understanding of the nature of the food and the microbes that could possibly be associated with it. Furthermore, it is very important that process parameters such as pressure, temperature, and time be optimized in order to ensure maximum destruction of microbes with minimum effect of the nutrients.

Table 17.1: Effect of HPP on microorganisms in meat and meat products.

Microorganism	Product	Pressure (MPa)	Time (min)	Temperature (°C)	Microbial log reduction	References
Lactobacillus sakei	Cooked ham	500	10	40	4	Hugas et al., (2002)
Staphylococcus carnosus	Cooked ham	500	10	40	1	Hugas et al., (2002)
Yersinia enterocolitica	Pork	300	10	25	6	Shigehisa et al. (1991)
Campylobacter jejuni	Pork	300	10	25	6	Shigehisa et al. (1991)
Campylobacter jejuni	Poultry	375	10	25	6	Solomon and Hoover (2004)
Escherichia coli O157:H7	Poultry	600	15	20	3	Patterson et al. (1995)
Staphylococcus aureus	Poultry	600	15	20	3	Patterson et al. (1995)
Listeria monocytogenes	Poultry	375	15	20	2	Patterson et al. (1995)
Listeria monocytogenes	Poultry	500	1	40	3.8	Chen (2007)

17.3.2 Mycotoxin control

Mycotoxins are toxic compounds that are naturally produced by certain types of fungi that present a potential hazard regarding food safety. They are usually associated with numerous foodstuffs such as cereals, dried fruits, nuts, and spices usually under warm and humid conditions. These toxic compounds are secondary metabolites of different fungus that have adverse health effects ranging from acute poisoning to long-term effects such as immune deficiency and cancer. Several hundreds of different mycotoxins have been identified, but the most commonly observed mycotoxins that present a concern to human health and livestock include aflatoxins, ochratoxin A, patulin, fumonisins, zearalenone, and nivalenol/deoxynivalenol. HPP has been reportedly used in the destruction of mycotoxins such as patulin in fruits (Woldemariam and Emire, 2019). Patulin (100 ppb) in apple juice was reported to have been reduced by up to 56.24% when treated with HPP (Avsaroglu et al., 2015). The operating conditions varied from pressures of about 30 to 500 MPa with temperature ranging between 30 and 50 °C. Higher operating pressure seems to require shorter processing time and levels of reduction appear to depend on the pressure, with higher pressure resulting in higher log reduction or destruction in foods. Avsaroglu et al. (2015) found that patulin concentrations in apple juice showed varying levels of reductions (0—51.16%) when processed with high pressure (300—500 MPa) and varying temperatures (20—50 °C). These authors reported no linear decrease in patulin concentrations for different pressure/temperature applications, and the highest reduction was obtained at 400 MPa/30 °C. Other authors found varying levels of patulin reductions of 42%, 53%, and 62% in apple juice treated by high pressure application (300, 500, and 800 MPa) at room temperature for 1 h. (Bruna et al., 1997). More research efforts may be required to further establish the control of HPP on other products, especially the dried foods with low water activity that are more vulnerable to attack by toxin-producing molds.

17.3.3 Effect of HPP on meat quality

The application of HPP to meat processing is mainly effective in postpackaging decontamination technology for ready-to-eat (RTE) meat products particularly in cases where heat treatment is not possible or convenient. Varying levels of pressure (400—600 MPa) have been applied for the pasteurization of meats and meat products (Bajovic et al., 2012). According to Bajovic et al. (2012), depending on the pressure level applied, the fresh meat quality can be remarkably modified. The meat becomes more gel-structured and paler losing the typical appearance of fresh meat. Protein structure is also modified and the sensory properties of the meat including color and texture may be significantly modified. For example, the application of HPP on minced beef in the pressure range of 200 and 350 MPa increased the lightness value (L*), which resulted in a pinkish discoloration (Carlez et al., 1995). At higher pressures (400–500 MPa), the meat turned gray brown. The variation in color of HPP meat seems to have limited the commercialization of HPP fresh meats since the high pressure results in dramatic changes in the color (Cheftel and Culioli, 1997). This seems plausible since the processed meat lack the typical color of fresh meat from the consumer's perspective. Although these changes may not be a serious concern when such products will be further processed into other meat products such as hamburger patties for the food service, it is important to control the level of changes through optimization studies, so that while there is microbial inactivation and reduction, color as well as other sensory properties are preserved or at least do not change significantly. With regards to protein structure in meat, HPP has been found to result in varying levels of modification as mentioned earlier. The four levels of protein structure are primary, secondary, tertiary, and quaternary. The different levels respond differently to varying pressures resulting in unfolding of the protein structure and in some instances denaturation depending on the pressure applied. Quaternary structure is the most susceptible to HPP since it is mainly held by hydrophobic interactions, but tertiary structure is altered at pressures above 200 MPa, while secondary structure will require higher pressures for significant changes (Rastogi et al., 2007). HPP can also modify the texture of meat, which may enhance tenderization. Meat tenderization can be achieved in several ways including the use of proteolytic enzymes. However, the degree of meat tenderness using HPP will depend on the rigor stage, pressure, and temperature level applied (Bajovic et al., 2012; Sun and Holley, 2010). Other impact of HPP on meat quality is with respect to inducing lipid oxidation, which may have detrimental effects in terms of keeping quality of the meat as well as may pose health threats since free radicals generated during oxidation are known to be problematic. At varying pressures (300–600 MPa), HPP has been reported to trigger lipid oxidation in beef (Beltran et al., 2004; Ma et al., 2007; Mcardle et al., 2010), poultry (Beltran et al., 2004; Bragagnolo et al., 2007; Kruk et al., 2011), and pork (Cheah and Ledward, 1995–1997). Future studies may be required to possibly apply antioxidants before or after HPP to reduce the posttreatment effect of HPP on meat quality, especially on lipid oxidation.

17.3.4 Effect of HPP on milk and dairy products

The application of HPP to the processing of milk is suggested as a cold alternative process to thermal pasteurization and has been applied in dairy food with mainly the advantage of shelf-life extension (2–3 times) relativity to nonpasteurized products (Dhineshkumar et al., 2016).

In dairy food, HPP has been applied in milk, fresh cheese, ripened cheese, whey, cheese, yogurt, ice cream, and butter. According to previous research, HPP can influence the physicochemical and technological characteristics of milk by modifying the structure of milk components (Cadesky et al., 2017; Patterson, 2005). For instance, HPP does not seem to affect lactose in milk suggesting that no Maillard or lactose isomerization reaction takes place in milk as a result of pressure treatment (Lopez-Fandino et al., 1996). Stratakos et al. (2019) investigated the application of HPP as an alternative to standard raw milk processing. Pressure was applied at different levels (400–600 MPa) and exposure times (1–5 min) against artificially inoculated pathogenic *E. coli*, *Salmonella*, and *L. monocytogenes*. The study showed that HPP effectively inactivated all the bacteria by 5 log CFU/mL. Furthermore, in comparison with pasteurized milk, HPP-treated milk at 600 MPa for 3 min reportedly prolonged shelf life of the milk by 1 week (Stratakos et al., 2019).

One important product in the dairy industry is cheese, which is produced by coagulation of the milk protein casein. Cheese comes in a wide range of flavors, textures and is primarily made from cow milk. Cheeses are generally classified according to length of fermentation, texture, methods of production, fat content, animal milk, and country or region of origin where it is produced. According to previous research, the effect of HPP on maturated cheeses depends on the age of the cheese when the treatment is applied (Arqués et al., 2007; Ávila et al., 2016; Delgado et al., 2013; Delgado et al., 2011). Delgado et al. (2013) found that HPP treatments induced important changes in raw goat milk cheeses, when applied before maturation compared to when the HPP treatment is applied at the end of maturation. The application of up to 600 MPa was found to prevent excessive bitterness and over-ripening in matured Torta del Casar cheese (Delgado-Martínez et al., 2019).

Yoghurt has also been processed using HPP technology. According to Trujillo et al. (2002), HHP has been applied in two ways to yoghurt. The first method is to produce yoghurt from HPP-treated milk and the other is to apply pressure to yoghurt in order to inactivate microorganisms.

Johnston et al. (1994) studied the properties of acid-set gels prepared from HPP-treated milk and found improvement in texture and syneresis resistance of the gels. Furthermore, authors reported viscosity improvement of stirred-style yoghurt-type product prepared from HPP-treated skim milk with pressure varying between 100 and 600 MPa for up to 1 h (Johnston et al., 1994). Other studies found that HPP-treated packaged yoghurt showed no significant changes in texture and did not reduce the number of lactic acid bacteria (Tanaka and Hatanaka, 1992). The operating conditions were 200 to 300 MPa at 10–20 °C for 10 min. However, higher pressures (>300 MPa) reportedly prevented over-acidification, but reduced the number of viable lactic acid bacteria (Tanaka and Hatanaka, 1992). These findings suggest that pressure-time combination may influence the number of microbes associated with the processed food, which may influence the quality of the food. Due to the variation in the properties of food packaging materials, future studies may be required to establish the impact of HPP on packaged dairy products.

17.4 Ionizing radiations

For food processing, radiation refers to gamma rays, X-rays, and electron beams. The processes associated with various types of radiations are capable of ionizing atoms and molecules within food (Fan, 2012). Irradiation has been successfully used as a physical measure for decontamination and inactivation of insect pests and some pathogenic microbes. Gamma rays are high-energy photons produced by radioisotopes such as cobalt-60 or cesium-137. Electron beams are often produced by a particle accelerator. The mechanism of activity of gamma rays or other forms of ionizing irradiation on pathogenic microbes is their disruptive, disintegrating effects on base pairs or phosphate sugar bonds of DNA molecules. The effectiveness of ionizing radiations on pathogenic microbes derives from the production of free radicals, ions, and other reactive species, which damage cell membranes and microbial DNA. Irradiation of postharvest produce inactivates food-borne pathogens with minimum damage to the quality of produce during the limited duration of exposure (Niemira et al., 2005).

Gamma irradiation has been used consistently to inactivate enteric pathogens such as *E. coli* O157:H7 and *Salmonella* spp. on leafy greens, fruits, and other foods (Niemira et al., 2002; 2005). The effectiveness of gamma rays depends on the source of radiation, the uniformity and duration of treatment, the type of produce being irradiated, as well as the total water content of the produce. Other factors that impact the sensitivity to radiation (D_{10} values) for enteric pathogens such as *E. coli* O157:H7 cocktails or *Salmonella* on fresh produce include the re-suspension medium and sample preparation conditions (Moreira et al., 2012). The design of irradiation units varies amongst commercial operations that use them for quarantine.

The effectiveness of irradiation for inactivation of food-borne pathogens has been documented in various publications (Moreira et al., 2012; and Morris et al., 2005). In research on the use of gamma radiation for inactivation of pathogenic microbes on cabbage, carrots, and celery, reductions of 4–5 logs, 4 logs, and 3.6 logs, respectively, were documented for *E. coli* O157:H7 (Table 17.2). Similarly, a 6.7 log reduction of *E. coli* O157:H7 was recorded when cilantro leaves were irradiated with gamma rays at 1.05 kGy (Niemira et al., 2005). Differences in the reductions of *E. coli* O157:H7 and *Salmonella* were obtained when the pathogens were subjected to gamma irradiation following inoculation on cabbage and cucumbers (Khattak et al., 2005).

At a dose of 0.8 kGy, *E. coli* O157:H7 populations were reduced by 4–5 logs for cabbage and cucumbers. On the contrary, the reduction of *Salmonella* on cabbage and cucumber by gamma rays ranged from 3 to 4 logs. This demonstrates the difference in the sensitivity of the two pathogens to gamma irradiation when inoculated on produce (Khattak et al., 2005). In another study, a 2 kGy irradiation was shown to result in a 5 log (CFU/g) reductions of *Salmonella typhimurium* in pineapple (Zhang et al., 1995). For *L. monocytogenes*, reductions in the range of 3.1–5.2 logs were documented on celery, carrots, and cabbage, respectively, by various researchers (Table 17.2). However, in a

Table 17.2: Examples of nonthermal intervention (gamma rays) technologies utilized for reduction of food-borne pathogens on leafy greens and fruits at postharvest.

Food-borne pathogens[a]	Intervention technology[b]	Produce	Pathogen reduction[c]	References
E. coli O157:H7	Gamma rays	Cabbage	4–5 logs	Khattak et al., 2005
E. coli O157:H7	Gamma rays	Cucumber	4–5 logs	Khattak et al., 2005
E. coli O157:H7	Gamma rays	Carrots	>4 logs	Lacroix and Lafortune, 2004
E. coli O157:H7	Gamma rays	Celery	3.6 logs	Lopez et al., 2005
E. coli O157:H7	Gamma rays	Cilantro	6.7 logs	Foley et al.,, 2004
E. coli O157:H7	Gamma rays	Carrots	2–4 logs	Kamat et al., 2005
E. coli O157:H7	Gamma rays	Lettuce (Boston, iceberg, red leaf, green leaf)	3.6–3.8 logs	Niemira et al., 2002
Salmonella paratyphi	Gamma rays	Cabbage	3–4 logs	Khattak et al., 2005
Salmonella paratyphi	Gamma rays	Cucumber	3–4 logs	Khattak et al., 2005
Salmonella typhimurium	Gamma rays	Pineapple	5 logs	Shashidhar et al., 2007
L. monocytogenes	Gamma rays	Cabbage	5.2 logs	Bari et al., 2005
L. monocytogenes	Gamma rays	Carrots	2.6 logs	Kamat et al., 2005
L. monocytogenes	Gamma rays	Celery	3.1 logs	Prakash et al., 2000
L. monocytogenes	Gamma rays	Capsicum	2–3 logs	Ramamurthy et al., 2004

[a]Pathogenic microbes exposed to postharvest control measures.
[b]Postharvest intervention technology utilized for treatment of produce.
[c]Reduction of food-borne pathogen based on the corresponding application of postharvest control measure.

related study, complete elimination of *Listeria* was documented when a 2 kGy irradiation was applied to pepper (Ramamurthy et al., 2004).

It should be noted that variations in irradiation sensitivity might occur for different bacterial strains or when the bacteria are inoculated on various produce types or cultivars. In research, differences in radiation sensitivities were attributed to the antioxidant or ascorbic acid content of the lettuce or spinach cultivars (Niemira et al., 2002). The authors suggested that the antioxidant content may protect the produce from ionizing radiation and contribute to variation in the radiation sensitivity of various cultivars.

The advantages of using irradiation on produce and other foods include its ability to eliminate food-borne pathogens or bacteria that are systemic or internalized within produce. Chemical washes such as chlorine dioxide, sodium hypochlorite, or other nonoxidizing agents do not readily inactivate internalized microbes (Ge et al., 2013; Gomes et al., 2009). In addition to its role in food safety, irradiation has also been reported to extend the shelf life of produce and reduce quality losses in food or produce, especially during short- or long-distance transport (Prakash et al., 2000). A drawback to the use of gamma irradiation is that sublethal injury to microbes may result if irradiation is not administered at the correct dosage and/or for adequate duration. Excessively high doses may however have some effects on the sensory qualities of food, the nutritional and chemical composition of some RTE foods. In general, proper calibration of dosage is essential for effective use of irradiation.

17.5 Ultrasound technology effects on pathogen inactivation in food processing

Ultrasound is a type of acoustic energy that is produced by longitudinal wave in which the vibration frequency normally exceeds 20,000 cycles per second (20 kHz). This frequency is beyond that which can be heard by human beings (Alarcon-Rojo et al., 2015). The ultrasound waves generally ranges from 20 kHz to 10 MHz and is categorized as follows: (i) high-power (>5 W cm^{-2} or 10–1,000 W cm^{-2}) and low-frequency (20–100 kHz); (ii) medium-power and frequency (100 kH to 1 MHz); (iii) low-power (<1 W cm^{-2}) and high-frequency (1–10 MHz) as previously described (Ashokkumar et al., 2008). Since sound waves may be transmitted in a one-dimensional manner, the speed of ultrasound or sound waves may depend on the acoustic properties of the sound transmitting medium. Therefore, the sound transmission speed would be expected to be greater in solid substrates than in aqueous or gaseous medium. Ultrasound energy is noninvasive, produces no residual effects, and pathogen inactivation (Alarcon-Rojo et al., 2018).

The properties of energy efficiency, simplicity, and its utilization as a potentially emerging, green technology have led to potential wide applications in the food industry (Alarcon-Rojo et al., 2018). Ultrasound technology has been applied for improvement of food processing and food quality as well as pathogen inactivation as a consequent of cavitation effects created by ultrasound treatment (Bilek and Turantaş, 2013). The inactivation of microbial populations by ultrasound treatments may result from the changes in pressure generated by ultrasonic waves. The acoustic cavitation generated by ultrasound particularly in aqueous medium may be attributed to physical forces and its consequent effects on bacterial inactivation. It has been documented that the impact of different ultrasound wave amplitudes on microbial populations may vary with the treatment temperatures, microbial types, food composition, and quantity as well as the contact time with microbial populations. Similarly, enzymes inactivation has also been an important utility of ultrasound technique, thereby reducing the detrimental effects of food-spoilage enzymes in food matrices. Ultrasound (low-power and high-frequency method) has also been applied to assess the composition and physicochemical properties of food and its components during food processing, resulting in enhancement of food quality. The application of ultrasound technology for process optimization may include utility in meat tenderization, crystallization, freezing, drying to improve other processes, and sensory characteristics as well as texture in meat products.

There are numerous nonthermal intervention measures such as ultraviolet (UV) light for pathogen inactivation which have also been studied. UV is part of the electromagnetic spectrum with wavelengths between 100 and 400 nm. The UV-C irradiation is the electromagnetic spectrum for inactivation of food-borne pathogens and ranges from 200 to 280 nm. The effects of UV light on bacteria are based on the formation of thymine (Giese and Darby, 2000), which results in the breakage of DNA strands of pathogenic microbes. The effects of UV-C irradiation on Gram-negative and Gram-positive food-borne bacteria, as well as on bacterial spores, have also been documented (Giese and Darby, 2000; Schenk et al., 2008).

17.6 Concluding remarks

The periodic occurrences of food-borne pathogens and outbreaks present challenges for pathogen control. Several nonthermal and innovative food processing techniques have been outlined that could be utilized for lethality or inactivation of food-borne pathogens, while simultaneously preserving the nutritional and sensory attributes of food. Amongst those covered in this chapter are PEF utilized for microbial inactivation and its application for extraction of plant-derived materials such as juice and bioactive compounds. PEF has also been applied against enzymes in the food systems and for reduction of food deterioration. Additionally, HHP processing which has also been extensively applied particularly on industrial-scale levels to provide lethal effects on pathogen and preserve foods are mentioned. HPP utility for mycotoxin reduction or control and their effects on meat quality and dairy products (milk and cheese) has been outlined and elaborated. Similarly, ionizing radiation (gamma rays)/electric beam effects applied for decontamination and inactivation of microbial agents and insect pests have been mentioned. In a related manner, ultrasound technology effects on pathogen inactivation in food processing environments have also been presented. Although mechanisms of pathogen or microbial inactivation may differ amongst the various nonthermal technologies outlined above, the ultimate goal has been microbial inactivation and enhancement of food safety. The variations in pathogen or bacterial reductions as a consequent of applications of nonthermal technologies may be expected as variations in application times, dosage, food matrices, and other conditions may occur. Overall, microbial contamination may be effectively controlled as innovative and nonthermal processing techniques continue to evolve.

References

Ade-Omowaye, B., Angersbach, A., Taiwo, K. and Knorr, D. (2001). The use of pulsed electric fields in producing juice from paprika (*Capsicum annuum* L.). *Journal of Food Processing and Preservation*, 25, 353–365.

Ade-Omowaye, B. I. O. (2019). *Healthy Food for All: Exploring New Horizons*. Ogbomoso: Ladoke Akintola University of Technology Press.

Alarcon-Rojo, A., Janacua, H., Rodriguez, J., Paniwnyk, L. and Mason, T. J. (2015). Power ultrasound in meat processing. *Meat Science*, 107, 86–93.

Alarcon-Rojo, A. D., Peña-González, E., García-Galicia, I., Carrillo-López, L., Huerta-Jiménez, M., Reyes-Villagrana, R. and Janacua-Vidales, H. (2018). Ultrasound application to improve meat quality. In *Descriptive Food Science*. IntechOpen.

Arqués, J., Garde, S., Fernández-García, E., Gaya, P. and Nuñez, M. (2007). Volatile compounds, odor, and aroma of La Serena cheese high-pressure treated at two different stages of ripening. *Journal of Dairy Science*, 90, 3627–3639.

Ashokkumar, M., Sunartio, D., Kentish, S., Mawson, R., Simons, L., Vilkhu, K. and Versteeg, C. K. (2008). Modification of food ingredients by ultrasound to improve functionality: a preliminary study on a model system. *Innovative Food Science & Emerging Technologies*, 9, 155–160.

Ávila, M., Gómez-Torres, N., Delgado, D., Gaya, P. and Garde, S. (2016). Application of high pressure processing for controlling Clostridium tyrobutyricum and late blowing defect on semi-hard cheese. *Food Microbiology*, 60, 165–173.

Avsaroglu, M., Bozoglu, F., Alpas, H., Largeteau, A. and Demazeau, G. (2015). Use of pulsed-high hydrostatic pressure treatment to decrease patulin in apple juice. *High Pressure Research*, 35, 214–222.

Bajovic, B., Bolumar, T. and Heinz, V. (2012). Quality considerations with high pressure processing of fresh and value added meat products. *Meat Science*, 92, 280–289.

Balasubramaniam, V., Martinez-Monteagudo, S. I. and Gupta, R. (2015). Principles and application of high pressure–based technologies in the food industry. *Annual Review of Food Science and Technology*, 6, 435–462.

Barba, F. J., Koubaa, M., do Prado-Silva, L., Orlien, V. and de Souza Sant'Ana, A. (2017). Mild processing applied to the inactivation of the main foodborne bacterial pathogens: a review. *Trends in Food* Science & *Technology*, 66, 20–35.

Barbosa-Cánovas, G., Pothakamury, U., Palou, E. and Swanson, B. (1998). *Non-thermal Preservation of Foods*. New York, NY: Macel Dekker. Inc.

Bari, M.I., Nakauma, M., Todoriki, S., Juneja, V.K., Kawamoto, S. (2005). Effectiveness of irradiation treatments in activating *Listeria monocytogenes* on fresh vegetables at refrigeration temperature. Journal of Food Protection. 68, 318–323.

Bazhal, M., Lebovka, N. and Vorobiev, E. (2001). Pulsed electric field treatment of apple tissue during compression for juice extraction. *Journal of Food Engineering*, 50, 129–139.

Beltran, E., Pla, R., Yuste, J. and Mor-Mur, M. (2004). Use of antioxidants to minimize rancidity in pressurized and cooked chicken slurries. *Meat Science*, 66, 719–725.

Bilek, S. E. and Turantaş, F. (2013). Decontamination efficiency of high power ultrasound in the fruit and vegetable industry, a review. *International Journal of Food Microbiology*, 166, 155–162.

Bouzrara, H. and Vorobiev, E. (2000). Beet juice extraction by pressing and pulsed electric fields. *International Sugar Journal*, 102, 194–200.

Bragagnolo, N., Danielsen, B. and Skibsted, L. H. (2007). Rosemary as antioxidant in pressure processed chicken during subsequent cooking as evaluated by electron spin resonance spectroscopy. *Innovative Food Science & Emerging Technologies*, 8, 24–29.

Bruna, D., Voldrich, M., Marek, M. and Kamarád, J. (1997). Effect of high pressure treatment on patulin content in apple concentrate. *High Pressure Research in the Biosciences*, 335–338.

Cadesky, L., Walkling-Ribeiro, M., Kriner, K. T., Karwe, M. V. and Moraru, C. I. (2017). Structural changes induced by high-pressure processing in micellar casein and milk protein concentrates. *Journal of Dairy Science*, 100, 7055–7070.

Caminiti, I. M., Noci, F., Muñoz, A., Whyte, P., Morgan, D. J., Cronin, D. A. and Lyng, J. G. (2011). Impact of selected combinations of non-thermal processing technologies on the quality of an apple and cranberry juice blend. *Food Chemistry*, 124, 1387–1392.

Carlez, A., Veciana-Nogues, T. and Cheftel, J.-C. (1995). Changes in colour and myoglobin of minced beef meat due to high pressure processing. *LWT-Food Science and Technology*, 28, 528–538.

Castro, A. J., Barbosa-Canovas, G. V. and Swanson, B. G. (1993). Microbial inactivation of foods by pulsed electric fields. *Journal of Food Processing and Preservation*, 17, 47–73.

Cheah, P. and Ledward, D. (1995). High-pressure effects on lipid oxidation. *Journal of the American Oil Chemists' Society*, 72, 1059.

Cheah, P. and Ledward, D. (1996). High pressure effects on lipid oxidation in minced pork. *Meat Science*, 43, 123–134.

Cheah, P. and Ledward, D. (1997). Inhibition of metmyoglobin formation in fresh beef by pressure treatment. *Meat Science*, 45, 411–418.

Cheftel, J. C. and Culioli, J. (1997). Effects of high pressure on meat: a review. *Meat Science*, 46, 211–236.

Chen, H. (2007). Temperature-assisted pressure inactivation of Listeria monocytogenes in Turkey breast meat. *International Journal of Food Microbiology*, 117, 55–60.

Crawford, Y. J., Murano, E. A., Olson, D. G. and Shenoy, K. (1996). Use of high hydrostatic pressure and irradiation to eliminate Clostridium sporogenes spores in chicken breast. *Journal of Food Protection*, 59, 711–715.

Cserhalmi, Z., Sass-Kiss, A., Tóth-Markus, M. and Lechner, N. (2006). Study of pulsed electric field treated citrus juices. *Innovative Food Science & Emerging Technologies*, 7, 49–54.

Daher, D., Le Gourrierec, S. and Pérez-Lamela, C. (2017). Effect of high pressure processing on the microbial inactivation in fruit preparations and other vegetable based beverages. *Agriculture*, 7, 72.

Delgado, F. J., Delgado, J., González-Crespo, J., Cava, R. and Ramírez, R. (2013). High-pressure processing of a raw milk cheese improved its food safety maintaining the sensory quality. *Food Science and Technology International*, 19, 493–501.

Delgado, F. J., González-Crespo, J., Cava, R. and Ramírez, R. (2011). Effect of high-pressure treatment on the volatile profile of a mature raw goat milk cheese with paprika on rind. *Innovative Food Science & Emerging Technologies*, 12, 98–103.

Delgado-Martínez, F. J., Carrapiso, A. I., Contador, R., and Ramírez, M. R. (2019). Volatile compounds and sensory changes after high pressure processing of mature "Torta del Casar"(raw ewe's milk cheese) during refrigerated storage. *Innovative Food Science & Emerging Technologies*, 52, 34–41.

Dhineshkumar, V., Ramasamy, D. and Siddharth, M. (2016). High pressure processing technology in dairy processing: a review. *Asian Journal of Dairy & Food Research*, 35.

El Kantar, S., Boussetta, N., Lebovka, N., Foucart, F., Rajha, H. N., Maroun, R. G., Louka, N. and Vorobiev, E. (2018). Pulsed electric field treatment of citrus fruits: improvement of juice and polyphenols extraction. *Innovative Food Science & Emerging Technologies*, 46, 153–161.

Fan, X. (2012). Ionizing radiation. In: V. Gomez-Lopez (Ed.), *Decontamination of Fresh and Minimally Processed Produce*. UK: John Wiley and Sons, Inc. Wiley-Blackwell.

Fellows. (2000). *Food Processing Technology; Principles and Practice (Second Edition)*. England: Woodhead Publishing Limited.

Foley, D., Euper, M., Carporaso, F. and Prakash A. (2004). Irradiation and chlorination effectively reduces *Escherichia coli* O157:H7 inoculated on cilantro (*Coriandrum sativum*) without negatively affecting quality. Radiation Physics & Chemistry, 63, 391-396.

Fonberg-Broczek, M., Windyga, B., Szczawinski, J., Szczawinska, M., Pietrzak, D. and Prestamo, G. (2005). High pressure processing for food safety. *Acta Biochimica Polonica-English Edition*, 52, 721–724.

Gachovska, T., Ngadi, M. and Raghavan, G. (2006). Pulsed electric field assisted juice extraction from alfalfa. *Canadian Biosystems Engineering*, 48, 3.

Gagneten, M., Leiva, G., Salvatori, D., Schebor, C. and Olaiz, N. (2019). Optimization of pulsed electric field treatment for the extraction of bioactive compounds from blackcurrant. *Food and Bioprocess Technology*, 12, 1102–1109.

Ge, C., Bohrerova, Z. and Lee, J. (2013). Inactivation of internalized Salmonella Typhimurium in lettuce and green onion using ultraviolet C irradiation and chemical sanitizers. *Journal of Applied Microbiology*, 114, 1415–1424.

Giese, N. and Darby, J. (2000). Sensitivity of microorganisms to different wavelengths of UV light: implications on modeling of medium pressure UV systems. *Water Research*, 34, 4007–4013.

Gomes, C., Da Silva, P., Moreira, R. G., Castell-Perez, E., Ellis, E. A. and Pendleton, M. (2009). Understanding E. coli internalization in lettuce leaves for optimization of irradiation treatment. *International Journal of Food Microbiology*, 135, 238–247.

Gómez, B., Munekata, P. E., Gavahian, M., Barba, F. J., Martí-Quijal, F. J., Bolumar, T., Campagnol, P. C. B., Tomasevic, I. and Lorenzo, J. M. (2019). Application of pulsed electric fields in meat and fish processing industries: an overview. *Food Research International*, 123, 95–105.

Grimi, N., Mamouni, F., Lebovka, N., Vorobiev, E. and Vaxelaire, J. (2011). Impact of apple processing modes on extracted juice quality: pressing assisted by pulsed electric fields. *Journal of Food Engineering*, 103, 52–61.

Heinz, V. and Knorr, D. (2002). *Effects of High Pressure on Spores. In Ultra High Pressure Treatments of Foods*. Boston, MA: Springer.

Ho, S. and Mittal, G. S. (1996). Electroporation of cell membranes: a review. *Critical Reviews in Biotechnology*, 16, 349–362.

Hodgins, A., Mittal, G. and Griffiths, M. (2002). Pasteurization of fresh orange juice using low-energy pulsed electrical field. *Journal of Food Science*, 67, 2294–2299.

Hoover, D. (1993). Pressure effects on biological systems. *Food Technology (USA)*, 47, 150–155.

Huang, H.-W., Wu, S.-J., Lu, J.-K., Shyu, Y.-T. and Wang, C.-Y. (2017). Current status and future trends of high-pressure processing in food industry. *Food Control*, 72, 1–8.

Hugas, M., Garriga, M. and Monfort, J. (2002). New mild technologies in meat processing: high pressure as a model technology. *Meat Science*, 62, 359–371.

Jambrak, A.R., Vukusic, T., Donsi, F., Paniwyk, L. and Djekic, I. (2018). Three pillars of novel non-thermal food technologies: Food safety, quality and environment. Journal of Food Quality, 2018, 1–18. https://doi.org/10.1155/2018/8619707.

Jeyamkondan, S., Jayas, D. and Holley, R. (1999). Pulsed electric field processing of foods: a review. *Journal of Food Protection*, 62, 1088–1096.

Johnston, D. E., Murphy, R. J. and Birksl, A. W. (1994). Stirred-style yoghurt-type product prepared from pressure treated skim-milk. *International Journal of High Pressure Research*, 12, 215–219.

Kamat, A.S., Ghadge, N., Ramamurthy, M.S., Alur, M.D. (2005). Effects of low dose irradiation on shelf life and microbiological safety of sliced carrot. Journal of Science in Food & Agriculture, 85, 2213–2219.

Khattak, A. B., Bibi, N., Chaudry, M. A., Khan, M., Khan, M. and Qureshi, M. J. (2005). Shelf life extension of minimally processed cabbage and cucumber through gamma irradiation. *Journal of Food Protection*, 68, 105–110.

Knorr, D. (2003). Impact of non-thermal processing on plant metabolites. *Journal of Food Engineering*, 56, 131–134.

Knorr, D. and Angersbach, A. (1998). Impact of high-intensity electric field pulses on plant membrane permeabilization. *Trends in Food Science & Technology*, 9, 185–191.

Kruk, Z. A., Yun, H., Rutley, D. L., Lee, E. J., Kim, Y. J. and Jo, C. (2011). The effect of high pressure on microbial population, meat quality and sensory characteristics of chicken breast fillet. *Food Control*, 22, 6–12.

Lacroix, M., and Lafortune, R. (2004). Combined effects of gamma irradiation and modified atmosphere packaging on bacterial resistance in grated carrots (*Daucus carota*). Radiation Physics & Chemistry, 71, 77–80.

Lopez-Fandino, R., Carrascosa, A. and Olano, A. (1996). The effects of high pressure on whey protein denaturation and cheese-making properties of raw milk. *Journal of Dairy Science*, 79, 929–936.

Lopez, V., Avendano, S., Romero, J., Garrido, S., Espinoza, J., Vargas, M. (2005). Effect of gamma irradiation on the microbiological quality of minimally processed vegetables. Archives of Latioam Nutrition, 55, 287–292.

Ma, H., Ledward, D., Zamri, A., Frazier, R. and Zhou, G. (2007). Effects of high pressure/thermal treatment on lipid oxidation in beef and chicken muscle. *Food Chemistry*, 104, 1575–1579.

Maggi, A., Gola, S., Rovere, P., Miglioli, L., Dall'Aglio, G. and Loenneborg, N. (1996). Effects of combined high pressure-temperature treatments on Clostridium sporogenes spores in liquid media. *Industria Conserve*, 71, 8–14.

McArdle, R., Marcos, B., Kerry, J. and Mullen, A. (2010). Monitoring the effects of high pressure processing and temperature on selected beef quality attributes. *Meat Science*, 86, 629–634.

Mohamed, M. E. and Eissa, A. H. A. (2012). Pulsed electric fields for food processing technology. In: A. E. Ayman (Ed.), *Structure and Function of Food Engineering* (pp. 275–306). London, UK: IntechOpen.

Moreira, R. G., Puerta-Gomez, A. F., Kim, J. and Castell-Perez, M. E. (2012). Factors affecting radiation D-values (D10) of an Escherichia coli cocktail and Salmonella typhimurium LT2 inoculated in fresh produce. *Journal of Food Science*, 77, E104–E111.

Morris, C., Brody, A. L. and Wicker, L. (2007). Non-thermal food processing/preservation technologies: a review with packaging implications. *Packaging Technology and Science: An International Journal*, 20, 275–286.

Muntean, M.-V., Marian, O., Barbieru, V., Cătunescu, G. M., Ranta, O., Drocas, I. and Terhes, S. (2016). High pressure processing in food industry – characteristics and applications. *Agriculture and Agricultural Science Procedia*, 10, 377–383.

Nakayama, A., Yano, Y., Kobayashi, S., Ishikawa, M. and Sakai, K. (1996). Comparison of pressure resistances of spores of six bacillus strains with their heat resistances. *Applied Environmental Microbiology*, 62, 3897–3900.

Niemira, B. A., Fan, X. and Sokorai, K. J. (2005). Irradiation and modified atmosphere packaging of endive influences survival and regrowth of Listeria monocytogenes and product sensory qualities. *Radiation Physics and Chemistry*, 72, 41–48.

Niemira, B. A., Sommers, C. H. and Fan, X. (2002). Suspending lettuce type influences recoverability and radiation sensitivity of Escherichia coli O157: H7. *Journal of Food Protection*, 65, 1388–1393.

Olajide, J., Adedeji, A. A., Ade-Omowaye, B., Otunola, E. and Adejuyitan, J. A. (2006). Potentials of high intensity electric field pulses (HELP) to food processors in developing countries. *Nutrition & Food Science*, 36, 248–258.

Patterson, M. (2005). Microbiology of pressure-treated foods. *Journal of Applied Microbiology*, 98, 1400–1409.

Patterson, M. F. and Kilpatrick, D. J. (1998). The combined effect of high hydrostatic pressure and mild heat on inactivation of pathogens in milk and poultry. *Journal of Food Protection*, 61, 432–436.

Patterson, M. F., Quinn, M., Simpson, R. and Gilmour, A. (1995). Sensitivity of vegetative pathogens to high hydrostatic pressure treatment in phosphate-buffered saline and foods. *Journal of Food Protection*, 58, 524–529.

Pereira, R. and Vicente, A. (2010). Environmental impact of novel thermal and non-thermal technologies in food processing. *Food Research International*, 43, 1936–1943.

Prakash, A., Inthajak, P., Huibregtse, H., Caporaso, F. and Foley, D. (2000). Effects of low-dose gamma irradiation and conventional treatments on shelf life and quality characteristics of diced celery. *Journal of Food Science*, 65, 1070–1075.

Praporscic, I., Lebovka, N., Vorobiev, E. and Mietton-Peuchot, M. (2007). Pulsed electric field enhanced expression and juice quality of white grapes. *Separation and Purification Technology*, 52, 520–526.

Qin, B., Zhang, Q., Barbosa-Cánovas, G., Swanson, B. and Pedrow, P. (1995). Pulsed electric field treatment chamber design for liquid food pasteurization using a finite element method. *Transactions of the ASAE*, 38, 557–565.

Ramamurthy, M., Kamat, A., Kakatkar, A., Ghadge, N., Bhushan, B. and Alur, M. (2004). Improvement of shelf-life and microbiological quality of minimally processed refrigerated capsicum by gamma irradiation. *International Journal of Food Sciences and Nutrition*, 55, 291–299.

Raso, J., Alvarez, I., Condón, S. and Trepat, F. J. S. (2000). Predicting inactivation of Salmonella senftenberg by pulsed electric fields. *Innovative Food Science & Emerging Technologies*, 1, 21–29.

Rastogi, N., Raghavarao, K., Balasubramaniam, V., Niranjan, K. and Knorr, D. (2007). Opportunities and challenges in high pressure processing of foods. *Critical Reviews in Food Science and Nutrition*, 47, 69–112.

Rivas, A., Rodrigo, D., Martinez, A., Barbosa-Cánovas, G. and Rodrigo, M. (2006). Effect of PEF and heat pasteurization on the physical–chemical characteristics of blended orange and carrot juice. *LWT – Food Science and Technology*, 39, 1163–1170.

Sale, A. and Hamilton, W. (1968). Effects of high electric fields on micro-organisms: III. Lysis of erythrocytes and protoplasts. *Biochimica et Biophysica Acta (BBA) – Biomembranes*, 163, 37–43.

Schenk, M., Guerrero, S. and Alzamora, S. M. (2008). Response of some microorganisms to ultraviolet treatment on fresh-cut pear. *Food and Bioprocess Technology*, 1, 384–392.

Señorans, F. J., Ibáñez, E. and Cifuentes, A. (2003). New trends in food processing. *Critical Reviews in Food Science and Nutrition*, 43, 507–526.

Shashidhar, R., Dhokane, V.S., Hajare, S.N., Sharma, A., and Bandekar, J.R. (2007). Effectiveness of radiation processing for elimination of Salmonella Typhimurium form pineapple (*Ananas comosus* Merr.). Journal of Food Science, 72, M98–M101.

Shigehisa, T., Ohmori, T., Saito, A., Taji, S. and Hayashi, R. (1991). Effects of high hydrostatic pressure on characteristics of pork slurries and inactivation of microorganisms associated with meat and meat products. *International Journal of Food Microbiology*, 12, 207–215.

Siemer, C., Toepfl, S., Witt, J. and Ostermeier, R. (2018). *Use of Pulsed Electric Fields (PEF) in the Food Industry, DLG Expert Report.*

Simonis, P., Kersulis, S., Stankevich, V., Sinkevic, K., Striguniene, K., Ragoza, G. and Stirke, A. (2019). Pulsed electric field effects on inactivation of microorganisms in acid whey. *International Journal of Food Microbiology*, 291, 128–134.

Solomon, E. and Hoover, D. (2004). Inactivation of Campylobacter jejuni by high hydrostatic pressure. *Letters in Applied Microbiology*, 38, 505–509.

Stratakos, A. C., Inguglia, E. S., Linton, M., Tollerton, J., Murphy, L., Corcionivoschi, N., Koidis, A. and Tiwari, B. K. (2019). Effect of high pressure processing on the safety, shelf life and quality of raw milk. *Innovative Food Science & Emerging Technologies*, 52, 325–333.

Sun, X. D. and Holley, R. A. (2010). High hydrostatic pressure effects on the texture of meat and meat products. *Journal of Food Science*, 75, R17–R23.

Syed, Q., Ishaq, A., Rahman, U., Aslam, S. and Shukat, R. (2017). Pulsed electric field technology in food preservation: a review. *Journal of Nutritional Health and Food Engineering*, 6, 1–5.

Tanaka, T. and Hatanaka, K. (1992). Application of hydrostatic pressure to yoghurt to prevent its after-acidification. *Nippon Shokuhin Kogyo Gakkaishi*, 39, 173–177.

Tiwari, B., O'donnell, C. and Cullen, P. (2009). Effect of non-thermal processing technologies on the anthocyanin content of fruit juices. *Trends in Food Science & Technology*, 20, 137–145.

Torres, J. A. and Velazquez, G. (2008). Hydrostatic pressure processing of foods. In *Food Processing Operations Modeling: Design and Analysis (Second Edition)* (pp. 173–121). Boca Raton, FL: CRC Press.

Trujillo, A. J., Capellas, M., Saldo, J., Gervilla, R. and Guamis, B. (2002). Applications of high-hydrostatic pressure on milk and dairy products: a review. *Innovative Food Science & Emerging Technologies*, 3, 295–307.

Tsong, T. (1990). Review on electroporation of cell membranes and some related phenomena. *Biochemical and Bioenergy*, 24, 271–295.

Wilson, D. R., Dabrowski, L., Stringer, S., Moezelaar, R. and Brocklehurst, T. F. (2008). High pressure in combination with elevated temperature as a method for the sterilisation of food. *Trends in Food Science & Technology*, 19, 289–299.

Woldemariam, H. W. and Emire, S. A. (2019). High pressure processing of foods for microbial and mycotoxins control: current trends and future prospects. *Cogent Food & Agriculture*, 5, 1622184.

Wouters, P. C., Alvarez, I. and Raso, J. (2001). Critical factors determining inactivation kinetics by pulsed electric field food processing. *Trends in Food Science & Technology*, 12, 112–121.

Yeom, H., Zhang, Q. and Chism, G. (2002). Inactivation of pectin methyl esterase in orange juice by pulsed electric fields. *Journal of Food Science*, 67, 2154–2159.

Zhang, Q., Barbosa-Cánovas, G. V. and Swanson, B. G. (1995). Engineering aspects of pulsed electric field pasteurization. *Journal of Food Engineering*, 25, 261–281.

Zhang, Q., Qin, B., Barbosa-Canovas, G. and Swanson, B. (1994). Inactivation of E. coli for food pasteurization by high-intensity short-duration pulsed electric fields. *Journal of Food Processing and Preservation*, 19, 103–118.

Sabiu Saheed, Pillay Charlene, Garuba Taofeeq, Ngcala Mamosa

18 Food scanners: applications in the food industry

18.1 Introduction

Globally, a significant number of the populace have one or more preventable, chronic diseases, many of which are related to poor eating patterns and physical inactivity. These diseases are also caused, in part, by food industries that promote processed foods packed with unhealthy ingredients, artificial sweeteners and flavors, preservatives and other additives (Donsky, 2017).

Till date, the prevalence of these chronic, diet-related diseases continues to rise, and this has occupied a central stage in the food industry. From the progressive rise in the prevalence of debilitating diseases like obesity and diabetes to severe food allergies, there is no denying the extent foods, and the ingredients in the foods consumed influence the lives of consumers. Ingredient lists of food items are usually long and nearly incomprehensible, and technological advancements in the way food is processed make it nearly impossible to determine the exact constituents of what is consumed each day. This fact has resulted in several cases culminating in either established ill-health or potential degenerating health conditions.

In view of the foregoing, many health-conscious dieters, health experts, and food advocates are embarking on a back-to-basics approach to diet to ascertain the exact composition of food items with a view to alleviating most of the modern-day despairs resulting from food consumption. As interesting as this approach could sound as a probable solution to these food woes, it may not be that simple as just eating natural, fresh ingredients etc. One of the very proposed strategies to this problem is technological advancement and development in the form of devices that could comprehensively reveal the entirety of the constituent(s) of a given food sample. Typical example of these devices is the food scanners.

Food scanners are hand-held technological devices that can determine the composition of food with a simple point and shoot (Jacobsen, 2018). They could reveal the nutrients, calories, allergens, microbial types/loads, and residual pesticides in food samples. Food scanners could also ascertain if the food is genetically modified and provide adequate information with precision to the consumers. Generally, accurate nutritional content analysis technologies, such as optical spectroscopy,

Sabiu Saheed, Pillay Charlene, Department of Biotechnology and Food Technology, Durban University of Technology, Durban, South Africa
Garuba Taofeeq, Plant Biology Department, University of Ilorin, Ilorin, Nigeria
Ngcala Mamosa, Department of Molecular and Cell Biology, University of Cape Town, Private Bag, Rondebosch, South Africa

https://doi.org/10.1515/9783110667462-018

radiation, acoustics, have been investigated in the food production industry for many years. Of these, the near-infrared (NIR) spectroscopy is widely used mainly because of its ability to accurately determine the exact chemical and physical component of individual food in a short time (Blanco and Villarroya, 2002), with no sample handling and manipulation requirements (Thong et al., 2017), as it is the case with most scanners today. Although the concept of food scanner is relatively new and seems to be a promising emerging technology in the food industry. There are ongoing efforts to ensure its comprehensiveness, universal acceptability, compliance, and sustainability. This chapter looks at its advent, advances, and applications in the food industry.

18.2 Advent and advances till date

Parallel to the increase in diet-related diseases, societies are now witnessing development of huge technological advances. Individuals are now more reliant on technology to perform many tasks of daily living (Gilmore et al., 2014). For instance, self-monitoring through the use of technology-based devices has been advocated as one of the most important strategies for weight management and lifestyle changes. This allows for an increase in self-awareness with regards to targeting behavior and outcomes in relation to food intake goals (Zaidan and Roehrer, 2016). Many start-up companies that are proposing the use of mobile devices that are able to test the quality of food and to determine its constituents are emerging due to the use of crowdfunding platforms such as Indiegogo and Kickstarter (Rateni et al., 2017). Over the past few years, spectroscopic techniques have evolved and developed dramatically, particularly with the use of fiber optics, laser, and solid-state components facilitating the miniaturization of spectrometers. The most important requirement was that advances were made to lower the costs of nondestructive, noninvasive spectroscopic techniques to a level that is more generally acceptable in the food industry (Scotter, 1997). The early adopters of food scanners were those battling with weight loss, diabetics and pre-diabetics, and families coping with allergies and food intolerance (Gander, 2015). The alliance between digital technology and the Web is empowering consumers. In conjunction with smartphones, a new portable technology is linking spectrometer-based chemical and nutrient scanning to online, real-time analysis, and date (Gander, 2015). Some of the existing competitive key players of the global food scanners market include TellSpec Inc., Scioscan, Spectral Engines OY, and Nima Labs Inc, while FOODSniffer, Koninklijke Philips N.V., and Samsung Electronics Co. Ltd. are other anticipated key potential companies that may embrace manufacturing of food scanners in no distant time (EIN Presswire, 2017). In March 2017, a competition organized by the European Commission awarded a €1 million Horizon prize to companies with the best innovative, affordable, and noninvasive device that enables users to measure and analyze their food intake (https://ec.europa.eu/info/research-and-innovation/funding/funding-opportunities/prizes/horizon-prizes/food-scanner). The winner of the Horizon prize was Spectral Engines with the runner up being SciOscan and TellSpec.

18.2.1 Spectral engines

Spectral Engines is a Finnish company founded in 2014 that developed a novel spectral sensing platform that offers unique benefits and many applications such as food sensing and analysis. Its core competence in miniaturized spectrometers can be converted into plug-and-play industrial grade smart sensors. A food scanner prototype was developed based on NIR spectroscopy along with Bluetooth connection to a mobile device and a data connection to a Cloud server with a vast material library to reveal the fat, protein, sugar, and total energy content of food items. This food scanner prototype is compact and provides real-time results at a comparatively low price (https://ec.europa.eu/info/research-and-innovation/funding/funding-opportunities/prizes/horizon-prizes/food-scanner).

The food scanner by Spectral Engines was awarded as the winner of the Horizon Prize organized by the European Union in March 2017. In February 2016, the scanner had received the Photonics Prism Award in the category of detectors and sensors at the Photonics West 2016 (https://www.spectralengines.com/products/nirone/scanner/foodscanner)

18.2.2 SciO scanners

Founders, Remy and Astrid Bonnasse, devised the concept of SciO scanner in 2014 when their 9-year-old daughter was diagnosed with Type 1 diabetes requiring her to constantly monitor the carbohydrate intake of each meal and measure insulin. The couple developed a nutrition coaching app and liaised with an Israeli company Consumer Physics for scanner development and engineering that uses science for process automation (Murphy, 2016). The Israeli-based start-up company introduced its hand-held molecular food scanner in May 2014. Within a month of its release, more than $2 million in crowdfunding on Kickstarter was raised. SciO is a tiny device that shines light onto a piece of food, and articulately processes the information through a tiny spectrometer connected with a smartphone (Albright, 2014). After about 10 s, a nutritional breakdown will appear on an accompanying app called DietSensor providing chemical information such as calorie content, ripeness, and sweetness (Albright, 2014; D'Estries, 2016). The French start-up company, DietSensor, launched at the 2016 International Consumer Electronics shows a pocket-sized, Bluetooth-connected molecular sensor called SciO using NIR spectroscopy to determine the chemical makeup of food and drinks (Murphy, 2016). Essentially, it is a miniaturized, inexpensive version of technology that lab scientists have been using for many years to determine the physical composition of various materials. The hand-held scanner comes with apps that can report the physical composition of food and pharmaceuticals (https://ec.europa.eu/info/research-and-innovation/funding/funding-opportunities/prizes/horizon-prizes/food-scanner). The SciO is described as the state of the art micro-spectrometer primarily aimed at helping diabetics as well as those with cardiovascular diseases (Murphy, 2016; https://www.consumerphysics.com/scio-for-consumers/). Compared with the laboratory NIR scanners, the SciO scanner records a narrower spectrum, with wavelength ranging only from 740 nm to 1070 nm. However, it is suitable for daily usage by consumers due to its compact size, ability to sync and upload measured data wirelessly, and automatic data model construction in the SciO cloud. In a study by Thong et al. (2017), the nutrient facts such as energy and carbohydrate can be predicted by SciO NIR scanner using NIR spectra reflected from liquid foods. The authors found that the scanner performed best on support vector regression on the experimental evaluation on the collected RBF kernel samples in revealing their comprehensive characteristic constituents.

18.2.3 TellSpec handheld scanners

Isabel Hoffman along with mathematician Stephen Watson created TellSpec, a handheld device that could provide information on chemicals, allergens, and ingredients to avoid with a simple point at a food item (Sandhana, 2013). It was her own daughter's complex allergies that prompted her to investigate the idea of a scanner-and-phone combination that was sensitive enough to identify the parts per million needed to detect allergens and pesticides (Gander, 2015). Hoffman moved with her family from Europe to North America in January 2011. It was shortly after that her daughter became ill with severe allergies to certain foods and mycotoxins (https://www.jameco.com/Jameco/workshop/RollCall/tellspec-spectroscopy-food-sensor-innovation.html). Initially, the first prototype developed in 2014 utilized a small Raman spectrometer, a unique cloud-based algorithm and a simple smartphone app. It beams a low-powered laser at the item and analyzes the reflected light waves to identify the chemical makeup of the food to detect chemicals, allergens, and ingredients that make up the food (Sandhana, 2013). The device assists consumers to track what they consume by giving readings on sugars, carbohydrates, and fat content (https://www.jameco.com/Jameco/workshop/RollCall/tellspec-spectroscopy-food-sensor-innovation.html). Besides identifying the makeup of a food, the app can warn about allergens like gluten or egg. Additionally, this app allows the user to achieve calorie targets, clarify inaccurately reported food labels, and monitor intake of essential vitamins and minerals (https://www.digitaltrends.com/cool-tech/tellspec-food-scanner/).

Currently, the scanner uses Texas Instruments' DLP technology based on NIR offering an advantageous dimension to the first-generation model as described above which used a laser. The shift away from laser implementation was largely dependent on consumer safety. However, there are some restrictions acknowledged with NIR spectroscopy as it will not work with clear liquids as achieved by the laser. Rather, this system scans through transparent plastics and glass but not opaque packaging (Gander, 2015), and its overall precision is dependent on transparency of the food surface. The data obtained are uploaded to the analysis engine comparing it to the reference spectra. Successful identification after scanning the food's surface results in 97.7% similarity to the reference data (Sandhana, 2013). The TellSpec can scan foods directly or through plastic or glass allowing easy usage of the scanner (https://www.jameco.com/Jameco/workshop/RollCall/tellspec-spectros copy-food-sensor-innovation.html). TellSpec has been previously used in a study to develop multivariate models for simultaneous prediction of melamine and urea in wheat gluten samples based on data acquired with handheld NIR scanners and a user-friendly mobile app. The results obtained in this study proved that Tellspec scanner is a rapid, cost-effective and user-friendly tool that can be used for simultaneous quantification of the most common food adulterants in wheat gluten powders at or lower than 1% adulteration level (Kovacs et al., 2017). Moreover, TellSpec scanner offers a potential for manufacturers to prevent contamination of their products, regulators to track contamination to its source and consumers to be confident in the quality of their food.

An advancement from the prototype of the TellSpec scanner is the development and installation of TellSpecopedia, which is a searchable bank of scientific and research data about different food components, ingredients, and their health impacts (Gander, 2015). The TellSpec team scanned 3,000 food items toward the end of 2013 to create the initial database to potentially identify unlimited ingredients present in food. This number and ability to identify is expected to increase exponentially as the number of users of the TellSpec scanner grow and add their own scans of different food items (Sandhana, 2013). The more food scanned by people, the larger the food database and the larger the public memory of food composition and consumption (Gander, 2015). With users of TellSpec around the world, this database can be continually updated. As the founder of TellSpec, Isabel Hoffman stated, "it is literally in the hands of the people. It is they who will truly participate actively in creating a global footprint of food data. The food database is an evolving number – the more people scan, the more the database grows and the more precise the scans become" (Sandhana, 2013).

The data available on the TellSpec scanner are open source, allowing anyone to use the data to create their own health-based apps. A diabetic app tracking blood sugar levels utilizes TellSpec data to track the sugars and carbohydrates consumed and identify the ingredients present in food that are converted into sugar (Sandhana, 2013). In 2015, the device only achieved 96.7% accuracy in detecting gluten in foods. This is an acceptable level for gluten but not for a peanut allergy. Thus, with more scans, 100% accuracy is achievable (Gander, 2015).

TellSpec's technology can be used to test the quality, ripeness, and flavor of fresh fruit in less than 15 s in a reliable, nondestructive manner. A further advancement in the technology of TellSpec scanners is the incorporation of the Fruit QC application that allows one to determine the quality of fruits in terms of Brix and Titratable acidity. Degree Brix or total soluble solids are a measure of the sugar content present in fruits. Brix can be traditionally measured using a refractometer. However, the disadvantage of this is that it is a time-consuming process that destroys the fruit because it does require manual preparation of the juice to be read and this might lead to inconsistencies in the measurements. The TellSpec Enterprise Scanner coupled with the Fruit QC application allows measurement of Brix without destroying the fruit and measured within a few seconds. The sensor emits light into the fruit and based on the interaction of the light with the fruit, the degree Brix can be measured. Similarly, titratable acidity can be measured in fruits determining the degree of maturity, the quality of the fruit and the amount of organic acids (citric, malic, lactic, tartaric, and acetic acids) present that can directly affect flavor, color, and stability (http://tellspec.com/food-retail/).

Food fraud is an international adulteration to mask the nutritional characteristics of the product. Fish fraud, the practice of misleading consumers about the fish they consume to increase profits, has a negative impact on marine conservation, consumer trust, and human health (Staffen et al., 2017). This practice of fraudulent mislabeling can negatively impact the health of the consumer especially if toxic fish species are substituted for nontoxic fish and has become an important problem in the seafood industry (Galal-Khallaf et al., 2014). In addition, these fraudulent activities are global issues resulting in approximately 20% of seafood that are mislabeled worldwide. Wholesalers, retailers, restaurants, catering businesses, and consumers deserve the facts of the fish supply chain, the species of fish, its nutritional profile, and its freshness. Fish fraud can result in severe economic consequences by tarnishing the national reputation of a country in the global food market. To combat these fraudulent activities, TellSpec has developed a fish quality and fraud mobile app to test the nutritional value of fish and determine the shelf life of the fish (http://tellspec.com/eit-food/fishproject/).

18.2.4 Nima

The San Francisco-based labs (formerly known as 6Sensor Lab) devised the world's first connected food sensor engineered and developed by people who have food sensitivities and allergies. Nima was developed by adapting antibody-based chemistry used for protein or allergen detection (https://www.nimasensor.com/science-nima-understanding-device/). This product is a handheld scanner with a high sensitivity and specificity to gluten and peanut allergens in food (Elgan, 2016). Nima subjected the device to a third-party tester that found the scanner able to detect 10 parts per million or more peanut protein with 99.2% accuracy. The gluten scanner has been in the market for more than two years (Comstock, 2018).

The usage of this scanner requires the food in question to be inserted into a disposable capsule and placed into the triangular scanner. Within 2 min, a reading is given on whether gluten or peanuts is present in the food (Elgan, 2016). While the results display on the device, the user can refer to a connected smartphone app to log tests, thus contributing to a database that other users can benefit from.

This portable food testing device for allergens has tried to incorporate spectrometry allowing smoother user experience, avoiding the process on inserting food samples into the capsule in the scanner. However, this was not optimal in detection of allergens and therefore, the Nima scanner is purely chemical-based, using special proprietary antibodies to detect peanuts and gluten (Comstock, 2018).

The Nima chemistry team developed a pair of antibodies specifically for the detection of gluten and peanuts. Nima's 13F6 and 14G11 antibody system was evaluated and determined to be more sensitive in a standard laboratory test than one of the well-recognized gold standard antibodies currently used in the market. Nima's gluten antibody is currently used in the Biofront gluten ELISA kit, been evaluated for excellent performance in sensitivity and specificity in a wide variety of foods. Nima's 20B10 and 16B1 antibodies bind to a peanut protein called Arah3. Although this is not highly antigenic, it is found abundantly in all types of peanuts and found to be stable under processing conditions such as heat during roasting (https://www. nimasensor.com/science-nima-understanding-device/).

18.2.5 Others

Besides disclosing the quality and inherent characteristics of the food items, some food scanners, such as the Origin Trail, can trace the origin of the food along agricultural value chain. This kind of application is germane to addressing the pertinent issue in the 2013 report of European Consumer Organization, where about 70% of consumers in Central Europe are doggedly insistent to get information on the source of food item they consume (Gay, 2016). Similarly, Inspecto, which will be commercially available by 2020 from FoodTech Startup Inspecto Company, can detect contaminants in food items at low concentration levels (Eagle, 2019). It is envisaged that the scanner will be launched in Italy, Spain, and France, followed by the USA. The advent of this device could address the 2017 report of WHO, which speculated that, one-tenth of people become sick because of eating contaminated food, which was associated with incapability of food-producing industries to perform mass testing of their products. The use of Inspecto will also cut the cost of laboratory expenses, time, and complexity of quality assurance process.

The Fraunhofer-Gesellschaft is the leading organization for applied research in Europe and has another important "pocket" food scanner development underway. Like most other scanners, its mode of operation depends on infrared light for determination of food/fruit freshness. The United Nations Foods and Agricultural Association reported that one-third of food produced in the world is wasted (Szondy, 2019). Similarly, 10 million metric tons of food is thrown in the garbage every year in Germany despite still being edible (Fraunhofer-Gesellschaft, 2019; Szondy, 2019). The reason accounted for the loss is premised within inability of consumers to determine edibility of the food. The Bavarian Ministry of Food, Agriculture and Forestry has initiated a way of combating waste by means of a food scanner. This scanner is designed to help reduce waste at the end of the value chain, namely in stores and in the homes of consumers. In the future, this inexpensive pocket-sized device will determine the degree of ripeness of the fruits and the actual freshness of both packaged and unpackaged foods (Fraunhofer-Gesellschaft, 2019). According to the company, the device is still at the demonstrator stage looking for techniques to introduce to overcome its inability to scan heterogeneous food. Hyperspectral imaging, spectral sensors, and fusion-based approaches are the prospective technologies to overcome the defects in future. Researchers at the Fraunhofer Institute of Optronics are developing a more closely related compact device to scan the rate of microbial spoilage and to determine the shelf life and ripeness of produce through NIR spectra (Szondy, 2019). With this device, it is envisioned that microbial level will be monitored by sending the scanned data via Bluetooth to a cloud database for evaluation and the outcome will be displayed on the mobile device screen; the rate of microbial spoilage, how much shelf life is left and tips on how to use the food if its sell-by date has expired.

Over the years, food scanners have also metamorphosed and presented in more convenient and simpler applications. For instance, smartphone applications such as the Fooducate and Shopwell applications (from Fooducate Ltd, USA, and Shopwell Lab. Inc., USA, respectively) for food scanning have evolved. These are iphone applications that are sensitive to select healthier foods especially for people with allergies. Another groundbreaking technology is the AIRO wristband. The waistband tracks not only sleep, stress, and heart beat but also reveals the chemical components of food that is being eaten by the user. The device measures the calories and displays the nutritional composition of food. The food scanner by AIRO is built on different light properties of the constituent nutrients of the food sample. The wristband can split food composition into fat, carbohydrate, and protein, and to the level of simple and complex starches. It also analyzes the overall nutritional status of the food using different wavelengths of light that detects metabolites as they are released into the blood stream (Lee, 2019).

18.3 Significance and applications

The prospect of a life-threatening allergic reaction is always a major public concern. It is generally accepted that the prevalence of food allergy has been increasing in recent decades, particularly in westernized countries, yet undoubtable evidence that is based on challenge confirmed diagnosis of food allergy to support this assumption is lacking because of the high cost and potential risks associated with conducting food challenges in large populations (Tang and Mullins, 2017). Over the years, available reports have supported consistent evidence for increasing prevalence of food allergy at least in western countries, while recent reports that children of East Asian or African ethnicity who are raised in a western environment have an increased risk of developing food allergy compared with resident Caucasian children suggest that food allergy might also increase across Asian and African countries as their economies grow and populations adopt a more westernized lifestyle (Tan and Mullins, 2017). Given that many cases of food allergy persist, logical principles would predict a continued increase in food allergy prevalence in the short to medium term until an effective treatment is identified to allow the rate of disease resolution to be equal to or greater than the rate of new cases.

Patients may present with urticaria or hives that are attributed to a food allergy. Common food allergic manifestations include oral symptoms such as itching and swelling of the mouth, tongue, and throat and even throat or laryngeal obstruction with breathing difficulties (https://www.allergyclinic.co.za/food-allergies/). Successful testing for allergies can be very difficult without a clear precise history implicating one or other food. To complicate issues, the available tests for allergy are often not 100% specific and the possibility of wrong diagnosis could be high. Be it the true food allergy, where there is usually an immediate catastrophic IgE immune system mediated reaction and requiring emergency adrenaline injections and medical resuscitation, food intolerance (which are not immune system mediated but may be enzyme deficiencies, and other mechanisms that mimic true allergic reactions), or food toxicity, that is primarily caused by biological contaminants or poison present in the food (https://www.allergyclinic.co.za/food-allergies/), the severity could be grave and life-threatening. Meanwhile, this scenario could be effectively tamed or prevented by mere use of "point and shoot" devices (food scanners) that will foretell the kind and comprehensive constituents of the food to be consumed. For individuals with allergies to specific foods, the food scanners could help save numerous lives. According to European Statistics, an average of 230 million people may suffer from food allergies worldwide and about 5% of children and 4% of teens and adults have clinically shown allergy to foods (The Medical Futurist, 2016). Similarly, the Food Allergy Research and Education reports that 1 in 13 children in the USA alone have food allergies. And about 30% of these are allergic to more than one food (https://www.foodallergy.org/). Also, due to food allergies, one person is sent to the emergency room every 3 min in the USA with approximately more than 200,000 persons similarly affected annually (https://www.foodallergy.org/). Again, there is no perfect cure except for avoiding these foods. Although total avoidance might be practically impossible, as it is usually the case when on vacation abroad or when eating out. However, the food scanners can provide valuable warning signs prior to ingesting food.

Furthermore, the food scanners could also be very handy and useful even for those without food allergies. For instance, the International Diabetes Federation estimated in 2017 that 451 million people are suffering from diabetes worldwide and this statistical fact is expected to reach 693 million by 2045, with half of them undiagnosed (Cho et al., 2018). Food allergies, obesity, and diabetes are the most striking examples where knowing calorie intake and ingredients are pivotal in managing medical issues (The Medical Futurist, 2016). The food scanners could potentially be used to count the calories to be ingested, especially for those on the verge of losing weight to look good as well as victims of cardiovascular diseases, diabetes, and other debilitating disorders.

The promotion and use of food scanners will not only be of utmost significance to health experts, international organizations, and food advocates, but also to many health-conscious persons seeking healthy eating patterns. For instance, a careful consideration of the dietary guidelines for Americans 2015–2020 8th edition guidelines, which advocates and promotes healthy eating patterns could suggest and dictate the relevance of food scanners to healthy eating patterns. The guidelines enjoins to: (1) follow a healthy eating pattern across the lifespan: this explains the need to choose a healthy eating pattern at an appropriate calorie level to help achieve and maintain a healthy body weight, support nutrient adequacy, and reduce the risk of chronic disease; (2) focus on variety, nutrient density, and amount: this encourages that, to meet nutrient needs within calorie limits, choose a variety of nutrient-dense foods across and within all food groups in recommended amounts; (3) limit calories from added sugars and saturated fats and reduce sodium intake: this advices to embrace an eating pattern characterized with low-added sugars, saturated fats, and sodium. Cut back on foods and beverages higher in these components to amounts that fit within healthy eating patterns; (4) shift to healthier food and beverage choices: this explains the need to choose nutrient-dense foods and beverages across and within all food groups in place of less healthy choices. Consider cultural and personal preferences to make these shifts easier to accomplish and maintain; and (5) support healthy eating patterns for all, that is, everyone has a role in helping to create and support healthy eating patterns in multiple settings nationwide, from home to school to work to communities.

To achieve the goals of the guidelines, it could be logically inferred that a crystal-clear ambient is created for food scanners as they could aid in making informed decisions on an appropriate calorie level to help achieve and maintain a healthy body weight, support nutrient adequacy, reduce the risk of chronic disease, and ultimately promote a healthy eating pattern.

Table 18.1 summarizes the different food scanners, their respective application (s), and limitations.

Table 18.1: Notable applications and limitations of food scanners.

Scanner	Company	Year of invention	Core principle	Uses	Limitation(s)
Spectral Engines	Finnish company	2014	Uses the world's smallest true NIR spectral sensing module, advanced algorithms, cloud connectivity, and a vast material library	It scans the fat, protein, sugar, and total energy content of food items at good accuracy	It can only analyze homogeneous foods and not heterogeneous. The result is based on the cloud database. It cannot work outside the cloud.
SciO scanner	Consumer Physics	2014	It is an NIR microspectrometer that absorbs light reflected from an object, breaks it down into a spectrum, and analyzes it to determine the food's chemical makeup	It reveals the chemical composition (calorie content, ripeness, and sweetness) of the food	It does not give information on exact functional group in the food. It works with smartphone and cloud database. It cannot work on information that is not in the cloud database. Only able to identify elements that compose more than 0.1% of the overall chemical makeup of a compound. Packaging and other obstacles can interfere with the sensor.
TellSpec	Tellspec Inc.	2014	Works on reflective NIR spectroscopy between spectral wavelength of 900 nm and 1,700 nm	It reveals allergens, chemicals, nutrients, calories, and ingredients in the scanned food	It is not a stand-alone device. It requires a smartphone. Food items must be "transparent" to obtain an accurate scan

(continued)

Table 18.1 (continued)

Scanner	Company	Year of invention	Core principle	Uses	Limitation(s)
Nima	Nima Lab (formerly known as 6 sensor Lab)	2016	The mixture from sample loaded in the cartridge and the solution at the bottom of cartridge react with test strip on the capsule and the result displays on OLED screen.	Use to quantify the amount of gluten in food items	There is a chance of cross-contamination from the plate, hand, or table while loading the food sample to the cartridge. Users must put a pea-sized food sample in the device, so any large meal or a sampling of a dish with multiple parts (e.g., burger) is impossible to analyze in full. Not tested on nonfood substances such as medications and make up. Unable to detect gluten in the following: Soy sauce, beer, and other fermented or hydrolyzed foods due to the chemistry composition. In addition, cannot test alcohol or pure xanthan/guar gum.
Origin Trail	Origin Trail Network	2013	Origin Trail app involves scanning the barcode on any food product with smartphone using blockchain technology	It reveals nutritional values, the origin of the food items along agricultural value chain, and other descriptive information. The data from suppliers are synced every day	Mainly designed to reveal the source of the food item

Inspecto	Inspecto Solution Ltd	2016	Surface-enhanced Raman spectroscopy. A sample is placed in a disposable capsule for detection, inserted into the device, and then activated with a simple press of a button. The sample is scanned and automatically processed within minutes	Detection of chemical contamination in the food items at low concentration levels	The device can currently perform survey testing and detect contaminants in fruits and vegetables only
Fraunhofer	Fraunhofer Institute of Optronics, System Technologies and Image Exploitation (IOSB)	2019	Uses high-precision NIR sensor	Determines the degree of freshness, edibility, and shelf life of food item.	It can only analyze homogeneous food
Fooducate	Fooducate Ltd	2012	It is an iPhone application that scans the barcode of foodstuffs and immediately produces a nutritional breakdown of the food item	Apart from nutritional analysis, it also provides information on the artificial flavoring and coloring agents in food samples. Scans and grades foods based on their ingredients	It is restricted to iPhone users. Adding extra nutrient (such as fat) tracking features requires premium subscription
ShopWell	Innit Inc.	2017	It is an iOS application that scans the barcode of a food item to see its nutritional information	It analyzes food nutrient based on the user's profile that has been previously created in the application. After compiling the shopping list, ShopWell can suggest similar items with higher scores	Though it can be used in English-speaking countries, but it was primarily designed for use in the USA

(continued)

Table 18.1 (continued)

Scanner	Company	Year of invention	Core principle	Uses	Limitation(s)
AIRO wristband	AIRO Health	2013	Uses optical spectroscopy to track nutrition. Spectrometer uses light wavelength, penetrate the blood stream to detect metabolites and measuring the calorie level. Food items are scanned by AIRO which is built on different light properties of the constituent nutrients of the food sample	It detects the metabolites in the blood and measures the caloric intake. Splits food composition into fat, carbohydrate, and protein, and to the level of simple and complex starches	It cannot be used to scan food items in grocery stores or pantry. It is also able to track stress levels but cannot warn the user if the stress levels rise significantly
"Talkable Vegetables"	Hakuhodo Inc.	2015	The device is activated by turning the voltage differential between the moisture in humans and vegetables into an audio signal. This happens when customers pick up the vegetable at the grocery store	It tells the consumer what makes them special and recommended recipes	It is designed for raw fruits and vegetables in stores.
MyFitnessPal	MyFitnessPal Inc.	2015	It is an iOS application using iPhone camera for scanning	It unveils proximate and mineral compositions of the food items. Allow users to track their nutritional values of their diets and fitness goals	It is restricted to iOS (11.0 or later) users

NIR, near-infrared.

18.4 Future prospects

The food scanners have come to stay due to their tremendous functions and values. With their invention, most food-related preventable, chronic diseases can be controllably checked. Before the introduction of this technology, the ingredients/compositions of most of the food products are externally placed to provide dietary information about the products. In most instances, it was practically impossible to verify such information until samples are taken to laboratory for proximate and other related analyses. With this technology, the future holds better consumer health-conscious approach to nutrition as consumers will no longer buy food products ignorantly without confirming the compositions. The usage of this technology is not restricted to people with allergies, it could be used for self-monitoring of body weight, food intake, and physical activity (Gilmore et al., 2014). In addition, these developing food scanners will be able to determine the geographic origin of the raw ingredients, the nutritional quality, and shelf lives of foods as well as the ethical considerations related to the processing of foods.

Consensus but low awareness is ongoing on the acceptability of some of the newly invented scanners. For example, a test phase was due to begin in supermarkets at the start of 2019 for Fraunhofer, which will investigate how consumers respond to food scanners. It is expected that the versatile technology will be used throughout the value chain, from raw material to end products. The ability to detect quality changes at an early-stage facilitates alternative uses and helps reduce waste. Therefore, a food scanner is more than just an instrument for testing food items, it is also a general-use, cost-effective scanning technology that can be easily adapted (Fraunhofer-Gesellschaft, 2019). Food producers as well as raw material suppliers should invest in having these scanners installed to check for contamination and out-of-specification ingredients (Gander, 2015). However, the lack of awareness toward such devices amongst consumers especially in developing countries is hampering the growth of food scanners market. Availability of these scanners in developing countries can be accessible to consumers if supermarket operators could invest in them for usage. Moreover, high cost of food scanners is a major challenge which could dampen the growth of global food scanners in the near future (https://www.researchnester.com/reports/food-scanners-market-global-demand-analysis-opportunity-outlook-2024/370).

Majority of consumers due to their fast-paced life do not understand the labels or foods that they consume. Therefore, these scanners are critical in analyzing unlabeled foods and purchased foods to allow tracking of calories and carbohydrate intake per meal (Gander, 2015). Current methods for ensuring food safety rely on routine, highly resource-intensive laboratory-based examination of chemicals and/or food-borne pathogens. In remote areas, where resources are scarce, sending specimens for analysis can often become difficult (Rateni et al., 2017). Technical developments with food scanners are also one of the important factors driving the growth of the market for global food scanners (https://www.researchnester.com/re ports/food-scanners-market-global-demand-analysis-opportunity-outlook-2024/ 370). The global food scanners market is segmented into end user such as restaurant, home, laboratory, and food industry (EIN Presswire, 2017).

The demand for food scanners is expected to rise in food industries for quality control operations and to ensure product safety (EIN Presswire, 2017). The ability of food scanners to examine the food with respect to its hygienic and unhygienic properties and its use in industries for food quality maintenance is expected to propel the demand for food scanners to reach at notable revenue by the end of the year 2024 (https://www.researchnester.com/reports/food-scanners-market-global-demand-analysis-opportunity-outlook-2024/370). The market is expected to expand during the forecast period due to rising disposable income and growing affordability of the consumers. In addition, the growing awareness amongst consumers toward health and fitness combined with maintaining proper balanced diets, high adoption rate of innovative and advanced technologies are expected to bolster the growth of global food scanners in the near future (EIN Presswire, 2017). Apart from this, growing health problems such as obesity, diabetes, various food allergies, and an increasing number of food-borne infections are expected to benefit from the expansion of the global food scanner market (https://markettalknews.com/food-scanners-market-analysis-by-growth-emerging-trends-and-future-opportunities-till-2024/). In addition, rules and regulations from governments and food industry bodies to prevent food frauds and scandals and to ensure food safety are also expected to increase the demand for food scanners (EIN Presswire, 2017). This is ultimately envisaged to eliminate the politicized and polarized gestures on the part of some Governmental Departments and Organizations that could trump science and the government's own health advisors.

The SciO Company offers a platform for developers to create new applications like monitoring of beer fermentation. The varying choices of foods are a complex combination of factors that are in constant flux, including stress, mood, and hunger. Future possibilities of application of SciO scanners will develop a food technology that leads to a more honest and a better food system for all (Albright, 2014). Although the DietSensor is one of the first apps to work with the SciO, developers around the world are in the process of creating additional applications. Besides being a food scanner, additional uses could be analyzing moisture levels in plants and perhaps even blood alcohol content (D'Estries, 2016).

A number of tests are underway for the TellSpec scanners to be able to identify heavy metals, pesticides, and by-products such as acrylamide. The goal is to be able to use the scanner for observation at a molecular level (Gander, 2015). Work is in progress to use the scanner to calculate the volume of food a person consumes (Sandhana, 2013). However, the biggest problem with the scanners is that most foods in their entirety cannot be scanned (Elgan, 2016). A limitation to the TellSpec scanner is that the food item requires transparency of the food item in order to obtain an accurate scan. In certain cases, two separate scans are required on the outer surface of the food item as well as in the center (Sandhana, 2013). Both the TellSpec and SciO scanners only reads homogeneous foods such as cheese, breads, chicken, and beef but cannot pick up all the ingredients in a sandwich or pizza. This app allows you to input more complex meals manually (Murphy, 2016). Therefore, this major limitation of food scanners must be overcome in order to obtain an informative scan of a hetero-geneous food item. However, scientists are investigating high-spatial-resolution tech-nologies such as hyperspectral imaging and fusion-based approaches using color images and spectral sensors (Fraunhofer-Gesellschaft, 2019; Szondy, 2019).

Food scanners can also determine the authenticity of a product and will be able to identify if a product has been adulterated (Fraunhofer-Gesellschaft, 2019). Comparison between the absorption spectrum from a food product with that of a known sample can determine if the food is edible, also its ripeness and can even detect if it is a coun-terfeit (Szondy, 2019). Another aspect under development is a machine-learning algo-rithm allowing for advanced pattern recognition. Experimental testing has been conducted with tomatoes and ground beef sing statistical techniques to match the NIR spectra with the rate of microbial spoilage and other chemical parameters, allowing measurement of microbial count and the shelf life of the meat (Szondy, 2019).

With advances in micro-manufacture, sensor technology, and miniaturized electronics, diagnostic devices on smartphones will be increasingly used to perform biochemical detections in healthcare diagnosis, environmental monitoring, and food evaluation in the near future (Rateni et al., 2017).

Another probable hiccup with some of the devices is the size. Food scanners should be portable and hand-held. The engineers must work on the technology to-ward achieving sensitivity and accuracy relative to size. Besides, the scanners are not taken cognizance of the state of the product. The kinetic energy of particles at a given temperature is higher in liquid than solid although maximum energy move-ment is obtained in gas. This is an important factor that may influence, interfere, or interrupt the outcome of the scanning. The technology needs to be improved upon to give desirable results and meet the purpose of its invention irrespective of the state of the food. Another pitfall of this technology is that most of the devices ana-lyze homogeneous food. For instance, DietScanner reads nutritional components of a slice of bacon not bacon cheeseburger and scanner from Fraunhofer can only ana-lyze a potato but not a pizza. For a balanced diet, people are keenly interested in assessment of heterogeneous food. With the trend of technological advancement, the existing scanners can be modified to answer the questions of majority.

Put together, the food scanners are potential revolutionary tools that might soon be used to generate host of data for research or to build databases. The technology will also move to the level of identifying the meat type incorporated in a food such as beef, mutton, pork, and horsemeat, and so on. However, the effect of radiation on the food sample needs to be considered in modifying the existing or developing new scanning devices. Radiation induces some changes that alter nutritional quality of foods depending on the radiation dose and irradiated food.

18.5 Conclusion

The advent and technological advancements of food scanners are welcoming developments that hold promising grounds for patients with specific allergies to food substances, health-conscious individuals, and the food industry. However, for universal acceptability and continued sustainability of this laudable technological concept, appropriate policies must be in place to allow for all-inclusive and unbiased ambient that is consumer-centeredness oriented. Efforts must also be geared toward improving on the existing technologies to design a one-size-fit-all device that will be void of the identified limitations and could reveal the entirety of the necessary information about a food item at a simple point and shoot.

References

Albright, M. B. (2014). *New Food Scanner Helps Calorie-Watchers Identify Ingredients* (pp. 12, 13). https://www.nationalgeographic.com/people-and-culture/food/the-plate/2014/06/09/food-scanner.html.

Blanco, M. and Villarroya, I. N. I. R. (2002). NIR spectroscopy: a rapid-response analytical tool. *Trends in Analytical Chemistry*, 21(4), 240–250.

Cho, N. H., Shaw, J. E., Karuranga, S., Huang, Y., da Rocha Fernandes, J. D., Ohlrogge, A. W. and Malanda, B. (2018). IDF Diabetes Atlas: global estimates of diabetes prevalence for 2017 and projections for 2045. *Diabetes Research and Clinical Practice*. DOI: 10.1016/j.diabres.2018.02.023

Chodosh, S. (2017). *I Tested 'Gluten-Free' Food with the New Gluten Sensor – Here's What I Found.* https://www.popsci.com/gluten-sensor-test-review/ (accessed on May 23, 2019).

Comstock, J. (2018). *Nima's Newest Smart-Phone Connected Device Scans Foods for Peanut Allergens.* https://www.mobihealthnews.com (accessed June 21, 2019).

D'Estries, M. (2016). *Real-Time Food Scanners have Arrived: DietSensor Utilizes a Wild New Gizmo that Analyses Your Food to Help You Hit Your Wellness Goals.* https://www.mnn.com/health/fitness-well-being/blogs/real-time-nutritional-food-scanners-have-arrived (accessed May 20, 2019).

Donsky, A. (2017). *Exploring the Association Between Eating a Whole Food Plant-Based Diet and Reducing Chronic Diseases: A Critical Literature Synthesis* (pp. 1–67). MSc Degree Thesis. Pennsylvania: University of Pittsburgh.

EIN Presswire. (2017). *Food Scanners Market: Global Demand Analysis and Opportunity Outlook 2024.* https://www.einpresswire.com/article/409965579/food-scanners-market-global-demand-analysis-opportunity-outlook-2024 (accessed May 20, 2019).

Eagle, J. (2019). *FoodTech Startup Creates Portable Scanner that Detects Chemical Contamination in Food.* https://www.foodnavigator.com/News/Food-Safety-Quality/FoodTech-startup-creates-portable-scanner-that-detects-chemical-contamination (accessedJune 6, 2019).

Elgan, M. (2016). *What's in Your Food? Tech will Tell!* https://www.computerworld.com/article/3104539/whats-in-your-food-tech-will-tell.html.

Findling, D. (2016). *Portable Sensors will Tell You if There is Gluten in Your Food.* https://www.cnbc.com/2016/01/22/6sensorlabs-nima-is-a-portable-gluten-sensor.html. (accessed June 28, 2019).

Fraunhofer-Gesellschaft. (2019). *Pocket-Size Food Scanner.* https://physic.org/news/2019-01-pocket-size-food-scanner.html (accessed May 12, 2019).

Galal-Khallaf, A., Ardura, A., Mohammed-Geba, I. K., Borrell, Y. J. and Garcia-Vazquez, E. (2014). DNA barcoding reveals a high level of mislabelling in Egyptian fish fillets. *Food Control*, 46, 441–445.

Gander, P. (2015). *Technology Case Study: Portable, Real-Time Nutrient Analysis for Consumers.* tellspec.com/wp-content/uploads/2015/04/NutriNB-APR07_Tellspec.pdf (accessed June 13, 2019).

Gay, J. (2016). *Interview: OriginTrail, an App that Tells You Where Your Food is from.* FoodBev Media. https://www.foodbev.com/news/interview-origintrail-the-app-that-tells-you-where-your-food-is-from/ (accessed June 28, 2019).

Gilmore, L. A., Frost, E. A. and Redman, L. M. (2014). The Technology Boom: a new era in obesity management. *Journal of Diabetes Science and Technology*, 8(3),596–608.

Hamilton, T. (2018). *Can this Device Actually Tell You What is in Your Food?* https://www.theglobeandmail.com/report-on-business/rob-magazine/this-tech-device-might-ruin-dessert-forever/article30147250/ (accessed June 20, 2019).

Jacobsen, B. (2018). *Food Scanners – Will They Change Your Diet?* https://www.futuresplatform.com/blog/food-scanners-will-they-change-your-diet-calorie-molecular (accessed May 13, 2019).

Kovacs, K., Bazar, G., Darvish, B., Nieuwenhuijs, F. and Hoffmann, I. (2017). Simultaneous detection of melamine and urea in gluten with a handheld NIR scanner, OCM 2017. In: *3rd International Conference on Optical Characterization of Materials* (pp. 13–23).

Lee, N. (2019). *AIRO Wristband Tracks Not just Sleep, Exercise and Stress but also what You Eat.* https://www.engadget.com/2013/10/28/airo-wristband/?guccounter=1&guce_referrer=aHR0cHM6Ly93d3cuZ29vZ2xlLmNvbS8&guce_referrer_sig=AQAAAGOCw5BjnQ7Jfi7FrZNF53jZGLt1CiW43BQruyQ918T-6UILK_x6l2VJhyzGiLRb09PGMar0gen4boiZ5i4mjmjZrZV7HlGmMwveZiWusfz0DfgeOCRArsGM9Ccy_SYdyBSNUhgr-ykUNx__16cjdpWoAKr7ftqLmRdSo-7xqLoE (accessed June 13, 2019).

Li, S. (2015). *Expanding Technological Frontiers to an Everyday Level.* http://princetoninnovation.org/magazine/2015/04/26/scio/ (accessed May 10, 2019).

The Medical Futurist. (2016). *The Fascinating World of Food Scanners.* https://medicalfuturist.com/food-scanners (accessed June 17, 2019).

Murphy, S. (2016). *Magical Gadget Scans Your Food to Reveal Its Nutritional Value.* https://mashable.com/2016/01/04/dietsensor-scio-scans-food/ (accessed June 13, 2019).

Rateni, G., Dario, P. and Cavallo, F. (2017). Smartphone-based food diagnostic technologies: a review. *Sensors.* 17, 1–22.

Sandhana, L. (2013). *TellSpec Hand-Held Scanner Identifies What's in Your Food.* http://newatlas.com/tellspec-food-source/30221/ (accessed June 13, 2019).

Scotter, C. N. G. (1997). Non-destructive spectroscopic techniques for the measurement of food quality. *Trends in Food Science & Technology*, 8, 285–292.

Staffen, C. F., Staffen, M. D., Becker, M. L., Lofgren, S. E., Muniz, Y. C. N., Ache de Freitas, R. H. and Marrero, A. R. (2017). DNA barcoding reveals the mislabeling of fish in a popular tourist destination in Brazil. *Peer J*, 5, e4006. DOI: 10.7717/peerj.4006

Szondy, D. (2019). *Pocket Scanner Blasts Food with Infrared Light to Determine Its Freshness.* https://newatlas.com/pocket-food-scanner-freshness/57866/ (accessed May 11, 2019).

Thong, Y. J., Nguyen, T., Zhang, Q., Karunanithi, M. and Yu, L. (2017). Predicting food nutrition facts using pocket-size near-infrared sensor. In *2017 39th Annual International Conference of the IEEE Engineering in Medicine and Biology Society (EMBC)* (pp. 742–745).

Zaidan, S. and Roehrer, E. (2016). Popular mobile phone apps for diet and weight loss: a content analysis. *JMIR Mhealth Uhealth*, 4(3),1–10. https://cdn2.hubspot.net/hubfs/4905262/Assets/

Application%20Notes/SE_ApplicationNotes_SmartHomes_FoodScanner_v04.pdf (accessed May 13, 2019).

Horizon Prize for a Food Scanner. https://ec.europa.eu/info/research-and-innovation/funding/funding-opportunities/prizes/horizon-prizes/food-scanner (accessed May 14, 2019).

Food Scanners Market Analysis by Growth, Emerging Trends and Future Opportunities till 2024. https://markettalknews.com/food-scanners-market-analysis-by-growth-emerging-trends-and-future-opportunities-till-2024/ (accessed May 14, 2019). https://nimasensor.com/shop/products/nima-starter-kit (accessed June 24, 2019).

Improving Trust in the Fish Chain: Rapid and Portable Monitoring Tools for Better Control of While Fish. http://tellspec.com/eit-food/fishproject/ (accessed May 14, 2019).

Mobile Application for the Enterprise Scanner. http://tellspec.com/food-retail/ (accessed May 14, 2019).

https://www.allergyclinic.co.za/food-allergies/ (accessed May 13, 2019).

https://www.consumerphysics.com/scio-for-consumers/ (accessed May 14, 2019).

SciO for the Consumers: It's Sci-fi at Your Fingertips. https://www.consumerphysics.com/scio-for-consumers/ (accessed May 14, 2019).

Tang ML. Mullins RJ. 2017. Food allergy: Is prevalence increasing? *Internal Medicine Journal,* 47 (3): 256–261

TellSpec Scanner Could Save Your Life by Identifying Food Allergen, Chemicals. https://www.digitaltrends.com/cool-tech/tellspec-food-scanner/ (accessed June 13, 2019).

https://www.foodallergy.org/ (accessed May 14, 2019).

TellSpec Spectroscopy: Food Sensor Innovation. https://www.jameco.com/Jameco/workshop/RollCall/tellspec-spectroscopy-food-sensor-innovation.html (accessed May 22, 2019).

Our Science – A Lab in Your Pocket. https://www.nimasensor.com/science-nima-understanding-device/ (accessed June 24, 2019).

Food Scanners Market Overview. https://www.researchnester.com/reports/food-scanners-market-global-demand-analysis-opportunity-outlook-2024/370 (accessed June 17, 2019).

https://www.spectralengines.com/products/nirone/scanner/foodscanner (accessed May 14, 2019).

Part VI: **Food business: entrepreneurship and regulation**

Oluwatoyin Oluwole, Ahmad Cheikhyoussef, Fatima Raji,
Olaide Akande, Oyedeji Ajibola Bamikole,
Oluwatosin Ademola Ijabadeniyi

19 Food entrepreneurship: principles and practice

19.1 Introduction

The United Nations Industrial Development Organization (UNIDO) (1999) defined entrepreneurship as "the process of using initiative to transform business concept to new venture, diversify existing venture or enterprise to high growing venture potentials." It is believed that entrepreneurship drives several benefits to national development by driving innovative ideation and improving productivity, encouraging critical thinking and competitiveness, and fostering socioeconomic development (Acs, 2008; Parker, 2009; Bolzani et al., 2015). It is becoming general consensus that identification and pursuit of opportunities in entrepreneurship is an integral part of business (Lans et al., 2017), nevertheless, there is yet to be a distinct definition for agricultural entrepreneurship because its scope is wider than development of agribusinesses, spanning to non-agro-based products and services as well as new product development from agricultural outputs, their distribution, and marketing (Pindado and Sánchez, 2017; Dias et al., 2019).

Agricultural entrepreneurship is directly linked to food sector as it is the main source for food products. Food sector is evidently a significant component for economic development globally and currently, challenges such as food supply (Moy 2018), food security (Cole et al., 2018), and food waste (Andler et al., 2018) are common in the food industry, presenting huge gaps and potential opportunities for entrepreneurs who may wish to innovatively combat these challenges (Kuckertz et al., 2019).

Oluwatoyin Oluwole, Fatima Raji, Olaide Akande, Food Technology Department, Federal Institute of Industrial Research Oshodi, Lagos, Nigeria
Ahmad Cheikhyoussef, Science and Technology Division, Multidisciplinary Research Centre (MRC), University of Namibia, Windhoek, Namibia
Oyedeji Ajibola Bamikole, Department of Biotechnology and Food Technology, University of Johannesburg, Doornfontein Campus, Gauteng, South Africa
Oluwatosin Ademola Ijabadeniyi, Department of Biotechnology and Food Technology, Durban University of Technology, Durban, South Africa

https://doi.org/10.1515/9783110667462-019

19.1.2 Who is a food entrepreneur?

Entrepreneurs are not only people starting new businesses or who invest in money. By definition, an entrepreneur is a person sensitive enough to identify a problem which could serve as a business opportunity and taking up the responsibility of creating a value chain system to transform the problem to opportunities. Such a person is comfortable taking informed and calculated risks, organized, determined, and confident of positive outcomes (Stephen, 2016). Entrepreneurs are often leaders, who are capable of innovations and are driven primarily by the desire to meet consumers' needs, while growing the business and expanding market opportunities (Stephen, 2016). This possibly implies that we are all entrepreneurs in our homes and most importantly in our kitchen because we ensure that the food consumed in our homes are hygienically prepared and contains important nutrients needed for healthy living. To have a standard definition for entrepreneur is a bit challenging as it could be inadequate to define an entrepreneur as someone who established a new organization without including the considerations of the distinct qualities such as identification of opportunities, confidence in taking risks, and determination to see set goals and execute them (Shane and Venkataraman, 2000).

19.2 Reasons for entrepreneurship

There are four categories of reasons for entrepreneurship in Figure 19.1.

Figure 19.1: Reasons for entrepreneurship diagram.
1. Being your own boss: self-management is the motivation that drives many entrepreneurs.
2. Financial success: entrepreneurs are wealth creators.
3. Job security: large companies have eliminated more jobs than they have created over the past 10 years.
4. Quality of life: starting a business gives the founder some choice over when, where, and how to work.

19.3 Starting a new venture in food entrepreneurship

Food entrepreneurs have to ask, plan, and consider many factors before starting any new project, these factors can be categorized under three steps:

19.3.1 Potential business ideas: these could involve the following

1. Food producer
2. Restaurant
3. Food processing
4. Frozen foods
5. Organic foods
6. Food on wheels
7. Weight loss or dietary foods
8. Baking or cooking lessons
9. Food blogging

19.3.2 Feasibility tests of some food business ideas: these should include

1. Ask your local farm shop owners what they would like to see on their shelves?
2. Use social media (e.g., Instagram, Twitter, and Facebook) and ask people what they are interested in? Upload photos and videos of different food ideas you have had and see which ones get the warmest response.
3. Use the media (e.g., television, radio, magazines, and newspapers) to seek and dish out information of current food trends.
4. Find out how existing products can be improved upon in terms of packaging, flavor, and nutrition. These are big selling points especially for novel food products that are just entering the market.
5. Attend food networking events such as food expos, food shows, festivals, farmers' markets, and trainings.
6. Make a list of everyday food and drink and make resolutions about how they could be improved upon.
7. Explore the science of nutrition in a short course.
8. Sign up for advanced-level cookery lessons for those to operate restaurants.
9. Ask family and friends about ideas on food for thoughts.

19.3.3 Familiarization with rules, regulations, and local laws

To start a cottage food business, an entrepreneur must contact the local government council/authority for necessary information and actions on permits, licenses, and local zoning laws for operating a business. In other climes, there is need to check with local health department about being a street vendor. In Nigeria for example, it is quite difficult to do that since almost every street, corners, and make-shift-shop are selling space for street vendors.

However, in order to receive a license to operate, preparation area will need to pass through inspection. This is done by staff of the Local Government Councils. But there still exist one thousand and one other vendors who keep their wares anywhere (near the toilet, by the sewage, on the drainage, etc.). These vendors only come out and sell food to desperate consumers who overlook these unhealthy environments. Regulatory inspection by responsible government department is necessary to prevent such vendors from operating.

In the case where zoning or cooking area regulations do not allow for indiscriminate food preparation, commercial cooking spaces are used while some other business tasks can be managed from home. Packaging of food items will require permission from the State's Health Department of the National Agency for Food and Drug Administration and Control (NAFDAC). Packaged items could be required to carry a standard recipe, nutritional information, expiry date, and NAFDAC registration number on the label.

19.4 Reasons to consider food entrepreneurship

The ever-increasing population and increased consumer awareness about quality and safety of foods makes food business a lucrative venture for the future. There are few reasons food businesses will continue to be relevant and become more popular in the coming years. These include the focus of consumers on health-related benefits of foods, the increasing need, and crave for minimal cooking with a very low level of inconvenience (tilt toward convenience foods), awareness about affordable organic food options, increase in life pace, leading to demand for fast and quickly delivered foods, consumers' drive to learn more about what they eat and their willingness to expose others to these knowledge.

19.5 Entrepreneurship investment to move food value chain forward

It is already known that food create business opportunities that drive the agricultural value chain forward thereby creating different opportunities for the producers and ensuring that the farmers get good returns from their operations. It implies therefore that a lot of investments should be directed toward food processing business development and food product marketing. Such investments will be more effective in job creation and expansion of the value-added products. It will also be beneficial for the earning of foreign exchange from the export of value-added food products. Food scientist and technologist have a critical role to play to ensure the success of some of the objectives of the new agriculture policy as indicated above. They are in possession of the scientific knowledge required for the conversion of raw agriculture produce into value-added food products through the know-how for preservation, storage, packaging, and handling of food materials. Kuckertz et al. (2017) emphasized that the concept of entrepreneurial opportunity which is central to entrepreneurship and economic theory could suggest number of sources of such opportunities that help to structure the food industry and to add value to the food chain.

These economic sources for entrepreneurial opportunities according to Kuckertz et al. (2019) are:

1. Whenever customer demand changes, opportunities for entrepreneurs to cater to these new demands emerge.
2. Changes in supply, such as newly developed enzymes or flavorants, offer food entrepreneurs opportunities to remodel their value chain.
3. Information asymmetries that expose entrepreneurial opportunities for those entrepreneurs to be able to address them.
4. Exogenous shocks to the market are likely to present the most interesting entrepreneurial opportunities.

19.6 Current food trends

As a food entrepreneur working with the current trend in food innovation is critical for product success. For instance, people are now health conscious such as in terms of calories, cholesterol content of the food they consume, consumers are more inclined to gluten-free, organic foods, sugar-free, reduced salt, less oil/fat, healthy nuts and seeds such as chia seeds, almond, walnut, pistachios, and so on. Ecofriendly packaging materials also attract consumers, and this is significant in product development (Aramouni and Deschenes, 2015).

Approach to food safety is from farm (during planting and harvesting) through logistics (transport and handling) and processing to storage and utilization (preparation and preservation) by consumers. (WHO, 2015). In 2018, some traditional food chains have given ways to technological advancement as seen in the current research and development on the use of stem cells to produce laboratory meats in developed countries, which in few years from now will fully flood the market (Zegler, 2018).

Globally, food delivery sector is estimated to be an €83 billion market, a mere 1% of the entire food value chain (Hirschberg et al., 2016). The food industry has attained maturity in many countries, with annual growths of about 3.5% projected for the next 5 years. Also, the upsurge of digital technology is expanding the scope of the food industry as consumers are exposed and comfortable with the use of online platforms and web applications for shopping food items, with the advantage of convenience, transparency (Hirschberg et al., 2016). In Nigeria, online entrepreneurs like Jumia (one of the largest online marketplaces) are taking up the orders and delivery of food for consumer's convenience. Now the sector is getting competitive as individual food entrepreneurs are taking up the challenge too. Another trend is the application of food biotechnology tools and techniques such as genetic engineering, enzyme catalysis, and system-based biochemical processes in the development of novel, genetically engineered and laboratory cultured foods. Food biotechnology is quickly gaining global impact, as many tradition food processing operations such as wine beer and cheese technologies have been replaced by food biotechnological approaches, leading to higher product yield in reduced time (Popa et al., 2019).

19.7 Stages in food entrepreneurship

The metamorphosis of stages involved in the development of a new food product from ideation to a physically sellable item requires that informed choices regarding the business' visions and goals be clearly stated and understood. The vital stages for a new food entrepreneurship include development of business plan, market survey and analyses, development of organizational structure and arrangement of business divisions, production and distribution plan, sales estimation, market and financial analyses. Business plan, among the various requirements of a new food entrepreneurship is very vital. It is a document which contains details of the goals and visions of the business and serves as a reference throughout the business establishment process. A business plan is based on facts and figures obtained from extensive market research and survey and not on opinions or feeling. It also helps to clearly define the business to regulatory agencies, as well as the required licenses for practice of the business. It is important to painstakingly develop a business plan and conduct thorough market survey, as this will reveal the profitability potentials and possible break-even points of the business and helps the entrepreneur make informed decisions. A business plan must be continually updated before, during, and after the establishment of the food business, in order to be in tune with the current realities.

19.8 Food product development

The essential reason for a new food product is the satisfaction of consumers both in appeal and nutrition. Hence, a new food product must be nutritious, safe, and satisfy the consumers' desires. These attributes of a new food product must be achieved without jettisoning food safety requirements and regulatory guidelines. Food product development entails a logical research to develop products and methods substantial to meet a consumer's need. It involves the combination of applied and social science skills for food processing and food marketing.

According to Cooper and Kleinschmidt (1986), four stages are key in product development and these include: product plan development, product design and development, product commercialization, and product unveiling and promotion (FAO, 2006). In the development of a new food product, the roles of the producer, distributor, and the consumers are very important. The producer has to be creative and innovative with the product, the composition, function, and benefits of the product and the packaging. While the distributor has to be able to exhibit great marketing skills to promote the products to meet all the food chain stakeholders and the requirement needed for its marketability. The consumer plays the ultimate role as the final user in the food value chain and thus the influence of the producer and distributor helps in determining the food choices.

In product plan development, initial idea screening will be carried out followed by preliminary market assessment to ascertain its market acceptability. Thereafter, a detailed market research will be conducted using questionnaires or by conducting mini interviews on consumers and distributors. Finally, the outcomes of these will lead to development of the product concept and the financial feasibility study. Feasibility consideration for a business includes regulations, policies, technology, and finances (Aramouni and Deschenes, 2015). During product design and development, bench-top or laboratory scale production will be done, which will serve as a prototype design. During this stage, the production preliminary trials will be done, analysis will be carried out and eventually the procedure will be standardized. Bench-top testing will be conducted afterwards a comprehensive sensory evaluation will be done. Shelf life and safety studies will be conducted on the developed product to ascertain its safety for human consumption. On successful completion of this stage, scaling up of the production process is done (FAO, 2006).

Product commercialization involves pilot-scale production of the developed product and consumer acceptability test. Product unveiling and promotion involves promotion run, logistics, product launch, sales and marketing promotion, and financial and operational analysis. According to Tetra Pak (2004), certain factors predict the success of a new product in the market. Such factors therefore can serve as guide in determining what product to develop. Consumers want well-detailed important product information, they prefer novel products which are better than what is currently in the market, good taste, color, packaging, nutritional value, and functionality. More specifically, during the last decade, consumer requirements in the field of food production have changed considerably, whereby consumers are increasingly believing that food contributes directly to their health (Mollet and Rowland, 2002).

19.9 Product packaging

In food entrepreneurship, the package of a food product is paramount for its safety, shelf life, and ease of distribution. For a food product, the producer should consider a primary, secondary, and tertiary packaging material as applicable to the type of the developed product. A food package should contain, store, present, and protect food products from contamination and mechanical damage from handling. Factors such as the mechanical strength of packaging material, its barrier properties, cost, and availability must be critically considered before adopting such material for food packaging (Hanlon et al., 1998)

The integrity of the product should be kept till it meets the final consumer thus packaging testing should be conducted before the appropriate material is chosen. In determining the appropriate packaging material, the food properties should be considered. Is the product in solid, liquid, free-flowing, powdery, or gel-like form? This helps to inform the right packaging material that will preserve the quality of the product. Second, determination of the functional properties of the product, its compatibility, its brittleness, the product's ability to break or squeeze or smear will assist in packaging type and choice (Brod et al, 2001). The packaging marketing design also dictates what packaging material is required. For instance, will the product be refrigerated, placed on shelf, or displayed on a hanger as in the case of sweets? Legal and regulatory requirements also infer packaging choice, for instance food products containing additives or fortificants need proper packaging to preserve the quality, nutritional composition, and aesthetic appearance of the product. Considering the cost estimated and incurred in packaging, the price of the product should suffice for it. Before the product is exposed to the market, the packaging integrity tests should be conducted to ascertain the strength and weakness of the packaging material (Harper et al., 1995). The commonly used packaging materials for food products include papers, pouches, cardboards, polypropylene, polyethylene, biaxially oriented polypropylene, stainless steel, aluminum, cans, metallic foils, metallic films, glass, and plastics (Aramouni and Deschenes, 2015).

A food package label creates an interface between the food business and consumers. A good label must be appealing and attractive to consumers, with respect to the kind of food product and its intended consumers. Labels contain vital information about the food product including its picture or image, ingredients' quantities, gross and net weights, nutritional information, peculiar processing operations, regulatory endorsements, and other information that will convince consumers of the quality of packaged products and may persuade them to buy. Labeling of food products on packaging materials must be appropriately carried out, in line with the guidelines of the local and international regulatory agencies responsible for such product. Information of food labels must not be false or misleading and should not misrepresent the product in the package. In countries where regulatory agencies approve food labels, food businesses must submit the designs of an intended label of a new food product for approval.

19.10 Food hygiene and safety

Food businesses must offer consumers with food of suitable quality which is safe to eat. Hygiene and food safety are critical issues for all food businesses. Food business should adopt the Hazard Analysis Critical Control Point (HACCP) during the product production process. This is to ensure that customer's satisfaction is achieved in terms of wholesomeness, safety, and food laws must be implemented strictly. Good manufacturing practices (GMPs) must be followed during production to ensure food safety. Food spoilage and food hazards causes can be evaluated so they can be avoided. Thus, controlling food safety and hygiene in each critical stage in any food production is crucial.

Through the food production flow process, the integrity of products can be compromised at different stages or points. To this end, food businesses must be proactive in identifying these possible hazard introduction points, identify these hazards and design plans to prevent, minimize, or critically control hazard the introduction of contaminants at these points (Sperber, 1998). For food safety considerations, hazards introduced into foods can be broadly classified into three; physical hazards which include dirts from poorly processed foods, broken glasses, sand, metal filings; biological hazards such as pathogenic microorganisms, pests such as rodents and insects, spoilage microorganisms; chemical hazards which are pesticide residues, excess ingredients, poorly washed cleaning agents, etc.

Since there is a huge risk of contamination in a food production process, a food business must anticipate these hazards and make plans to prevent their occurrence, in order to produce safe foods. Van Kleef (2006) highlighted the considerations for reduction of food safety hazards in the predevelopmental stage of food businesses. They include the correct choice of production facility, process flow for desired product, ingredient formulations to use, and regulatory procedures.

Also, throughout the production process, processing conditions must be regulated and strictly controlled to prevent the introduction and proliferation of food safety hazards. Among the factors to be considered, the following are included: regulation of acidic environment of food production process, physical conditions such as time, temperature, water activity, and the hygiene of handlers must be closely monitored and constantly checked to ensure compliance to quality and safety requirements.

19.10.1 Good Manufacturing Practices

GMPs are carefully written guiding principles by regulatory agencies regarding sensible practices that must be followed by food manufacturers to produce quality, wholesome, and safe foods. GMPs are required to be followed, irrespective of the scale of the food business and the product being manufactured. Government quality and safety audit agencies are always on the look out for compliance to these practices in food manufacturing business and any deviation poses huge business risks including closures and forfeiture of licenses. A good GMP program must give guidelines for buildings and facilities, personnel, sanitary operation, production processes, and equipment. These sections must be adjudged satisfactorily by a quality assurance personnel before production could be allowed on a processing line and this must be done routinely, depending on the sensitivity of food products (Dias et al., 2012).

To ensure adequate preparation and readiness of these various sectors of food manufacturing process, documents detailing the step-by-step instructions to be followed for a process should be made available to personnel, to serve as reference document. Log books for operational stages, specified processing equipment, and analytical instruments must also be made available to track access and enhance trouble-shooting. Proper routing records of the production process should also be adequately taken to easily identify areas of critical control and quickly employ preventive or control measures. Virtually all food categories including heat-treated foods, cold or frozen foods, ready to eat foods or foods prepared for vending equipment and genetically modified foods require that these requirements be adequately monitored (Blanchfield, 2005).

19.10.2 Hazard Analysis and Critical Control Points

The Codex Alimentarius indicated the HACCP principles as the most effective tool in ensuring food safety worldwide (EC, 2016; FAO/WHO, 2004). HACCP requires the identification of possible stages or points of contamination or cross-contamination in a food process flow, the potential hazards, and the control measures that could be applied to prevent the occurrence of such hazards. A well-designed HACCP plan I crucial to the successful implementation of the program (Dzwolak, 2019; Wallace, Halyoak, Powell and Dykes, 2014).

Ingredients of an HACCP design are put together in an HACCP plan which is a process flow chart that helps to identify, control, and properly document the results of hazard analyses. The ingredients or steps required for an HACCP plan include hazard analysis; determination of critical control points; establishment of critical limits for hazards, establishment of monitoring procedures, establishment of corrective actions, design, and implementation of verification procedures for corrective actions, record keeping, and documentation mechanisms.

Although, after several decades of implementing HACCP principles and successes recorded worldwide, there are some challenges militating against the application of HACCP plans in many food businesses. These challenges include management commitment, financial constrains, regulatory lapses, educational, knowledge, and psychological demands for effective implementation.

19.10.3 Food Safety Modernization Act

Food Safety Modernization Act (FSMA) was signed into law in January of 2011 and marked a major shift in the Food and Drug Administration's (FDA) approach to food safety from outbreak response to prevention-based controls (Adalja and Lichtenberg, 2018). This Act enable the agency authority to require the use of sanitation measures in growing, harvesting, packing, and holding of fresh fruits and vegetables in order to reduce the incidence of food-borne illness, of which a large share have been attributed to fruits and vegetables (Painter et al., 2013). FSMA is an act intended to be more proactive rather than reactive to food safety issues. Through FSMA food processors and producers who currently do not fall under categories additional regulations such as HACCP or canned food guidelines would be required to formulate a system or plan called preventative controls (PC) for their food (Grover et al., 2016). The details of what is expected or required for such PC plan are not yet known as they are still being proposed and amended. One proposal for the regulation includes a Food Safety Plan. Controls during the production process serve to eliminate or reduce widespread illness by reducing the possibility of pathogen survival in the product, eliminating allergen exposure, ensuring and recording proper sanitation practices, and preparing for the event of a recall. The legislation is still being amended, however, businesses are encouraged to stay astride or ahead of these proposed changes as to prevent shock when they are finally installed by starting to think about food safety planning (Grover et al., 2016). FDA (2015) introduced officially what so known as standards for the growing, harvesting, packing, and holding of produce for human consumption. The Produce Rule was finalized in November of 2015 and became effective from January of 2016. These standards established the agricultural production to cover the following areas: (1) agricultural water; (2) biological soil amendments of animal origin; (3) health and hygiene; (4) intrusion of domesticated and wild animals; and (5) sanitation of equipment, tools, and buildings (Adalja and Lichtenberg, 2018).

19.11 Logistics and distribution

With a viable business idea, an entrepreneur can prepare food from home or solely in a business space, investment, and delivery in supplies and equipment can then be executed.

Logistics concentrates on the flow of goods and services along the value chains linking primary producers, processors, manufacturers, retailers, and consumers. Logistics encompasses getting the right product, in the right quantities, to the right place, at the right time and cost, to meet the demands of successive customers and product sustainability. Supply chain has recently received heightened attention because of its potentials for greater utilization of resources and lighter adverse effects on the environment (Gružauskas, Gimžauskienė and Navickas, 2019). Sustainability involves a combination of ecological, social, and financial responsibilities for future generations to satisfy their needs (Amui et al., 2017). Logistics activities are unified to provide the necessary quality and customer service for the minimum possible cost. Logistics and distribution involve organization, execution, and monitoring efficient and operational movement and storage of products, including the return of materials for reuse, recycling, or disposal. The improvement of alignment of demand and supply would not only result in the improvement of inventory levels, but also in more stable agricultural processes (Gružauskas, Gimžauskienė and Navickas, 2019). Thus, the first in first out (FIFO) principle should be adopted for proper product monitoring, recording, and traceability.

Financial demands for manufacture of food products vary. They include labor demands in terms of man-hour rates to pay and the time invested in business and technical logistics; direct of ingredients, inputs and packaging materials; and operation costs, including those incurred by buying machineries and equipment, production wastages, rent, storage, and distribution fees and insurance.

Quality and safety of foods must not be jeopardized in the food distribution chain, irrespective of other constrains that may be imminent in the food business. Sustainability in agro-food supply chain also involves that the employees' health and safety must be ensured (Mangla et al., 2019).

19.12 Consumer preferences

High-quality and safe food are always the top priorities in food industry. To determine consumer preferences, the target of the product should be known. To maintain the consistency and accuracy of consumer's preferences for quality and safe foods has become the focal point of research, in order to guide policy framework formation, implementation, and business strategies (Goldstein and Einhorn, 1987; Seidl, 2002; Chen et al., 2020). There are several factors that infer consumer choices such as age (children, adults, and the elderly), religion, income, and ethnicity. Consumers' choices of what they consume have become more informed and systematic, guided by increased knowledge of food and their benefits, societal norms, awareness of the environment, and ethical considerations. (Ladhari and Tchetgna, 2017).

19.13 Customer relationship

An entrepreneur should learn to build and develop buyer and seller relationships and apply the various analytical techniques to assess their relevance in different market and competitive environments. The effective management of suppliers and buyers is seen to be critical to the entire operation of the business. The quality, technology, cost, and delivery performance have ability to meet the challenges of competitors and the demands of customers. An entrepreneur could make a list of local companies, organizations, and events where complimentary samples of products, gift baskets, or small catered meals can be distributed. Testing of the food is a function of the entrepreneur getting close to the consumers. And once the sample food is accepted, the entrepreneur can be contacted for orders. Orders might come in a bit slowly but with time there will be boost in sales. Furthermore, an entrepreneur will get more sales by contacting groups of people rather than individuals. This will make it possible for more people to be reached at a time furthermore larger orders can be secured in future. Furthermore by building a good customer relationship, you will be effectively able to market your products. According to Fried and Hansson (2010), marketing is something everyone in an organization should do 24 h, every week and 365 days a year. A company is indirectly marketing herself by seemingly little things such as the way the phone is answered or an email is sent (Fried and Hansson).

19.14 Challenges of food entrepreneurs

The reason why food entrepreneurs sometimes have challenges with their proposed innovative product or probably encounter product failure could be due to overly high prices because they want to breakeven just at the start of business, poor, or insufficient marketing strategies and planning, unrealistic expectations, technical difficulties with machineries, logistics bottlenecks, tight market competition, and poor product market penetration and distribution (Ilori et al., 2000, FAO, 2006). Kapinga and Montero (2017)'s work also revealed that women food entrepreneurs in Tanzania face the problems of capital shortage and limited market access. However, they were able to overcome the identified challenges through creation of economic groups and entrepreneurship clubs.

19.15 Succeeding as a food entrepreneur

Menrad (2004) indicated that innovations are understood as new products, processes, and services, which are recognized as an important instrument for companies to stand out from competitors and to satisfy their consumer expectations. To succeed as a food entrepreneur, product development cycle time must be reduced that is, the producer should have a competitive advantage over the market in order to be dominant and have an increased market share value as well as investment turnover (Ref). Furthermore, increased product development innovation should be adapted so that the product gains consumer base by increasing production line extensions through successive product innovations. Entrepreneurs can also reuse existing knowledge, work on product errors, and rebrand (Moskowitz et al., 2009). The drive for development of new food product is usually birthed by the drive to develop more nutritious and functional products or the quest to create replacement or competitive food products for already existing ones. Newly introduced food products are often improved, with greater consumer appeal due to improvement in food processing technologies such as product formulation, engineering, and recent information about consumer needs and preferences. (Bigliardi and Galati, 2013). A classification of the food innovations was proposed by Bigliardi and Galati (2013) categorizing innovations food industry as the following:
1. New food ingredients and materials
2. Innovations in fresh foods
3. New food process techniques
4. Innovations in food quality
5. New packaging methods
6. New distribution or retailing methods

Apart from staying relevant through innovation, a food entrepreneur needs market strategy to continue to grow in leap and bounds because it helps to promote the business, target the right clients, and allocate resources wisely. An effective market strategy entails defining the product and highlighting the benefits; clarity about the strength and weakness of the product; knowledge of the competitor; determination of target market; development of a vision for the sales of the product; and finally, establishment of a budget for marketing, promotion, and advertising.

19.16 Conclusion

Most food entrepreneurs will not be food scientists and technologists or nutritionists but entrepreneurs who will need the knowledge base of food scientists and nutritionists for research results, technology out-scaling, recipes formulation, food safety controls, food quality management, and process yield/performance optimization. Without the provision of these capabilities, the food entrepreneurs cannot operate profitably and find their ways in food industry. Coupled with that food entrepreneurs must be leaders and innovators who are striving to remain curious, creative, competent, courageous, and confident marketers.

References

Acs, Z. J. (2008). Foundations of high impact entrepreneurship. *Foundations and Trends in Entrepreneurship*, 4(6), 535–620.

Adalja, A. and Lichtenberg, E. (2018). Implementation challenges of the food safety modernization act: evidence from a national survey of produce growers. *Food Control*, 89, 62–71.

Amui, L. B. L., Jabbour, C. J. C., de Sousa Jabbour, A. B. L. and Kannan, D. (2017). Sustainability as a dynamic organizational capability: a systematic review and a future agenda toward a sustainable transition. *Journal of Cleaner Production*, 142, 308–322.

Andler, S. M. and Goddard, J. M. (2018). Transforming food waste: how immobilized enzymes can valorize waste streams into revenue streams. *npj Science of Food*, 2, Article 19.

Aramouni, F. and Deschenes, K. (2015). *Methods for Developing New Food Products. An Instructional Guide*. Lancaster, PA: DEStech Publications, Inc.

Bigliardi, B. and Galati, F. (2013). Innovation trends in the food industry: the case of functional foods. *Trends in Food Science & Technology*, 31, 118–129.

Blanchfield, J. R. (2005). Good manufacturing practice (GMP) in the food industry. In *Handbook of Hygiene Control in the Food Industry* (pp. 324–347). Elsevier Inc. DOI: 10.1533/9781845690533.3.324

Bolzani, D., Carli, G., Fini, R. and Sobrero, M. (2015). Promoting entrepreneurship in the agri-food industry: policy insights from a pan-European public–private consortium. *Industry and Innovation*, 22(8), 753–784.

Brody, A. L., Strupinsky, E. P. and Kline, L. R. (2001). *Active Packaging for Food Applications*. CRC Press.

Chaoniruthisai, P., Punnakitikashem, P. and Tajchamaha, K. (2018). Challenges and difficulties in the implementation of a food safety management system in Thailand: a survey of BRC certified food productions. *Food Control*, 93, 274–282.

Chen, E., Flint, S., Perry, P., Perry, M. and Lau, R. (2015). Implementation of non-regulatory food safety management schemes in New Zealand: a survey of the food and beverage industry. *Food Control*, 47, 569–576.

Chen, X., Gaob, Z. and McFadden, B. R. (2020). Reveal preference reversal in consumer preference for sustainable food products. *Food Quality and Preference*, 79, 103754. https://doi.org/10.1016/j.foodqual.2019.103754.

Cole, M. B., Augustin, M. A., Robertson, M. J. and Manners, J. M. (2018). The science of food security. *npj Science of Food*, 2, Article 14.

Cooper, R. D. and Kleinschmidt, E. J. (1986). An investigation into the new product process: steps, deficiencies, and impact. *The Journal of Product Innovation Management*, 3(2), 71–85.

Corlett, D. A. (1998). *HACCP User's Manual*. Springer Science & Business Media.

Dias, C. S. L., Rodrigues, R. G. and Ferreira, J. J. (2019). Agricultural entrepreneurship: going back to the basics. *Journal of Rural Studies*, 70, 125–138.

Dias, M. A., Sant'Ana, A. S., Cruz, A. G., José de Assis, F. F., de Oliveira, C. A. and Bona, E. (2012). On the implementation of good manufacturing practices in a small processing unity of mozzarella cheese in Brazil. *Food Control*, 24 (1–2), 199–205.

Dzwolak, W. (2019). Assessment of HACCP plans in standardized food safety management systems – the case of small-sized Polish food businesses. *Food Control*, 106. DOI: 10.1016/j.foodcont.2019.106716.

EC, European Commission. (2016). Commission notice on the implementation of food safety management systems covering prerequisite programs (PRPs) and procedures based on the HACCP principles, including the facilitation/flexibility of the implementation in certain food businesses. *Official Journal of the European Union*, C278, 30.7, 1–32.

FAO. (2006). *Food Product Innovation: A Background Paper*. Winger, R and Wall, Gln: Agricultural and Food Engineering Working Document.

FAO/WHO. (2004). *FAO/WHO Guidance to Governments on the Application of HACCP in Small and/or Less-Developed Food Businesses*. FAO Food and Nutrition Paper 86. http://www.fao.org/3/a-a0799e.pdf (accessed October 23, 2019).

FDA, Food and Drug Administration. (2001). *HACCP: A State-of-the-Art Approach to Food Safety, 2001*. http://vm.cfsan.fda.gov/~1rd/bghaccp.html (accessed September 12, 2007).

Food and Drug Administration. (2015). *Standards for the Growing, Harvesting, Packing, and Holding of Produce for Human Consumption*. 80 FR 74353. https://www.fda.gov/media/117414/download (accessed October 25, 2019).

Fotopoulos, C. V., Kafetzopoulos, D. P. and Psomas, E. L. (2009). Assessing the critical factors and their impact on the effective implementation of a food safety management system. *International Journal of Quality & Reliability Management*, 26(9), 894–910.

Fried, J. and Hansson, D. H. (2010). *Rework*. London: Vermilion.

Goldstein, W. M. and Einhorn, H. J. (1987). Expression theory and the preference reversal phenomena. *Psychological Review*, 94(2), 236–254.

Grover, A. K., Chopra, S. and Mosher, G. A. (2016). Food safety modernization act: a quality management approach to identify and prioritize factors affecting adoption of preventive controls among small food facilities. *Food Control*, 66, 241–249.

Gružauskas, V., Gimžauskienė, E. and Navickas, V. (2019). Forecasting accuracy influence on logistics clusters activities: the case of the food industry. *Journal of Cleaner Production*, 240, 118225. https://doi.org/10.1016/j.jclepro.2019.118225.

Hanlon, J. F., Kelsey, R. J. and Forcinio, H. (1998). *Handbook of Package Engineering*. CRC Press.

Harper, C. L., Blakistone, B. A., Litchfield, J. B. and Morris, S. A. (1995). Developments in food packaging integrity testing. *Trends in Food Science & Technology*, 6(10), 336–340.

Hirschberg, C., Rajko, A., Schumacher, T. and Wrulich, M. (2016). *The Changing Market for Food Delivery*. https://www.mckinsey.com/industries/technology-media-and-telecommunications/our-insights/the-changing-market-for-food-delivery (accessed October 29, 2019).

Ilori, M. O., Oke, J. S. and Sanni, S. A. (2000). Management of new product development in selected food companies in Nigeria. *Technovation*, 20, 333–342.

Jouve, J. L., Stringer, M. F. and Baird-Parker, A. C. (1998). *Food Safety Management Tools*. Chicago.

Kanothi, R. N. (2009). *The Dynamics of Entrepreneurship in ICT: Raphael Ngatia Kanothi* (Vol. 466). The Hague: Institute of Social Studies (ISS).

Kapinga, A. F. and Montero, C. S. (2017). Exploring the socio-cultural challenges of food processing women entrepreneurs in Iringa, Tanzania and strategies used to tackle them. *Journal of Global Entrepreneurship Research*, 7, 17. https://doi.org/10.1186/s40497-017-0076-0.

Kuckertz, A., Hinderer, S. and Röhm, P. (2019). Entrepreneurship and entrepreneurial opportunities in the food value chain. *npj Science of Food*, 3, Article 6.

Kuckertz, A., Kollmann, T., Krell, P. and Stöckmann, C. (2017). Understanding, differentiating, and measuring opportunity recognition and exploitation. *The International Journal of Entrepreneurial Behavior & Research*, 23, 78–97.

Ladhari, R. and Tchetgna, N. M. (2017). Values, socially conscious behaviour and consumption emotions as predictors of Canadians' intent to buy fair trade products. *International Journal of Consumer Studies*, 41(6), 696–705.

Lans, T., Seuneke, P. and Klerkx, L. (2017). Agricultural entrepreneurship. In: E. G. Carayannis (Ed.), *Encyclopaedia of Creativity, Invention, Innovation and Entrepreneurship* (pp. 1–7).

Leon, I. (1999). *A Guide to Small Business Investments*. Lagos, Nigeria: Impressed Publishers.

Mangla, S. K., Sharma, Y. K., Patil, P. P. and Yadav, G. (2019). Logistics and distribution challenges to managing operations for corporate sustainability: study on leading Indian diary organizations. *Journal of Cleaner Production*, 238, 117620. DOI: 10.1016/j.jclepro.2019.117620

Menrad, K. (2004). Innovations in the food industry in Germany. *Research Policy*, 33, 845–878.

Mohamad, A. S. (2004). The challenges in building a safe and sustainable food industry. Policy and development: the multi-disciplinary bulletin focusing on managing and development and perspectives on national issue. *INTAN*, 1, 29–35.

Mollet, B. and Rowland, I. (2002). Functional foods: at the frontier between food and pharma. *Current Opinion in Biotechnology*, 13, 483–485.

Moskowitz, H. R., Saguy, I. S. and Straus, T. (2009). *An Integrated Approach to New Food Product Development*. Boca Raton, FL: CRC Press Taylor & Francis Group.

Moy, G. G. (2018). The role of whistleblowers in protecting the safety and integrity of the food supply. *npj Science of Food*, 2, Article 8.

Painter, J. A., Hoekstra, R., Ayers, T., Tauxe, R., Braden, C. and Angulo, F., et al. (2013). Attribution of foodborne illnesses, hospitalizations, and deaths to food commodities by using outbreak data, United States, 1998–2008. *Emerging Infectious Diseases*, 19(3), 407–415.

Parker, S. C. (2009). *The Economics of Entrepreneurship*. Cambridge: Cambridge University Press.

Pavičić, N., Rešetar, Z. and Toš Bublić, T. (2013). Recommended contents of business plans and feasibility studies at home and abroad. In: *DIEM: Dubrovnik International Economic Meeting* (Vol. 1, No. 1, pp. 0–0). Sveučilište u Dubrovniku. Organization of the Business-Leadership Division/Structure.

Pindado, E. and Sánchez, M. (2017). Researching the entrepreneurial behaviour of new and existing ventures in European agriculture. *Small Business Economics*, 49(2), 421–444.

Popa, M. E., Mitelut, A. C., Popa, E. E. and Matei, F. (2019). Creating products and services in food biotechnology. In: F. Matei and D. Zirra (Eds.), *Introduction to Biotech Entrepreneurship: From Idea to Business*. Cham: Springer.

Roos, Y. H. (2003). Water activity; principles and measurement. In: M. F. Sancho-Madriz (Ed.), *Encyclopaedia of Food Sciences and Nutrition* (pp. 6089–6094). Elsevier. DOI: 10.1016/B0-12-227055-X/00968-8

Schlimme, D. V. (1995). Marketing lightly processed fruits and vegetables. *HortScience*, 30(1), 15–17.

Seidl, C. (2002). Preference reversal. *Journal of Economic Surveys*, 16(5),621–655.

Shane, S. and Venkataraman, S. (2000). The promise of entrepreneurship as a field of research. *Academy of Management Review*, 25(1), 217–226.

Sperber, W. H. (1998). Auditing and verification of food safety and HACCP. *Food Control*, 9(2–3), 157–162.

Skripak, S. J., Cortes, A. and Walz, A. (2016). *Entrepreneurship: Starting a Business*. Pamplin College of Business and Virginia Tech Libraries. http://hdl.handle.net/10919/70961.

Tetra Pak. (2004). *Company Magazine Number 89*. Lund, Sweden.

United Nations Industrial Development Organization (UNIDO). (1999). *Report*. www.unido.org.

van Kleef, E. (2006). *Consumer Research in the Early Stages of New Product Development: Issues and Applications in the Food Domain*.

Wallace, C. A., Halyoak, L., Powell, S. C. and Dykes, F. C. (2014). HACCP – the difficulty with hazard analysis. *Food Control*, 35, 233–240.

Williams, C. C. and Nadin, S. (2012). Entrepreneurship in the informal economy: commercial or social entrepreneurs? *International Entrepreneurship and Management Journal*, 8(3), 309–324.

World Health Organization Regional Office for South East Asia. (2015). *Food Safety: What You Should Know*. World Health Organization. https//apps.who.int/iris/handle/10665/160165.

Zegler, J. (2018). *Global Food and Drinks Trends 2018*. https://gastronomiaycia.republica.com/wp-content/uploads/2017/10/informe_mintel_tendencias_2018.pdf (accessed December 9, 2019).

Aruwa Christiana Eleojo

20 Food regulations and governance

20.1 Background – a United Nations perspective

In the early parts of first century America, food farming/production was mainly subsistent, and purchases were made from persons whose produce were well-known such that food safety laws were not required. However, over time and with the advent of industrialization, there was a significant reduction in the number of people that could produce foods for themselves. This led to the development and establishment of the food industry to serve the purpose of production and distribution of foods. The initial attempts at providing food regulations and governance were done at the state level from 1850 and beyond and were difficult to execute. The first significant federal food governing law was the Federal Food and Drug Act of 1906 which serve as the foundation for food regulations. Under this law, it was unlawful to sell misbranded or adulterated foods in interstate trading. It also listed illegal chemicals such as formaldehyde or borax, which were not to be used as food additives. The law was, however, weak in defining punishments when the law is broken and had weak techniques to enforce its implementation (AAE, 2010). Food producers are however bound to comply with jurisdictional food quality specifications and standards through practice of food quality control and assurance activities to ensure they meet global standards (Rahmat et al., 2016).

Current Food and Agricultural Organization (FAO) food regulations are derived from international standards and principles channeled toward the regulation of food production (food control), trade, and handling (food safety) of foods. To safeguard global public health, it is imperative that foods consumed remain untampered with and are not subjected to illicit practices aimed at deceiving the consumer (FAO, 2019). Food regulations are built on topics such as food quality, hygiene, and safety laws. So, food laws and regulations require regular reviews since these embedded topics are constantly evolving globally. The reevaluation of regulations at all levels of government ensures they remain valid and functional in line with international agreements such as the Codex standards and World Trade Organization (WTO). Again, and more important, are the mechanisms put in place to ensure food regulations are appropriately implemented (Aruwa and Akinyosoye, 2015 and 2017; FAO, 2019).

Regulatory laws laid down by food societies and governing bodies at all levels are channeled toward ensuring that various food classes, depending on their nature, are handled, prepared, stored, distributed, and sold to the respective final consumer in an acceptable and hygienic manner. On the other hand, food governance is a

Aruwa Christiana Eleojo, The School of Sciences, Department of Microbiology, Main (Obanla) Campus, Federal University of Technology, Akure (FUTA), Ondo State, Nigeria

https://doi.org/10.1515/9783110667462-020

broad terminology closely linked to global food security. Food governance is a facilitating mechanism that allows debate, the convergence of views, and unification of actions to help improve food security at all levels of governance (FAO, 2019). Nevertheless, food regulation and governance is a global phenomenon with the sole objective of delivering safe foods/food products to ensure public health and food security. Within food safety regulations, complex governance interactions exist which involve stakeholders in the public, private, and hybrid/emerging sectors/economies. To better understand food governance in relation to food safety, food models have been utilized to try understand the overall outlook/concept. Each model, however, falls short one way or the other. Recently, the regulator-intermediary-target (RIT) model has been used to appropriately understand the reality and existence of emerging governing relationships in modern food safety regulation. For example, the Global Food Safety Initiative (GFSI), a private corporation and meta-regulator setting food safety standards for both farm assurance and manufacturers, is an intermediate actor in food safety regulations in the European Union (EU). Within the RIT model the regulator makes the rules, the target takes the rules, while the intermediary does one or both functions. This model was put forward and analyzed in-depth by Havinga and Verbruggen (2017) and argued that

1. In the context of private and public regulatory bodies/functions, more than one intermediary participates, and each may act independently. An intermediary may be organized in form of a hierarchy or linked to another in a more complex form.
2. Intermediary functions lie between those of several regulators and targets/consumers.
3. Intermediary actors may act as targets or regulators in relation to other participants in the governing hierarchy. In other words, their function may change/evolve depending on the inter-relationship studied (Havinga and Verbruggen, 2017).

Unlike the "command and control" model of food regulation, the RIT model allows the separation of functions into regulatory subunits in which every participator stalls or contributes to the regulatory process. It is a useful scheme for understanding and analyzing complicated food governing interrelationships in food safety regulatory systems. However, the model may not fully portray the complicated and dynamic links between participators in the food safety regulatory government. Given the outlined contentions of the model, it is essential that the model be viewed in light of other theories which may identify, map, and expatiate on the functions of each participator in a food safety regulatory government (Havinga and Verbruggen, 2017).

Interestingly, intermediate actors in the food regulatory chain have taken on new forms/functions especially in the agricultural food (agrifood) or food retail sector. Also, due to the lack of clarity between the power (the structural power/function of retail corporations) and authority (the perceived legitimate function of retail organizations as political participators in the global food governing scheme) to govern, private food governance (PFG) emerged. Worldwide, food/agriculture governance has faced significant problems since the beginning of a new millennium (Fuchs and Kalfagianni, 2010). The influence of the food supply chain on the environment is often overlooked. The supply chain encompasses the processing and production of unprocessed materials, as well as product transportation and sale. Still, environmental standards are pushed aside in private food retail (PFR) governance. PFR governance has little advantage for food safety and adversely impacts food security, such as improved choices for consumers which only benefits a small section of the world population. At the same time, food retailers also have the power to rearrange social networks and grassroot markets, thereby imposing unnecessary hardships on subsistent farmers, which worsen their food insecurity. Retail corporations base the action and emergence of PFR governance on improved efficiency values, expertise, and the perceived appropriate distribution and delegation of functions especially at the most basic level. PFR governance structure and power is backed by certain supportive public policies which enhanced free trade and movement of funds and capital. However, PFR governance sacrifices food security and environmental sustainability, two key features of global food governance, for selfish gains (Fuchs and Kalfagianni, 2010).

Despite the associated complexities, food governance has the capacity to contribute effectively to planning and formulation, monitoring and implementation of food security, and nutrition programs and policies which include the worldwide rights to food principles and good food practices as they align with the guidance provided by governing bodies such as the Committee of World Food Security (CFS). The CFS is an inclusionary global platform were stakeholders combine efforts to ensure food security and delivery of nutritive foods to all. It is tasked with the revision and further scrutiny of policies that can impact worldwide food security. CFS reviews cover production, as well as physical and economic access to food (FAO, 2019). While food governance has a wider far-reaching goal of achieving food security to address related causes of hunger, it admits that meeting this goal is greatly dependent on a strong consistency and unity among all relevant entities that essentially contribute to food schemes and guidelines of nations and stakeholders in the food industry (FAO, 2019). In addition, food governance also recognizes that food safety and food security are basic human rights. Considering this, it is expected that all food businesses should be appropriately registered with the respective food governing or regulatory bodies within their operational jurisdiction. Some food industry operators may require formal approval from councils such as the Barnsley Metropolitan Borough Council in the UK and National Agency for Food and Drug Administration and Control in Nigeria, before they can begin operations (BMBC, 2019).

20.2 Food regulation/policy implementation

Effective implementation of food regulations at all levels is dependent on a variety of factors.

1. There must be an enabling economic environment for food laws and policies to thrive within relevant institutions which then extend to the grass roots. This would ensure effective participation in actions channeled toward furthering the right to safe food. There is usually a wide political and social divide between national policy and legislative intent and what is practiced at the grassroot by the most vulnerable segments of the population. Nevertheless, laws with immense equity orientation which emphasize good governance is vital in prioritizing the most food insecure and vulnerable sectors of society, especially in resource creation and allocation, social and political awareness about conditions at the grass roots, and with inclusion of nongovernmental participators including local groups.

2. Second, there is need for government institutions and nongovernmental and community-based organizations (NGOs) to have sufficient capacity and resources to support local actions. The mandates and capacity of each body should be explicit enough to ensure compliance and good food regulation/governance practices to support grass roots actions.

3. Third and most importantly the population should be sensitized and empowered enough to see themselves as worthy subjects of human rights that have a lucid and simple understanding of their right to good food principles and good governance as these should apply to them in practice. The public should also be alert to their own responsibilities and choices to the extent that they can evaluate real options that affect their development. This includes the means with which they seek recompense in cases where regulations are bypassed.

20.3 Why regulate foods? – An outlook on food hygiene and safety

Food hygiene and safety regulations are aimed at safeguarding food consumers by ensuring that food is handled, prepared, stored, sold, and distributed under hygienic conditions (Aruwa and Ogundare, 2017; FAO, 2019). Food regulations are increasingly becoming more important given the growing world population, annual increments in incidences of foodborne diseases and outbreaks worldwide, and increased consumer awareness on their right to safe foods (Rahmat et al., 2016). In 2014, the World Health Organization (WHO) reported that more than 90% of human exposure to disease is through food, mainly seafoods, dairy, and meat products. A good food regulation policy should incorporate mechanisms that permit consumers and food production outlet feedback, and evaluation and dissemination of food

industry participants to seek advice on food safety laws based on location and juris-
dictions. It would also facilitate the prompt reporting of food safety concerns, challenges, and problems to aid in the rapid reduction of incidences from foodborne
outbreaks (BMBC, 2019).

20.3.1 Insight into the EU and US food safety laws

20.3.1.1 The Food Safety Act 1990

This Act gives certain powers to food inspectors and allows them where necessary
to detain, seize, or recall any food inspected that is found suspicious or unsafe.

20.3.1.2 General food regulations of 2004

1. Here, food ventures and businesses are expected to meet the requirements of
   the European commission (EC) regulation No. 178/2002. Article No. 14 of the
   law expanded on food safety requirements and states that food containing
   harmful components, for example, heavy metals, should not be offered for sale.
2. Article No. 18 under the same regulation touched on traceability and states that
   food operators should be able to show place of purchase of their food raw materials and who the products are supplied to. This touches essentially on record
   keeping right from the source of food raw materials supplied and their suppliers to where the processed or unprocessed food is going to be delivered, to consumers or otherwise. In other words, transaction receipts and invoices and
   other relevant forms of documentation should be appropriately kept.
3. Article No. 19 on the other hand modulates and controls decisions on food
   product recalls and states that where a food operator has distributed products
   which are unsafe, it must make immediate and urgent attempts to call back the
   food product from the market. In case the products already reached the target/
   consumer, the operator should inform the retailer/consumers or other suppliers
   of the reason behind the recall.

20.3.1.3 The food hygiene regulations of 2006

Under this regulation, food ventures are expected to meet the requirements of EC regulation No. 852/2004 on the hygiene condition of foods materials. Some specific and
general hygiene requirements include food premise layout, structure, and design, as
well as maintenance conditions, provision of adequate lobbies and toilet facilities,
food handlers' personal hygiene and regular training, periodic arrangements for

pests control, and appropriate modalities for food wastes disposal (BMBC, 2019; Aruwa et al., 2017).

20.3.1.4 The Food Safety Modernization Act (FSMA) of 2011

As part of a drastic focus shift in worldwide food supply systems from food safety response (palliative) to food safety prevention (anticipatory and precautionary), the FSMA was created by the US Food and Drug Administration (USFDA) and enacted by the US Congress into law in January 2011 under the Obama administration. The Act stemmed from the significant burden posed by foodborne illnesses globally, and has better metamorphosed the food safety system in the United States. The USFDA recognized that reducing the incidence of preventative foodborne diseases would have a tremendous impact on global public health and should be focused on and strengthened, besides the palliative measures already in place (USFDA, 2019a). The Act applies to both US and non-US food industry actors and is therefore globally accepted. Training, audit, technical, and supplier verification services such as those organized by the Société Générale de Surveillance (SGS) are still being offered at all levels of government worldwide to aid compliance with the underpinning laws of the Act and expand knowledge on its benefits to food consumers, importers, and world economies (SGS, 2019). The Act brings the global community closer to achieving a preventative food safety system. The implementation of standards in the Act may be achieved through an open process that gives all stakeholders the opportunity for meaningful inputs (USFDA, 2019a).

The Act lays out main actions, rules, or programs for prevention of food contamination which significantly affect the animal and human food supply chains. The programs covered include authorization and third-party certifications (TPP) and Voluntary Qualified Importer Program (VQIP). Other inherent preventive rules within the Act cover mandatory good manufacturing practice (GMP) and hazard analysis and critical control points (HACCP) controls for human and animal foods, affirmation programs for international or foreign food suppliers, strategies that mitigate against the intentional adulteration of foods, hygienic transfer of animal and human foods, and mandatory requirements for harvesting, growing, storage, and packaging of foods for human consumption. On the other hand, palliative measures focused on extra record-keeping requirements for high risk foods, obligatory food recalls, suspension of food facility registration when necessary, improved product traceability, and expanded detention of suspicious/illicit products (USFDA, 2019a). In addition, the Act holds both domestic and imported foods to same standards and is integrated in collaboration with actors at local, state, national, and international governance strata. The improved partnership and reliance on other inspection agencies, as well as frequent and compulsory inspections and access to records and food testing by accredited laboratories cannot be overemphasized regarding ensuring compliance with the Act. In terms of flexibility, the law also provides produce safety standards exemptions for local, subsistent farmers who sell their farm produce through a community supported agriculture program or directly to consumers (USFDA, 2019a).

Since inception in 2011, FSMA rules and programs have undergone several reevaluations which apply to food industry participants. Some of these reviews include the new user fee rates for the year 2020 for importers participating in the VQIP (which took effect from July 24, 2019), certification bodies involved in the TPP program and a revised guide for food protection strategies to mitigate against deliberate adulteration of foods in 2019. In 2017, a sixth chapter draft guidance on preventive control for human foods was released to aid food facilities adhere to standards such as the development of a documented food safety plan, as well as a guidance on control of *Listeria monocytogenes* in ready-to-eat (RTE) foods (docket No. FDA-2008-D-0096) for affected food industry stakeholders (USFDA, 2019b).

20.3.2 Hazard Analysis and Critical Control Points (HACCP) role in food safety

As part of food safety regulations, food merchants are expected to apply the HACCP principles as a globally accepted food safety management framework. HACCP principles help food operators/vendors:

1. Identify hazardous points in a production system/line which must be reduced, removed, or prevented to satisfactory levels.
2. Identify the points at which control is required, that is, critical control points (CCPs) in order to achieve the first point above.
3. Establish critical cut-off points at each CCP that defines the level of acceptability or otherwise for the reduction, removal, or prevention of hazards identified.
4. Implement and establish efficient monitoring protocols at CCPs.
5. Establish corrective measures when there are indications that a CCP has not been effectively controlled.
6. Map out a process to be followed regularly to ensure that the actions outlined in points one through five function effectively.
7. Maintain and store relevant business records and documents to depict efficient application of measures from points one to six (BMBC, 2019; Rahmat et al., 2016).

20.4 The implications of food regulations on consumers and businesses

Globally, most governments and societies have laid down rules that govern businesses involved in processing, production, storage and transportation, marketing, and preparation of food products. This is done to reduce the disadvantageous effects of unsafe foods. However, while food laws/regulations DO NOT guarantee safe food; achieving safe food practices or GMP to reduce the risk of unsafe food is attainable. Food laws also cover the prohibition of "adulterated" and "misbranded" foods for consumer protection and health (AAE, 2010). Food regulations serve to:

1. Ensure food vendors do what they can to assure safety of their products.
2. Prohibit food operators from engaging in activities that will render their product unsafe.
3. Ensure consumers have the required information to make informed choices, for example, food product labeling.
4. Proscribe food vendors from providing confusing or false product details to consumers.
5. Ensure the right of consumers to learn about food processing techniques and nutrition, as well as how to read, comprehend, and utilize food product information.
6. Assure public health about unpropitious effects of unsafe foods such as foodborne outbreaks.
7. Disseminate information to target market, for example, food product labeling guidelines (labeling aid consumers in making informed decisions) and awareness campaigns
8. Ensure consumers get value for money for products advertised and purchased while protecting them against fraud.
9. Assure lawful and fair trade practices among market actors, while preventing misrepresentation and misleading information on food product quality.
10. Ensure environmental protection through the recognition that food production processes affect the quality of water and air.
11. Assure consumer protection against intentional or malicious attacks on the food system (AAE, 2010).

20.5 Some food safety organizations worldwide

The prevalence of foodborne illnesses increased drastically, and this became a global concern at the start of the twenty-first century and initiated the need for food supply and protection through food regulations. Food regulatory agencies are set up at all levels of government to assure that food supplied to the public is safe from harmful ingredients (additives) and agents (microbial, chemical, and other hazardous components) to reduce the risk and incidence of foodborne illnesses globally. This is essential because at all stages of food preparation/production; harvesting, packaging, processing, cooking and storage, and food contaminants may be introduced. Some food safety bodies are herein discussed.

The FAO is an international regulatory and governing agency (Table 20.1) saddled with the responsibility of food security for the global population, while ensuring access to quality foods for healthy living. Every nation's health is its wealth, and the other side of unhealthy living leads to increased morbidity and mortality rates worldwide. The FAO tasks itself with three major functions that include eradicating malnutrition, food insecurity, and hunger, poverty elimination and increase in economic drive and social progress, and lastly the sustainable utilization of natural bioresources such as air, water, land, and genetic resources for the benefit of all. The organization in collaboration with the WHO developed a standard operating procedure (SOP) as an intervention for guidance on food safety exigencies, thereby reinforcing preparedness, for example, in case of food related outbreaks or illnesses (MSU libraries, 2019). Other food regulatory agencies around the world whose offshoot functions are derived from these international agencies are listed in Table 20.1. They carry out their food safety tasks by ensuring GMP and SOPs are utilized in food production and

Table 20.1: Food safety regulatory organizations around the world.

S/N	Region	Food safety organizations within countries
A	America	– United States Department of Agriculture (USDA) – Centre for Food Safety and Applied Nutrition (CFSAN), US – Food and Drug Administration (FDA) division of the US Department of Health and Human Services (DHHS) – Centre for Disease Control and Prevention (CDC) Food Safety Initiative – World Health Organization (WHO) Food Safety Department (FOS) – The Food and Agriculture Organization of the United Nations (FAO) – The Codex Alimentarius Commission – The World Food Safety Organization – Guelph Food Technology Centre (Canada) – Canadian Food Inspection Agency (CFIA) – National Food Safety and Quality Service (SENASA), Argentina
B	Europe	– European Union Food Safety Policy Committee – Food Safety Promotion Board (Multinational) – Food Safety Agency (United Kingdom) – Department for Environment, Food and Rural Affairs (UK) – Committee on the Environment, Public Health and Food Safety (EU) – Federal Ministry of Food, Agriculture and Consumer Protection (Germany) – Agencia Española de Seguridad Alimentaria y Nutrición (AESAN, Spain) – Belgian Federal Agency for the Safety of the Food Chain – Global Food Safety Initiative (GFSI), Belgium – Economic and Food Safety Authority (Portugal) – Food Standards Scotland – Norwegian Food Safety Authority – Ministry of Economic Affairs, Agriculture and Innovation (Netherlands)

(continued)

Table 20.1 (continued)

S/N	Region	Food safety organizations within countries
C	Africa	– Ghana Food and Drugs Authority – National Agency for Food and Drug Administration and Control (NAFDAC) (Nigeria)
D	Asia	– Food and Drug Administration (Philippines) – Saudi Food and Drug Authority (SFDA) – Minister for Health, Welfare and Family Affairs (South Korea) – State Food and Drug Administration (China) – Centre for Food Safety (Hong Kong) – Food Safety and Quality Division (FSQD) (Malaysia) – Food Safety and Standards Authority of India
E	Oceania	– Food Standards Australia New Zealand (FSANZ) – Australian Minister for Agriculture, Fisheries and Forestry – Australian Quarantine and Inspection Service (AQIS) – New South Wales Food Authority – Biosecurity Australia – New Zealand Food Safety Authority (NZFSA) – Minister for Food Safety (New Zealand)

Compiled from CSCERSDC (2012), MSU libraries (2019), Schmelzer (2004).

distribution. Other far reaching functions of food regulatory agencies include regular food sampling, testing, and monitoring to ensure they meet with stipulated jurisdictional food standards, specifications, and guidelines. Also, the reporting and carrying out of public awareness campaigns for continued education of the populace on the need for food safety practices remains an integral function.

Conclusively, it is generally expected that the federal/national government food regulatory bodies would work closely with state and local governments, nonprofit corporations, individuals, and private operators to monitor and ensure the safety of foods supplied. The FAO and WHO function in a policymaking capacity for its 192 member countries, and thus provide an overall intercontinental presence. Global food security and safety cannot be overemphasized. Food trade has become more open and international and novel food products are regularly introduced into the food chain/market through biotechnological, bioengineering, and other scientific and technological means. The constant monitoring and regulation of the continuous flow of foods and food products from farm to mouth calls for extensive global networking and collaboration in order to be successful (Schmelzer, 2004).

20.6 International food standard organizations and agreements

1. **Codex Alimentarius Commission (CAC):** The WHO and FAO formed the CAC in the year 1963. The commission was tasked to establish food specifications, standards, guidelines and codes of conduct/practice, among other functions. The Codex Alimentarius (CA) essentially ensures good practices in the food market, achieves consumer health protection and enhances coordination of all food standard works carried out by international governmental organizations and NGOs.

2. **Sanitary and Phytosanitary Measures (SPS):** This agreement lays out fundamental guidelines which aim to aid governments on the application food safety, as well as stipulates plant and animal health measures under the SPS measures.

3. **FAOLEX:** FAOLEX is an updated and comprehensive legislative library/database collated under the FAO. FAOLEX is one of the largest compendiums of national food regulations, as well as renewable natural resources and agricultural laws. FAOLEX database users have full text access to most legislations contained therein. Within the FAOLEX are subthematic databases for specified sectors such as the pastoralism legislation, fishlex, aqualex, and agroecologylex databases.

4. **The International Plant Protection Convention (IPPC) agreement:** The IPPC was created in 1952 and is a global agreement on plant health which serves to conserve and protect both wild and cultivated plant species by preventing the spread and/or introduction of pests (IPPC, 2011).

5. **World Organization for Animal Health (OIE):** The OIE is an integral reference organization with about 178 member countries as of 2013 and recognized by the WTO. It is essentially an intergovernmental organization, with the responsibilities of improving animal health globally. The OIE maintains permanent collaborations with about 45 regional and international corporations.

6. **International Trade Centre's Standards Map:** Standards Map is transparent and verified compendium of information on voluntary sustainability standards and food safety and quality initiatives. The map's major goal is to strengthen the ability of food consumers, merchants, and policymakers to participate more in sustainable food trade and production.

7. **Organization for Economic Co-operation and Development (OECD) in agriculture and fisheries:** The OECD was set up with the central aim to promote laws that will improve the well-being and socioeconomic status of populations globally (MSU libraries, 2019).

8. **United States Pharmacopeia Convention (USP):** The USP was established in 1820 and produces reference standards to ensure the consistency and quality of food ingredients and dietary supplements, as well as medicines (includes their identity and purity) (USP, 2008 and 2011).

9. **International Organization for Standardization (ISO):** Since its establishment in 1947, the ISO has been a vast network of standards institutes from 162 nations (ISO, 2011a) and sets market standards and fosters standardization activities among member nations (Giovannucci and Purcell, 2008). ISO standards and activities cover good management practices and food products, as well as management systems for food safety.

10. **International Commission on Microbiological Specifications for Food (ICMSF):** The ICMSF was established in 1962 and has since provided evidence-based criteria to industry and government on control and evaluation of the microbiological safety of foods (ICMSF, 2011).

11. **Association of Official Agricultural Chemists (AOAC) International:** The AOAC provides science-based expertise for validation studies, food microbiology, single laboratory validation for botanicals, and antibody characterization used in immunochemical procedures (AOAC, 2009). It was founded in 1884 (National Academy of Science, 2012).

12. **The Africa Continental Free Trade Agreement (AfCFTA):** This agreement was signed in 2018, and in 2019 it has been signed by 54 African Union member states. Several sections of the AfCFTA remain under negotiation and implementation would be in phases. Nevertheless, the major role of the AfCFTA is to serve as an umbrella body to which annexes and protocols will be added.

20.7 Food safety at the center of food regulations and governance

Achieving the reduction in risk of foodborne illnesses involves practice and application of basic food production principles, that is, cleaning, separation, cooking, and chilling. These four basic principles are consumer education oriented. About one in six Americans are affected by foodborne illnesses with mortality rate of 3,000 deaths per year. Along the food production chain, foods are handled, transported, and transferred several times with a likelihood of introduction of contaminants at any point in the chain. Such contaminants may be reduced by consumers through the application of safe food handling principles such as the clean, separate, cook, and chill principle. Also, the practice of cleaning-in-place during food production cannot be overemphasized (Dietary guidelines, 2015).

1. **Clean:** Microbial contaminants can be transferred within the food production areas through utensils, hands, foods, and cutting boards, also known as cross contamination. Prevention here would involve regular hand washing especially after handling raw animal products, for example, seafoods, meat, fish and encountering sick persons, for example, people with flu and Staphylococcal infections. Surfaces and countertops also require frequent cleaning to prevent

cross contamination. Nevertheless, all produce/grocery irrespective of where they were purchased or grown should be thoroughly washed/rinsed/disinfected to reduce microbes and other impurities in foods. Again, hand washing using soap and water before and after food preparation, especially after handling raw foods and before eating is recommended. The continued use of antimicrobial soaps/agents may not be needed for consumers since long term use may result in the development of antimicrobial resistance to the agents. Hence, the recommendation of rinse-free alcohol-containing (≥ 60%) sanitizers. Also, food equipment and appliances need to be cleaned, and overstayed food, expired and cooked food leftovers should be discarded on a regular basis. Food produce also often require thorough rinsing prior to being cut or peeled to prevent transfer of microorganisms to the inner parts of the produce to be consumed. The produce may then be dried using paper towel or clean kitchen cloth to achieve further reduction of microbial contaminants which may be present. As a rule of thumb, raw animal products are not rinsed off as microbes in the raw extracts can be transferred and cross-contaminate other kitchen surfaces, utensils, and foods resulting in foodborne illness.

2. **Separate:** The separation of raw from RTE foods is integral to the prevention of foodborne illnesses. Foods separation should be the norm at all stages of food processing and handling. It is of utmost importance that raw animal products be separated when shopping and placed in separate packs away from other foods during refrigeration, but under RTE foods. Separate clean chopping boards also should be used for raw animal products and fresh farm/food products.

3. **Cook and chill:** Poultry and other raw animal foods like meat, eggs, and seafoods should be processed at the recommended minimal temperatures that kill inherently opportunistic and pathogenic microorganisms. The use of food thermometers is encouraged to ensure food is safely cooked and thereafter stored at the recommended temperature until ready for consumption. This is because the safety of food cannot be determined by its look. Producer's instructions for use should be adhered to and for the appropriate period required to achieve the required internal temperature in foods. Again, due to uneven distribution of heat, microwave cooking often leave "cold or uncooked areas" where disease-causing bacteria may grow and thrive. Microwave processing instructions on food product packages should always be adhered to. While hot foods are maintained at 60 °C or above, cold foods on the other hand are usually held at 4 °C or below. Again, if thawing frozen foods are allowed to become warmer than 4 °C, the growth and multiplication of microbes present prior to freezing can be initiated. This may be prevented by thawing foods in a leak proof bag in cold water, but with the change of water every half hour. Foods may also be thawed in a microwave or refrigerator (the refrigerator temperature should be kept relatively stable at 4 °C or below) (Dietary guidelines, 2015).

20.8 Safe minimal internal temperatures for food processing

The application of safe minimal cooking temperatures is recommended to consumers and can be monitored and measured using a food thermometer. Also, due to individual preferences, consumers may decide to process meats at even higher temperatures. According to dietary guidelines of 2015, the following list gives some recommendations on temperatures at which various food categories are to be processed to mitigate health risks and foodborne illnesses.

1. Beef, pork, veal, and lamb at 72 °C.
2. Steaks, roasts and chops, and fresh/raw ham at 62 °C.
3. Poultry, roasts, whole chicken, turkey, and pork to be cooked at 75 °C minimum.
4. Reheat precooked ham at 60 °C.
5. Eggs are to be cooked until yolk and white become firm.
6. Egg dishes at 72 °C minimum.
7. Finfish cooked at 62 °C minimum. It must also be cooked till it readily flakes and becomes opaque.
8. Shrimps, scallops, and lobsters should be cooked until they reach their appropriate color, that is, milky white for lobsters and shrimp, and firm and opaque for scallops.
9. Food leftovers are recommended to be cooked at a minimum of 75 °C (Dietary guidelines, 2015).

Studies have shown and emphasized the need to keep foods, including RTE foods, at safe minimum temperatures and in line with the earlier recommendations, prior to their purchase and consumption (Aruwa and Akinyosoye, 2017; Davis et al., 2008).

20.9 Risky eating habits and individuals at risk of foodborne illnesses

It is usually quite impossible for consumers to know what signs of food contamination to look out for especially in cases of harmful parasites and microorganisms that usually do not change the appearance or aroma of foods. Again, the risk of foodborne illness increases with intake of improperly cooked or raw animal food products, for example, ground beef in undercooked hamburgers, eggs with runny yolks, and raw oysters/seafoods and cheese made from unpasteurized milk. The consumption of pasteurized dairy products only and food processing at minimal safe temperatures are highly recommended to diminish the risk of foodborne diseases. Despite the potential risks, seafood lovers are advised to choose previously frozen seafoods over raw seafoods because prior freezing would at least kill parasites but not harmful microbes. Also, it is advised that in the production of foods which require raw eggs as ingredients, for example, smoothies, certain sauces, eggnogs, other drinks and ice cream, the utilization of pasteurized eggs is best during their preparation (Dietary guidelines, 2015). Furthermore, certain members of the population such as the immunocompromised are at risk of contracting foodborne illnesses. Persons with weak immune systems include pregnant women and their unborn, the elderly, young children, people undergoing organ transplant, and those living with HIV and undergoing cancer treatment are more prone to foodborne diseases like salmonellosis and listeriosis compared to other members of the population. In other words, for these classes of individuals extra caution must be taken to ensure the safety of foods consumed as foodborne illnesses may be fatal for them (Dietary guidelines, 2015).

20.10 Food labeling regulations and food safety

Food packaging and labeling comes under regional legislation with the main aim to promote food safety and fend off false advertisements. A food label is an informative mark/tag or brand, which may be pictorial, stencilled, printed, impressed on, marked or attached to a food container. Food labels are attached to and accompany foods and food products in order to promote their disposal or sale (Codex Alimentarius, 2007). While food labels are controlled by jurisdictional legislation, they are expected to be accurate, safe, and not misleading. Food labeling applies to all foods sold to establishments and consumers by food merchants and producers and must contain key details about the food product. Labels educate the consumer, assists them in making informed decisions, and essentially informs on how the food product(s) should be safely stored before, during, and after use (NHS, 2019).

Food labeling policies may be categorized based on the following:

1. A wide variety of label laws and foods, for example, CA which outlines the international voluntary food labeling standards
2. Veracity, for example, food product health claims and false advertising
3. Ingredients and nutrition information, for example, nutrition facts label and calorie count laws
4. Food handling materials, for example, food safe symbols like the wine glass and fork
5. Specific food, for example, food grading labels and olive oil regulation and adulteration
6. Framing practice, for example, the 2002 UTZ sustainable farming label laws for coffee and cocoa, free range, and organic certifications

7. Vegan and vegetarian, for example, the globally recognized "V-label" for consumer ease, clarity, and transparency for vegan and vegetarian food products, and the "Certified Vegan" brand by Vegan Awareness Foundation trademark for vegan companies (GMFs) and organizations (European Vegetarian Union, 2019)
8. Genetic and commercial origin, for example, laws governing the labeling of genetically modified foods (GMFs) and ability to trace products to their origin
9. Religious certification, for example, Islamic dietary laws covering halal foods and Kashrut covering kosher foods in Jewish law
10. Pricing, for example, dine and dash
11. Safety information, for example, International Food Safety Network, African Food Safety Network (AFoSaN, 2019), warning labels, food allergy, alcohol abuse, and the International Standard Organization's (ISO) 22000 for food safety standards derived from ISO 9000 regulations
12. Region, for example, international and global FAO and WHO policies, local (grassroot) and national (countries)

Food labeling laws are often covered under the consumer protection law/act (CPA) in different countries and regions around the world. For example, the food labeling law published in 2010 as Notice No. R146 in South Africa sought to tighten labeling loopholes and strengthen the consumer's ability to make more informed choice. The CPA simply emphasizes that a consumer has the right to truthful, plain, understandable, and nonmisleading information on all goods sold to them. It also enforces a consumer's right to quality and safe goods, as well as the right to responsible marketing. CPA applies to all trades involving supply of good and services to consumers and includes every food product/good marketed for human consumption. It is expected that good food labels should exhibit certain characteristics.

1. They should be legible enough to read.
2. Label should clearly give the food product name.
3. List of ingredients must be shown and the quantity (known as "Quantitative Ingredient Declaration – QUID") of certain ingredients.
   a. The food label list of ingredients must have the "ingredients" heading.
   b. Ingredients are also usually listed in descending order based on mass/weight used during product processing.
4. The net quantity of products should be indicated either in weight (grams and kilos) or volume (liters, milliliters, and centiliters) excluding the package weight,
5. Instructions for use (if needed) should be part of the food tag.
6. Food product durability period should be indicated, for example, expiry date or "best before."
7. Food product conditions of storage and use should be clearly stated.
8. Food business/operator address and name, and/or supplier should be provided. This information affords unsatisfied consumers the opportunity to contact the food manufacturer.

9. Food product place of provenance is mandatory if the product was made in a different country, for example, "Nigerian mackerel smoked in Ghana" is a correct and more truthful label compared to the incorrect "Ghanaian smoked mackerel" label.

10. Possible food allergens within the product (including their derivatives) need to be clearly stated and should stand out from other ingredients.

    Note that for '"gluten free" labeled products, it implies that products contain less than 20 mg/kg gluten, while "very low gluten" products contain less than 100 mg/kg gluten. There are 14 specific allergens or allergen sources which have been listed and must be highlighted in food labels. The allergens include eggs, fish, peanuts, milk, mustard, soybean, sesame, lupin, celery and celeriac, nuts, molluscs, crustaceans, and sulfite preservatives found in some dried fruit (but only when used above 10 mg/kg or 10 mg/L benchmark). In several regions and jurisdictions, allergen information is also required for foods not prepacked or sold loose, for example, in retail and takeaway outlets, restaurants, hotels, and cafes. Such information may be passed across verbally or through menus, blackboards, or separate allergy list sheets (NHS, 2019). Several food additives may also be allergenic and need to be identified in food labels, for example, food preservatives added to products to maintain food quality and extend product shelf life. Certain factors, including regulatory factors, determine the use of a food additive (Fig. 20.1) and must be adhered to. Of note are conditions under which food additives must not be used (Povea-Garcerant, 2017).

11. Nutritional content information must be present. This is usually in the form of calories/energy content or reference intake (RI). RI is an approximate measure of the daily energy or nutrients needed for a healthy diet and represented as a percentage. While an average person is said to require about 8,400 kJ, 70 g in total fat, 20 g in saturated fats, 260 g carbohydrate, 90 g total sugars, 50 g in protein, and 6 g of salt as part of the daily diet, these values vary with individuals and body weight.

12. In the case of beverages, alcohol content/strength of more than 1.2% should be shown (NHS, 2019; Safefood, 2019).

In the case of pre-packaged foods, the acceptable global standard principle for food labeling set by the FAO and WHO states that:

Prepackaged foods shall not be described or presented on any label or in any labelling in a manner that is false, misleading or deceptive or is likely to create an erroneous impression regarding its character in any respect. Secondly, pre-packaged foods shall not be described or presented on any label or in any labelling by words, pictorial or other devices which refer to or are suggestive either directly or indirectly, of any other product with which such food might be confused, or in such a manner as to lead the purchaser or consumer to suppose that the food relates to such other product (Codex Alimentarius, 2007).

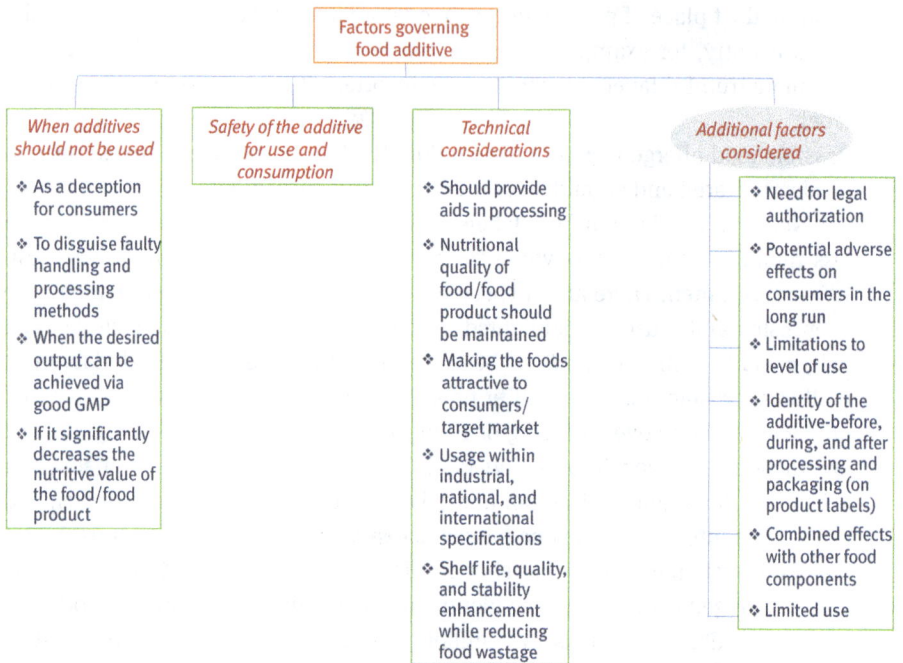

Figure 20.1: Factors governing food additives. Adapted from Povea-Garcerant (2017).

20.11 Challenges of food safety regulation and implementation

1. **Quality assurance and adherence to international standards:** A major challenge is when foods supplied fail to meet set safety and quality standards, especially where international trade is involved. Adherence is particularly challenging in low and middle-income nations, as well as in agricultural and food industries (Rahmat et al., 2016). For example, stringent international regulations that check for the presence of minimum pesticide residues in import and export produce to/from developing countries should not be overlooked. Food safety should never be sacrificed for food supply/production. Also, level of involvement of local and national agencies from the developing and least developed countries in international standard organization such as the Codex is poor/insignificant. A solution would be for these nations to improve compliance and involvement in international trade practices/standards and maintaining harmonized standards across markets may help reduce bureaucratic bottle necks.

2. **Supply chain control:** Given the food supply chain complexity, the inability to control a supply chain makes it difficult to trace or recreate the chain. In a bid to tackle this, the US 2002 Bioterrorism Act was enacted and required participants in a

supply chain to state the immediate previous source and recipient of their product. This is called the "one-up, one-back" traceability report (Gessner et al., 2007). This report pathway allowed for easier recreation and traceability of products even when not one participant has a total grasp/picture of it. While product tracking remains a daunting challenge for developing nations, several intercontinental companies have embraced the use of bar codes and radio frequency tags to enhance traceability of products within their supply chains.

3. **Poverty and poor infrastructure:** Lack of infrastructures such as good road, telecommunication systems, adequate storage facilities, stable and reliable electricity, and good surveillance system, as well as laboratory, manufacturing, and market infrastructures, as obtainable in many developing countries, pose a great deal of challenge; thereby the food supply chain suffers immensely. When the supply chain suffers, foods are lost, food safety is mortgaged, and food security may be hampered. Poverty and poor infrastructure are closely linked and constitute two major problems facing implementation of food safety regulations. In some African and Asian countries, food safety laboratories had no equipment or personnel to properly collect and analyze food samples, while other countries transport samples at regular intervals to regional laboratories which is slow and expensive. Therefore, an increased interest from governments in improving infrastructure and reducing poverty is essential in addressing this problem (Hao, 2012; World Bank, 2009).

4. **Prevailing laws:** At the heart of food regulations must be relevant and enforceable laws which foster fair trade, considerably reduces fraud and ensures public health. This is important as laws set the political, cultural, and economic climate of a nation/region (WHO, 2007). While some developing countries have no food safety governing laws, others have contradictory laws. Poorly coordinated regulations create confusion in the assignment of responsibilities to agencies. Governments therefore need to make regular monitoring and reevaluation of existing laws and make attempts to put laws in place where none exists to safeguard the life of their citizens. Enforcement means to implement stated punitive and corrective measures to violators of food regulations. Enforcement is important if laws are to be taken seriously and strictly adhered to. Not in all cases do regulatory agencies participate in enforcement. In most cases, the regulatory agencies are quite different from the enforcement agencies and varies with region/jurisdiction. Enforcement must however be transparent in order to be effective.

5. **The regulatory agency workforce:** The challenges associated with the regulatory agency workforce is mostly seen in the public sector and in developing countries. It includes inability to retain and/or sufficiently pay available staff, too few staff, staff with insufficient technical training, and/or little or no sense of pride or loyalty (that is, "*espirit de corps*") among regulators. Tackling corruption in developing countries as well as careful monitoring and disbursement of funds for agencies and staff training may help manage the workforce challenge.

6. **Institutional fragmentation:** Fragmentation implies assigning different responsibilities to different regulatory agencies. However, this is not always the case as the same functions may be borne by different agencies and this problem cuts across both developing and developed nations. For example, about 12 federal regulators enforced 35 different food safety laws in the US (Martin, 2007). However, a new fragmentation pattern now exists in certain emerging world economies such that there are no clearly defined responsibilities and established enforcement programs or command chain.

7. **Surveillance and poor feedback mechanisms:** Surveillance is an integral systematic function in food regulation and public health systems. It involves the continuous collection, collation, and assessment of public and global health reports, and public health response analysis where necessary (WHO, 2008). Good surveillance practice requires good infrastructure, a relevant and committed workforce, established techniques, and avenues for efficient diffusion of findings. Regulatory surveillance systems that deliver reliable data are difficult to attain and maintain, especially in developing countries. Poor surveillance limits risk assessment. For example, if a foodborne disease or outbreak incident is not reported into a central repository, surveillance functions such as laboratory identification and epidemiological investigation of the causative pathogen responsible for the incident, market withdrawal of adulterated products where necessary, and ability to trace back to the contamination source would be significantly and negatively affected. A good and effective feedback and surveillance system also requires the trust and consent of indigenous people/consumers. Therefore, all stakeholders including respective regulatory agencies have a responsibility to develop and determine the level of failure or success of response and surveillance systems in each region/country (Kimball et al., 2008).

8. **Communication and political will:** Communication here implies the need to share information appropriately and strategically, and in a way that supports an open dialogue culture. Communication challenges with regulatory agencies include communication with international counterpart agencies, communication to the public, communication problems across agencies with overlapping regulatory roles and within a regulatory agency, and communication between regulatory authorities and those they regulate (National Academy of Science, 2012). These challenges negatively impact both developed and developing economies and may be improved through the signing of confidentially agreements and building trust over time. In terms of political will to enforce food safety laws, established economies have greater political will compared to emerging economies. Nevertheless, the political will of governments may be affected at any time depending on prevailing economic situations such as in recessive economies (Newport, 2010). In a democratic society, public opinion can also drive political will.

In summary, some international strategic schemes have been put in place to ensure food safety, nutrition and security, and improve implementation. Similar schemes are required at all levels of food governance since they recommend regulatory benchmarks for states, intercontinental, private actors, and other stakeholders, as well as the CFS. Such global strategies give techniques on how to foster unified policies within the rights-based framework which is channeled toward the complete awareness of the right to adequate food for all. Last, the world recognizes that climate change has introduced the emergence of new agricultural/food pests and diseases coupled with increased demand for quality/safe foods due to the growing global population and changing international trade policies. These evolving motivating factors call for the creation and/or strengthening of food safety regulatory chains where all relevant stakeholders play a role.

References

African Food Safety Network (AFoSaN). (2019). *Food Safety in Africa*. https://www.africanfoodsafety network.org/?page_id=234 (accessed September 20, 2019).

Agribusiness and Applied Economics (AAE). (2010). *Purpose of Food Law (Sec 2)*. https://www. ndsu.edu/pubweb/~saxowsky/aglawtextbk/chapters/foodlaw/470-6704.htm (accessed August 30, 2019).

Aruwa, C. E., Akindusoye, A. J. and Awala, S. I. (2017). Socio-demographic characteristics and food hygiene level assessment of food handlers in cafeterias around a Federal University in Nigeria. *Journal of Scientific Research and Reports*, 14, 1–9.

Aruwa, C. E. and Akinyosoye, F. A. (2015). Microbiological assessment of ready-to-eat foods (RTEs) for the presence Bacillus species. *Journal of Advances in Biology and Biotechnology*, 3(4), 145–152.

Aruwa, C. E. and Akinyosoye, F. A. (2017). *In vitro* efficacy of temperature and preservatives on fast food bacilli, and their antibiotic susceptibility profile. *World Journal of Pharmaceutical and Medical Research*, 3(6), 43–51.

Aruwa, C. E. and Ogundare, O. (2017). Microbiological quality assessment of pupuru and plantain flours in an urban market in Akure, Ondo State, South Western Nigeria. *Library Journal*, 4, e3783.

Association of Analytical Communities International (AOAC). (2009). *Validation Guidelines*. http://www.aoac.org/vmeth/Validation_Guidelines.htm (accessed September 23, 2011).

Barnsley Metropolitan Borough Council (BMBC). (2019). *Food Hygiene Regulations*. https://www. barnsley.gov.uk/services/business-information/food-businesses/food-hygiene-and-safety/food-hygiene-regulations/ (accessed August 30, 2019).

Codex Alimentarius. (2007). *Food Labelling*. http://www.fao.org/tempref/codex/Publications/Booklets/Labelling/Labelling_2007_EN.pdf (accessed August 10, 2019).

Committee on Strengthening Core Elements of Regulatory Systems in Developing Countries (CSCERSDC). (2012). *Ensuring Safe Foods and Medical Products Through Stronger Regulatory Systems Abroad* (Board on Global Health, Board on Health Sciences Policy, Institute of Medicine, J. E. Riviere and G. J. Buckley, Eds.). Washington, DC: National Academies Press; Core Elements of Regulatory Systems. https://www.ncbi.nlm.nih.gov/books/NBK201160/ (accessed September 9, 2019).

Davis, J. R., Lawley, R., Davis, J. and Curtis, L. (2008). *The Food Safety Hazard Guidebook* (p. 17). Cambridge, UK: RSC Pub.

Dietary Guidelines. (2015). *Appendix 14. Food Safety Principles and Guidance.* https://health.gov/
    dietaryguidelines/2015/guidelines/appendix-14/ (accessed September 7, 2019).
European Vegetarian Union. (2019). *The Seal of Quality for Vegan and Vegetarian Products.*
    https://www.v-label.eu/en (accessed September. 12, 2019)
Food and Agricultural Organization (FAO). (2019). *The Right to Food: Governance.* http://www.fao.
    org/right-to-food/areas-of-work/governance/district-level-implementation-rtf-zanzibar/en/
    (accessed August 29, 2019).
Fuchs, D. and Kalfagianni, A. (2010). The causes and consequences of private food governance.
    Business and Politics, 12(3), 1–34.
Gessner, G. H., Volonino, L. and Fish, L. A. (2007). One-up, one-back ERM in the food supply chain.
    Information Systems Management, 24(3), 213–222.
Giovannucci, D. and Purcell, T. (2008). *Standards and Agricultural Trade in Asia.* Tokyo, Japan:
    Asian Development Bank Institute.
Hao, Z. (2012). *Substandard Cardiac Drugs Claim over 100 Lives in Pakistan.* http://english.cntv.
    cn/20120206/113321.shtml (accessed February 12, 2019).
Havinga, T. and Verbruggen, P. (2017). Understanding complex governance relationships in food
    safety regulation: the RIT model as a theoretical lens. *The ANNALS of the American Academy of
    Political and Social Science*, 670(1), 58–77.
International Commission on Microbiological Specifications for Foods (ICMSF). (2011). *Purpose.*
    http://www.icmsf.iit.edu/main/home.html (accessed September 20, 2019).
International Plant Protection Convention (IPPC). (2011). *What We Do.* https://www.ippc.int/index.
    php?id=what&no_cache=1&L=0 (accessed September 23, 2011).
Kimball, A. M., Moore, M., French, H. M., Arima, Y., Ungchusak, K., Wibulpolprasert, S., Taylor, T.,
    Touch, S. and Leventhal. A. (2008). Regional infectious disease surveillance networks and
    their potential to facilitate the implementation of the international health regulations. *Medical
    Clinics of North America*, 92(6), 1459–1471.
Martin, D. S. (2007). Lawmakers push for change in food safety oversight. *CNN.com*, May 17.
Michigan State University (MSU) Libraries. (2019). *International Food Law and Regulations:
    International Organizations.* https://libguides.lib.msu.edu/c.php?g=212831&p=1404071
    (accessed September 6, 2019).
National Academy of Science. (2012). *Ensuring Safe Foods and Medical Products Through Stronger
    Regulatory Systems Abroad. Consensus Study Report.* https://www.nap.edu/catalog/13296/
    ensuring-safe-foods-and-medical-products-through-stronger-regulatory-systems-abroad
    (accessed September 13, 2019).
National Health Service (NHS). (2019). *Food Labelling.* https://www.nhsinform.scot/healthy-living/
    food-and-nutrition/food-packaging/food-labelling (accessed September 6, 2019).
Newport, F. (2010). Americans Leery of Too Much Gov't Regulation of Business. Princeton, NJ: Gallup.
Povea-Garcerant, I. (2017). *International Food Law and Regulations: A Review.* https://www.re
    searchgate.net/publication/317098565_International_Food_Law_and_Regulation_A_Review
    (accessed September 13, 2019).
Rahmat, S., Cheong, C. B. and Hamid, M. S. R. B. A. (2016). Challenges of developing countries in
    complying quality and enhancing standards in food industries. *Procedia-Social and
    Behavioral Sciences*, 224, 445–451.
Safefood. (2019). *Food Labelling Requirements.* https://www.safefood.eu/SafeFood/media/
    SafeFoodLibrary/Documents/Education/Whats%20on%20a%20label/GCE/What-s_on_a_
    label_GCE-Classroom-Slides-Topic-1.pdf (accessed September 9, 2019).
Schmelzer, C. D. (2004). *Regulatory Agencies – Nutrition and Well-Being A to Z.* https://www.ency
    clopedia.com/sports-and-everyday-life/food-and-drink/food-and-cooking/regulatory-agencies
    (accessed September 6, 2019).

Société Générale de Surveillance (SGS). (2019). *Training Services – FSMA Training*. https://www. sgs.co.za/en/training-services/industry-based-training/agriculture-and-food/fsma-training (accessed September 20, 2019).

United States Food and Drug Administration (USFDA). (2019a). *Food Safety Modernization Act (FSMA)*. https://www.fda.gov/food/guidance-regulation-food-and-dietary-supplements/food-safety-modernization-act-fsma (accessed September 18, 2019).

United States Food and Drug Administration (USFDA). (2019b). *What's New in FSMA*. https://www. fda.gov/food/food-safety-modernization-act-fsma/whats-new-fsma (accessed September 20, 2019).

United States Pharmacopeia (USP). (2008). *USP Standards Development*. http://us.vocuspr.com/ Newsroom/ViewAttachment.aspx?SiteName=USPharm&Entity=PRA set&AttachmentType= F&EntityID=97277&AttachmentID=6026e674-9b0f-4b23-aced 85f8853d7da4 (accessed September 23, 2011).

United States Pharmacopeia (USP). (2011). *Food Chemicals Codex – An Overview*. http://www.usp. org/fcc/ (accessed September 26, 2011).

World Bank. (2009). *Implementation, Completion and Results Report on a Credit in the Amount of SDR 39.7 Million (US$54 Million Equivalent) to India for Food and Drugs Capacity Building Project*. Washington, DC: The World Bank.

World Health Organization (WHO). (2007). *Countries urged to be more vigilant about food safety*. http://www.who.int/mediacentre/news/releases/2007/pr39/en/

World Health Organization (WHO). (2008). *International Health Regulations (2005)*. Geneva, Switzerland: WHO.

World Health Organization (WHO). (2014). *Dioxins and Their Effects on Human Health*. http://www. who.int/mediacentre/factsheets/fs225/en/.

Part VII: **Outlook**

Oluwatosin Ademola Ijabadeniyi

21 Conclusion and outlook

This is a body of work produced by experts and professionals from different fields on diverse food science and technology subjects. Stakeholder collaborations, such as this, are needed to significantly reduce world hunger, food waste, empower smallholder farmers, and mitigate against climate change. Policy makers also need to remember that poverty, inequality, and conflicts play significant roles in world hunger.

Poverty remains a major challenge in many countries in sub-Saharan Africa and other developing countries. According to Hazell and Haddad (2001), about 1.2 billion rural people live in poverty (defined as living on less than $1 per day). Nearly 9% of these people live either in Asia or sub-Saharan Africa. Although, Asia dominates with two thirds of the total poor who are concentrated in the South Asia. Poverty is, however, declining in China, East China, and Middle East and North Africa. Unfortunately, this is not the case with South Asia and sub-Saharan Africa. While the global number of people living below the extreme poverty line of $1 per day decreased between 1981 and 2004 from 1,470 million to 969 million (i.e., 40–18%) more than a billion people continue to live in poverty in sub-Saharan Africa leading to economic inequality growing exponentially (Kates and Dasgupta, 2007). According to World Bank, people living in poverty in sub-Saharan Africa grew from 278 million in 1990 to 413 million in 2015, that is, most of the global poor live on the continent as of 2015 (World Bank, 2018).

The plight of extremely poor people can be seen in this poverty description by Kates and Dasgupta (2007): In the world of the poor, people don't enjoy food security, don't own many assets, are stunted and wasted, don't live long, can't read or write, don't have access to easy credit, are unable to save much, aren't empowered, can't insure themselves well against crop failure or household calamity, don't have control over their own lives, don't trade with the rest of the world, live in unhealthy surroundings, suffer from "incapabilities," are poorly governed.

Hunger which is linked to poverty is a universal problem. One in nine people do not eat adequately to live a healthy and active life. Hunger and malnutrition are the top two risks to health worldwide – greater than AIDS, malaria, and tuberculosis combined (FAO, 2019a). Sub-Saharan Africa has the highest burden of hunger and malnutrition. According to the Food and Agriculture Organization (FAO), Africa remains the world's most food insecure continent, with relatively low levels of agricultural productivity, low rural incomes, and high rates of malnutrition, despite important economic progress and agricultural successes (FAO, 2014; Jimoh, 2014). Although progress has been made in few African countries, there is still a long way to go to overcome food insecurity. For example, even though South Africa

Oluwatosin Ademola Ijabadeniyi, Department of Biotechnology and Food Technology, Durban University of Technology, South Africa

https://doi.org/10.1515/9783110667462-021

is improving its food system, 38% of South Africa's children, under five, are stunted, wasted, and overweight (Sylla, 2019). A new report by Lancet (2019), however, shows that no country is immune to malnutrition. One in three low- and middle-income population are battling extreme obesity and malnutrition at the same time. Nutrition and malnutrition therefore need to be tackled from multiple perspectives (Lancet, 2019).

At the 28th session of the Regional Conference for Africa in Tunis from 24 to 28 March 2014, FAO called on African ministers of agriculture to take action in priority areas to accelerate investment and broad-based transformation in support of smallholder farmers, including rural youth and women (Jimoh, 2014). There is evidently an opportunity for using agroprocessing in addressing and solving the problem of food insecurity through empowerment of smallholder farmers. Food security however cannot be achieved without an effective partnership between farmers, public, and private sector in support of smallholder farming. This was evident from our funded project sponsored from Research and Technology Fund/Department of Agriculture Fisheries and Forestry (DAFF), South Africa. Smallholder farmers' majority who are women were lectured on basic food processing and trained on adding value to pulses. Such initiatives show to farmers that they are relevant food stakeholder for the achievement of food security in Africa. Our interactions with smallholder farmers also revealed the necessity of training the smallholder farmers on good communication, management, and entrepreneurial skills so as to achieve production efficiency and sustainability. Also, addressing gender inequality and lack of support for women farmers, unlike their male counterparts, is probably one of the best ways to reduce poverty and solve the problem of food insecurity.

Another panacea to the problem of food insecurity and hunger is reduction of food waste. According to FAO (2016), about 1.3 billion tons of food is lost or wasted every year around the globe, that is about one third of all food produced for human consumption. Without question, food waste needs to be stopped. Tackling food waste demands that focus is shifted from the food to the system. A report from a year-long research project funded by the Walmart Foundation revealed where food loss and waste (FLW) is occurring across the entire value chain, that is, from farm level (production and post-harvest handling and storage), processing and manufacturing, distribution and retail (including hotels, restaurants, and institutions) to consumption in Global North (Nudds, 2019). Although FLW varied across the value chain, the greatest total FLW (by weight in tonnage) occurred during manufacturing and processing (47% of the total waste), followed by production (24%). While, households accounted for 14%, the remaining 15% of the total FLW comes from retail, distribution, hotels, restaurants, and institutions (Nudds, 2019). In the Global South however, FLW occurs principally at farm level. According to United Nations Regional Information Centre, UNRIC, 2019, inadequate harvest techniques, poor-harvest management, lack of suitable infrastructure, processing and packaging, and lack of marketing information are the main causes of waste in developing countries. Better shelf life dating has been suggested to reduce food wastage in South Africa, where out of 31 million tons of food produced annually, about 10 million tons of food is wasted (Krige, 2019). FAO (2016) has also suggested 9 easy tips to help with food waste reduction and they include (1) ask for smaller portions, (2) love your leftovers, (3) shop smart, (4) buy "ugly" fruits and vegetables, (5) check your fridge, (6) practice FIFO: first in, first out, (7) understand dates on your food, (8) turn waste into compost, and (9) sharing is caring: give to help.

Rodin (2015) described the three distinctly modern phenomena – urbanization, climate change, and globalization – responsible for disruption. The three are interrelated and they can compound the problem of food insecurity in the absence of proper planning, innovation, leadership, research, and collaboration among the stakeholders. Climate change particularly can lead to drought and shortage of water for irrigation. Humid conditions influenced by climate change may cause higher mycotoxin production in groundnut and maize. More research and strategies are needed to mitigate the effect of climate change. One of such strategies is the use of biotechnology or genetic engineering. For example, genetically modified stress tolerant and high yielding transgenic crops that can withstand the effect of climate change are needed. In addition, the use of biofertilizer and regenerative agriculture also have roles to play in mitigating the climate change effects (Testahun, 2018).

Incidence of persistent foodborne pathogens and foodborne outbreaks is on the rise (Whitwort, 2019) and this trend is expected to continue as the number of older persons is expected to double by 2050, projected to reach about 2.1 billion (UN, 2017). Adults aged 65 and older are more susceptible to foodborne illnesses as a result of their immune systems not combating like before (CDC, 2019). Other at-risk people include children younger than 5 years, people with weakened immune system, and pregnant women (CDC, 2019). Food safety will therefore continue to be relevant now and in the future. Although, it was not a priority in developing countries especially African countries, unlike in developed countries, things are now changing because of the impact of multinational food companies, social media, export, and growing middle class. While developing countries try to focus on food accessibility, it should be emphasized that there is no food security without food safety. Food safety is very important for achievement of food security and especially during food product development. Khoza (2019) recently wrote about a feeding scheme in South Africa that resulted in the death of two pupils after they had food poisoning. Again, the unfortunate incident signifies that there cannot be food security without food safety. Whether in developing countries or developed countries, it is necessary to be aware of primary drivers of existing and emerging food safety risks because awareness is the first step toward an effective mitigation strategy. The following have been suggested as primary drivers of food safety risks by Kendall et al. (2018):
- Demographic change
- Economic driving forces
- Resource shortages

- Environmental driving forces
- Increased complexity of the food supply chain (e.g., globalization)
- Water security
- Malevolent activities (e.g., food fraud, terrorism, and vandalization)

Food fraud will remain a challenge going forward. It is crucial to understand the drivers of food fraud and find a solution. Example of such solution is the National Food Crime Unit (NFCU) established in 2015 by the Food Standards Agency, UK (New Food, 2019). The main objectives of NFCU include improvement in the understanding of the food crime threat and identification of specific instances of dishonesty within the food supply chains and working with partners to address it. The seven types of food frauds that the unit tackles according to its head (Darren Davies) are theft, unlawful processing, waste diversion, adulteration, substitution, misrepresentation, and documentation fraud (New Food, 2019). FSMA is the best piece of regulation or food safety improvement plan since hazard analysis and critical control point. Developing countries should be encouraged to implement the US Food Safety Modernization Act (FSMA). FSMA shifts the focus from responding to foodborne illness to preventing it (FDA, 2019a).

Digital technology has the potential to improve food safety and reduce production costs in addition to assisting farmers to make precise decisions and improve access to information, knowledge, and markets (World Bank, 2019). Digital technology, however, will not be the solution to infrastructure challenges such as better roads, uninterrupted power supply, postharvest storage facilities, and improved logistics that connect farmers to markets (World Bank, 2019). A lot of investment will continue to be necessary in these areas.

According to FAO (2019a), the future of food depends on food safety. In the article, the importance of whole-genome sequencing (WGS) in decoding risks to food was also described. For example, WGS can single out with more precision the ingredient in a multiingredient food responsible for an outbreak. It can also determine the source of contamination or which illnesses are part of an outbreak. And lastly, it can show the linkages between multinational outbreaks (FAO, 2019b). A new era of smarter food safety has also been suggested by FDA (2019b). This will entail changing food traceability from largely paper-based system to the use of digital technologies such as blockchain, internet of things, and artificial intelligence (FDA, 2019b).

Blockchain could help the food industry to have better traceability and transparency invariably giving rise to what has been described as smarter food industry. Coupled with big data, the source of outbreak can be determined real time and economic cost reduced. It has also been suggested that blockchain has a role to play in food waste reduction. Currently, one-third of food produced globally end up being wasted. In the United Kingdom alone, an approximately £20 billion of food is wasted annually. In Canada, 35.5 million metric tons of food – 58% of national production – is lost and wasted (Nudds, 2019). The blockchain technology is able to achieve waste reduction because of its ability to improve visibility across the entire food supply chain in addition to generating business intelligence which invariably leads to better decision making and reduced waste (The Grocer, 2019).

There is an increase in blockchain adoption in the food industry signifying that it has relevant applications far beyond bitcoin and cryptocurrency. A total of 23 product lines have also been tested and launched on Walmart China Blockchain Traceability Platform. The platform is expected to scale by another 100 product lines by the end of 2019. More than 10 product categories including fresh meat product, rice, mushrooms, and cooking oil are covered on the platform (Huo et al., 2019). A recent research showed that blockchain will assist the food industry to save about $31 billion by 2024 principally from reducing food fraud (Frangoul, 2019).

Another way technology and innovation can assist to achieve a significant leap in food safety and quality is the implementation of virtual reality. Researchers are creating a virtual reality experience to train food service employees on good hygiene practices (Hyde, 2019). Virtual reality can be used to train restaurant workers and other food handlers.

Food will continue to evolve due to changing consumer preference and innovation trends. Meal replacement beverages, 3-D printed food products, meat grown in laboratories, and nanofood are examples of types of foods that will be common in few years to come. Some of these foods are necessary because of the current traditional food system. Since food production contributes to climate change, it is advantageous for global consumption of beef to be reduced. Companies like Beyond Meat and Impossible Foods have an important role to play in developing meat analogues and plant-based substitutes.

Also, consumers will use different proven methods to achieve caloric restriction so as to overcome the international malady of obesity and overweight. One of such method is personalized nutrition. Hamaker (2019) noted that improved understanding of how food carbohydrates intersect with the body is opening up the possibility of personalized nutrition for health outcomes. Personalized nutrition is relevant because individuals are unique regarding their genetic background, gut microbiota (or microbiome), and health status. These unique parameters are therefore considered for nutritional strategies (Zeng et al., 2016).

System approach is also imperative to solve food global challenges because of its capability for prediction. It is an approach that recognizes that the understanding of complex interactions between multiple processes and stakeholders is the foundation of solving the problems of food insecurity, food waste, and climate change (Ericksen et al., 2012). Marvin and Bouzembrak (2020) showed how such model could support risk assessors and risk managers in their understanding of the impacts of a given factor on food and feed safety and inform risk managers decisions to mitigate potential risks. According to Hermansson and Lillford (2019), a food-based system approach should have defined missions and should be coupled with interdisciplinary research. The following are the seven missions that the authors came up with:

1. to introduce more diverse and sustainable primary production;
2. to design sustainable process and system engineering including novel processes using reduced water and energy;
3. to eliminate waste in production, distribution, and consumption;
4. to establish complete traceability and product safety;
5. to provide affordable and balanced nutrition to the malnourished;
6. to improve health through diet;
7. to integrate big data, information technology, and artificial intelligence through the food chain.

As wonderful as these missions are, they cannot be achieved without conscious and decisive actions. It should be understood that the task ahead of food scientists and stakeholders in the food system is feeding the world's future 9 billion inhabitants with nutritious, safe, and wholesome food while making sure that food production, food processing, and distribution do not negatively impact the environment while also maintaining the value of food products for as long as possible as well as minimizing food waste through adoption of circular economy.

References

CDC. (2019). *People with a Higher Risk of Food Poisoning*. https://www.cdc.gov/foodsafety/people-at-risk-food-poisoning.html (accessed December 17, 2019).

Ericksen, P., Stewart, B., Dixon, J., Barling, D., Loring, P., Anderson, M. and Ingram, J. (2012). *The Value of a Food System Approach in Food Security and Global Environmental Change* (384pp.).

FAO. (2014). *FAO calls for action to accelerate economic transformation and development in Africa*. http://www.fao.org/news/story/en/item/218238/icode/. Accessed May 22 2020.

FAO. (2016). *Do Good. SaveFood*. www.fao.org/save-food (accessed November 2019).

FAO. (2019a). *The state of food security and nutrition in the world*. http://www.fao.org/state-of-food-security-nutrition. Accessed 22 May 2020.

FAO. (2019b). *The Future of Food Depends on the Future of Food Safety*. http://www.fao.org/fao-stories/article/en/c/1187077/ (accessed April 19, 2019).

FDA. (2019a). *Food Safety Modernization Act (FSMA)*. https://www.fda.gov/food/guidance-regulation-food-and-dietary-supplements/food-safety-modernization-act-fsma (accessed September 18, 2019).

FDA. (2019b). *Statement from Acting FDA Commissioner Ned Sharpless, M.D., and Deputy Commissioner Frank Yiannas on "Steps to Usher the U.S. into a New Era of Smarter Food Safety."* https://www.fda.gov/news-events/press-announcements/statement-acting-fda-commissioner-ned-sharpless-md-and-deputy-commissioner-frank-yiannas-steps-usher (accessed May 1, 2019).

Frangoul, A (2019). *Blockchain to Save Food Industry $31 Billion*. www.cnbc.com/2019/11/27/block chain-to-save-food-industry-31-billion-new-research-says.html (accessed December 3, 2019).

Hamaker B. R. (201). *Concept of tailoring fiber-based prebiotics for personalized gut health*. Presented at 33rd European Federation of Food Science and Technology Conference, Rotterdam, Netherlands.

Hazell, P. and Haddad, L. (2001). *Agricultural Research and Poverty Reduction*. http://www.ifpri.
    org/sites/default/files/publications/2020dp34.pdf (accessed March 27, 2014).
Hermansson, A. M and Lillford, P. (2019). Global challenges and the critical needs of food science
    and technology. In: *33rd EFFoST International Conference: Sustainable Food Systems –
    Performing by Connection, November 12–14,2019, Rotterdam, The Netherlands*.
Huo, N., Xia, Y., Zheng, R. and Li, W. (2019). *Walmart China Takes on Food Safety with VeChain Thor
    blockchain Technology*. https://medium.com/vechain-foundation/walmart-china-takes-on-
    food-safety-with-vechainthor-blockchain-technology-b1443e0e079c (accessed June 28, 2019).
Hyde, R. (2019). *How Virtual Reality Training Could Reduce Your Chance of Food Poisoning*.
    https://theconversation-com.cdn.ampproject.org/v/s/theconversation.com/amp/how-
    virtual-reality-training-could-reduce-your-chance-of-food-poisoning-95195?fbclid=
    IwAR0eAYYs6DNYGk65NOPFWWCbnSJNXE2lgZpVEtlhzySbYaZZW2D1hxlGpcU&amp_js_v=0.1
    (accessed May 2, 2019).
Jimoh, B. (2014). *"Africa Still Most Food Insecure Continent" – FAO*. http://www.vanguardngr.com/
    2014/03/africa-still-food-insecure-continent-fao/?utm_source=dlvr.it&utm_medium=twitter
    (accessed April 1, 2014).
Kates, R. W. and Dasgupta, P. (2007). African poverty: a grand challenge for sustainability science.
    PNAS, 104, 16747–16750.
Kendall, H., Kaptan, G., Stewart, G., Grainger, M., Kuznesof, S., Naughton, P., Hubbard, C.,
    Raley, M., Marvin, H. J. P. and Frewer, L. J. (2018). Drivers of existing and emerging food safety
    risks: expert opinion regarding multiple impacts. *Food Control*, 90, 440–458.
Khoza, M. (2019). *Feeding Scheme Grounded after Two Pupils Die of Food Poisoning*. https://www.
    sowetanlive.co.za/news/south-africa/2019-07-31-feeding-scheme-grounded-after-two-pupils-
    die-of-food-poisoning/. (accessed August 30, 2019).
Krige, N. (2019). *Tiger Brands Foundation Calls for Review of Food Sell-by Date Legislation*. https://
    www-thesouthafrican-com.cdn.ampproject.org/v/s/https://www.thesouthafrican.com/news/
    tiger-brands-foundation-calls-for-review-of-food-sell-by-date-legislation/amp/?fbclid=IwAR1-
    HSJtcoPw2Iq5XoYFdfBWyn-Qp6zCgcMN11eGvBZwfRogGoYNL8L_Gck&amp_js_v=0.1#referrer=
    https://%3A%2F%2Fhttps://www.google.com&amp_tf=From%20%251%24s&ampshare=https://
    %3A%2F%2Fhttps://www.thesouthafrican.com%2Fnews%2Ftiger-brands-foundation-calls-for-
    review-of-food-sell-by-date-legislation%2F (accessed September 8, 2019).
Lancet. (2019). *A Future Direction for Tackling Malnutrition*. https://www.thelancet.com/journals/
    lancet/article/PIIS0140-6736(19)33099-5/fulltext (accessed December 20, 2019).
Marvin, H. J. P. and Bouzenbrak, Y. (2020). A system approach towards prediction of food safety
    hazards: impacts of climate and agricultural use on the occurrence of food safety hazards.
    Agricultural Systems, 178. DOI: 10.1016/j.agsy.2019.102760
New Food. (2019). *Addressing Food Crime*. https://www.newfoodmagazine.com/article/96583/ad
    dressing-food-crime/ Accessed October 23 2019.
Nudds, K. (2019). Tackling food waste: ground-breaking study determines the root causes; much is
    lost in manufacturing and processing food in Canada. 79(3), 20–22.
Rodin, J. (2015). *The Resilience Dividend, Managing Disruption, Avoiding Disaster, and Growing
    Stronger in an Unpredictable World* (p. 398). London: Judith Rodin. Profile Books Ltd.
Sylla, M. (2019). *Too Many of SA's Children Are Stunted, Wasted or Obese*. https://mg.co.za/arti
    cle/2019-11-06-00-too-many-of-sas-children-are-stunted-wasted-or-obese?fbclid=
    IwAR3tYsOJ4PMH5E7gtHBPg70cq0QW22Ir622XSlWkoROCFvZazp1Na4iulYM
    (accessed November 7, 2019).
Tesfahun, W. (2018). Climate change mitigation and adaptation through biotechnology approaches:
    a review. *International Journal of Agriculture, Forestry and Life Sciences*, 2, 62–74.

The Avoidable Crisis of Food Waste. (2018). https://secondharvest.ca/research/the-avoidable-crisis-of-food-waste/ (accessed May 7, 2019).

The Grocer. (2019). *How Blockchain is Turning Back the Tide on Food Waste*. https://www.the grocer.co.uk/promotional-features/how-blockchain-is-turning-back-the-tide-on-food-waste/590970.article (accessed April 29, 2019).

United Nations. (2017). *World Population Ageing*. https://www.un.org/en/development/desa/popu lation/publications/pdf/ageing/WPA2017_Highlights.pdf (accessed December 17, 2019).

United Nations Regional Information Centre (UNRIC). (2019). *One Third of All Food Wasted*. https://www.unric.org/en/food-waste/27133-one-third-of-all-food-wasted (accessed October 23, 2019).

Whitwort, J. (2019). *STEC Climbs into Third while Listeria Continues to Rise in Europe*. https://www.foodsafetynews.com/2019/12/stec-climbs-into-third-while-listeria-continues-to-rise-in-europe/?fbclid=IwAR3-s5HOpmXGfodLV2NPR5xnEEVSMRK_30e6zmhTiNe2zBPIOAmWA7SOuoA (accessed December 17, 2019).

World Bank. (2018). *Poverty and Shared Prosperity 2018: Piecing Together the Poverty Puzzle*. https://www.worldbank.org/en/publication/poverty-and-shared-prosperity (accessed October 23, 2019).

World Bank. (2019). *Future of Food: Harnessing Digital Technologies to Improve Food System Outcomes*. http://www.worldbank.org/en/topic/agriculture/publication/future-of-food-harnessing-digital-technologies-to-improve-food-system-outcomes?cid=EXT_WBSocialShare_EXT&fbclid=IwAR1tb48rAFkQBFk09FAclLzfiF1uYoRouOZe2myqQ_nrYol03SNbey4R0ss (accessed April 29, 2019).

Zeng, B., Li, W., Yan, F., Zhao, Y., Pang, X., Zhang, X., Fu, H., Chen, F., Zhao, N. and Hamaker, B. R. (2016). *Nutrition Challenges Ahead*. https://efsa.onlinelibrary.wiley.com/doi/pdf/10.2903/j.efsa.2016.s0504 (accessed October 22, 2019).

Index

https://doi.org/10.1515/9783110667462-022

www.ingramcontent.com/pod-product-compliance
Lightning Source LLC
Chambersburg PA
CBHW060953210326
41598CB00031B/4808